Handbook of Research on Computational Arts and Creative Informatics

James Braman
Towson University, USA

Giovanni Vincenti
Gruppo Vincenti, S.r.l., Italy

Goran Trajkovski
Laureate Higher Education Group, USA

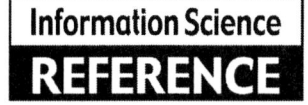

INFORMATION SCIENCE REFERENCE

Hershey · New York

Director of Editorial Content: Kristin Klinger
Senior Managing Editor: Jamie Snavely
Managing Editor: Jeff Ash
Assistant Managing Editor: Carole Coulson
Typesetter: Chris Hrobak
Cover Design: Lisa Tosheff
Printed at: Yurchak Printing Inc.

Published in the United States of America by
        Information Science Reference (an imprint of IGI Global)
        701 E. Chocolate Avenue
        Hershey PA 17033
        Tel: 717-533-8845
        Fax: 717-533-8661
        E-mail: cust@igi-global.com
        Web site: http://www.igi-global.com/reference

and in the United Kingdom by
        Information Science Reference (an imprint of IGI Global)
        3 Henrietta Street
        Covent Garden
        London WC2E 8LU
        Tel: 44 20 7240 0856
        Fax: 44 20 7379 0609
        Web site: http://www.eurospanbookstore.com

Library of Congress Cataloging-in-Publication Data

Handbook of research on computational arts and creative informatics / James Braman, Giovanni Vincenti and Goran Trajkovski, editors.

        p. cm.

    Includes bibliographical references and index.

    Summary: "This book looks at the combination of art, creativity and expression through the use and combination of computer science, and how technology can be used creatively for self expression using different approaches"--Provided by publisher. ISBN 978-1-60566-352-4 (hardcover) 1. Technology and the arts. 2. Creation (Literary, artistic, etc.) 3. Computer science. I. Braman, James, 1981- II. Vincenti, Giovanni, 1978- III. Trajkovski, Goran, 1972- NX180.T4H36 2009

    700.285--dc22

                        2008043767

British Cataloguing in Publication Data
A Cataloguing in Publication record for this book is available from the British Library.

Editorial Advisory Board

List of Contributors

Table of Contents

Section I
Intersections of Art, Science and Technology

Adérito Fernandes Marcos, University of Minho, Portugal
Pedro Branco, University of Minho, Portugal
João Álvaro Carvalho, University of Minho, Portugal

Salah Uddin Ahmed, Norwegian University of Science and Technology (NTNU), Norway
Letizia Jaccheri, Norwegian University of Science and Technology (NTNU), Norway
Guttorm Sindre, Norwegian University of Science and Technology (NTNU), Norway
Anna Trifonova, Norwegian University of Science and Technology (NTNU), Norway

Joseph William Pruitt, Slingshot Product Development Group, USA

Jim Bizzocchi, Simon Fraser University, Canada
Belgacem Ben Youssef, Simon Fraser University, Canada

Section II
Creativity Unleashed

Section III
Implications of Technology, Social Dynamics and Culture

Section IV
Creativity in Virtual Worlds and Artificial Spaces

Detailed Table of Contents

Section I
Intersections of Art, Science and Technology

In this section, the book takes an in-depth look at the intersection between the arts and the sciences and investigates its implications. In these normally separate domains, one begins to see the creative potential when they are combined.

Chapter I

Adérito Fernandes Marcos, University of Minho, Portugal
Pedro Branco, University of Minho, Portugal
João Álvaro Carvalho, University of Minho, Portugal

This chapter discusses various concepts behind digital art, emphasizing how the computer medium is itself the tool and the raw material of creativity. Digital art differs from conventional art pieces by the use of computers and computer-based artifacts that manipulate digitally coded information, inheriting the almost unlimited possibilities in interaction, virtualization and manipulation of information the computer medium offers. A framework is also presented here for digital art creation that consists of a common design space where digital artists can smoothly progress from concept to the final artifact, while pushing the computer medium to its maximum potential.

Chapter II

Salah Uddin Ahmed, Norwegian University of Science and Technology (NTNU), Norway
Letizia Jaccheri, Norwegian University of Science and Technology (NTNU), Norway
Guttorm Sindre, Norwegian University of Science and Technology (NTNU), Norway
Anna Trifonova, Norwegian University of Science and Technology (NTNU), Norway

The interaction between art and technology, especially computing technology, is an increasing trend in the recent days. The number of artists participating in multimedia software or game development projects is continuously increasing and so is the number of software engineers participating in art projects like interactive art installations. As this intersection of art and technology grows, it involves people from different disciplines with varying interests creating a milieu of interdisciplinary collaborations. In this context at the Norwegian University of Science and Technology, this chapter explores the intersection of software and art to understand different entities that are involved in the intersection.

Engineering Design focuses on the physical world. Industrial design focuses on the human experience. The marriage of these two design philosophies can often turn into a daytime soap opera. This chapter exposes the differences between the engineer and the designer and how they might, through the help of focused management, actually create ideas that not only function, but have meaning to the people that use them.

The authors describe a shared research agenda that successfully integrates three separate domains of practice: humanities scholarship, scientific research, and artistic creation. They work primarily within their respective specialties, but have also aligned their research and creation activities within a larger context. This shared work leads to joint research outcomes, but also enriches their individual work. The chapter analyzes the details of each individual research or creation strand, and identifies the dynamics of mutual support and synergy. The critical factors that support the success of cross-disciplinary collaboration are identified and explicated.

Section II
Creativity Unleashed

Here we see many diverse ideas, projects, applications and descriptions of research bordering the areas of technology and art, where much potential of creativity resides. The synergistic approach in these projects presents innovative ideas in these areas.

This chapter explores various definitions of randomness and how they might be applied to contemporary art. Special attention is paid to how randomness is used in computer science and how it is viewed in informational theory. A loose distinction is made between art that uses elements that are unpredictable, but meaningful, and art that uses randomness that conveys little or no meaning.

'Holography: Re-defined' is a side-ways look at our journey into the age of Photonics. Using historical parallel, its theoretical base describes holography's integration within digital media and how this breakthrough will affect our grasp of reality to the point where our objectivity is submerged in a sea of imaginings, blurring the "real" and the "un-real".

The combination of computer science and music has far reaching potential. This chapter introduces the reader to spatial hearing and to the binaural spatialization technique. The modeling of existing objects of sound and other natural artifacts can be enhanced through simulation where we can also see the creation of new designs. The author discusses various multi-dimensional auditory worlds where the authors see the advantages of simulating objects far beyond the scope of naturally occurring objects through the investigation of various "soundscape dimensions". Finally they get a glimpse into the idea of a Fourth auditory dimension by extending the idea of soundscapes.

This chapter serves as a guide for working with a visual art form using a digital time-based medium. An overview of the necessary theories; processes, concepts and most important the "elements," needed to create expressive visual artworks through the technologies associated with this visual art form is presented. An examination of various details of how we can effectively visually communicate and express our artistic ideas and intentions through a digitally time-based medium is discussed along with the artist's insight and experience in this domain.

Recently digital art performs in multi sensory forms of knowing and communicating. There are investigations of perceptual and emotional mechanisms of involuntary synesthetic experiences. Artistic experiments have predominantly been transferred through the human sensorium in interlaced approaches. The synesthetic experimentation by artists is arguable. The functions and interrelations of the synesthetic approaches in new media arts and neurological researches are discussed separately in this chapter.

Here in this chapter we discuss the characteristics of net art according to different categories, such are e-mail art, non-linear narrative, online performance, information art, game art, collaborative creation, Internet community, and physical interaction, attempting to emerge a breakthrough of net art aesthetics from the phenomenon of its chaos.

Recent developments of digital technology for artists have lead to the creation of a graphics tablet from Wacom Technologies. It is claimed that the graphics tablet is more favorable to creativity than other existing digital technologies. This chapter addresses this issue through a qualitative study of five artists using the Wacom graphics tablet. In particular the artist's own experience using the graphics tablet is explored. The outcome of this study indicates that the graphics tablet is a more useful artistic tool. However, there are still several improvements required to advance the graphics tablet to a stage suitable for fine artists.

This chapter seeks to contextualize the history and discourse surrounding information visualization. Here the author discusses his visual culture and the evolution of the interface as a ubiquitous tool and the aesthetics for understanding the organization of information. A selection of several recent visualization projects are discussed to illustrate the variety of areas enriched by contemporary information design.

"Memory Association Machine" (also known as "Self-Other Organizing Structure #1") is the first prototype in a series of site-specific responsive installations inspired by cognitive processes. The artist provides a mechanism that allows the structure of the artwork to change in response to continuous stimulus from its context. "Memory Association Machine" relates itself to its context using three primary processes: perception, the integration of sensor data into a field of experience, and the free-association through that field. "Memory Association Machine" perceives through a video camera, integrates using a Kohonen Self-Organizing Map, and free-associates through an implementation of Liane M. Gabora's model of memory and creativity.

    Stefano De Luca, Evodevo, Italy
    Eugenia Benelli, Evodevo, Italy
    Francesco Altarocca, University of Rome, Italy
    Dario Dussoni, University of Rome, Italy

Designing good and sound architectural projects is difficult and often involves many stakeholders. We can make this process easier using an "autonomous genetic design facilitator" and "collective subjective designer". These components are able to generate landscape designs that meet needs of every part involved. The PIGA system supports stakeholders expressing their preferences and improves collective abilities in term of creativity and design. This chapter describes a methodology as a process of urban parks design and can be adapted to other situations considering many variables (and consequently a huge amount of possible solutions) and many specific needs to satisfy.

    Sergiy Rakov, G.S. Skovoroda National Pedagogical University, Ukraine
    Viktor Gorokh, G.S. Skovoroda National Pedagogical University, Ukraine
    Kirill Osenkov, G.S. Skovoroda National Pedagogical University, Ukraine

Modern IT opens new perspectives and possibilities for mathematics and its applications to real life, in particular to Art. The authors show how creativity and exploration can lead to discovering new artifacts and how modern mathematical software (computer algebra and dynamic geometry) can guide and enrich the creative process. To illustrate the research approach and the competency approach we show the discovery of magic curves (parametric equations visualized on-the-fly), the famous Falling Ladder, building an interactive Triangle Expert System and exploring how a chain of generalization can lead from a statement about quadrilaterals to interactive morphing and animation of a sad face into a smiling face using interpolation - automated constructing of caricatures. The authors build interactive models of real-world-like linkages that can be moved and animated on screen, each of which performs a geometric transformation (dilation, rotation, translation, reflection). They also show the synergy of computer algebra and dynamic geometry while proving a newly discovered geometric statement using Derive and visualizing it using the dynamic geometry package DG. Finally, they conclude with some philosophical thoughts on new tools and a new form of education: creativity, competency paradigm, research approach, computer support, distributed educational communities.

    Jim Barta, Utah State University, USA
    Ron Eglash, Rensselaer Polytechnic Institute (RPI), USA

Students who may typically view mathematics as a sterile and disjointed subject are learning complex mathematical concepts incorporating culturally inclusive art and activities while exploring Native American beadwork, wampum, and Navajo rug weaving. Students explore artful and mathematical expressions within the context of Native American artwork and traditions via computer technology using a suite of virtual design tools within the context of Native American artwork and traditions. The ideas and activities shared within this chapter can provide teachers other options for incorporating art, mathematics, and technology in a blend of purposeful use. Students can acquire deeper knowledge and understanding as they create artful expressions of mathematical beauty through their use of these virtual design tools.

Chapter XVII
Mia Kalish, Diné College, USA

Knowledge, as it is used and reused, becomes progressively more abstracted and details become more assumed and consequently more invisible. Examples are ubiquitous and range from the ease with which we speak and drive cars to the understandings and interpretations of surds and integral signs that are actually mathematical visualizations. Every visualization must have at least one organizing story, for while the form provides structure and organization, it does not supply meaning. Meaning derives from the story that uses the form, and from the integrated cultural understandings and expectations. The Diné visualization has one organizing story, following the rays of the sun as the Sun-God carries it across the sky; many, small, spatial stories simultaneously enrich the organizing story and establish the cultural foundation and is represented culturally by four small dots arranged in a circular sequence at 90°, 0°, 270°, and 180°. Each position is associated with a time of day, a season, a color, a type of stone, a time in the lifecycle, and a process of living and learning. This chapter explores Conceptual Blending Theory to explore this complex information space of small spatial stories that combine to form an "information system of information systems."

Section III
Implications of Technology, Social Dynamics and Culture

As technology has grown into parts of our everyday lives and become meshed with many aspects of society and our culture, we often fail to realize its overall impact and implications. Not only has technology influenced us directly, it has also changed our views of various artifacts of culture, society and expression. In this section we discuss the various impacts of technology.

Chapter XVIII
Lindsay Grace, University of Illinois and Illinois Institute of Art, USA

Enculturation is the act of passing cultural ideologies from one person to the other. It is what breeds innovation instead of new creation. It is the disease of derivation, instead of the birth of creativity. This chapter assumes the practical perspective of critical anthropological distance to understand the culture of art. Such critical evaluation should illuminate the distinct characteristics that encourage patterns. In the tradition of anthropological and sociological study of existing culture, this chapter seeks to illuminate the distinguishing characteristics of contemporary art production and offer perspective on the critical creative process. It takes new media art as its case study because it serves as a cross-cultural intersection of scientific invention and artistic innovation.

Chapter XIX

Lindsay Grace, University of Illinois and Illinois Institute of Art, USA

Software is philosophical. It is designed by people who have been influenced by a specific understanding of the way objects, people and systems work. These concepts are then transferred to the user, who manipulates that software within the rules set forth by the software developer. This chapter diagnoses and constructs a critical perspective through which these philosophies are made evident. It outlines the ways in which software philosophies affect creative problem-solving and compares the scientific approaches to artistic approaches. This diagnosis is centered upon three axes, the use of analogy, reductivism, and transferred agency.

Chapter XX

Judson Wright, Pump Orgin Computer Artist, USA

In this chapter the author examines an approach to understanding, incorporating and utilizing aspects of our environment. From a routinely logical point of view, peoples' mental image of technology does not seem a perfect fit with their needs, nor do their uses of it. It is informative to delve into technological issues as they relate to this logical stance. Applying alternative perspectives, particularly Linguistics, Musicology and Sociology, reveal yet another aspect of our needs. These needs are often obscured by hype, and though it is easy to blame "the industry", Group Behavioral studies show us that it is a function of our psychology to displace real problems. Despite our having created insurmountable walls of smoke, people can now begin to ask: what are constructive ways to step out of this mirage-trap.

Section IV
Creativity in Virtual Worlds and Artificial Spaces

Computer generated virtual worlds provide a unique atmosphere for creativity, expression and design. In this section several projects and ongoing research is presented where we can see the potential of these artificial spaces. With recent growing interest in these mediums, many are turning to these types of environments to experiment in new ways of interaction.

Chapter XXI

Stephen A. Schrum, University of Pittsburgh at Greensburg, USA

As creative people inhabit virtual worlds, they bring their ideas for art and performance with them into these brave new worlds. While at first glance, virtual performances may have all the outward trappings of theatre, some believe they don't adhere to the basic traditional definition of the theatre: the live interaction between an actor and an audience. Detractors suggest that shared physical presence is required for such an interaction to take place. However, studies have shown that computer mediated communication can be as real as face-to-face communication, where emotional response is concerned. Armed with this information, we can examine how performance in a virtual world such as Second Life may indeed be like "real" theatre, what the possibilities for future virtual performance are, and may require that we redefine theatre for presentation in online performance venues.

This chapter provides a brief history and discussion of machinima (films created by computer users within virtual worlds). Several examples are highlighted here as well as a discussion on the creation of machinima in the social virtual world of Second Life. This chapter also connects user-produced content, like machinima, with the openness and rules of the platform in which it is created. It concludes with discussions centered on legal thinking surrounding these issues and points to the rise of the player-producer in these systems.

In this chapter the author discusses the role of massively multiplayer online role-playing games (MMOR-PGs) in a social and recreational context. Following the rise and evolution of games and technology over the years, he takes a look at how these virtual spaces have changed into a complex arena for social interaction. Looking at key areas of player motivation following research of Bartle and Yee, this chapter directs its focus on player behavior and game dynamics.

Foreword

The days of "traditional" computing have long past; we are no longer bound to programs that only process spreadsheet data, business operations, or calculations alone. Computer technology has changed everything around us, including our daily routines and the society in which we live. Computers now can do so much more and contain far greater computational power compared to many years ago. We now have the ability to assemble rich interactive multimedia content, create and play digital music, art, design, Web pages and other highly expressive content. Many times in the computer sciences, the creative side and potential of technology is ignored. Many times in areas of art and design, heavily technical information is left out of the details where the focus is on implementation and the design itself. This book aims to include both sides where technology and creativity come together on equal terms. In a creative array of discussions, projects and examples, combined with the range of experience and expertise of many areas of discourse, the *Handbook of Research on Computational Arts and Creative Informatics* presents a unique research focus. This book aims to give readers a unique experience by showing both the creative spirit of technology while looking at new and interesting uses of computers, software and our digital society.

Section I of this book focuses on the intersection of Art and Technology. Many discussions in this area arise throughout the book, but here we deal with many of these issues head-on. Examples of projects that can be found at the intersection of Art and Technology as well as how both scientific and artistic methods can be used in the design process are just a few topics. This section explores how computer technology can be used as an artistic tool to unlock the creative potential of these mediums.

The next section titled "Creativity Unleashed" presents a wide variety of unique examples, projects and discussions. The creative array of ideas presented in this section is difficult to describe. Readers will have a vast snapshot of some of the concepts being explored in this research area. Many of the authors have very diverse backgrounds bringing it all together to explore what technology and art research has to offer. The chapters span over several granularities of technical expertise, design related to the arts, trends and issues, creating an eclectic creative thread. One simply cannot read through this section without getting some new ideas for their own future projects.

Section III is concentrated around the idea of how Art, Culture and Technology have undoubtedly influenced each other ultimately impacting society. We can ask ourselves how computer interaction and software processes have shaped our view of understanding certain concepts of realty. How have culture and self concept become reinvented with changes in our ability to express ourselves and communicate through our new inventions and technical artifacts? The "culture" within Art and Science also is an interesting phenomenon where we can see how our perception and abilities to understand these issues can lead us to a better understanding of the need to strive for creativity. Many interesting ideas and topics are discussed in this section.

In the final section of the book, we see how computer generated realties can spark our creative potential. These spaces create new ways in which we can interact and express ideas and concepts. Here we are no longer bound by physical limitations of time, space or physics, freeing us from normal constraints. This section presents film, theater and other forms of creativity that exist and reach their full potential in the virtual realm. We also get a closer look at how the virtual world can impact real life interaction and vice versa.

The Handbook of Research on Computational Arts and Creative Informatics is a unique collection of creative and innovative ideas and projects. This book will be sure to leave a lasting impression on those interested in art or computer science (or both!) and serve to demystify some of the boundaries between the disciplines.

Gabriele Meiselwitz
Baltimore, USA
September 2008

Preface

Stop and think for a moment about some of your recent ideas. Many times these ideas (and projects) fail to fit nicely in the arts box or in the computing one. They fall *somewhere* in the *middle of* this spectrum. We need to think outside of our preconceived notions of "boxes" and what they are supposed to contain. Blurring the edges of both domains and joining them into a larger set allows for more flexibility and creativity. And, of course, this process helps us get out of disciplinary containers and think in new, innovative ways. This cross-disciplinary volume originated from our frustration in teaching computing disciplines in the "mainstream" way, lacking creative expression brought about by the very technology it so created. Trying to encourage students and other individuals to use technology in new meaningful ways that are outside the norm of mainstream thinking has been indeed a formidable challenge. *The Handbook of Research on Computational Arts and Creative Informatics* attempts to push limits and blur boundaries within these domains.

The arts and the computer sciences are traditionally not seen as having a significant intersection. Despite the fact that many of our colleagues would call programming an art, the cross-disciplinary tangents do not seems to go further than that. Arts and computing, therefore, are commonly seen as two disciplines that do not have much in common, apart from the fact that artists use computing systems and applications as tools in fleshing out their creative ideas. As software and computer systems have grown more powerful, so has their potential for performing tasks other than data analysis, spreadsheets and word processing. Even while surfing on the Internet we are faced with a barrage of images, multimedia and hyperlinked structures creating a unique new medium. Through technology we can create spaces that bypass the normal boundary of self concept, time, physics and space. These technological artifacts become extensions to our creative minds, allowing us abilities of expression that are not always possible through traditional means. It is important to understand the underlying foundation of technology in order to grasp the creative potential of these systems. The importance of technology in our everyday lives is often trivialized and taken for granted since it has become meshed into daily routine. We fail to realize the totality of its impact on how we think and interact with the world around us.

We often ask our students at the beginning of the semester to characterize what it means to be "creative". After a brief discussion of the various and very different answers (many of which are often conflicting), they are asked to characterize the "Sciences" and the "Arts". They interject with different lists of the typical stereotypes that seem mainstream. Generally they say that Art is more expressive, free and emotional while science is colder, factual and restrictive. It is when we tell the class to list the similarities between these areas that things become interesting. We have had many debates over the apparent differences and the possible existence of many similarities. Some have even argued (not only the students) that the purpose of a particular project is what should be considered either as artistic or scientifically based; they feel no middle ground can exist. We clearly feel that this is not the case. An artifact can fall into a gray area, in the middle, not defined by either side of the coin as it were. A piece can exhibit great artistic quality and still boast great scientific and computational achievements. Often, when very different ideas come together very great things occur.

We decided to launch the initial call for chapters for this book within forums accessible to a majority of computer scientists, and then to forums accessed by a majority of artists. Being on the computing side of the coin, our colleagues in the art world helped us spread the word around. The most interesting observation about the initial stages of this project was the amount of emails and phone calls we received. Colleagues were coming back to us with many questions regarding the scope of the book. We heard and read a lot of "wow's", as colleagues were challenging themselves to get out of their disciplinary boundaries. Many were apprehensive in submitting their particular proposals in fear of rejection due to their topic. Many were appreciative however to be able to explore their unique projects and ideas in an open medium. We wanted a book which allowed authors to express themselves in an open and friendly atmosphere where new ideas were abound. Closing the lid on certain "boxes" of ideas by limiting the book's scope and range would be a tragedy to both fields. The volume that you have before you targets the creative nature of technology itself. As you will see through the chapters composing this book, the collaboration of ideas yields both fascinating and thought-provoking concepts.

Technology is here to stay. It has shaped events in the past and will clearly continue to shape the future. Through this handbook we are addressing novel concepts from creation, interaction, communication, to the interpretation and emergence of art through various technological means and media. The book itself is divided into four main categories. Section one focuses on the overlap of these domains where the discussion of collaboration and intersection remains the overall theme. Section two presents a plethora of current creative projects spanning multiple disciplines, artists and researchers from many diverse backgrounds. The third section focuses on the impact of various aspects of culture and society as influenced by technology and art. This section also deals with our interpretation of these artifacts and how we are shaped by certain concepts and philosophies. In section four we take a look at the creativity offered through MMORPGs, online environments and virtual worlds. We hope that these chapters will serve to inspire new ideas and shed light on topics previously unknown to many readers.

Section I: Intersections of Art, Science and Technology, highlights how these areas overlap. One of the most obvious intersections that many envision as the concepts of "Computers" and "Arts" are mentioned is the idea of visual representations of mathematical concepts. The masses are often exposed to films created with the help of computer experts, with special effects that corroborate the importance of catastrophic repercussion of a series of events the story's main character seems to run into. Or the more fantastic stories that are quickly replacing classic-style cartoons, heavily relying on ray-tracing engines that produce funny ogres or families of superheroes. There are other applications that instigate one's artistic side more, as we review the visual representation of Mandelbrot and Julia sets. But this is exactly what those are, visual representations. These are not art. The art lays in the perception of the human who explored a particular area of a mathematical universe, or created an animation character. Artists are mathematicians, musicians, painters, tailors, and writers. Sometimes such artist can also be an artificial agent capable of producing a medium that is perceived as art.

Art is a concept that is often relative and misinterpreted. It is usually personal. Beauty is in the eye of the beholder, many popular folks sayings state through the world. This is particularly true for innovative streams of artists that push the envelope in order to progress in a particular niche, shifting ever so slightly, or sometimes significantly, the rules that govern it. This is the spirit behind the works that are reported in this handbook. We are not trying to describe works of art like the David or the Sistine Chapel. We are also not trying to illustrate how such impressive masterpieces can be digitized in one form or another. We are not applying computer science as a tool available to artists or computer scientists. We want to project the true intersection of computers and arts, scientists and artists, pixels and canvases.

Section II: Creativity Unleashed. The diversity and depth contained in this section presented a challenge to classify by one unifying theme. This difficulty is a direct consequence of the vast overlap between the domains of computing and arts. It is in the eclectic collection of projects reported here that we truly visualize the blurred boundary. We want to give room to projects that are truly innovative in their nature.

Vision is one sense that most people associate directly with arts. Hearing is a close second. Monet, Michelangelo, Beethoven and Brahms are all masters at their crafts, focusing on the direct communication of their visions and feelings to us through one sense. The innovation lies in exploring the indirect communication, the blending of the senses where the full effect can be achieved only by interpolating our feelings with our senses, and, at last, with the intellect. This section illustrates many projects just as worthy that explore the breaking of generally static physical and conceptual boundaries.

Section III: Implications of Technology, Social Dynamics and Culture focuses on slightly different aspects. Art is often an individual experience, but it can be rarely isolated from a social and cultural perspective. This section analyzes these very aspects, where social and individual factors can be visited and revisited with an innovative magnifying lens.

We often become so self-involved that we think of computing and sometimes even arts as an individual process. Although this idea is many times true, these two domains create social and cultural waves that cannot be underestimated. And especially, as we look beneath the surface of routine actions and reactions, we can identify underlying causes and mechanisms that offer a brand new viewpoint to the usual set of notions and parameters that we use to evaluate certain actions and situations.

Section IV: Creativity in Virtual Worlds and Artificial Spaces focuses on the creative side of computer-mediated environments. Virtual worlds provide a unique medium for both personal and collaborative expression. This is especially true in some of the newer persistent virtual worlds were we can share our virtual lives and identities with others around the globe. Some of us even spend much time and energy (and sometimes even money) in creating an alternative self that inhabits these worlds. Not only can we shape "physical" objects here, but we use our virtual identity as a creative medium for expression and self representation. Many are realizing the creative power locked in this technology, where we are no longer bound to sharing information, text, images and files like the flat 2D Internet. In these three-dimensional spaces we can view informational artifacts as visual or "physical" manifestations. In their own right, we could also consider many of these virtual worlds as four-dimensional; their dynamics changing over time, affected by the various visits and interactions of users, evolving into its own emergent society.

By exploring the relationship between "worlds" we have evolved our concepts of reality that extends creativity into spaces that are merely represented by pixels on a screen while a human audience perceives them as real. One needs not be a machinimator or avid game player to realize the potential of these spaces once you are immersed inside any of these environments, meeting the people that "live" there. As higher education, artists, researchers and social scientists (to name a few) are embracing these technologies, we are starting to see just how interesting and creative these environments can be made to be. These synthetic environments are a unique medium ready for exploration.

It is to be expected that the number of questions you might have after reading this Handbook significantly surpasses the amount of solutions and answers given across the chapters. But that happens in cross-disciplinary efforts like this anyway. If we managed to challenge you to think outside your disciplinary bounds and boxes that you are comfortable in, we would declare we had reached the goal of this project. We hope that through this book we can reach a broad range of individuals and blur boundaries.

The domains of arts and computing have never been closer than now. We hope that our readers realize that such interaction goes well beyond the simple diffusion of art through Information Technology solutions. Arts involve you, make you feel alive, and let you express yourself. Computers are perhaps one of the best allies. Do not be afraid of exploring new frontiers, break beyond them and become aware of how it feels.

Only as we stand on the shoulders of giants we can look forward and see what is ahead. This concept is particularly important in this work. And it is for this reason that we would like to shift the focus from this handbook to the people who were giants and on whose shoulders our friends and contributing authors stand. It is their tireless efforts that brought us to this handbook. Without visionaries we would still have to adapt to what nature gives us. Such visionaries, added over time, lead us to today, and to the ones who contributed to this book, more than the ones who humbly assembled it. In turn, it is the visionaries who made this book possible that will become the platform onto which others will stand and dare to build.

James Braman, Towson University, USA
Giovanni Vincenti, Gruppo Vincenti, S.r.l., Italy
Goran Trajkovski, Laureate Higher Education Group, USA

Acknowledgment

The editors of this book would first and foremost like to thank all of the authors who contributed to this handbook. It has been a pleasure working with such a great team of talented individuals. Not only did this book emerge from all of our collaborative efforts, but many new alliances and innovative ideas and projects were formed. We appreciate everyone's effort, support, insight and dedication over the last year of this project.

We would also like to thank the editorial review board who significantly assisted in many phases of this project. The reviews and constructive comments were much appreciated and essential to the outcome of the handbook. Many anonymous colleagues also assisted in the distribution of the initial call for chapters and helped to spread the word of this project's creation, to which we also would like to acknowledge and thank.

The publishing team at IGI Global has been a tremendous asset as well. Their guidance and support has been truly appreciated. We would like to personally thank Julia Mosemann, Christine Bufton and Jan Travers for their personal assistance with this project. Many times they stepped in and answered our questions and provided support.

We sometimes forget the amount of time, energy and commitment projects like these take out of us all, and would like to thank all of our families, friends and colleagues for their encouragement and support throughout the duration of this project.

James Braman, Towson University, USA
Giovanni Vincenti, Gruppo Vincenti, S.r.l., Italy
Goran Trajkovski, Laureate Higher Education Group, USA

September 2008

Section I
Intersections of Art, Science and Technology

Chapter I
The Computer Medium in Digital Art's Creative Process

Adérito Fernandes Marcos
University of Minho, Portugal

Pedro Branco
University of Minho, Portugal

João Álvaro Carvalho
University of Minho, Portugal

ABSTRACT

Art objects might be described as symbolic objects that aim at stimulating emotions. They reach us through our senses (visual, auditory, tactile, or other). They are displayed by means of physical material (stone, paper, wood, etc.) and combine some patterns to produce an aesthetic composition. They convey some message, normally to suggest some state of mind or to induce an emotion and the consequent feeling on the side of the viewer. Digital art differs from conventional art pieces by the use of computers and computer-based artifacts that manipulate digitally coded information, inheriting the almost unlimited possibilities in interaction, virtualization and manipulation of information the computer medium offers. In this chapter the authors propose to analyze and discuss the concepts and definitions behind digital art, emphasizing how the computer medium is itself the tool and the raw material in its creation, especially if we stress the fact that the conception and design of artistic information content is at the heart of any artistic work. Furthermore the authors present a framework for digital art creation that consists of a common design space where digital artists can smoothly progress from the concept until the final artifact while exploring the computer medium to its maximum potential.

INTRODUCTION

Arts and culture are social phenomena, consequential of the social interaction, of the individual and collective imaginary manifestations, that together establish a common communicational and informational space embracing artifacts said to be cultural and artistic. These artifacts, where some are possibly non-tangible, constitute, in fact, the resulting product from the artistic and cultural phenomenon. They are expressions of our imagination.

In this respect, the common communicational and informational space is created by the process of collaboration among a group of people who communicate and operate together by sharing the same interests and goals. Information or information content, meaning the intended message of each artifact, is a central constituent of the common communicational and informational space. Accordingly, artistic artifacts, may these be of digital or physical nature can be defined as informational objects.

Art objects might be described as symbolic objects that aim at stimulating emotions. They reach us through our senses (visual, auditory, tactile, or other). They are displayed by means of physical material (stone, paper, wood, etc.) and combine some patterns to produce an aesthetic composition. Like any art object, digital art objects are informational in nature; they are symbolic and purposeful built. Their creator intends to convey some message, normally to suggest some state of mind or to induce an emotion and the consequent feeling. They differ from conventional art pieces by the use of computers and computer-based artifacts that manipulate digitally coded information, what opens unlimited possibilities in interaction, virtualization and manipulation of information.

The computer medium is defined here as the set of digital technologies ranging from digital information formats, infrastructures to processing tools that together can be observed as a continuum art medium used by artists to produce digital artifacts.

When we consider the creative process itself, we can establish its beginnings when the creator gets an hold of the first concept or idea resulting from his/her subjective vision, gradually modeled into a form of (un)tangible artifact. It constitutes the message, this *about* something, the artist wants to transmit to the world. When digital content is used in this process, it can be both the means and the end product. On one hand, the digital content can be explored as the means to create non-digital artifacts, as for instance, digitally altered paper-based photography, and, on the other hand, be the end-result intended as it is the case in animated comics (Marcos, 2007).

In fact, digital art applies the computer medium both as raw material (e.g. the digitally coded information content) and as a tool of enhancing creativity. Notice that raw material is related here to unprocessed (or in minimally processed state) material that can be acted by the human labor to create some product. Similarly, digitally coded information content can be manipulated by digital artists to create artistic objects. When in the creative process, digital artists apply information content along with technologies from multimedia, virtual reality, computer vision, digital music and sound, etc. as also the information and communication infrastructure available such are the internet, presentation devices, and storage arrays, among others, to create interactive installations and generate digital artifacts. Therefore, the computer medium traverses effectively all the stages of the creative process, from concept drawing until the final artifact production and exhibition. Today's powerful editing and programming tools make it possible to an artist to modify, correct, change and integrate information content as valuable raw material in the creative process, that may be presented in several digital formats such are text, image, video, sound, 3D objects, animation, or even haptic objects.

Moreover, artistic communities need to have access to common technological infrastructures that facilitate collaboration (collaborative editing, annotating, etc.), communication and sharing of work experiences, of materials, being these, unprocessed digital content or final artifacts, activities that are essential for a soft progress from the starting concept to the final artwork. We argue here that as in other human activities, artistic creation benefits from the collaboration within a community of equals while having access to materials and tools. Such common information space is in effect a creative design space; thought design (in the sense of shaping) is the fundamental activity in the creative process of digital art.

In this chapter we propose to analyze and discuss the concepts and definitions behind digital art; its evolution from its begin until today; the technological issues related with artistic applications, emphasizing how the computer medium represents itself the raw material and the tool of any digital artwork, especially if we stress the fact the conception and design of artistic information content is at the heart of any artistic work. A proposal for a creative design space for digital art – the artech framework - is presented and described in detail. It allows for a smooth progress from the concept/idea until the final product (artwork) while exploring the computer medium to its maximum potential.

This chapter is divided in the following sections: first we give an overview of the background of digital art in terms of its history, fundamental concepts and developing vectors from its beginning until nowadays. Next we present and discuss some digital art fundamental concepts and key technological issues related with design and development constraints that artistic creativity has to deal with. In the third section we present some key aspects related with development of digital artifacts. In the fourth section we describe the artech framework for digital creation, embracing the creative design process, the creative design space architecture while presenting concrete examples. The future trends, that digital art is facing is the subject of the fifth section. Finally we draw out some conclusions.

BACKGROUND

Evolution of Digital Art

The history of digital art has been formed as much by the development of science and technology, as by artistic-historical influences. This does not mean that artists were unaware of the potential of the technology, but on the contrary, they have been among the pioneers who experimented with the computer medium and reflected on the cultural implications of technology of their time, some of them even decades before the digital revolution became a reality. The technological progress that influenced decisively the digital art evolution is tightly coupled with the development of the computer technology, a process that has been initially promoted by the military industry sector, on one hand, and on the other by the research centers as well as by the consumer market. Today, research centers still continue to play a major role in the design of founding concepts and production of digital art.

Digital art has its roots within the first decades of the twentieth century with isolated experiments created by a few visionaries whose results where mostly exhibited in art fairs, conferences, festivals and symposia devoted to technology or electronic media. These first artworks have been mostly classified as marginal to the mainstream art world. Alike in the Dadaist art movement some of these artworks were seen as a form of *anti-art*.

The world's first general purpose computer, the ENIAC (Electronic Numerical Integrator and Computer) is presented in 1946, by the University of Pennsylvania. While it is not yet a machine exploitable for artistic creation it represents one of the first steps in the set up process of the today's digital medium. The UNIVAC (Universal Automatic Computer), the first patented commercial digital computer, appeared in 1951, was already able to process textual data (besides numerical). The science of cybernetics appeared also in the 1940s, with a special mention to the work of Norbert Wiener (1894-1964) who created the term based on a comparative study of different communication and control systems, such as the computer and the human brain. This grounded the work of several digital artists who later explored the Wiener concept of man-machine symbiosis.

The 1960s was a rather important decade for the history of technologies, as it embraces most of the foundations of today's technology and its artistic utilization. In this decade we saw the emergence of the hypertext/hypermedia concepts by Theodore Nelson with his "docuverse", a kind of the pre-World Wide Web. Also the ARPANET, the first network that laid the basis for today's Internet, was created. In 1968, Douglas Engelbart from the Stanford Research Institute, introduced the notion of bitmap window and invented the mouse for its manipulation. These two concepts constituted a revolution of the way people interacted with computers. Computers started to be seen as entries into *communicational and informational spaces* whose content could be visualized and manipulated as bitmaps on the screen and exchanged over the network among a group of machines. The later creation of the Graphic User Interface (GUI) by Ivan Sutherland (a colleague of Engelbart) and Alan Kay from Xerox PARC in Palo Alto, California along with its integration into a "desktop" metaphor with layered windows brought to life the high popularized personal computer, launched first by Apple Macintosh in 1983. The following decades of the 80s and 90s have witnessed an accelerated expansion of digital art due to the remarkable development of the technologies. Computers became personal and linked together over common networks that turned out to be the today's global Internet; technologies developed gradually to be more intuitive and easy to use as well as less expensive an affordable for common people. Especially considering the last fifteen years, digital technologies became ostensibly pervasive and highly expressive. Along with the advancement of the collaboration and communication tools individual artists can now cooperate with their counterparts in order to pursue a common goal in a specific creative process.

As stated before, the development of science and technology has been the principal engine of the evolution of digital art, but has digital art been influenced by existing and previous art movements? Yes, indeed, among others, Fluxus, Dada, and Conceptual Art have influenced decisively the progress of digital art as we know it today. These movements brought into digital art the emphasis on formal instructions, the focus on concept, on the event *per se*, and also, the emphasis on the viewer's participation, contrasting to the art based on unified static material objects. From the Dadaism specifically, digital art inherited the concept of creating art by using precise predefined rules, i.e., a finite set of instructions generates the final artwork (a poem, a painting). The rule' or algorithm' instruction was adopted as the conceptual central element in the creative process. *Instruction-based* art is a fertile soil of today's digital art. Similarly, the Fluxus art movement has also extensively explored the idea of instruction-based generated art along with the immersion of the audience in the event, forcing an *interaction* between the spectator and the artworks. Influences from the Conceptual art, a movement emerged in the 1960s, came from its central statement "the idea or concept is the most important aspect of the work". This is still a way of thinking and practice common to many digital artists in all over the world. The concept or idea is the leitmotif for the shaping of the digital artifact. It means that "all of the planning and decisions are made beforehand and the execution is a perfunctory affair, i.e., the idea becomes a machine that makes the art", by artist Sol LeWitt (1967).

When we go back to the first half of the twentieth century, two artists justify a special attention for their seminal work in the integration of *interaction* and *virtuality* (in the sense of the immaterial) in art, as they explored pivotal notions that laid the foundations of the digital art of today: Marcel Duchamp and László Moholy-

Nagy. In 1920, Marcel Duchamp and Man Ray presented their "Rotary Glass Plates", an interactive machine that invited users to interact with concentric rotating glass plates to generate visual effects. In 1933, László Moholy-Nagy created "kinetic sculpture moving," a device for creating light sculptures by projecting silhouettes in movement. The innovative aspect here was the shift from object to *concept* in the form of the "virtual object" that was seen as a structure in the process, sometimes dynamic and volatile, that created expressive effects on the part of the observer, as an active player when *interacting* with the artwork itself.

Artworks created by the OULIPO (Ouvroir de Littérature Potentielle) group, a French literary and artistic association founded in 1960 by Raymond Queneau and François Le Lionnais; and the work of the American digital artist Grahame Weinbren, to mention few, have also explored the concept of *combinatorial* and *strict rule-based* process of the Dadaism poetry, in conjunction with another notion that emerged during the pioneering era of digital artworks – the *random access*. It is a form of *controlled randomness* to generate and activate instructions for information access and processing. We find the notion of randomness and instruction-based information process in the work of the American composer John Cage, whose

work carried out in the 1950s and 1960s, explored extensively these concepts. Cage described music as a structure divisible into successive parts that could be filled by means of automatically controlled randomness and instruction-based algorithms. The work of Cage anticipated numerous experiments in interactive art. Also, the German composer Karlheinz Stockhausen explored the concept of controlled randomness in his work in electronic music and uncertainty (controlled chance) in serial composition.

In the 1960s and 1970s there were a considerable number of artists and art movements as well art initiatives that have promoted definitely the development of digital art. We refer here the works of John Whitney, Charles Csuri and Vera Molnar from the 1960s as they remain influential today for their experimental research in computer-generated transformations of visuals through mathematical functions. Works such as *Permutations* (1967) by Whitney or *Hummingbird* (1967) by Csuri are milestones of computer-generated animation. Deserving a special mention is the EAT (Experiments in Art and Technology) group, founded by Billy Klüver, started collaborations among artists, engineers, programmers, researchers and scientists that would become and remains a differentiate characteristic of today's digital art communities, i.e., they usually involve the

Figure 1. In the left: Rotary Glass Plates (Precision Optics [in motion]), 1920, by Marcel Duchamp. In the right: Hummingbird, 1967, by Charles Csuri (courtesy of the authors).

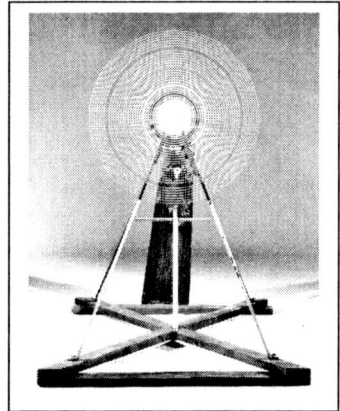
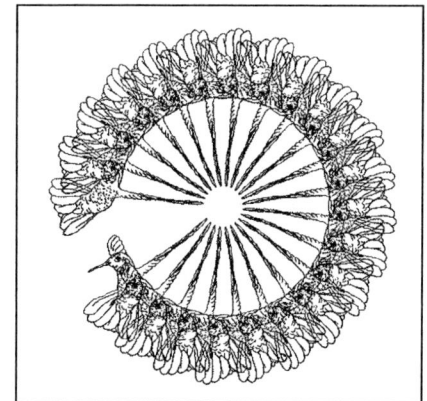

collaborative work of a multidisciplinary team. Some of the works created by EAT artists have been presented at the World Expo 1970 in Osaka, Japan as also in institutions such are the Institute of Contemporary Arts London and the Jewish Museum of New York, that have hosted exhibitions of digital art installations since the late 1960 and all through the 1970s. Artists such are Douglas Davis, Charlotte Moorman and Joseph Beuys or Robert Adrian have participated in shows that represented milestones in the development of digital art, namely, *Documenta VI* art show in Kassel, Germany, 1977 and the experience *Send/Receive Satellite Network*, a fifteen-hour, two-way, interactive satellite transmission between New York and San Francisco.

In the decade of 80s, painters, sculptors, architects, printmakers, photographers, and video and performance artists increasingly began to experiment with new computer audio and imaging techniques. Digital art evolved into multiple fronts of practice, ranging from more object-oriented work to artifacts that integrated dynamic and interactive dimensions, embracing also aspects of process-oriented virtual objects.

Digital art, as it is known nowadays, entered the world art in the late 1990s when museums and art galleries started increasingly to incorporate digital art installations in their exhibitions. The Intercommunication Center (ICC) in Tokyo, Japan; the Center for Culture and Media in Karlsruhe, Germany; the Ars Electronica Festival in Linz, Austria; the EMAF - European Media Arts Festival, Osnabrück, Germany; the VIPER (Switzerland); the International Art Biennale of Cerveira, Portugal; and the DEAF - Dutch Electronic Arts Festival are examples of initiatives that have supported and initiated digital art consistently all over the last two decades.

Today's digital artifacts range from virtual life as it is the case of *A-Volve* (1994) from Chris Sommerer and Laurent Mignonneau, a virtual environment where aesthetic creatures try to survive; to Internet art such is *Conversation Map* (2001-present) by Warren Sack, a mapping of the communication displayed in a browser, where results of a large-scale content analyzes from online email exchanges are used to create a graphical interface showing different social and semantic relationships. A good example of artificial life robotics installation is *Autopoiesis* (2000), by Kenneth Rinaldo that presents sculptures with sensors that react to the visitor by moving their arms towards the person provoking attraction or repulsion. Virtual Characters (usually called Avatars), Internet art and Cyborgs are topics where digital artists are active nowadays. A more comprehensive overview of the today's aesthetic

Figure 2. In the left: A-Volve, 1994, by Sommerer and Mignonneau (supported by ICC, Japan). In the right: Autopoiesis, 2000, by Kenneth Rinaldo (courtesy of the authors).

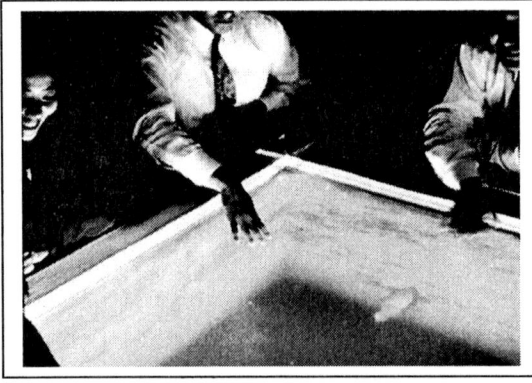 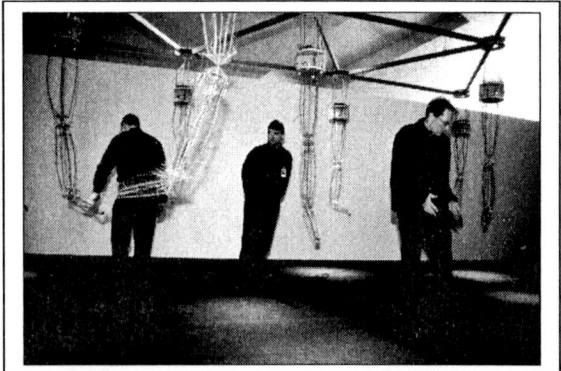

digital artifacts can be obtained from Paul (2005) and Greene, (2005).

It is worth mention that digital art is becoming part of several educational programs at graduation and post-graduation level in applied arts and technology (multimedia) all over the world. Digital art is also one of the main themes in dozens of conferences and workshops world wide. International institutions like UNESCO or the European Commission have incorporated digital art as one of their interest areas.

Several Arts of Digital Art

Digital art is in fact a recent term that became a general designation for several forms of computer-supported art, from *computer art* (since 1970s), *multimedia art*, *interactive art*, *electronic art* and more recently, *new media art*. Under the definition of digital art there are several art branches commonly connected to the specific media or technology they are based on.

Digital art can be defined as *art that explores computers (tools, technologies and digitally coded information content) as a tool and material for creation*.

In the course of this definition digital art has to incorporate the computer medium in its creative process, even if the final artifact does not visibly integrate computer or digital elements.

In Figure 3 we present an overview of the different artistic areas related to digital art. As we can observer, digital art embraces, by definition, all type of computer-supported art.

Computer art term appeared for the first time in 1960s mainly through the work of the digital artists and researchers at Bell Laboratories, New Jersey, as for instance, Michael A. Noll one of the founders of computer-generated images. This art expression was born directly from the fact the new aesthetic experiments where fundamentally supported by computers.

Multimedia art or *multimedia performance* relates to all type of art expressions mainly based on the exploitation of multimedia technologies to support installations and public performances. Its first appearance was in the early 1960s with the experiments of artist who applied a kind of

Figure 3. A general categorization of digital art

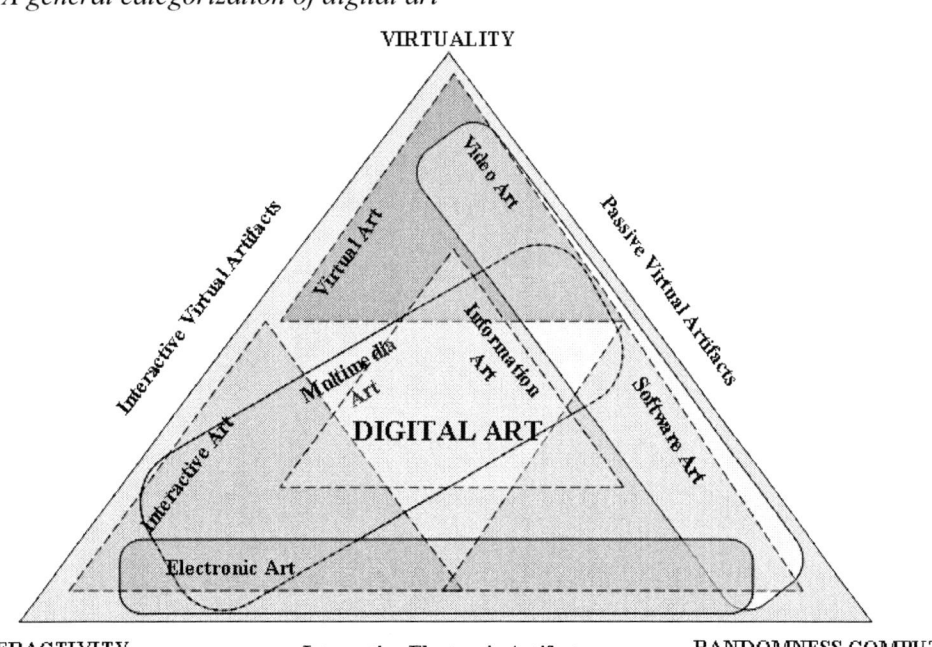

cross-fertilisation of between theatre, dance, film, video, and visual art, among others, to set up a multimedia (several media) show.

Video art emerged in the mid 1960s connected with experiments with television or captures made with half-inch video equipment when it became available in 1965. It relates nowadays to all type of art expressions mainly based on video (dynamic image) technologies where the emphasis lies on the exploitation of video as information content to enhance artistic audio-visual narratives.

Electronic art relates to artistic work strongly connected with the assembly of electronic components. It first appeared in the early 1960s connected with art installation incorporating electronic parts, where the emphasis lies here on the combination of electronic, mechanical, and other materials together with technologies, but not necessarily digital since they could even not embrace computational means.

Interactive art is a general classification of any type of art mainly interactive, where the viewer turns out to be an active player dialoguing with the artifact, possibly changing it.

Net.Art or *Internet art* is an art movement born in the 1990s that explores exclusively the World Wide Web for presentation of digital artifacts, benefiting from the extraordinary promotional potential and the substantive audience this communication channels allows. Internet holds the singularity to host many different aesthetic activities, most of them of chaotic, dematerialised and ephemeral nature. It resides in the world most open zone, the cyberspace, i.e., can be viewed anywhere in computer desktops, mobile phones, palmtops, etc., but rarely in museums halls and galleries (Greene, 2005).

Software art or *instruction-based art* is a recent term that relates to all type of art that explores the concept of controlled randomness of instruction sequences (algorithms) in the generation of artistic content. It is a classification mostly used by researchers working in *computational aesthetics* concentrated in the design of algorithms for

aesthetical manipulation of text, image and sound, among other media. Software artists often define the code itself as a form of creative writing, as the medium, the "paint and canvas" of the digital artist. Since the tools can also be rewritten, re-programmed, the artist can also re-invent the "paintbrush" and "palette".

Virtual art relates to a general classification of any type of art that implements forms of perceptual immersion (visual, auditory, haptic, etc.) of the viewer in the artifact, applying virtual and augmented reality technologies. It is a type of art that was born and developed in the mid of the 1990s following the emergence of the virtual reality technologies. Virtual Life is a form of virtual art, for instance (Grau, 2003).

Information art relates to an emerging field of electronic art that includes interaction with computers that generate artistic content based on processing large amounts of digitally coded data and information content (Wilson, 2002).

New media art is a global classification of any type of art embracing several media where the emphasis lies on the exploitation of new and recent media. It is a term employed today for any type of digital art that embraces informational media.

There are several other digital art branches; such are bio-art, games art, immersive art or digital poetry, among others, whose detailed description is beyond the scope of this chapter.

Core Concepts of Digital Art

We observe today a growing number of creative people who are exploring innovations in information and communications technology in global communication networks, as well as in media tools to create new aesthetical representational forms of information. Since the use of digital technologies is transversal to all daily-life activities, a process vastly increased during the last decade, we can argue that to a certain extend all forms of artistic media (and activities) will eventually be absorbed in the computer medium in the years to come, either through digitization or through the use of computers in a specific aspect of processing or production within the creative process.

Nowadays, more and more artists are converting themselves into computer medium practitioners, falling down to the enormous creative potential the digital technologies offer. The incorporation of the computer medium in each specific artifact or in the creative process is not always clear. In some cases the use of digital technologies and digitally coded information content is rather evident. But in other cases, for instance, in some mechanical-based installations the level of incorporation of the computer medium is not at all clear.

As described previously, digital art is mainly based on three grounding concepts: *controlled randomness access*; *presentational virtuality* and *interactivity* that have been behind emergent artwork from the 1960s to today's digital art installations. They can be described as follows:

- **Randomness access:** (pseudo) Non-deterministic instruction-based algorithms open the possibility of instant access to media elements that can be reshuffled in seemingly infinite combinations;

- **Virtuality:** The physical object is migrated into a *virtual* or *conceptual object*. The *concept* itself becomes perceptible through its virtualization;

- **Interactivity:** The viewer may assume an active role in influencing and changing the artwork itself.

The artwork is often transformed into an open structure in process that relies on a constant flux of information and engages the participant in the way a performance might do. The audience becomes a participant in the work, resembling the components of the project that may display information of a specific perceptive nature (visual, auditory, tactile, or other). The artist plays usually the role of facilitator for the participant's interaction.

The creative process in digital art relies often on collaborations between an artist and a team of programmers, technicians, engineers, scientists and designers, among others. This collaboration implies the multidisciplinary work involving art, science, technology, design, psychology, etc., that form a common communicational and informational space. Due to the widespread of the digitally coded information content that is increasingly

Figure 4. The continuum art medium

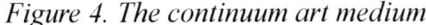
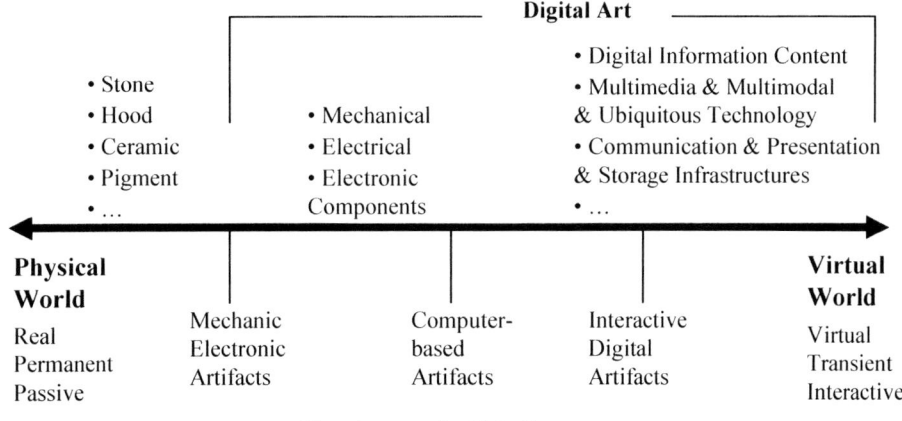

Continuum Art Medium

available in high expressive multimedia formats, the creative process is becoming more and more based on the manipulation and integration of digital content for creation of artworks. This is particularly the case in artifacts displayed by means of digital devices (e.g. Internet, DVD).

Accordingly, we need to devise a common creative design space where digital artists can smoothly progress from the concept/idea until the final product (artwork) while exploring the computer medium to its maximum potential. This common creative design space incorporates necessarily a communicational and informational space beneath, where digitally coded information content of different nature and level of processing is available for the artists' use. Furthermore, tools for editing, design or for any specific processing and composing have to be offered along with facilities for communication and collaboration among the community members. The creative design space shall also provide tools to support all the activities at all phases of the creative design process, ranging from the drafting phase, passing through the artifact's implementation phase until the artifacts exhibition preparation (exhibition space design) as also the access to physical and/ or digital exhibition space. This way, the creative design space will facilitate the establishment of communities of interests in art, where people from different backgrounds share materials (raw material), and digital collections while collaborating throughout common goals. In fact, digital art is about multidisciplinary team work rather than sole individual engagement. This has been proven to be a fact in several artist communities since the 1960s until today. For instance OULIPO, EAT or even more recent artist communities such are the ones linked to the Ars Electronica or to Cerveira Biennial of Art festivals, have established in their work, even if informally, common creative spaces where collaborative facilities, share of materials or simple exchange of ideas are seen as beneficial.

Consequently, digital artists have to learn how to extract the greatest potential from the computer medium (tools, technologies, digitally coded information content) if they expect nothing but the best results. For instance, artwork based on controlled randomness access usually requires artists to possess programming skills.

The meaning of design in this context, appoints to a conscious effort to create something that is both functional and aesthetically pleasing. Design is here taken from both the perspective of design in engineering and from a more inventive view as it is the case in applied arts.

As Löwgren and Stolterman (2007) state design is always carried out in a context (p.45). In digital art, design of digital artifacts is mainly based on the conceptualism's aphorism where the initial "idea or concept becomes a machine that makes the art" (Sol LeWitt, 1967). Thus the design process itself evolves from a vision or idea (even if it is not aware for the creator) until the final digital artifact is released. The message the viewer can obtain from the artifact in terms of a personal or group experience is the central issue the digital artifact holds.

From this point of view the digital artifact is nothing but a designed thing built around a core of digital technology. In digital art context, the artifact is an object embracing information content displayed by means of digital media or a combination of digital and physical components. The artifact acts as a materialization of a message, a piece of information, throughout the presentation of information content intended to stimulate emotions, perceptive experiences on side of the user. Thus, artistic digital artifacts, being these of pure digital or a combination with physical constituents are more adequately defined as *informational objects*.

Digital content is defined as informative material of digital nature that holds the ability to be acted to transmit a message. Some authors, as for instance Robert Musil, refer to digital technology and by legacy, digitally coded information content, as the material without qualities due to its pervasive characteristics and constantly development. These are, however, characteristics that open, almost on a daily basis, new challenges and possibilities for aesthetical experiments since the computer medium can constantly wear new presentational facets.

Preservation in Digital Art

Digital art, being based on the computer medium, faces a number of unprecedented challenges, if we compare with the traditional art, because of the characteristics of the medium itself in terms of its presentation, collection and preservation. Digital art often requires user engagement, which can be highly volatile when its message is not understandable at a glance. It also poses very specific challenges in its preservation and collection due to the pervasive characteristic of the computer medium along with the daily advancement in the development of software and hardware. This makes digital art, in general, unstable and very difficult to deal with. For instance time-based, interactive digital artworks imply numerous open issues regarding their presentation and preservation along time. How can we digitally preserve an artifact that incorporates mechanical components? A video capture of the artifact can hardly preserve all of its significance for future reuse or visit.

Digital preservation is therefore a challenging matter thought large parts of our civilization's cultural heritage is widely going digital but risking obsolescence if is not adequately preserved. The rising awareness of this urgency led to a number of research initiatives over the last decades. This research has undergone two main strategic lines: *migration* and *emulation*.

Migration means here the ability of the conversion of a digital object to another representation. It has been successfully used in documents (mainly text) and image conversion. Problems arise when the artifact to be migrated embraces such different features based on hardware/software specific or proprietary characteristics and time-based issues that over time are not replicable due to the obso-

lescence of the grounded technology. This also happens in the emulation of the artifact.

Emulation aims at mimicking a certain environment that a digital object needs, e.g., a certain processor or a certain operating system. The emulation turns to be unfeasible when dealing with closed proprietary hardware/software whose inner functioning remains unknown or not fully replicable.

As a result, digital art is usually very difficult to preserve due to its digital characteristics: it is inherently interactive, virtual and temporary, aggravated with the heterogeneity of its employed media. Projects such as the Variable Media Network, a joint effort of several institutions ranging from the Guggenheim Museum New York to the Berkeley Art Museum/Pacific Film Archives (Depocas, Ippolito, and Jones, 2004) or the PANIC project (Hunter and Choudhury, 2006) have investigated properties and developed preservation strategies for digital artworks, including tools, methods and standards for preservation of unstable mixed-media objects. The open challenges are mainly related to the diversity of and complexity of obsolete file format that are used in the field of digital art. (Depocas, Ippolito, and Jones 2004) argue that efforts to preserve born-digital media art always have to be based on structured documentation. He also adds that often this documentation is the only thing that remains from the artwork. However, criticism has been raised against any type of prescription of formats that might oblige the artists to conform their artworks to pre-defined submission policies. This reduces to a great extend the creativity freedom and cannot, therefore, be somehow accepted.

The success of the digital preservation initiatives will depend largely on the standardization of formats, tools, archiving methods for long-term preservation of digital objects. It is worth mentioning here the Reference Model for an Open Archival Information System (OAIS), published in 2002 as a standard model for general purpose digital information archiving and preservation

(OAIS, 2002). It has been developed and extended since some five years now in order to implement concrete preservation solutions for cultural, scientific and art digital content. Examples of research projects working with the OAIS reference model are CASPAR, devoted to cultural and scientific digital content (Caspar, 2005) and PLANETS project focusing in interactive multimedia art (Planets, 2004).

IMPLEMENTATION OF DIGITAL ARTIFACTS

The implementation of the digital artifacts, being these complex systems or more simple applications, involves an increased importance due to its direct dependency of the technology potential, limits and constraints. An adequate use of the technology will grant artists the chance to explore its maximum potential in generating aesthetic pleasant and technological innovating results. The correct choices of input/output devices, facing and solving maintenance or on-site issues are aspects to be considered in the implementation of digital artifacts.

Input Devices

The input devices are often (together with the output devices) the only hardware the user may recognise. Especially non-technical and technically unversed users do not want to know how the artifact is functioning, since they are mainly interested in the idea or concept behind. How this concept is technically achieved may be hidden to the user (Linaza, 2003).

In general, input devices can be classified with respect to two different goals:

- Devices for user interaction; and
- Devices for the creation of digitized exhibits (like scanners).

Devices for user interaction should be as easy to use and as intuitive as possible in order to address a wide range of different human beings and even technically unversed users (unless the difficult of use if part of the concept of the installation). As opposed to interaction, devices for the creation of digital exhibits like digitized painting or 3D models of sculptures may be more difficult to use, due to the fact that the task of creating digital exhibits mainly addresses expert users or users that are able to spend time in a learning phase to get familiar with a special device.

Furthermore, input devices have to be classified with respect to some constraints indicating, if the current state of the art is sufficient, whether or not a device can be used for an artifact:

- **Usability of the input device:** Raises the questions that determine if the device is mainly usable with or without training, if it can be handled by a wide range of different users like children and adults, or if cables may reduce the freedom of interaction of the user.
- **Evaluation of user acceptance:** New technological devices that are not widely used, such as data gloves, have to be especially evaluated with respect to their acceptance by different user groups. Users should not quit their interaction due to an insufficient understanding of the device behaviour.
- **Need for further development:** Clearly, some innovative input devices or interaction techniques that are still under research and only available as prototypes, being, therefore, only partially usable.
- **Need for expert support:** Input devices go beyond the mouse or a keyboard, they can as also be a tracking system or a video cameras to track user's position or movement which normally need regular maintenance by an expert.

Content Delivery and Management

Even though content delivery and content management systems are normally hidden from the user, they play an important role for artifacts in the field of arts and cultural heritage. We propose to classify content delivery and management systems with respect to the following items:

- **Content dimension:** The main task is to clarify whether 2D content like digitized paintings or 3D content such as digitized sculptures should be rendered and subject of interaction. The overall dimension of the content objects has obviously a direct influence on the choice of input and output devices.
- **Data-types:** There is a wide range of different data formats available for both 2D and 3D rendering. During the creation of a new artifact, it is important that all involved modules (from the creation of the digitized copy to the final output rendering module) "understand" the same data formats. Import filters and conversion tools should be identified in order to translate different data types without loss of quality or artifact performance if required. Exchangeability of content between similar artifacts has to be considered in order to ensure data usage and transfer instead of single application solutions.
- **Expert support:** Even for content delivery and the management system, expert support can play an important role. Currently available databases for storing single content objects are especially difficult to handle, while easy to use and intuitive user interfaces to store and retrieve information are not commonly used.

Output Devices

Output devices are, together with the input devices, the direct interface with the user and, therefore,

an important choice for the design of an artifact. We propose to classify the existing output devices in the following way:

- **Usability of the output device:** The choice of the output device depends upon several artifact constraints. For example, digital artifacts that render high detailed digitized objects should prefer large screens instead of standard computer monitors; applications addressing a group of users should obviate flat screens; and applications displaying 3D objects could take advantage of stereo capable output devices.
- **Need for further development:** The increasing resolutions and falling prices of new output devices like Head Mounted Displays, binocular, beamer point ups or even CAVEs (Cave Automatic Virtual Environment) are subject to fast research and development. Some devices are sufficient for prototypical installations but will not be accepted in the current development stage for artistic artifacts addressing a wide range of interested users at public places like a museum.
- **Acceptance evaluation:** As for the classification of the input devices, evaluation of user acceptance is necessary for several output devices. It is obvious that Head Mounted Displays are not favoured by a large percentage of users that have to wear them for a long time.
- **Expert support:** Even output devices need technical support by experts. For instance, stereoscopic video projectors displaying content onto a large-scale screen need to be carefully calibrated.

Maintenance

During the research and development stages of new technology (both hardware and software), maintenance often goes unattended. Nevertheless, when bringing new artifacts out of the laboratories and into public places like museums, art galleries or cultural heritage institutions, the maintenance of the application plays an important role in the usability of the system. From the technological point of view, applications have to be designed and implemented with respect to:

- Personal costs of eventually necessary expert support:
  - How many times and how often is an expert necessary to keep the application running?
  - How often must hardware or software be recalibrated or reconfigured?
- Cost of maintenance of the equipment;
- Robustness of hardware (e.g. life time of the equipment);

Environment Constraints

There are general technological constraints in the development of the artifact if we take into consideration the conditions of the site and environment where the artifact is expected to be used, among others. Technological constraints can be summarized as follows:

- **Site usage:** The choice of the input, output and computational hardware depends mainly on the prevailing environmental conditions. In mobile artifacts for an outdoor scenario, for instance, only light weight hardware such as light pocket PCs and binoculars are normally the ones used.
- **Environmental conditions:** New input devices like optical tracking systems impose illumination constraints, which restricts their usage to indoor scenarios with controllable lighting.
- **Mobility:** Mobile artifacts need light weight hardware, while stationary installations often use large-scale screens for displays.
- **Single/multi-user applications:** Clearly, the number of individuals acting on an artifact at the same time possibly requires different

input and output devices. For instance, small displays with a restricted viewing angle can not be used for multi-user applications.

- **User experience:** The range of different individuals using an artifact plays an important role in the choice of input and output devices, so it is important to know if the system will be used by experts or beginners and whether it is mainly addressing children or adult users.
- **Addressees:** A typical constraint for the design of a new artifact is the target group. Such constraints are less technical but have to be dealt equally.

THE ARTECH FRAMEWORK FOR DIGITAL CREATION

The Artech (Art and Technology Industries in the frontier Region Minho–Galicia) Framework for Digital Creation is a framework for digital and interactive art creation that has been designed and implemented in order to promote a community of digital artists in the frontier region of Minho-Galicia. The main concept behind this framework is to implement a common creative design space where digital artists can smoothly progress from the concept/idea until the final artifact while exploring the computer medium to its maximum potential.

In the following subsections we start by proposing a model of a creative design process, describing its different phases and their interconnections. After, the Artech main architecture is described in detail. Finally, in the third subsection we illustrate the implementation of Artech with some concrete examples.

THE CREATIVE DESIGN PROCESS

As previously discussed the creative process in digital art is mainly based on the design of the artifact's message and its development. The computer

Figure 5. The creative design process

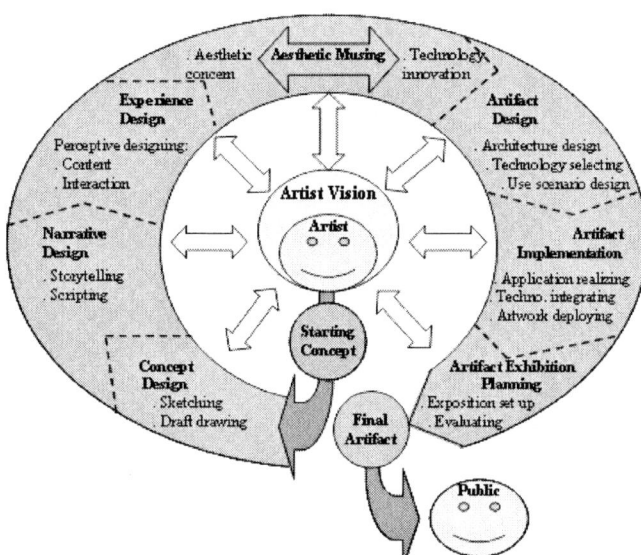

medium in the form of editing, communication and collaboration tools as well as digitally coded information content is likely to be always present and traversing the overall creative process.

As depicted in Figure 5 the creative design process is launched when the artist gets hold with an initial idea/concept. Then, the artist starts to design the concept, entering a process that will lead into the final artifact. This process is not a linear process, on the contrary, artists may go back and further in the activity sequence, skipping one or focusing the work in another. The process is usually highly dynamic, yet, the artist's vision is always present. The creative process involves the following phases:

Message Design phase:

- **Concept design:** In this activity the artist gets involved in converting his/her idea/ concept or vision into a set of sketches, informal drawings, i.e., the abstraction is concretized in a perceptive structure. The artist does exploratory drawings that are not intended as a finished work. The outcomes of this activity are, thus, sketches, draw-

ings that allow the artist to try out different ideas and establish a first attempt for a more complex composition.

- **Narrative design:** Here the artist takes the drawings resulting from the concept design activity and designs a composition, a construct of a sequence of events that set up the message that will allow its recount to the future users/viewers. The narrative of the message behind the initial concept is designed taking into consideration aspects such as the structure of its constituent parts and their function(s) and relationships. The narrative or story is *what* is narrated assuming usually a form of a chronological sequence of themes, motives and plot lines. The outcome of this activity can be resumed as the design of the message as a story.

- **Experience design:** This activity embraces the process of designing the message, taking into account its related concept and narrative, to design and conceptualize specific characteristics of each narrative event from the point of view of the human experience it shall provide. This design or planning of the human experience is made based on the consideration of an individual's or group's needs, desires, beliefs, knowledge, skills, experiences, and perceptions. The experience design attempts to draw from many sources including cognitive and perceptual psychology, cognitive science, environmental design, haptics, information content design, interaction design, heuristics, and design thinking, among others.

Aesthetic Musing: this is a central activity in the creative design process, it represents the moments of contemplation where the artist revise his/her vision against the decisions made (to be done) during the design and development of the artifact. We identify two guiding vectors in aesthetic musing of artifacts:

- **Aesthetic concern:** Process of integrating characteristics in the artifact that eventually provide a perceptual experience of pleasure, meaning or satisfaction, arising specifically here from sensory manifestations of the artifact such are shape, color, immersion, sound, texture, design or rhythm, among others. Beauty here relates almost exclusively to the aesthetic dimension of the perceptive nature of the artifact components.
- **Technology innovation:** Process of integrating novelty in the reshape, use, combination and exploitation of digital technology. This appoints to the computer medium dimension of the beauty creation, i.e., the technology is a driven force to set up new aesthetic dialogues. Taken the fact of the digital technology is under accelerated development; integration of high levels of technology innovation in digital art is commonly desired.

Artifact Development phase:

- **Artifact design:** This activity relates to all aspects concerned with the design of the computer system or application that will support the final artifact. This includes the design of the system architecture, interface and interaction, as well as the selection of technology to implement them. Since the artifact is to be acted usually by an audience of viewers, we have also considered in this activity the design of the use scenario from the technological point of view. Design adopts here an hybrid perspective mixing aspects from applied arts and engineering. It applies principles from a more rigorous design based on exploitation of technology, science and even mathematical knowledge along with the aesthetical concerns.
- **Artifact implementation:** In this activity the artist proceeds to the implementation of the artifact itself. This incorporates tasks as programming, testing and debugging, as well as, technology integration and the final artifact deployment. This demands from the artist to hold programming and technological skills if he/she wants to have a more direct control over the implementation process. The artist can even be assisted by a team of programmers and technologists; however, to be in command of the artwork, the artist has to be skilled in technology to a certain level.
- **Artifact exhibition planning:** This activity joins together all aspects related with the setting up of the artifact exhibition. This represents the final stage of the overall creative design process, where the artifact is brought into the world, i.e., the art object meets the audience. The success of this meeting will depend increasingly on the attractiveness of the artifact, the way the exhibition space is organized, how the logistic of its different components are managed and supported and also on the contextualization of the artifact

in the overall exhibition. Notice this activity will be based on the decisions made before in terms of the message design, the artifact implementation, and above all, on the use scenario configuration. Artifacts may be presented in museums, art halls, art clubs or private art galleries, or at some virtual place such is the Internet.

THE CREATIVE DESIGN SPACE ARCHITECTURE

The creative design space is the local, physical and virtual, where the creative design process is realized. As previously defined, a creative design space is a digital communicational and informational space that enables the generation of artistic content, the storage, transmission and exchange of digital data while providing the exhibition and presentation space for access to information and content by both specialists and the public.

The creative design space aims at supporting an artistic community by enabling all the main activities of the creative design process by providing tools for design, shaping, planning, collaboration, communication and sharing of information as well as giving access to digitally coded information content of diverse nature. Usually, such a space has also to provide exhibiting facilities for presentation of final artifacts to the audience.

As a whole, the creative design space as depicted in Figure 6 is not entirely affected either by technological advances or the needs of users and creators. The flow of work from one activity to another remains conceptually the same.

As previously noticed, the computer medium is likely to traverse all the stages of the creative design process, from concept drawing until the final artifact production and exhibition. As we can observe in the Figure 6 the computer medium can be divided in two main lines of contributions, namely:

Figure 6. The creative design space architecture

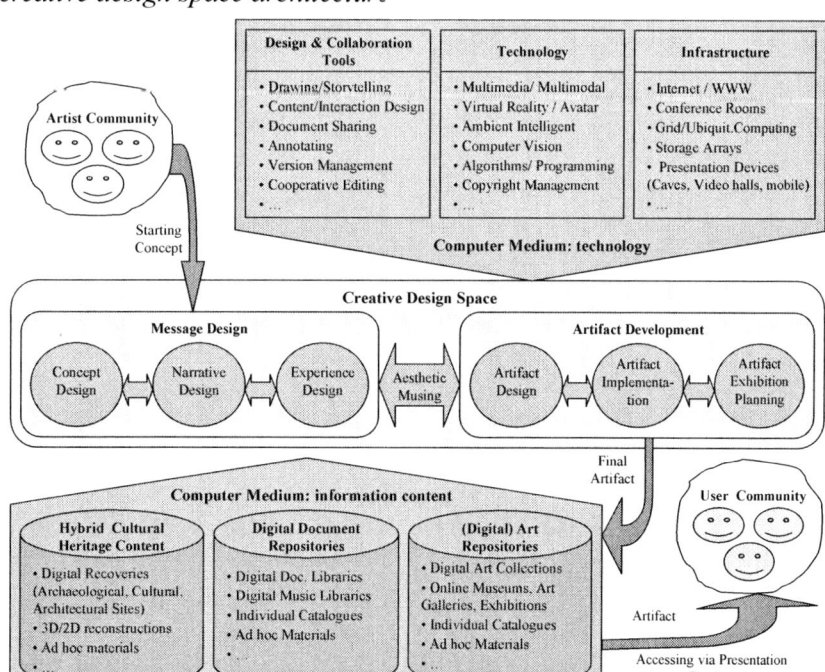

- Computer medium as technology: we identify here three principal types of tools:
  - **Design & Collaboration Tools:** They include all type of tools and applications that support activities related with design, drawing, planning, etc. as well as those allowing the collaboration among groups of artists to happen throughout communication, sharing of files, joint editing and annotating, etc.
  - **Technology:** We consider here all the computer technologies that are offered not only as tools or applications but principally as technological areas whose knowledge, procedures and techniques can be exploited in benefit of the creative design process. Programming languages, toolkits, specific algorithms, concepts and architectures, scripting techniques or procedures in areas such are virtual reality, computer vision or ambient intelligent are good examples of the technology mentioned here.
  - **Infrastructure:** This relates to all supporting infrastructures that make the computer medium to happen, in terms of communication, conferencing, storage facilities, computing capacity, presentation devices, etc.
- Computer medium as digitally coded information content: we identify here three principal types of information content:
  - **Hybrid cultural heritage content:** This relates to all kind of content, partial or full digital, collected from different cultural heritage sources such are archeological sites, museum, 2D and 3D digital recoveries of architectural and historical findings, etc. Cultural heritage content has been serving as raw material for the shaping of digital artifacts that aim at transmit specific cultural messages. For instance, digi-tally altered photography is exploiting to a great extend digital photographs of famous paintings.
  - **Digital document repositories:** These relate to the more formal document repositories ranging from text and image documents, digital music databases, from institutional or personal catalogues and collections. This type of information content is adequate, for instance, to be applied in artifacts that explore more official information sources, as for instance, the ones based on narratives referring to historic, real-life elements (dates, names, events).
  - **Digital art repositories:** These relate to digital-born art objects, media, documents, etc. owned by art galleries, museums, festivals, art houses, individual or ad hoc collections that are accessible online. Under this classification we consider also all the artifacts generated within the creative design space that can be digitally stored.

Artists enter the creative design process by providing a starting concept. Then, all along the message design and artifact development phases, the artist may bring into play several types of tools, by a single manner or collaborating with colleagues, while using digitally coded information content. Incorporated in this information content we might have also parts of or complete artifacts. They can be, possibly, reused as simple musing objects or be even transformed into new forms. Thus, the management of copy rights in the accessing and re-use of digital content is a mandatory requirement for a successful development of the community of interest over the common creative design space.

Notice that the final artifact is released into the digital repositories and not directly to the audience. This is because the access to the digital artifacts has to be done by presentation devices within an exhibition space, being this physical such is a museum room or virtual like the Internet.

THE ARTECH IMPLEMENTATION

The Artech creative design space is being constructed based on three locations: Guimarães at the School of Engineering of University of Minho (www.uminho.pt) and at Computer Graphics Center (www.ccg.pt); Vila Nova de Cerveira at the premises of International Art Biennale of Cerveira (www.bienaldecerveira.org); and Pontevedra at the Faculty of Arts of the University of Vigo (belasartes.uvigo.es).

The creative design space infrastructure has been gradually implemented along with all the last five years and includes the following main components:

1. An experimental lab at each location, integrating hardware/software tools, specific experimental infrastructures (e.g. a virtual reality lab in Guimarães; audio lab located in Cerveira; and multimedia lab in Pontevedra) as well as printing facilities to support creative work;
2. A common Information and Communication infrastructure based on:
    o   The Torga.net - a trans Portugal-Galicia communication network connecting the different locations of Minho and Galicia;
    o   One access grid room at each location. The access grid is an ensemble of resources including multimedia large-format displays, presentation and interactive environments, and interfaces to grid middleware and visualization environments. These resources are used to support group-to-group interactions across the creative design space.
    o   Shared multimedia content repositories accessible through access grid.

3. Exhibition spaces at each location where artists can present their final artifacts, namely: the space of the International Art Biennale of Cerveira, at Vila Nova de Cerveira; Exhibition space (room X) at Faculty of Arts in Pontevedra; and a future art exhibition space at Computer Graphics Center in Guimarães at the campus of Azurem of the University of Minho.

The local artistic community is living around different art fairs, conferences, workshops and training actions where an artist can exchange ideas, present their work and establish connections within the Artech community. Two events deserve special mention here: the International Art Biennale of Cerveira, organized since two decades ago; and the Artech – International Conference on Digital and Interactive Art that has been organized alternatively in Portugal and Galicia, starting in 2004.

The usage level of the common creative design space is growing since the Torga.net network has been brought into life four years ago. Since it started we had up to one hundred direct and indirect users of the infrastructure. The number of artifacts produced within the Artech community is also reaching the number of tens each year.

The Artech community also promotes postgraduate courses in the field of digital arts and cultural heritage. In this respect the University of Minho is offering a new Master Course on Technology and Digital Art, which aims at forming a professional profile in the fields of digital art, cultural expression, education and entertainment.

Even admitting we are still in an implementing phase of the Artech creative design space, a process that is never ending, the feedback received so far, the number of artworks presented, as well the increasing number of users of the infrastructure along with the rising number of digital art related initiatives in the region, altogether, allow us to conclude that the adoption of common creative design space enhances the artistic production and promotes collaboration among the local community of artists.

In the next pictures we present some examples of artworks presented at events organized within the Artech community, along with the relevant application constraints dealt during their setting up within the creative design space.

Example 1: WAVE – A Virtual Audio Environment

WAVE is an immersive musical instrument based on the exploitation of the concept of micro and macro gestures of the users; of sound/musical objects with embedded sound sources and synthesizer sounds and, finally, the facilities behind the visual and sound immersion. It integrates the real-time displacement of several musical categories of sound objects together with the performance control of external MIDI synthesizers. (Figure 7).

Main Application Constraints

- **Site usage:** *Wave* is for indoor usage. Minimal room dimension (4x4x4m) had to be

Figure 7. User performing in a WAVE installation. It was presented at Artech 2004

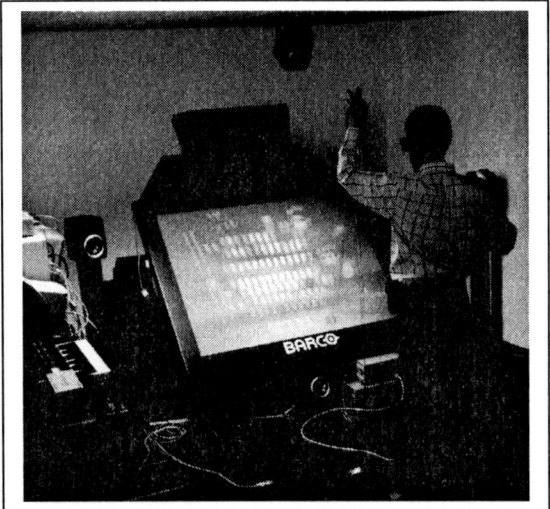

considered in order to allow the configuration of the set of loudspeakers' physical positions necessary for the aural immersion implementation.

- **Single/multi-user applications:** *Wave* has been primary designed for single users. However a relative large visual display (preferably stereoscopic) had to be provided in order to permit the audience to follow the performing actions.

- **User experience:** *Wave* addresses different kind of users, experts and beginners as also adults and children. The system' interaction was mainly based on gestures, i.e., hand and arm displacements, as also fingers movements.

- The I/O devices have been constructed over a 6DOF (six degrees of freedom) tracking system with enough amplitude to cover all the displacements of the open arms of a common user.

- **Addressees:** *Wave* addressees ranged from experimented musicians to common people without music background. There was no evidence of specific constraints appearing here. Based on empirical evidence resulted from usability studies and users' surveys realized, it has been concluded that immersive musical instruments have a significant acceptance, since users have demonstrated a positive intention to use them (Valbom and Marcos, 2005, 2007).

Example 2: Displacement

Displacement is an interactive installation applying Augmented Reality technology. It explores computer vision to capture the image of the performing user in order to generate its geometrised contour while projecting it over a large display. The image projected changes along with the user displacement within the installation area. The user can interact with his/her dynamic contour as also with some other projected objects, where collisions produce sound effects. This permits to create an audiovisual dialogue between the user and the installation. Visual and auditory elements are tightly linked and cannot be dissociated within each other.

Main Application Constraints

- **Site usage:** *Displacement* is for indoor usage. Minimal room dimension (4x4x4m) had to be considered in order to allow the configuration of an adequate immersion of the users within the audiovisual space.
- **Environmental conditions:** *Displacement* requires specific illumination constraints in order to allow the visual tracking of the users' contours. Users image have to be capture against a contrasting background.
- **Single/multi-user applications:** *Displacement* has been designed for both single/multi-users. The aforementioned space constraint applies here also (Quintas and Dionísio, 2005).

Example 3: Dynamic Topography

Dynamic Topography consisted of a cylindrical interactive multimedia installation. Three projectors spread the information onto three retro-projection screens – a transparent, a plain and a rough/creased one – to express through these supports the historical axis of the work. These three screens are surrounded by another one – cylindrical and transparent – whose purpose is to represent the globalisation of information in both space and time.

The creased screen is linked to the moment of the birth of writing (stone), the plain one to the invention of printing (paper) and the transparent one represents, symbolically, the screen.

This visitor is given a wireless mouse that enables him/her to freely interact with the projections throughout the installation where blocks of combining texts are displayed.

The visitor will be guided right into the centre of the installation by a set of fluorescent letters placed on the floor. The purpose of these letters is to guide the visitor across the space and to motivate him/her to explore the installation. Sound

Figure 8. (Quintas and Dionísio, 2005) Displacement, installation for body performance presented at Artech 2005

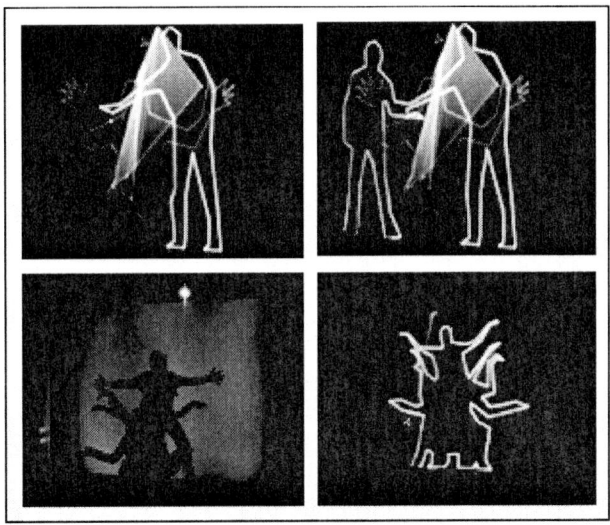

is employed to emphasise different aspects of the projections.

Main Application Constraints

- **Site usage:** *Dynamic Topography* is for indoor usage. Minimal room dimension (15x25x5) had to be considered in order to allow the configuration of an adequate immersion of the users within the interactive space. The structure of the installation consists of two concentric cylinders with diameters of approx. 15 and 25 metres, respectively. The three projection screens (3.0 by 2.8 metres) are placed within the inner cylinder.

- **Environmental conditions:** *Dynamic Topography* requires specific illumination constraints in order to allow the adequate environment of light and shadow, opaqueness and transparency, sharp or out-of-focus images in an attempt to create a magical and ritualistic atmosphere.

- **Single/multi-user applications:** *Dynamic Topography* has been designed primarily for both single-users. The aforementioned space constraint applies here also.

- **User experience:** *Dynamic Topography* addresses different kind of users, experts and beginners as also adults and children. The system' interaction was mainly based on manipulating an wireless interactive device (the gyromouse). It does not only translate the physical movements of the visitor/user within the space but it also controls the movements of the cursor during the projection. The gyromouse uses the technologies of radio frequency and gyroscope. The gyroscope allows the mouse to be used without a firm support (in the air) and the radio technology makes it possible to keep control over the cursor even at a great distance from the CPU (Pimentel and Branco, 2005).

Figure 9. Dynamic Topography – a photograph of the installation. It has been presented at Artech 2004.

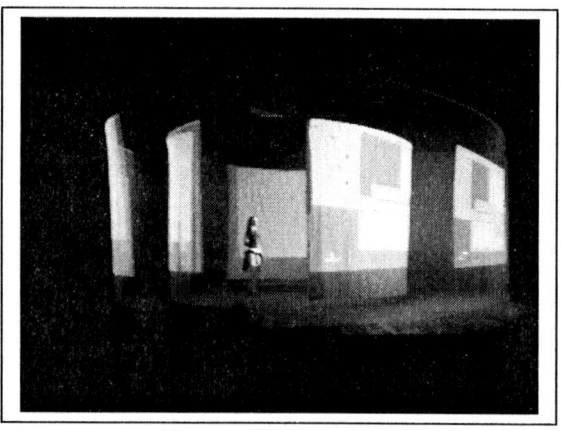

FUTURE TRENDS

The future of digital art will be shaped by the evolution of the creative potential of the computer medium. Art will, however, reflect on the specifics of cultural change, and the computer medium in the broadest sense has always been an important engine for the cultural transformation. Due to the increasingly accelerated evolution of the computer medium it is not easy (almost impossible) to predict what exactly the future of digital art will be. Even so, there are lines of evolution of the digital technologies that allow us to devise some of the eventual development challenges in the near future. In following we present some of the most representative of these lines on our view:

- **Ambient Intelligence environments:** The technology will be part of our surrounding environments, embedded in objects, halls and buildings as also in nature. The technology becomes an invisible, thought acting, integral part of life and art in general. There is a need for research and development of new interface and interaction models, agent technology, devices and integrated systems, which will be ubiquitous and traversal to all aspects of our life. Interaction will become more and more multimodal based on gestures, speech, gaze motion and even biological signals will be the channels to communicate with computers. Artifacts developed in such environments will be indistinguishable from other daily-life objects, however, allowing perceptive experiences to happen.

• **Intelligent Interfaces and Virtual Entities:** More and more intelligent interfaces and entities have being developed in the last years leading to the emergence of the technologies of the avatars or virtual characters, widely attracting the attention of the public. The Second Life (http://secondlife.com/) and the related social phenomenon it generated is a good example of the potential of this new type of technologies. One important line of research is the development of virtual actors. These are virtual characters who are intelligent enough to be able to understand directing instructions and behave autonomously with each other in a drama sequence that can be further recorded as a movie. Authors will be able to shape digital personalities and engage them into different interactive artifacts. The artifact, as a virtual character, may thus act and interact with the human viewer as a digital person or intelligent entity. The digital artifact becomes alive and intelligent.

• **Intersection of Digital, Bio- and Nanotechnology:** We are already living in a world where bioengineering and cloning of life-forms are a reality. The usage of genetic altered life-forms in art is not new (eg. *Telegarden* by Ken Goldberg and Joseph Santarromana, 1995-present; or *OneTree* by Natalie Jeremijenko, 2000); however, it will receive an increased attention as bioengineering technology is becoming common-place. The challenges here will lie on opening the door of the extreme small to the world of art. It is not easy to devise how

this will be explored by digital artists, but one can imagine tiny machines that projects biologic signals of our body to a screen or image, in real-time. The human body will become part of the digital medium, thus, also a constituent of the artifact.

Furthermore, the intensive development of the information society will have implications in the widespread of huge volumes of rich multimedia content and their usage in shaping digital artifacts. One way or other, our civilization' heritage will be in digital format and, to a great extent, available for free. Design and processing tools will become common place and increasingly trouble-free though they will integrate artificial intelligence in order to facilitate the creative process. Art will be a prerogative of every body, granted the access to the computer medium.

CONCLUSION

Digital art objects are informational in nature; they are symbolic and purposeful built. Their creation aims at convey some message, normally to suggest some state of mind or to induce an emotion and the consequent feeling. They differ from conventional art pieces by the use of computers and computer-based artifacts that manipulate digitally coded information.

We have defined here the computer medium as the set of digital technologies ranging from digital information formats, infrastructures to processing tools that together can be observed as a continuum art medium, the set of digital materials, techniques and tools used by artists to produce digital artifacts.

In this chapter we have analyze and discussed ground concepts and definitions behind digital art, emphasizing how the computer medium is itself the tool and the raw material in its creation, especially if we stress the fact the conception and design of artistic information content is a central issue of any artistic work. Furthermore we have presented a framework for digital art creation that consists of a creative design process implemented by means of a common design space where digital artists can smoothly progress from the concept until the final artifact while exploring the computer medium to its maximum potential.

We have seen the creative process in digital art is essentially about design of the message and experience the artifact will transmit and allow, as also its implementation as a computational system or application.

The computer medium affects here the role as the tool to enhance the creative process; as also as the raw material when the digitally coded information content and computer components are primarily explored in the shaping of the artifact. We have also stated the activity of digital art creation is mostly about collaboration among a multidisciplinary team. It requires a common communicational and informational space where the different activities of the creative process can be realized along with communication and collaboration facilities, as also, the access to digital information content and exhibition spaces have to be provided.

Finally we have presented some future trends in digital art and some lines of further research.

REFERENCES

CASPAR. (2005). *European Project*. http://www.casparpreserves.eu

Depocas A., Ippolito J., & Jones C. (Ed.). (2004). *Permanence Through Change: The Variable Media Approach*. New York: Guggenheim Museum Publications.

Grau, O. (2003). *Virtual Art – From Illusion to Immersion*. Cambridge, Massachusetts: The MIT Press.

Greene R. (2005). *Internet Art*. London: Thames & Hudson Ltd.

Hunter J., & Choudhury Sh. (2006). PANIC – an integrated approach to the preservation of complex digital objects using semantic web services. *International Journal on Digital Libraries: Special Issue on Complex Digital Objects, 6*(2), 174-183.

Linaza, T. (2003). *Artnouveau project: Recommendations and Generic Framework*. Brussels: European Commission, Project ID: artnouveau IST-2001-37863, Deliverable ID: D.5.

Löwgren, J., & Stolterman, E. (2007). *Thoughtful interaction design – a design perspective on information technology*. Cambridge, Massachusetts: The MIT Press.

Marcos, A. (2007). Digital Art: When artistic and cultural muse and computer technology merge. *IEEE Computer Graphics and Applications, 5*(27), 98-103.

OAIS - Open Archive Information System (2002). http://ssdoo.gsfc.nasa.gov/nost/isoas/us/overview.html.

Paul, Ch. (2005). *Digital Art*. London: Thames & Hudson Ltd.

Pimentel, T., & Branco, V. (2005). *Dynamic and Interactive typography in digital art*. Computer & Graphics Journal, *6*(29), 882-889.

PLANETS. (2004). European project. http://www.planets-project.eu

Quintas, R., & Dionísio, T. (2005). Displacement: Instalação Musica-Visual Imersiva que Analisa e Retracta a Expressividade Corporal. In A. Marcos, L. Valbom, & M. Meira. (Eds) *Proceedings of Artech 2005 – International Conference on Digital and Interactive Art*. Vila Nova de Cerveira, Portugal: Computer Graphics Center Press.

Valbom L., & Marcos A. (2005). WAVE: Sound and music in an immersive environment. *Computer & Graphics Journal, 6*(29), 871-881.

Valbom, L., & Marcos A. (2007). Presenting a prototype of an immersive musical instrument.

IEEE Computer Graphics and Applications, *4*(27), 14-19.

Wilson, S. (2002). *Information Arts: Intersections of Art, Science, and Technology.* Cambridge, Massachusetts: The MIT Press.

KEY TERMS

Computational Aesthetic: Is not a proper term for a digital art branch. This is a term widely used in the art and science communities devoted to the exploitation of information and communication technologies in the creation of new aesthetic forms. Computational aesthetics bridges the analytic and synthetic and integrates aspects of computer science, philosophy, psychology, and the fine, applied and performing arts and seeks to facilitate both the analysis and the augmentation of creative behavior.

Common Creative Design Space: In digital art is defined as the collection of infrastructures, tools and technologies that enables the generation of artistic content, the storage, transmission and exchange of digital data while providing the exhibition space for access to information and content by both specialists and the public.

Information Digital Content: Is defined as informative material of digital nature that holds the ability to be acted to transmit a message.

Artistic Digital Artifact: Is a designed thing built around a core of digital technology, embracing digitally coded information content displayed by means of digital media or a combination of digital and physical components. The artifact acts as a materialization of a message, a piece of information, throughout the presentation of information content while enhancing a perceptive experience on side of the user. *Artistic digital artifacts* can also be seen as *informational objects.*

Digital Art: Is defined as art that explores the computer medium (tools, technologies and digitally coded information content) as a tool and material for creation.

Design in Digital Art: Is the process that is arranged within existing digital resource constraints to create, shape, and decide all message-oriented aspects (concept, narrative, experience, technology, and aesthetic) of a digital artifact.

Chapter II
Conceptual Framework for the Intersection of Software and Art

Salah Uddin Ahmed
Norwegian University of Science and Technology (NTNU), Norway

Letizia Jaccheri
Norwegian University of Science and Technology (NTNU), Norway

Guttorm Sindre
Norwegian University of Science and Technology (NTNU), Norway

Anna Trifonova
Norwegian University of Science and Technology (NTNU), Norway

ABSTRACT

The interaction between art and technology, especially computing technology, is an increasing trend in the recent years. The context of this intersection is growing in numbers, size and aspects, each year. The number of artists participating in multimedia software or game development projects is continuously increasing and so is the number of software engineers participating in art projects like interactive art installations. As this intersection of art and technology grows, it involves people from different disciplines with varying interests creating a milieu of interdisciplinary collaborations. In this context at the Norwegian University of Science and Technology, we explore the intersection of software and art to understand different entities that are involved in the intersection. This is done by literature review and inspired by our previous experiences from participation in art projects. The objective is to conceptualize the framework of the intersection between software and art and develop a knowledge base at this interdisciplinary domain.

INTRODUCTION

Art finds expression in numerous products in society, where the development of products is complex, competitive, global and intercultural in scope. The literature is full with examples of artists applying mathematics, technology, and computing e.g. genetic art, algorithmic art, artificial intelligence to the creation of art. With the rapid development of technology, software is being used in almost every sector of life. The interconnection between art and computer science has a long history that dates back to the early sixties (1970) and it has interested many artists, researchers, art critics and theorists in the recent years. As the intersection is drawing attention of people from a diverse background and growing in size and scope, it is beneficiary for people interested in software and art to know each other's background and interests well. In a multidisciplinary collaboration, the success depends on how well the different actors in the project collaborate and understand each other. Both researchers and artists report problems regarding collaboration in multidisciplinary projects involving technologists and artists (Meyer, Staples, Minneman, Naimark, & Glassner, 1998). Thus understanding each other's interests and background knowledge is important for having a smooth collaboration and successful cooperation with all actors during an interdisciplinary project. The objective of this chapter is to provide a basic understanding of the interdisciplinary domain of software and art through a literature review. We describe the intersection through a conceptual framework which is represented by the different entities that we have encountered in our literature review as part of *SArt* project at the intersection of software and art. The framework is described by several entities such as *who, why, where, what* which stand, respectively, short for *who* are the people involved, *why* the people are interested to the intersection, *where* the intersection takes place, *what* tools and software are used in this intersection of software and art.

BACKGROUND AND RESEARCH METHOD

Background

The SArt project is conducted inside the Software Engineering group at the Department of Computer Science in the Norwegian University of Science and Technology. The focus of the project is the exploration of research issues in the intersection between software engineering and art. Our final objective is to propose, assess, and improve methods, models, and tools for software development in art context while facilitating collaboration with artists. Oates (2006) looks at computer art as an information system and proposes to extend Information Systems research agenda to include computer art. Similarly we regard the software developed in the context of art as to be considered for software engineering research and thus extend the scope of software engineering research to include research issues found in the intersection of software and art.

Since 2006, members of SArt have taken part in three interdisciplinary projects involving both artists and software engineers: *Flyndre* (http://www.flyndresang.no), *Sonic Onyx* (http://www.soniconyx.org) and *Open Digital Canvas* (http://mediawiki.idi.ntnu.no/wiki/sart/index.php/Main_Page). In the first two projects the artworks are sculptures with interactive sound systems. *Flyndre* takes as input parameters from the environment such as the local time, light level, temperature, water level and depending on these parameters creates music by exploiting algorithmic composition techniques. *Sonic Onyx* takes as input audio files and text files which are sent by users from their handheld devices such as mobiles or PDAs through Bluetooth technology. It converts those files into sound files which are later played by the sculpture. The third project, *Open Digital Canvas*, aims to embellish a white wall with a number of main boards with LEDs (Light Emitting Diodes) on them, creating a big matrix of light pixels. The project creates a platform that allows freedom of artistic expression and holds the concept of openness by keeping the hardware, software and behavior as open as possible.

Our experience shows that software engineering can play an important role in such interdisciplinary projects. For example, often the developing time and budget are limited in an art project which might lead to neglecting the design of the software and lead to a quick trial and error based solution. As a consequence, the software is often created without proper architecture, thus becomes difficult for later modification and upgrade; the documentation is poor or missing which makes software reuse hard and complex. We believe that software engineering perspective can help increasing awareness in these issues. We have also observed some common interests of the artists, for example they tend to use latest technologies in their art work. For example, Bluetooth technology is used in Sonic Onyx and a variety of sensors are used in Flyndre. Artists also want publicity of their works, in all of the mentioned projects, artists wanted websites to publish their artworks. Common interest of using open source software is also noticeable from the projects.

The experience gained from the projects give us insights about some of the issues in the development of software based artworks and provides us interesting information about the artists, their ideas and interests. Furthermore, a preliminary literature investigation showed that art's relationship with software involves people from diverse background and interest, for example art critics, software developers, educators and each has his/her own interests and attitudes. In order to get a wider picture of the intersection and capture issues related to software engineering beyond **software dependent art** projects, we extended our focus and conducted the review on the intersection of software and art. The objective of our literature review here in the context of SArt is to answer to the questions: where and how software and art intersect each other? Who are the important

actors / entities in this intersection and how can we conceptualize the intersection with respect to these entities?

Research Method

The conceptual framework presented in this chapter is the result of our literature review which has been conducted following the systematic review process suggested by Kitchenham (Kitchenham, 2004). The details of the process have been published in our article (Trifonova, Ahmed, & Jaccheri, 2007). In this section we present the criteria that we have used to select the articles in order that the readers get an overview of the scope and the focus of our literature review. We have set in advance and agreed on common relevance criteria for inclusion/exclusion of articles in the literature review. We have also identified the search strategies, including a list of searchable electronic databases of scientific publications and a starting list of keywords. Criteria were cleared and polished in several iterations at the beginning of the literature review. Relevant articles are those that address one or more of the following:

- Artists attitude towards software
- Software engineers attitude towards art
- Influence and usage of software in art
- The influence of arts in computing
- Artists' and software developers' joint works such as art projects, multimedia installations, multidisciplinary courses.
- Working process related issues such as collaboration problems, communication problems between artists and software developers.
- Software development related issues such as maintenance, requirements, development process and CASE (Computer Aided Software Engineering) tools in context of software development projects involving artists.

- Features of artistic software, their development history, usage and evolvement including both the communities of artists and software developers.

Articles that are not published in any scientific journal or conference are excluded from the review, for example, online articles, articles accompanying art festivals and workshops were excluded on account of this reason. We want to mention here that the conceptual framework presented later in this chapter includes not only reviewed articles from our literature review, but also includes relevant books and knowledge gained from our participation into several art festivals, art projects and conferences.

The Intersection of Software and Art

The intersection between software and art includes people with varying backgrounds such as artists, developers, critics who involve themselves into the intersection with purposes that vary for each of them. This leads to the people at the intersection having different viewpoints on different issues in this interdisciplinary domain. For example, a software engineer has a different purpose/interest than that of an art critic when he/she involves him/her self in the intersection. In the following section we describe the **intersection of software and art** in the view of the entities that exist in the intersection, such as actors (who are involved?), interests/viewpoints (Why are the people interested?), places (where this involvement takes place?), tools and technologies (what tools and technologies has bound this relationship?).

Who (Are the People Involved in this Domain)

The people involved in the intersection of software and art can be identified as artists, designers, software developers, software engineers, theorists, critics and researchers. It should be mentioned

here that these categorizations are based on the roles of the individuals involved and they are not mutually exclusive. This list is neither complete nor exhaustive; rather it covers the roles that we have found in our literature review. One person can have many roles at the same time, for example, when the interactive filmmaker Florian Thalhofer creates interactive documentary software Korsakov; he is both an artist and a software developer (Blassnigg, 2005).

Artists

Artists here refer to those artists who are using software for realizing some part of their artwork. This includes new media artists, photographers, filmmakers, curators, animators and all others who utilize software to accomplish their work. Many artists use software technology not directly for their work, but for publishing their work or communicating with others through web tools and technologies. It is interesting to note that some of the artists coin the term 'info-architect' to refer to the artists working with new media technologies (Blassnigg, 2005).

Artists, together with theorists and **art critics** (see the following section) are more involved in discussions about the **intersection of software and art** compare to the software practitioners. For example, according to Bond (2005), it is the new media artists, not the software practitioners that take part in the movement of **software art**. This movement of **software art** refers to the viewpoints of the artists who believe that software itself, that is, the raw code of software that works behind every software application should be treated as a medium or material of art. Bond (2005)states, "Casting software as an artistic medium might strike many readers as odd, or even objectionable, but there is a growing body of evidence to show that it is perceived and utilized in just this way" (p. 118). Theorists Florian Cramer and Ulrike Gabriel extend the view of **software art** by focusing on the underlying code (Cramer & Gabriel, 2001).

Theorists and Art Critics

Development of digital, information and communication technologies has built a complex corpus called cyber-culture. "Media and multimedia information and communication technologies generate new promises, problems and threats; and artists undertake efforts to examine this emerging area that has been repeatedly considered as a post biological syndrome. In other words artists do not only use media technologies but also scrutinize and challenge them" (Kluszczynski, 2005, p. 124). Thus, in the **intersection of software and art** we find a number of theorists and **art critics** as well. As an example, Erkki Huhtamo, Mathew Fuller, Florian Cramer, Jeffery Cox, Lev Manovich are a few of the names to list in this category. Many of the people mentioned here have several roles, varying from artist, teacher, theorist and programmer. For example Erkki Huhatamo is a lecturer, researcher, writer and curator all by the same time. Manovich is a lecturer and writer of many articles and books. His book, "The Language of New Media" is considered by many reviewers to be the first rigorous and far reaching theorization of the subject. Even though there might not be a person who can be termed as only theorist, we mention them as a separate category here as we find a significant portion of research articles that we have reviewed are contributed by these theorists and **art critics**.

Software Engineers

In context of the intersection between software and art, by software engineers we mean any person who is involved in the development of what we call '**artwork support tools**', i.e. software developed for the artists and intended to be used for some art purposes. These software engineers or developers might be students who follow some interdisciplinary courses and/or work together with artists in art projects. In many cases, software developers take part in art projects and work with artists for a short period. Sometimes artists working with digital media recruit their personal developers, for example, Paris based artist Christophe Bruno employs his personal software developer to realize his concepts with the help of software (http://www.christophebruno.com). However, there are new media application developing groups or companies run by artists that recruit software developers for long time collaboration, example of which might be the Mediamatic Lab in the Netherlands (www.mediamatic.nl) and Soundscape Studios in Norway (www.soundscape-studios.no).

In addition, there are some software engineers who do not have direct involvement with artists, but are involved with the debate of the role of art versus science in software engineering. The debate focuses on the creativity and innovativeness in software engineering discipline. For example, referring to the process of solving complex problems where there is no perfect solution achievable through rigorous methods of science, Bollinger (1997) states "the artistic part of this process lets science move unexpectedly into new currents and previously unmapped understanding" (p. 125). It is interesting to see that the panel discussion of OOPSLA 2004 was titled "Software Development: Arts & Crafts or Math & Science?" But there are also others who think different, for example, McConnell says that this debate was appropriate at the beginning, at least 30 years back when it started, as at that time software engineering didn't have an established body of knowledge. But now, the author recommends that it should be called engineering as there is significant body of knowledge (McConnell, 1998). A viewpoint combining the both parties could be "Building software is a complex and exciting task that is a unique confluence of engineering, mathematics and artistic insight, and it is important we resist the tendency to view it in a single dimension" (Wei-Lung, 2002, p. 29).

There are other software engineers who refer some software as "art for art's sake". Knuth includes in this categorization programs which are appreciated in light of the challenge the programmers face in creating it, for example one line programs, programs that output themselves, etc. (Bond, 2005). Knuth also compares the beauty of a program to the beauty in literature or music – programs whose tasks are stated elegantly and whose parts come together symphonically are beautiful programs.

Researchers

Many people are involved with software and art for the purpose of research in areas such as user interface, collaboration, creativity and innovation. These researchers may come both from art and computing disciplines. Human Computer Interaction (HCI) community is interested in aesthetics of user interface. Bertelsen and Pold assert that interacting with the interface is a cultural activity. They propose an approach to interface design in which the interface should be 'criticized' by someone with knowledge in aesthetics and ideally some experience with art and literature. This approach might be used for increasing the aesthetic value in the user interface (Bertelsen & Pold, 2004).

Creativity and Cognition Studios research group is an active group interested and involved in the intersection of software and art. Researchers like Linda Candy, Ernest Edmonds and many others working in the Creativity and Cognition Studios have contributed in numerous collaborative research works involving art and technology. In fact, creativity and cognition conference which was instigated by Ernest Edmonds and Linda Candy has turned into a regular event in Association of Computing Machinery's SIGCHI calendar.

There are also many art researchers who have addressed significantly the collaboration between artists and technologists. For example, UIST 98 panel discussion was about artists and technologists working together where artists like Michael Naimark, Loretta Staples were in the panel with the technologist-researchers Jon Meyer, Andrew Glassner and Scott Minneman (Meyer et al., 1998). Individual researchers and researchers whose main stream research is not focused on 'software and art' are also found to be interested in the intersection. An example is Briony J. Oates who proposes to extend the scope of information systems' research by including **computer art** (Oates, 2006).

Why (Art and Software Need Interaction)

Software engineers and artists work together for many reasons, varying from personal interests to the more general interests of their respective disciplines. In this section we highlight some of the reasons that we have found in the literature review that brings artists together with the developers.

Artists Need Tools Support

One of the main reasons artists seek help from the technologists is to get support with the tools that they need for the realizations of their artwork. "… they [artists] are often ill equipped to work with complex technologies. It is this factor that may incline any artist wishing to work within the domain of contemporary Art & Technology and digital art towards collaboration as a necessary means for realising their intentions" (Jones, 2005, p. 76).

Supporting artists with necessary tools invoke many questions, such as: What kind of tools artists want? What desired features might be included in these tools? For what purposes the available tools are used? and so on. Many researchers have addressed issues related to the creation of tools to support artists, designers or creative people in general. Here are some examples: (Machin, 2002) aims at reducing the gap between developers and artists and enhancing creativity; Warr and O'Neill (2007) focus on visualization at early stage of collaborative work, supporting individual creativity; Mamykina et al. (2002) discuss the importance of the ability to track progress and revisit design decision. Referring to Macromedia Director Machin (2002) states "this kind of software provides the artists with a powerful tool which assist in visualising the piece even at a very early stage of its design" (p.3). Biswas et al. (2006) described that a development toolkit (for new media content design) should support separate design (by the artists) and implementation (by software developers) of the final application. Thus, it should be able to decrease the semantic gap between artists and technologists and assist their dialogue. Gross (2005) believes that when it comes to build software tools for artists, pragmatics analysis (context and behaviour) is crucial because art is heavily immersed in practice and action, and because art is valued on its ability to communicate.

Art Projects Need Software Engineering

Software developers who have the experience of working together with artists in artistic projects have realized the need of software engineering knowledge in those projects. This is what we have also realized from our participation into the art projects. Machin (2002) states "What is required now is to carry out further work in order to establish methods that can be utilized in future requirements capture software. By working with artists in the installation art field, we can seek out ways to enable the capture of the artist's ideas without inhibiting the artistic process." (p. 8). Many researchers/software engineers are also interested in comparing the software development method and analyze which ones suit better an art project in a certain context. Furthermore, Candy and Edmonds (2002) investigate what are the most appropriate evaluation methods in

software intensive art and if the evaluation should be done only by the artists or should include the software engineers as well. Where the artworks are implemented in limited time and budget and where artists leads the project, the maintenance and upgrading issues are often overlooked. Thus the maintenance and upgrade of these kinds of software supported artworks becomes one of the prime sectors where art projects need software engineering help.

Computing Needs Aesthetics

Fishwick (2005) reports the result of a survey on the usefulness of aesthetic methods on several areas of computer science. The result shows that data structure, algorithms, digital logic, computer architecture was chosen by the respondents as some of the fields where **aesthetic computing** can be used. Information visualization and software visualization are other fields that can contribute to bringing art/aesthetics inside of computing (P. A. Fishwick, 2007).

Adams (1995) addresses the importance of teaching aesthetics in engineering education and the role of aesthetics in engineering. Paul Fishwick has coined a term "**Aesthetic Computing**" to refer to a new area of study, which is concerned with the impact and effects of aesthetics on the field of computing. As an example, the discrete models found in computing can be transformed into visual and interactive models which might increase the understanding of the students. Fishwick et al. (2005) represents a method for customizing discrete structures found in mathematics, programming and computer simulation. Based on the method, they transform some discrete models (for example finite state machine) to geometric models and assess students' perception on how customized models affected their understanding and preferences regarding visual and interactive model representations.

Technology Changes Art's Medium, Audience And Business

In the recent years more artists are exploring the Internet as a medium to reach their audience/spectators. When Internet is used as a way to transmit art the focus falls on the individual spectator, in contrast from the museum type of art presentation (Nalder, 2003). The Open Source Software tools developed in cooperative spirit and distributed via the internet, enable artists to experiment and create low-cost **artworks**. Besides the new way of reaching the target audience, Internet and Web has created new business models as well. For example, softwareARTspace (http://www.softwareartspace.com) is a website which provides digital **artworks** commercially. Usage of information technologies implies further improvements of tools and technologies to address the issues related to business and publicity, such as protection of artworks and copyrights. Thus, it brings artists and technologists together to solve these issues. As an example, small granularity technique has been proposed by Shoniregun et al. (2004) for the protection of artwork. The potential reciprocal interaction between artists and technologists in the form of demands of the user (artists) which stimulate an engineer to extend the technology in some way thus extends the possibility of its use (Jones, 2005). This helps both technology and art to co-evolve in a configuration of mutual interdependence.

Where (The Intersection Happens)

The influence of technology on art has also created a number of new genres of arts, such as internet art, generative art, new media art, net art, **software art** and many more (Nalder, 2003). Many of these fields might still not be well defined or still at an early stage of evolvement. Professor Peter Weibel coins all of these fields by a single term "**Computer art**". For example, he states in the jury statement at the Ars Electronica '92,

"**Computer art** is concerned with artworks that 'were impossible to produce before the invention of the computer … even unimaginable" (Oates, 2006). Some sectors of art where interaction between software and art happens extensively are computer generated sound and electronic music, visual arts, virtual reality, performance arts etc. Interdisciplinary events and activities are often supported by institutions or organizations. Thus, even though software is accessible to individual artists, it is not unusual to find most of the cases of intersection described in the reviewed articles have occurred in the context of some institutional support. Here we present some major places/sectors that we have identified from the literature where art and software meet each other.

In Educational Institutes

In art schools and computing schools interdisciplinary courses are conducted which include students from both art and computing discipline. Besides these interdisciplinary courses, there are also cases where the need of computing education is realized in the art discipline, and the need of art education in the computing discipline.

In Art Schools

Donna (1991) mentions "It is a fundamental mistake for a university not to provide such training [computer training] to non computer science majors because of the rapid growth of computing in arts and humanities" (p.2). The importance of the inclusion of computer graphics courses in the fine arts syllabus is recognized by art institutes (Garvey, 1997; Sardon, 2006). However, keeping the balance between the traditional fine arts skills (e.g. drawing, painting, sculpture) and digital skill mastery is of a great importance. Designing and conducting multidisciplinary courses need consideration of special issues such as ensuring good team work, having common language and understanding each other's skills and abilities as well as having respect for each other's domain.

While presenting their experience of organizing and conducting an interdisciplinary computer animation course in Ebert and Bailey (2000) mention the importance of effective teamwork and learning from each other's disciplines.

In Computing Schools

Parberry et al. (2006) describe their mainly positive experiences from a course and projects in game programming offered to the two groups (art and computer science) of students. Argent (2006) describes two game development courses offered at the University of Denver which were built upon a four way partnership between computer science, digital media studies, electronic media arts design and studio art. The role of teaching aesthetics to engineering students was addressed by Adams almost 12 years earlier (Adams, 1995). Zimmerman et al. (2001) describe a course on Virtual Reality and its "implications for computer science education". Jaccheri et al. (2007) reports their experience of running an interdisciplinary course for three years where they have reported that such an interdisciplinary course gives software students learning outcomes that are quite different from what they get from more traditional software engineering team projects, in particular concerning interdisciplinary skills and self insight. Fishwick (2003) describes the Digital Art and Science (DAS) curricula provided at the University of Florida where he discusses the importance of combining computer science and art in the educational phase and the positive experiences obtained. He suggests that such practices will stimulate creativity in Computer Science students.

In Software Industry

Art meets with software in the software industry as well, for example in building entertainment and leisure related applications, games, and demo-scenes. Computer based entertainment and leisure related applications range over a wide spectrum evolving from console based ones to augmented/mixed reality, pervasive, ubiquitous, immersive, context aware and social computing experiences. The involvement of artists and developers takes place in the industry in the following way: technologists develop more intelligent interactive context-aware systems while designers, artists and design-art researchers use the existing technologies as a "new media" to design with, and deliver "values" and "experiences" which go beyond the notional functionalities of technology (Biswas et al., 2006). The major areas where the collaboration takes place can be identified as Game Development, Human Computer Interaction, User Interface Design and Web Design. Game development is an interdisciplinary field which requires both technical and creative appreciation. Games are considered as a rich field for art not only because of the design, but also because of the cultural issues and perspectives that might be identified within the games (Blais & Ippolito, 2006).

In Research Institutes

Apart from the industry and the art or computing schools, there are also many research institutes where a research setting is intentionally created for artists and technologists to work closely together. Even though these research institutes may be a part of a university or an industry, the objective of research setting is different from general purpose art projects. Often, this kind of collaboration is done through "Artist-in-residence" programs, for example, the Xerox PARC artist-in-residence program (Harris, 1999), COSTART project (Candy, 1999). The objectives of these programs are different, in a way that they are aimed for innovation, creativity or elegant solutions of a complex problem, whereas in a general art project, the objective is to realize an art work without any explicit intention of innovation. Many industries also arrange artists in residence program from time to time. This situation differs from the one where artists work in the same enterprise with other technologists as part of regular production work.

In Art Projects

In public art especially new media art projects, art meets software for the realization of the projects. Sometimes artists learn to use or code the software by themselves; sometimes they work together with the technologists to get the software developed. The art projects can be creation of simple piece artwork such as internet art, **software art**, and digital art and so on where a single artist works with the software. On the other hand it can be an interactive art installation, or a multimedia presentation, film or electronic music where one or more artist work with software and collaborates with other technologists.

In Art Centres and Festivals

Art is promoted by many festivals where artists get chance of exhibiting their works to the audience, communicating ideas and concepts, and evaluating and/or criticising artworks. Many art festivals along with art institutes promote new media art which lies at the intersection of art and software. Examples include Ars Electronica (http://www. aec.at/de/index.asp), PixelACHE (http://www. pixelache.ac/), Read_me (http://readme.runme. org/), Transmediale (http://www.transmediale. de), Piksel (http://www.piksel.no/), Make Art (http://makeart.goto10.org/2007/), Trondheim Matchmaking (http://matchmaking.teks.no/). Competitions like PrixArs (International competitions for Cyber Arts organized by Ars Electronica) offers prizes on different categories such as computer animation, films, interactive art, digital music, hybrid art, digital communities and media art research. ARCO BEEP new media art awards targets the goal of advancing the production and exhibition of new media art and art linked to new technologies (www.arco.beep.

es). PixelACHE presents projects experimenting with new media and technology with a goal to act as a bridge between the traditional creative discipline and the rapidly developing electronic subcultures. Other institutes and festivals mentioned here also stress importance on media art and use of technology.

What (Software Tools are Used in this Domain)

After we have identified people (who), reasons behind their interest at the intersection (why) and the places/sectors (where) art intersects with software, here we present some practical examples of what (tools and technologies) binds the relationship between software and art. Artists tend to use software for different purposes. Quite often they use commercial software; often they are interested in open source software as a cheap alternative. In few cases, artists develop their own software. Most of the time they use the software as it was intended to be used by the creator of the software but sometimes they can be creative and use it in a different way which was not intended. For example the artist Jen Grey used the proprietary software *Surface Drawing* in a unique way to draw live models, a purpose which was not intended (Grey, 2002). Some software is used as a tool to develop artwork; some as a media to support artists' activities indirectly (for example collaboration) while others are general purpose programming languages used to build applications. Besides these, there is also customized software i.e., software that is built for a specific artistic purpose. Several articles mention this kind of software (example, Datareader, Korsakov mentioned in Table 1) which was developed by either artists alone, or with the help of programmers as part of an art project. In this section we list the software/tools that were mentioned in the literature. These tools provide the reader an overview of what type of software and tools are used or required by the artists.

Artwork support tools, i.e. tools used to develop artworks are mainly special purpose artistic software which specializes on some tasks such as visualization, sound manipulation or animation. The following table gives a list of these kinds of software that were mentioned in different articles found in our literature review along with their purposes/ functionalities. The list gives the readers an idea about the tools that artists are using/ interested in using for different art purposes. The list also contains customized tools that were developed for some specific artwork or art purpose, for example Datareader was developed for taking as input meteorological data.

Apart from the **artwork support tools** there are other tools and software that artists use for supporting other activities such as communication, publicity, sharing works, ideas etc. Internet and Web tools has become not only a medium for the artist to publish and present their work and activities but also a medium for communicating and collaborating with other artists. "The digital arts site Rhizome is recognized for the crucial role it plays enabling exchange and collaboration among artists through the network" states Walden in his review on the book Net_Condition: Art and Global Media (Walden, 2002). The tendency of publishing artists work through websites to reach

Table 1. Software tools mentioned in different articles

Category	Software Name	Description	Referenced in
1. Graphics Manipulation	Illustrator	Vector based drawing program	(Garvey, 1997)
	Freehand	Tool to create layouts and illustrations for print /Web designs	(Garvey, 1997)
	CorelDraw	A vector graphics editing software	(Garvey, 1997)
	Photoshop	A graphics editor software	(Grey, 2002)
	Painter 6.0	Painting tool for graphic designers and fine artists.	(Grey, 2002)
2. Multimedia Authoring	Flash	Animation and interactivity	(Sung-dae, Jin-wan, & Won-Hyung, 2006) (Sardon, 2006) (Marchese, 2006)
	Director	Creating interactive content for fixed media and internet	(Garvey, 1997) (Sardon, 2006) (Machin, 2002)
3. 3D graphics Manipulation	SoftImage 3D	3D animation software for games, film and television	(Jennings, Giaccardi, & Wesolkowska, 2006)
	3DStudio Max	Professional 3D modeling, rendering and animation software	(Garvey, 1997) (Strömberg, Väätänen, & Räty, 2002)
	Mini CAD, Auto CAD	A suite of CAD (Computer Aided Design) software products for 2- and 3-dimensional design and drafting	(Garvey, 1997)
	Alias / Wavefront	3D graphics software	(Garvey, 1997)
	WorldUp	Software development environment for building 3D/VR applications	(Zimmerman & Eber, 2001)
	Maya	High end 3D computer graphics software	(Zimmerman & Eber, 2001) (Steinkamp, 2001)
	Breve swarm simulation	A package for building 3D simulations of multi-agent systems and artificial life	(Boyd, Hushlak, & Jacob, 2004)

continued on following page

Table 1. continued

4. Sound Manipulation	Pure Data	Tool for creating interactive computer music and multimedia works	(Jennings et al., 2006)
	Max/MSP	A graphical programming environment for music and multimedia	(Jennings et al., 2006) (Marchese, 2006) (Edmonds, Turner, & Candy, 2004)
	GigaStudio 160	Software for music and sound effects	(Strömberg et al., 2002)
5. Video Manipulation	SoftVNS video toolkit	A real time video processing and tracking software for MAX/MSP	(Edmonds et al., 2004)
	Jitter	Extends MAX/MSP to support real time manipulation of video data	(Polli, 2004)
	Korsakov	Interactive Documentary software	(Blassnigg, 2005)
6. Other Applications	ELE (Expressive Lighting Engine)	Control Lighting for a virtual environment	(Gross, 2005)
	Mobile Bristol toolkit	A tool to create and share mobile, location–based media	(Biswas & Singh, 2006)
	Mobile Experience Engine	A software development platform for creating advanced context-aware applications and media-rich experiences for mobile devices	(Biswas & Singh, 2006)
	Sculpture simulator	A sculpture simulator with its own programming language	(Machin, 2002)
	Particle dynamics	A software tool set for simulating natural phenomena.	(Steinkamp, 2001)
	Datareader	A Max/MSP object to read text data and convert them into sound	(Polli, 2004)
7. Programming languages /API	VRML	Virtual Reality Modeling language	(Nalder, 2003)
	General purpose programming languages such as python, C++, Java etc.	To create wide range of applications	(Nalder, 2003)
	OpenGL	Industry standard and API for high performance graphics	(Fels et al., 2005)

audience was also observed during our participation in the three projects described in section 0. The other purposes of website includes, publishing artworks, selling art products, virtual tour of museums and creating online communities, discussion groups or forums, and blogging.

Domain specific programming language is preferred by the artists compared to the general purpose programming languages unless the artist does not aspire to be a professional programmer. This is because general languages can be daunting due to the steep learning curve associated with learning programming. Besides, artists often prefer to work with intermediate tools where the need for programming is reduced. But that does not make any limitations for artists to learn the general purpose programming languages. In some of the articles that we have reviewed mentions a number of general purpose languages which was used to realize **artworks** or some artistic software, for example, C++, ActionScript, UML, 2D OpenGL.

Last but not the least, the role of open source software has to be mentioned as an important factor for making artists more interested to software. Artists tend to move towards using open source technology not only because they are cheap, even free of charge, but also because many artists believe in the open source ideology. Halonen (2007) mentions that new media art is based on cooperation to a greater degree than many art forms that can be created alone. He identified four groups with diverse motives: i) using open source network as an important reference for professional image, ii) using open source projects as a platform for learning, iii) an opportunity to seek jobs and iv) enrich professional networks. From our project experience, we identified that some artists want to have open source projects so that they can build an interested community around the project which might assist in the further development, upgrade and maintenance of the project at a low cost (Flyndre and SonicOnyx). Open source and free software usage in artists community is also encouraged by different art festivals such as piksel (http://www.piksel.org), makeart (http://makeart.goto10.org/). The interest is also visible by the activities of different art organizations/institutes such as APO33 (http://apo33.org) ap/xxxxx (http://1010.co.uk/) Piet Zwart Institute (http://pzwart.wdka.hro.nl/).

The Framework

In the earlier sections we have presented the framework of the intersection in terms of different entities in the intersection, i.e. people and views, place and tools. Here the framework is presented through a table (table2). In the table, rows represent where (software and art meet) column represent who (are the actors), entry in each cell represent the reason why the actors participate in a given place.

Here we present the list of reasons (Why):

1. Design and conduct interdisciplinary courses
2. Conduct research and disseminate knowledge of conducting interdisciplinary courses
3. Foster innovation and creativity
4. Learn technology
5. Learn to work in **multidisciplinary projects**
6. Apply aesthetics in computing
7. Do research and development of products
8. Design user interface and enhance/improve human computer interaction
9. Enhance user experience
10. Use technology in artworks
11. Realize the artwork through software
12. Provide tools and technology support
13. Share and exchange knowledge

Table 2. Where, who and why dimension of the intersection of software and art.

Who / Where	Artists	Software Engineers	Researchers	Theorists & Art Critics
Educational Institutes (Computing and Art schools)	4, 5, 10	5	1, 2, 3	X
Research Institutes	3, 17	3, 6, 17	3, 18, 17	X
Software Industry	8, 9	7, 8, 9	7	X
Public Art, Art Projects	10	11, 12	18	16
Festivals	13, 14, 15	12, 15	17	16

14. Exhibit art work in public
15. Extend collaboration between artists and software engineers
16. Follow the changes of art and their social and cultural effects
17. Present research works
18. Conduct research on artists-technologists collaboration and reduce the gap between them

Different people have different reasons to involve themselves with the interdisciplinary domain. The fact that different actors have different viewpoints is largely due to their varying roles and background knowledge. That is why, while we see artists picking up different software tools and technologies and even the naked codes as a material for artworks, software engineers are seen debating over the role of art in software engineering.

FUTURE TRENDS

The interaction between software and art is an increasing trend followed by many artists, software developers and technologists. As the technology advances, more and more sectors of life are influenced by the use of technology. Art finds its way to reflect on every artefact of life, so the intersection of art with software takes place naturally as it happens with other technologies. The application of different computing tools and technologies has already been established in arts. The trend to include the aesthetics in computing, especially in aspects of design is a recent but growing trend. In future when the computing speed and network speed will increase further, the value of aesthetics will be even more recognized in computing (Hoffmann & Krauss, 2004). Until now aesthetics is recognized mostly in human computing interaction, but in future other fields of computing such as data structure, algorithms, digital logic, computer architecture might be also using more aesthetics as anticipated by Paul Fishwick (2005).

As time moves on and the interdisciplinary domain matures, this collaboration between art and computing attracts not only more people but also more disciplines and it gives birth to new issues and new perspectives. For example, the need of software technology for artists leaded to the inclusion of different computer courses in art and computing institutes. This in turn has raised the pedagogic perspectives of conducting interdisciplinary courses involving art and computing students. In future this might also raise the importance of learning technology through art and fun especially for children. **Software dependent arts** such as installation arts that are placed in public space attract the attention of many viewers including children and old people. Thus it involves social and cultural aspects which include perspectives of learning, consciousness in the society through computational arts. In future when there will be more **software dependent arts** placed in public space, it will be interesting to consider these findings, in relation to the different people and different perspectives of life surrounding computational arts as well as building a knowledge base in this growing interdisciplinary domain.

CONCLUSION

In this article we present an outline of different entities (who, why, where, what) at the interdisciplinary research domain of software and art. The interaction between art and software happens naturally as technology finds its way to influence every sphere of our life and artists reflect, scrutinize and challenge technology at the same time as they use it for extending their expressions. Artists working with technology are faced with multiple tasks that demand them to perform different roles such as researcher, engineer, programmer etc. In many cases artists have background knowledge of technology from their previous career or education, in other cases they learn a bit of the technology while working with it. But in general case, artists require support from the technologists to realize their visions. This brings artists together with technologists and the whole phenomenon brings together other professionals who are interested or get involved with this intersection of technology and art. The conceptual framework that we present in this chapter provides a detailed insight of how the intersection between software and art is. This intersection is mostly based on the articles from our literature survey and our experience from art projects. So it might not include all the actors and entities in this interdisciplinary domain. We have limited our literature review to scientific publications, but many recent trend and relevant information in the intersection of art and software is available in sources such as websites, online articles, and artists' biographies. The framework might look different if we include these sources.

Even though art has connection with software since a long while, but the intersection between art and software is a very recent trend and it is continuously changing and growing. Thus, in future this framework will have more entities to include. The conceptual framework adds value to the knowledge base of the interdisciplinary domain by structuring the information about the intersection. A knowledge base is a necessary element for an interdisciplinary domain based on which the domain can prosper and grow. The conceptual framework will thus act as basis for getting into detailed understanding of this domain. Future work regarding this conceptual framework might include extension and improvement which will further enhance the knowledge base of this interdisciplinary domain.

REFERENCES

Adams, C. C. (1995). *Technological allusivity: appreciating and teaching the role of aesthetics in engineering design.* In Proceedings of the Frontiers in Education Conference, 1995. (pp. 3a5.1-3a5.8). Washington, DC, USA: IEEE Computer Society.

Argent, L., Depper, B., Fajardo, R., Gjertson, S., Leutenegger, S. T., Lopez, M. A., et al. (2006). Building a game development program. *Computer, 39*(6), 52-60.

Bertelsen, O. W., & Pold, S. (2004). *Criticism as an approach to interface aesthetics.* In Proceedings of the third Nordic conference on Human-computer interaction (pp. 23-32). New York, NY, USA: ACM Press.

Biswas, A., Donaldson, T., Singh, J., Diamond, S., Gauthier, D., & Longford, M. (2006). *Assessment of mobile experience engine, the development toolkit for context aware mobile applications.* In Proceedings of the 2006 ACM SIGCHI international conference on Advances in computer entertainment technology (pp. 8). New York, NY, USA: ACM Press.

Biswas, A., & Singh, J. (2006). *Software Engineering Challenges in New Media Applications.* In the Proceedings of Software Engineering Applications (~SEA 2006~) (pp. 7). Dallas, TX, USA: ACTA Press.

Blais, J., & Ippolito, J. (2006). *At the Edge of Art.* London: Thames & Hudson Ltd.

Blassnigg, M. (2005). Documentary Film at the Junction between Art and Digital Media Technologies. *Convergence-The International journal of New Media Technologies, 11*(3), 104-110.

Bollinger, T. (1997). The interplay of art and science in software. *Computer, 30*(10), 128, 125-127.

Bond, G. W. (2005). Software as art. *Communications of the ACM, 48*(8), 118-124.

Boyd, J. E., Hushlak, G., & Jacob, C. J. (2004). *SwarmArt: interactive art from swarm intelligence.* In Proceedings of the 12th annual ACM international conference on Multimedia (pp. 628-635). New York, NY, USA: ACM Press.

Candy, L. (1999). *COSTART Project Artists Survey Report: Preliminary Results.*: Loughborough University.

Candy, L., & Edmonds, E. (2002). *Modeling co-creativity in art and technology.* In Proceedings of the 4th conference on Creativity & cognition (pp. 134-141). New York, NY, USA: ACM Press.

Cramer, F., & Gabriel, U. (2001). Software Art and Writing. *American Book Review, 22*(6).

Donna, J. C. (1991). Interdisciplinary collaboration case study in computer graphics education: "Venus & Milo". *SIGGRAPH Computer Graphics, 25*(3), 185-190.

Ebert, D. S., & Bailey, D. (2000). A collaborative and interdisciplinary computer animation course. *ACM SIGGRAPH Computer Graphics, 34*(3), 22-26.

Edmonds, E., Turner, G., & Candy, L. (2004). *Approaches to interactive art systems.* In Proceedings of the 2nd international conference on Computer graphics and interactive techniques in Australasia and South East Asia (pp. 113-117). New York, NY, USA: ACM Press.

Fels, S., Kinoshita, Y., Tzu-pei Grace, C., Takama, Y., Yohanan, S., Gadd, A., et al. (2005). Swimming across the Pacific: a VR swimming interface. *Computer Graphics and Applications, IEEE, 25*(1), 24-31.

Fishwick, P. (2003). Nurturing next-generation computer scientists. *Computer, 36*(12), 132-134.

Fishwick, P. (2005). Enhancing experiential and subjective qualities of discrete structure representations with aesthetic computing. *Journal of Visual Languages & Computing, 16*(5), 406-427.

Fishwick, P., Davis, T., & Douglas, J. (2005). Model representation with aesthetic computing:

Method and empirical study. *ACM Trans. Model. Comput. Simul., 15*(3), 254-279.

Fishwick, P. A. (2007). Aesthetic Computing: A Brief Tutorial. In F. Ferri (Ed.), *Visual Languages for Interactive Computing: Definitions and Formalizations*: Idea Group Inc.

Garvey, G. P. (1997). Retrofitting fine art and design education in the age of computer technology. *ACM SIGGRAPH Computer Graphics, 31*(3), 29-32.

Grey, J. (2002). *"Human-computer interaction in life drawing, a fine artist's perspective".* In Sixth International Conference on Information Visualisation (IV'02) (pp. 761-770). Los Alamitos, CA, USA: IEEE Computer Society.

Gross, J. B. (2005). *Programming for artists: a visual language for expressive lighting design.* In IEEE Symposium on Visual Languages and Human-Centric Computing, 2005 (pp. 331-332). Los Alamitos, CA, USA: IEEE Computer Society.

Halonen, K. (2007). Open Source and New Media Artists. *Human Technology - An interdisciplinary journal on humans in ICT environments, 3*(1), 98-114.

Harris, C. (Ed.). (1999). *Art and innovation: the Xerox PARC Artist-in-Residence program.* Cambridge, Massachusetts: MIT Press.

Hoffmann, R., & Krauss, K. (2004). *A critical evaluation of literature on visual aesthetics for the web.* In Proceedings of the 2004 annual research conference of the South African institute of computer scientists and information technologists on IT research in developing countries (pp. 205-209). South Africa: South African Institute for Computer Scientists and Information Technologists.

Jaccheri, M. L., & Sindre, G. (2007). *Software Engineering Students meet Interdisciplinary*

Project work and Art. In Proceedings of the 11th International Conference on Information Visualisation (pp. 925--934). Washington, DC, USA: IEEE Computer Society.

Jennings, P., Giaccardi, E., & Wesolkowska, M. (2006). *About face interface: creative engagement in the new media arts and HCI.* In CHI '06 extended abstracts on Human factors in computing systems (pp. 1663-1666). New York, NY, USA: ACM Press.

Jones, S. (2005). *A cultural systems approach to collaboration in art & technology.* In Proceedings of the 5th conference on Creativity & cognition (pp. 76--85). New York, NY, USA: ACM Press.

Kitchenham, B. (2004). *Procedures for Performing Systematic Reviews*: Keele University Technical Report TR/SE-0401 and NICTA Technical Report 0400011T.1.

Kluszczynski, R. W. (2005). Arts, Media, Cultures: Histories of Hybridisation. *Convergence-The International Journal of New Media Technologies, 11*(4), 124-132.

Machin, C. H. C. (2002). *Digital artworks: bridging the technology gap.* In Proceedings of the 20th Eurographics UK Conference, 2002. (pp. 16-23). Washington, DC, USA: IEEE Computer Society.

Mamykina, L., Candy, L., & Edmonds, E. (2002). Collaborative creativity. *Communications of the ACM, 45*(10), 96-99.

Marchese, F. T. (2006). *The Making of Trigger and the Agile Engineering of Artist-Scientist Collaboration.* In Proceedings of the conference on Information Visualization (pp. 839-844). Washington, DC, USA: IEEE Computer Society.

McConnell, S. (1998). The art, science, and engineering of software development. *IEEE Software, 15*(1), 120, 118-119.

Meyer, J., Staples, L., Minneman, S., Naimark, M., & Glassner, A. (1998). *Artists and technologists working together (panel).* In Proceedings of the 11th annual ACM symposium on User interface software and technology (pp. 67-69). New York, NY, USA: ACM Press.

Nalder, G. (2003). *Art in the Informational Mode.* In Proceedings of the Seventh International Conference on Information Visualization (pp. 110). Washington, D.C, USA: IEEE Computer Society.

Oates, B. J. (2006). New frontiers for information systems research: computer art as an information system. *European Journal of Information Systems, 15*(6), 617-626.

Parberry, I., Kazemzadeh, M. B., & Roden, T. (2006). *The art and science of game programming.* In Proceedings of the 37th SIGCSE technical symposium on Computer science education (pp. 510-514). New York, NY, USA: ACM Press.

Polli, A. (2004). *DATAREADER: a tool for art and science collaborations.* In Proceedings of the 12th annual ACM international conference on Multimedia (pp. 520-523). New York, NY, USA: ACM Press.

Sardon, M. (2006). *Books of sand.* In Proceedings of the 14th annual ACM international conference on Multimedia (pp. 1041-1042). New York, NY, USA: ACM Press.

Sedelow, S. Y. (1970). The Computer in the Humanities and Fine Arts. *ACM Computing Surveys (CSUR), 2*(2), 89-110.

Shoniregun, C. A., Logvynovskiy, O., Duan, Z., & Bose, S. (2004). *Streaming and security of art works on the Web.* In IEEE Sixth International Symposium on Multimedia Software Engineering (ISMSE'04) (pp. 344-351). Washington, DC, USA: IEEE Computer Society.

Steinkamp, J. (2001). My Only Sunshine: Installation Art Experiments with Light, Space, Sound and Motion. *Leonardo, 34*(2), 109-112.

Strömberg, H., Väätänen, A., & Räty, V.-P. (2002). *A group game played in interactive virtual space: design and evaluation.* In Proceedings of the conference on Designing interactive systems: processes, practices, methods, and techniques (pp. 56-63). New York, NY, USA: ACM Press.

Sung-dae, H., Jin-wan, P., & Won-Hyung, L. (2006). *Designing Audio Visual Software for Digital Interactive Art.* In 16th International Conference on Artificial Reality and Telexistence--Workshops, 2006. ICAT '06. (pp. 651-655). Washington, DC, USA: IEEE Computer Society.

Trifonova, A., Ahmed, S. U., & Jaccheri, L. (2007). *SArt: Towards Innovation at the intersection of Software engineering and art.* Paper presented at 16th International Conference on Information Systems Development, Galway, Ireland.

Walden, K. L. (2002). Reviews : Peter Weibel and Timothy Druekrey (eds), Net_Condition: Art and Global Media. *Convergence-The International journal of New Media Technologies, 8*(1), 114-116.

Warr, A., & O'Neill, E. (2007). *Tools to Support Collaborative Creativity.* Paper presented at Tools to Support Collaborative Creativity workshop held as part of Creativity and Cognition conference 2007, Washington D.C., USA.

Wei-Lung, W. (2002). Beware the engineering metaphor. *Communications of the ACM, 45*(5), 27-29.

Zimmerman, G. W., & Eber, D. E. (2001). *When worlds collide!: an interdisciplinary course in virtual-reality art.* In Proceedings of the thirty-second SIGCSE technical symposium on Computer Science Education (pp. 75-79). New York, NY, USA: ACM Press.

KEY TERMS

Art Projects: In the context of this chapter, with the word art projects we mean Art projects that use technology for the development of the final product which is usually an artwork. These projects include at least one artist and a number of technologists. In context of this chapter, technologists are software engineers. But in general, besides the software engineers there can be many other technologists such as sound engineers, electrical engineers and so on. Often art projects are multidisciplinary projects which involve many people including the sponsors and public administration, researchers and so on.

Artwork Support Tools: Software used to realize a certain artwork by implementing the core functionalities of the artwork. This software can be commercial off the shelf (COTS) software or open source software or even customized software that is developed only to realize a particular art project. By artwork support tools we mean the whole application that works behind the artwork.

Computer Art: Wikipedia defines Computer art as any art in which computers played a role in production or display of the artwork. Such art can be an image, sound, animation, video, CD-ROM, videogame, web site, algorithm, performance or gallery installation. Often computer art is used in general to refer to artworks that were impossible to create before the invention of computer.

Conceptual Framework: Conceptual framework can be described in many ways: (1) a set of coherent ideas or concepts organized in a manner that makes them easy to communicate to others. Or (2) an organized way of thinking about how and why a project takes place, and about how we understand its activities or (3) An overview of ideas and practices that shape the way work is done in a project. Wikipedia defines it as "A conceptual framework is used in research to outline possible courses of action or to present a preferred approach to a system analysis project. The framework is built from a set of concepts linked to a planned or existing system of methods, behaviours, functions, relationships, and objects."

Installation Art: Merriam Webster defines Installation as "a work of art that usually consists of multiple components often in mixed media and that is exhibited in a usually large space in an arrangement specified by the artist". In wikipedia we find following definition of Installation Art: "Installation art uses sculptural materials and other media to modify the way we experience a particular space. Installation art is not necessarily confined to gallery spaces and can be any material intervention in everyday public or private spaces. Installation art incorporates almost any media to create an experience in a particular environment. Materials used in contemporary installation art range from everyday and natural materials to new media such as video, sound, performance, computers and the internet. Some installations are site-specific in that they are designed to only exist in the space for which they were created."

Software Engineering: According to IEEE Standard Glossary of Software Engineering Terminology, Software engineering (SE) is the application of a systematic, disciplined, quantifiable approach to the development, operation, and maintenance of software. According to the Software Engineering Body of Knowledge, the discipline of software engineering encompasses knowledge, tools, and methods for defining software requirements, and performing software design, computer programming, user interface design, software testing, and software maintenance tasks. It also draws on knowledge from fields such as computer science, computer engineering, management, mathematics, project management, quality management, software ergonomics, and systems engineering. In industry software engineers can have several specialized roles such as analysts, architects, developers, testers (according to Wikipedia). In context of this chapter, we call all people who are involved in the development of software for art projects as software engineers irrespective of their specialization.

Chapter III
The Design of Engineering

Joseph William Pruitt
Slingshot Product Development Group, USA

ABSTRACT

The purpose of this chapter is to define the roles of engineering and design within the product development cycle looking at both the scientific and artistic methods used by the creators of new ideas. With the vastly different philosophies of product development between the engineer and the designer, the production manager is often faced with an ubiquitous tension that is frequently misdirected and mismanaged. The disparate design philosophies tend to force companies to pick either "science" or "art" in their development cycles which in turn creates either products that have no connection with human beings or products that cannot conceivably be produced on this planet. This chapter addresses these concerns and suggests methods in managing the creative insanity of successful product design.

INTRODUCTION

In the last fifty years, the developed world has seen a considerable increase in not only material wealth, but also in the choices and lifestyles provided by products and services that are a staple of our everyday lives. The increase in product choices, specifically "designer" product choices, has bred a new type of selective consumer who uses the goods and services in their life as a medium to convey a "personal brand" to the world through their selected experiences with the products they buy (*Brettell, 2007*).

This increase in the **self brand** has left many companies scrambling to connect with people on deeper levels that go far beyond function and possession. In the 21st century, many high-tech, engineering based companies have come to the realization that, if they want to compete in the global market, they have to integrate **human centered** design somehow into their core competency (Hoover, 1991). Basically, they realize that they need design, but don't know why. Business is just starting to grasp the emotional value of life and this shift from the analytical to the artistic is especially showing up in product and user-interface (UI) design.

The focus of this chapter is to define the engineer and the designer, highlighting their strengths and weaknesses in the product development cycle and then outline a way in which the two different personalities can be combined together to create a collective genius within companies through the role of a manager. The first part of the chapter discusses the history of the development of both engineering and design and then goes on to discuss the current state of design development and highlights several case studies in relational psychology between the artistic and the scientific with concern for product generation.

The methods in which integrated design are managed and addressed in the second part of the chapter followed by the roles of the designer, the engineer and the production manager within this assimilation. These areas of product development are discussed in order to better understand the key shortcomings in the synthesis of the workplace.

Overall, product development comes down to art and science, engineering and design. So far in product design, science has dominated the playground. The science physically creates the development, but the art creates the story and in the end, stories are all we really have.

BACKGROUND

It's four in the morning on a muggy Sunday night in North Jacksonville and I'm sitting in a Wal-Mart® parking lot with a few friends playing a game of Tidily Winks™, waiting for the Tire and Lube Express® to open. My buddy Theron is dominating all of us by a solid 50 points. We had driven the two hour stretch down I-95 from Savannah, Georgia earlier that night in hopes of finding adventure before taking Theron to the airport where he would catch a seven o'clock flight out to the west coast. Ka-thump, thump, thump! That's the sound of a shredded tire. As we sat there waiting for the tire center to open, it came to me: "This is what it's all about…the story." It

is the stories derived from experience that unite humans and give importance and understanding to our relationships with the world around us.

In the age of mass consumerism and knowledge at the press of a finger, people are searching for meaning in life beyond the stainless steel espresso maker that matches the kitchen sink and the lemon scented dish soap that "leaves hands feeling soft." People want to have a connection with their belongings that transcends the functional and dives into the experiential (Norman, 2003). As extreme as it sounds, people want to create stories with the products they buy and the services they purchase, because stories are how people remember and remembering is where the true value lies. If all that was written in the last paragraph was a monologue about the pros and cons of oblique engineering in the automotive industry, probably half of the people reading this chapter would stop and the other half would fall asleep. It was the essence of a shared experience, a story, which kept the reader intrigued and fascinated with the outcome. The story was where the value was maintained and the experience understood. This value in the experience must be conveyed within business, between and among departments, if it is to be successfully implemented.

This value of the story is further elaborated by designer Scott Klinker, who says

Wordsmiths have long been able to deliver complex ideas about the human condition with simple emotional stories. Language is the starting point for any design process; words define your goals and often determine the results. Inventive keywords can help you reposition the product to breach new categories. Literature is full of theoretical tools like deconstruction for the designer who learns to play with language to build meaningful forms, intellectual positions, and experiences. (Klinker, 2007)

The story is where the value is, but the product is where the business is.

So who, in business, is responsible for new product development? Many different departments may have a good deal of impact on production within any corporation, but for the sake of this paper, I have narrowed it down to 3 main roles: the manager (or production manager), the engineer and the designer. To best understand how these roles interact and communicate today, it is important to understand where they come from and how each has developed their own method and philosophy toward the generation and production of new ideas.

The Engineer

Engineering is possibly one of the oldest professions known to man. Starting with the early control of fire and stones, humanity has always been driven to understand nature and apply that understanding to improve life. Perhaps this is what engineering truly is, the functional control of the universe to provide utility towards the betterment of mankind. This can maybe be summed up to: *Engineers use physics and science to design for people.*

Carl Mitcham explains in more technical terms the role of science and design within engineering in his thesis entry on *The Importance of Philosophy to Engineering.* Mitcham (1998) states:

As for the engineering sciences, these 'have their roots in mathematics and basic sciences but carry knowledge further toward creative application.' Such rootedness is what justifies course requirements in mathematics and the basic sciences. In the words of the ABET [Accreditation Board for Engineering and Technology] *criteria: 'The objective of the studies in basic sciences is to acquire fundamental knowledge about nature and its phenomena, including quantitative expression.'*

As for engineering design, this is defined as the process of devising a system, component, or process to meet desired needs. It is a decision-making process (often iterative), in which the basic sciences and mathematics and engineering sciences are applied to convert resources optimally to meet a stated objective. (Mitcham, 1998)

One of the most notable engineers of this type, Sir Isaac Newton, is responsible for bridging many observations in nature and translating them into mathematics. Engineers and physicists had been quantifying the world for centuries before Newton, but Newton developed a large foundation for the methods in tackling the analytical modeling of natural phenomena. Understandably, when Newton derived equations correlating the dynamics of force, mass and acceleration, other engineers were able to use his formulas to predict the outcomes of their hypothesis on paper long before actually building any prototypes.

Other highly notable physicists and engineers followed, creating a somewhat unified practice of engineering that pursued a path of first defining the problem, second defining the variables, constraints and assumptions of the problem, third mathematically modeling the problem, refining the design as appropriate, and then finally applying the mathematical model to real world, functional design. This method of exploration through mathematics did two things: first, it allowed the engineer to quickly and efficiently solve complicated problems on paper without every having to construct a physical model and second, it created a recordable and repeatable solution for others to follow and build upon.

Over the last few centuries, the **engineering method** has been perfected and refined to such an extent that the mathematical theories of many new ideas in physics are beyond the realm of current observation, either because the ideas involve particles that are too small, or are too large and too far away. Nevertheless, the engineering mind is limited to the affect that its backbone is composed of logic and its soul is made of math. A very large part of engineering is quantifying the amount that an engineering analysis cannot be quantified. This is known as accuracy and precision. The engineer knows that nothing can be nailed down 100%, but he accepts it knowing that he can only get so close to nature, hoping that he is close enough. There is always the unknown and somehow the engineer tries to define the indefinite element with numbers and percentage signs. With human behavior, this is rarely possible. In steps the designer.

The Designer

Some people have defined design as applied art. I once heard that designers are people who create experiences. I agree with these statements to an extent, but I think that fundamentally, when you get down to it, design is more than that and at the same time, less. If engineering is the applied understanding of nature, then design is the applied understanding of humanity.

Each person has a unique set of needs, wants and desires that are attributed to his or her genetic makeup and to societal and cultural upbringing. These emotions that a human being experiences in life are perhaps best understood as a reflection of our own mortality. Beauty, pain, anger, joy; these are all indications of the delicacy and survival of life. We want the emotions and experiences that enhance and continue life and generally reject and avoid experiences and emotions that hinder and destroy it. Most life experiences are very mild in the emotional spectrum, but nevertheless can have tremendous impacts into our acceptance or rejection of them. The designer is the person who tries to understand the emotional ties between humans and their surroundings and then applies that knowledge to the objects they design.

Donald Norman touches on this theme repeatedly throughout his many books on the subjects of design and human interaction. I had a chance encounter with Mr. Norman at the Industrial Design Society of America (IDSA) conference in San Francisco this last year and began speaking with him about design and humanity not really knowing who he was or where his knowledge of expertise lay. I was humbled to learn that he coined the phrase "user centered" in a book he wrote in the 1980's entitled *The Design of Everyday Things* only after I began explaining the term to him and how I was applying it in my work. Needless to say, his work on the subject is phenomenal, particularly his book *Emotional Design* where he states:

Where as emotion is said to be hot, animalistic and irrational, cognition is cool, human and logical. This contrast comes from a long intellectual tradition that prides itself on rational, logical reasoning. Emotions are out of place in a polite, sophisticated society. They are remnants of our animal origins, but we humans must learn to rise above them. At least that is the perceived wisdom.

Nonsense! Emotions are inseparable from and a necessary part of cognition. Everything we do, everything we think is tinged with emotion, much of it subconscious. In turn, our emotions change the way we think, and serve as constant guides to appropriate behavior, steering us away from the bad, guiding us toward the good. (Norman, 2003)

The emotional connection with objects is fundamentally the vehicle that categorizes design as either "good" or "bad." Since often there is no quantitative way to determine how well objects "work" with people, we must rely on the emotional or "gut" reaction that people have with the products they use. This emotional connection can range from respect and appreciation of an object (That screw driver really made it easy for me to fix the radiator hose) to reliance and dependency (I would rather be dead than spend a week without my cell phone). As psychologist Sherry Turkle points out:

The first thing designers as well as consumers should think about are the human purposes being served by the design. It is more crucial now more than ever before to consider how to respect, not exploit, our human vulnerabilities... (Turkle, 2007)

Tapping into the emotional drive of well designed products is what designers pride themselves on and is one of the key separating factors in the differences between design and engineering; between the art and science of product development.

Joseph Pine and James Gilmore back up this emotional state in their book *The Experience Economy; Work is Theater & Every Business a Stage*. Pine and Gilmore explain that the change in customer expectations over time has gone from commodities to products to services and that our current society is now shifting from the service centered approach to the "Experience" driven approach. Kevin Mullet (2003) elaborates on this theme by explaining how *"We are well into the transition from a service to an experience-based economy in which sensation becomes a conscious source of consumer demand. Technology is not driving this transition, but the impact of new technologies will accelerate the emergence of rich, expressive, client-side platforms that deliver a more compelling sensory experience as well as a richer, more satisfying level of interactive control to the software user."* So what is an experience and what makes one experience better than another?

An experience is *"direct observation of or participation in events as a basis of knowledge and the fact or state of having been affected by or gained knowledge through direct observation or participation."* (Webster, 2007) Meaningful experiences occur when our sensations and emotions evoke a positive response either directly or at a later date in our lives. This is not to say that all meaningful experiences are free from negative feelings or emotions, but from a product develop-

ment standpoint, they generally should be. When designers engage the emotional side of experiences, they must have a strong understanding of human nature, cultural signatures and absolute beauty.

The emotions of human nature and understanding of beauty and culture are things that social scientists and psychologists have been discussing and debating for years. Where do the emotions come from? Which ones are universal? Which ones aren't? How do math and proportion tie into human perception of beauty? All of these are questions that the designer must ask to gain a better understanding of who we are emotionally as a human race. But in order to integrate the emotional side of a product experience, the designer must sell his idea to the analytical engineer, who ultimately, it seems, controls the destiny of making the product fly. But are analytical engineers really that important?

According to Daniel Pink (2004), in the next ten years, white-collar jobs are expected to loose $136 billion in salaries to Asia. These jobs, which can be performed as effectively and as timely as jobs in the U.S., are almost all from the analytical sector: accounting, law, engineering, computer programming, etc... The step by step knowledge based work is being outsourced in the same way that labor and manufacturing jobs were shipped over seas in the last century.

This change is not necessarily a bad thing; it just requires adaptation and adjustment. As Pink (2004) states in his book *A Whole New Mind, "As left brained work gets sent to Asia and to computers, engineers and programmers will have to master different aptitudes, relying more on creativity than competence, more on tacit knowledge than technical manuals and more on fashioning the big picture than sweating the details."* But the first step is for industry to acknowledge a need for improvement; which usually doesn't happen until significant market share has been lost to the competition (who changed their design philosophy years ago).

Though Daniel Pink is correct in stating that many people will need to adapt in order to conform to the progressive demands of our new global economy, he and many others within the creative revolution do not touch on the very important aspect of managing and working amongst the **creative divide** that separates the problem solving styles of many within the workplace.

The Manager

Most people can agree that the role of management is, by definition, a form of logically controlling the day to day activities within an organization. Management works on linear progression modeling with cause and effect problem solving to the answers that may arise each day. Therefore, many managers and executives are hired to oversee and control a work environment due to their strong analytical skills. Naturally, management hires people that they can understand and who have somewhat predictable behavior in respect to their job performance. Thus it goes to show that management, without an understanding of creative intelligence or non-linear production methods, fails to hire creative people who have seemingly random product development and idea generation techniques.

As Bruce Nussbaum (2006) says, "*What was once thought of as great design is now seen merely as good, what was once exceptional is now standard. Managers everywhere are turning to rapid ethnography, usability, special materials, and aesthetics—the tools of design—to innovate. The differentiation of products and services increasingly requires a much higher level of execution. In design, the bar is raised.*"

In 1991, the National Research Council published a report entitled: *Improving Engineering Design, Designing for Competitive Advantage.* The NRC outlined that creativity and human focused design were being placed on the back burner to analytical thought and logic based design solutions. Universities believed that associative and user centered design had little intellectual value in research or the classroom. This trend was carried out in industry with a majority of technical and analytical companies hiring only people with logic and reasoning skills and weeding out associative and creative individuals. Way back in 1991 the NRC thought this discrimination toward creative, user-centered design was a bad thing, a very bad thing.

Seventeen years later, this purely objective thinking has left companies with serious flaws. According to the NRC, "*Poor engineering design capabilities are leading US companies to design and produce products that are more expensive, of lower quality and slower to reach market than their top foreign competitors.*" (Hoover, 1991) (Not to mention product's that have little connection with people emotionally or experientially).

It seems to be very difficult for the current culture of big business to change from its analytical think tank status. For one, if things can be sequenced and quantified, they can be counted and analyzed (which are what top execs do best). In the scientific community, if it cannot be observed, it probably isn't real (unless you are a theoretical physicist, but that's a different story). Everything in science is ordered and logical and the processes implemented in all scientific problem solving areas, by definition, can be recorded and documented. In many respects, this is absolutely necessary to prevent chaos in order to cultivate and grow new ideas as well as maintain existing ones. This holds true for all aspects of human behavior when large groups of people are interacting for a common purpose. Politics, the military, business, education; without a defined structure with well defined measurements for success or failure, the human machine breaks down.

So it makes sense that business doesn't handle the creative mind well; it can't easily quantitatively measure it. Perhaps the final outcome of creativity, the product or service rendered, *can* have a gauge attached, but the process to achieving successful results within the **creative method** are unpredictable and unorganized to the scientific mind. There are no sequential steps with Boolean operators to determine if the creative method is on track or if the direction is even within the bounds of the "predetermined objectives."

I was speaking with my friend Jeff Walter a few weeks ago about this tension between the scientific and creative methods of design. Jeff is a computer programmer and user-interface designer for a web-based company in Savannah, Georgia. He's working on a widget that allows companies to put audio sound bites from multiple sources (cell phones, radio, and microphones) onto their website seamlessly. The idea is similar to You Tube™, only with audio. Jeff said that in his industry, programmers need to completely understand both the user and the computer intimately in order to be successful. If they know just the computer side of the equation, they may have exceptional software that serves no human purpose and if they know only the user and are excellent at mapping the needs, wants and desires into an effective design solution, they have little understanding of how to execute the concept from a programming standpoint and the idea has little chance of success.

In the world of computer programming, many in industry don't appreciate both ends of the design process," said Jeff. "This is because the programmers have little respect for the UI [user-interface] designers because [the programmers] think they know everything there is to know about programming. The designers, though they have the user figured out, can't program to save their lives. I'm trying to bridge the gap and be competent in both worlds. That's the only way to get things done.

Jeff's approach is a bit extreme in the sense that he is striving to become all things in his industry, but that is not necessarily what is needed for success in the ever changing world of design and commerce. More importantly, an understanding of the varying design processes and a respect for the different methods of design generates a productive, unified approach within a company instead of an atmosphere filled with skepticism and tension. This approach, weather it's between two branches of the same company or between an internal department and an external consultancy, will boost efficiency and launch companies into a new age of global management: the age of the **poignant dictator**. Sounds scary.

Perhaps some of the most creative people to ever live were those that fully understood both the artistic and scientific approaches to their expertise in life. A poignant dictator is the control of this conglomeration of the scientific and the artistic. A genius is someone who has the ability to fully understand the scientific enough to creatively defy the logic and step into the previously determined inconceivable. (Anderson, 2005) A poignant dictator is someone who has the ability to let this happen. In many cases, it is not a sole person that constitutes genius within a company, but the collaboration of disparate entities that creates genius-like behavior from the whole. So why a dictator?

Like in many geniuses of history, the mesh of scientific expertise and artistic foresight creates insanity. (Anderson, 2005) This is most definitely the case in business. Dictators are synonymous with control. If the management cannot control, then pandemonium ensues. Perhaps this comes back again to why management hires groups of people who are all the same: they work well together and therefore are easy to control and direct. Basically, no matter how many people a company takes on, if they all think the same way, then the cost of figuring out how to control them is reduced to unity. Oh, the efficiency of an analytical mind!

Nevertheless, on the other hand, the creative mind must be grounded in knowledge or impracticality will never generate growth and the creative soul will then become a homeless and starving soul with brilliant ideas that only work if physics is optional. So the poignant dictator must oversee these two adversaries, listening, sympathizing and controlling the many faceted monster of a workforce into a viable process that ultimately is named "Genius" by those less informed of the carefully directed internal bedlam.

The poignant dictator has two groups of people in his control which are constantly at odds: the designer and the engineer. A designer is someone who looks at how things could be, while an engineer looks at how things are. Designers, like I said previously, create experiences for people. Engineers perhaps make those experiences actually happen. So they must work together harmoniously to actually accomplish anything; this is a given. But they usually don't. Why not?

When I was finishing up graduate school, I was involved as a teaching assistant and project leader for some design work that was sponsored by a world wide equipment manufacturing company. One of our final concepts dealt with changing the manufacturing approach from tubular steel to stamped steel. The stamped steel had a much higher startup cost for tooling, but a lower material cost since the amount of steel was dramatically less than the tube-constructed approach and had a much shorter manufacturing time. Nevertheless, the idea had never taken off in the company due to production volume and implementation feasibility. Once we, on the design team, figured out how to manufacture the parts so that each stacked nicely into a condensed package, we proved that we could increase shipping efficiency from two units per shipping container to nearly 20; an increase of almost 1000%. We also demonstrated how we could cut down on manufacturing time by minimizing the number of spot welds but maintaining structural integrity as well as decreasing the instrumentation and control panels to critical displays and a unified, singular system. By cutting down on complexity and increasing quality, **rational design** won the engineers hearts and they were more than willing to change directions and work with us, rather than against us.

Re-read that last sentence. Engineering, as a discipline, prides itself in creation of new technology. Spending five years in engineering school taught me that the thought pattern and problem solving abilities of the engineer were logical, reasonable, scientific and, above all, superior to any other cognitive capabilities. If that truly were the case, than an emotional connection with the engineer, speaking to his "heart", would bear no ground in his or her decision making. Nevertheless, as Robert Pool (1999) says in his book entitled: *Beyond Engineering*:

...modern technology is not simply the rational product of scientists and engineers that it is often advertised to be. Look closely at any technology today, from aircraft to the internet, and you'll find that it truly makes sense only when seen as part of the society in which it grew up.

In our society, we are logical and rational human beings. But we are also sympathetic, poignant and introspective. We don't select birthday cards for loved ones based on price or size (hopefully), but on the sentiments that the card represents; the expression and emotion relayed beyond the words on the page. In the same way, the engineer decides the direction and scope of research and development based ultimately on the needs, wants and desires of society as a whole. As an engineer, one must be empathetic in understanding people, logical in the design process and considerate towards the varying disciplines of design in the creation of new forms and experiences. The poignant dictator is the future. The rational designer and sentimental engineer are his friends.

The sentimental engineer and the rational designer must work together, but how? The answer: inter-departmental dating. In any dating relationship, emotions are typically a large part of decision making and each party is trying to entice and attract the other party while simultaneously reaching and striving for compatibility and common ground within the relationship. Flattery and charm are usually successful methods in attracting a partner and are good grounds for building the more meaningful bonds of trust and respect. Many engineering and design departments seem to skip the dating phase and jump directly into married life. Without the foundation of respect and understanding toward the other party, little can be done to mend the gaps between the completely different methods of design thinking, so what typically results between these two disparate departments is lack of understanding and lack of appreciation, which yields unproductive decision making and poor quality in the end product or service. So how can respect be grown between these groups of different thinking individuals? Let's explore.

CURRENT TRENDS

The current state of the education system in the United States is heavily focused on mathematics and science. People seem to be concerned with our apparent inability to "keep up" with Asian countries in the scientific areas of education and so our government, in order to combat the problem, pushes the analytical areas of knowledge and leaves much to be desired in the arts and humanities. (Stewart, 2006) This correlation is also drawn in many businesses in the United States. At the moment, I am in the middle of a design project with a company who is feeling a tremendous impact from competition in China and Southeast Asia due to the dramatically lower cost of Chinese manufacturing. My economic design solution for the company is to not continue competing with the manufacturing efficiency of the Asians, but to understand their product and their consumer and ultimately generate a sense of what is valuable in the designs they currently have versus what is not valuable. This seems an obvious solution, but when one considers that their product has over 20 parts with half of those parts being manufactured to hide other parts, it

becomes blatantly obvious that a majority of their energies are focused on the wrong thing. If you can't beat them, then become better.

It has been a very painful process pulling them away from their engineering driven approach of designing for cost and function, not caring much that they are following every other trend that is currently on the market while left dazed and confused as to why customers don't want to pay extra for something that does not speak to them emotionally. They know that they need to change and adapt in order to survive. They know that design is the answer to their problems (otherwise they wouldn't of hired me), but they still can't seem to pull away from their old habits of "what's worked for the last 50 years should work for the next 50." Unfortunately, that's just not the case anymore.

Another notable example that I'll bring up involves a project I was heading up last year for one of the world's leading heavy equipment manufacturers. Part of their design team flew in from overseas to meet with us and discuss project scope and direction as well as what they expect at delivery time. The company brought in the head of the design department as well as the head of engineer for that particular division.

In discussing the project in detail, I asked one of the engineers what kind of departmental dynamic existed between the design groups and engineering groups at his company. "All out war," he said. He went on to explain that the designers wanted to put these fancy curves and bent-angle glass into everything and that once the engineering department received concepts they always tried to turn them back into functional boxes with tires on them. "So you don't understand each other," I said. He laughed. "It's getting better," he said. I assumed he was talking about the fighting. Situations like this are all too common in the workplace with engineers not understanding the role of the designers and managers resisting to how it all fits together. Engineering is crucially important, but it's not the end of the story. So, science, move over.

There's a new kid on the bus who has crayons and he's not afraid to use them.

FUTURE TRENDS

In the last 20 years, significant advances have been made in the area of neuroscience and mapping out the functions and problem solving capabilities of the human brain. It has only somewhat recently come to light how extremely complex, yet extremely efficient the human brain is at understanding the world and creatively generating new ideas. It has also been found in recent years that the "left brained," "right brained" phenomena that so many people adhere to so diligently isn't exactly true at all. (Anderson, 2005; Pink, 2004)

Up until the early 1990's it was thought in many public and scientific circles that people are either left brained or right brained, that they either think logically and reasonably or they rely on emotions and creativity to navigate through life. (Anderson, 2005) This theory came about largely due to a mid 19th century scientist named Paul Brocca, who found, through a patient that had suffered a stroke and was left speechless, that upon examining the brain in an autopsy, the front part of the left hemisphere of the brain had been affected and that this was the area responsible for communication. (Anderson, 2005) It was later hypothesized that left brained people are linguistically intelligent and therefore logical and rational, with the right half of the brain responsible for emotions, and so forth.

With the ability to now map the cognitive process of the brain using magnetic resonance imaging (MRI) and computed axial tomography (CAT) scans, it has been found that people generally use their entire brain simultaneously in many areas of creative and subconscious thought that were initially attributed to only one specific part of the brain. (Anderson, 2005) For example, in order to communicate verbally with other people, we must use one part of our brain for generating words, another part for grammatically putting those words together and yet another completely separate part to ensure that the words we say have meaning and context within the environment that we are using them (this is called the *language circuit* within the brain and uses both the left *and* right hemispheres). (Anderson, 2005) I bring this up because language is the ultimate widely used creative act that we experience each and every day. All of us have the ability to be creative, insightful and introspective with a profound sense of what is beautiful and we all each possess the ability to be logical and analytical of our environments and problems that we face on a daily basis.

Nevertheless, many people, through societal pressures and upbringing feel that they are stuck within either the left or right hemispheres of the brain. Perhaps in elementary school one was told that they had poor math skills or perhaps they were not good at art or music. With these preconceived notions of ones skill sets, one embarks on a life long journey of developing the "good" areas of their life and completely rejecting the areas that they do not excel in. It is this mind set that has created the ultimate divide within the work place and has generated an environment that is creating *"Poor engineering design capabilities…leading US companies to design and produce products that are more expensive, of lower quality and slower to reach market than their top foreign competitors."* (Hoover, 1991)

But there is hope. Actually, much more than just hope, there's science to back up that we can continuously develop new and connective abilities within the brain. As the acclaimed neuroscientist Nancy Anderson (2005) states in her book *The Creative Brain, the Science of Genius*:

…the brain is marvelously responsive, adaptive and eternally changing. Its adaptations and changes occur in response to the demands and pressures of the environment that it encounters. Sigmund Freud and the psychoanalytical movement gave us awareness that early life experiences

*affect emotional development and attitudes in later life. Neuroscience adds a new dimension: it makes us aware that experiences **throughout** life change the brain throughout life. We are literally remaking our brains – who we are and how we think with all our actions, reactions, perceptions, postures and positions – every minute of the day and every day of the week and every month and year of our entire lives.*

So get out there and redevelop that which you thought you never had. If you're into mathematics and science yet are repulsed by the idea of emotional creativity, then paint a picture, start a sketch book, learn to play the piano or the guitar. Express yourself artistically, you may be surprised. If you hate the idea of doing math problems or learning how a solid rocket booster works, then check out the science section at your local book store and buy a book on logic and work through some of the problems. You may be surprised at what you find you can do and understand. Ultimately, the point I am driving at is, don't conform yourself to what you *think* you can do or what you *think* that you're good at doing. Push the envelope of what you know and it may surprise you.

To sum this thought up, if your job is to calculate compressive stress loads on the roll strength of a fork lift cage, then you will probably never make money (or need to make money) playing piano. Stay strong at what you do know. But by struggling to learn the piano, you invest part of yourself into an area of weakness and from that you develop empathy which in turn helps you to appreciate why the designers at your company want to add a large mermaid emblem flowing from the roofline of the forklift chassis. Yes, it makes the computations three hundred times harder, but you also know that deep down, the designers are right: it *will* sell more forklifts. And by empathizing with the designers needs, you have opened up an opportunity to share with them why a solid gold body frame probably doesn't help the bottom line. And that makes the management happy.

CONCLUSION

Overall, this chapter has focused on the differing roles of the engineer and the designer within the construct of product development. It has looked at the role of the manager and has addressed the varying concerns of dealing with the artistic mind versus the analytical mind.

The engineer must be sentimental and empathetic with his decision making in the product development cycle, taking into account the value of human emotion and interaction in the design process. A respect for the designer's role is also crucial for the engineer since the engineer must fully understand the varying aspects of design and thus integrate them into the functional, working end-product.

The designer must be rational and respectful of the constraints placed on product development, taking into account the fact that physics and science are an important consideration within product development and cost as well as manufacturing techniques are viable concerns of the engineering department and should be taken seriously at some point within the designer's process.

The manager must fully understand both the design and the engineering approaches to problem solving and must direct and control the process with sincere empathy towards both parties. He must fully understand the importance of both the science and the art that goes into new product development and must act as the facilitator between the analytical and the creative to ensure products that are functional as well as meaningful to people in their lives.

Ultimately, all of us possess the ability to be both creative and logical. Modern science has shown us that we each enjoy the capacity for a full range of skills and attributes that go far beyond the stereotypical left brained/right brained mindset that we have endured throughout our lives. In the end, we all can create things that are beautiful, and to me, that makes sense.

REFERENCES

Anderson, N. C. (2005).*The Creative Brain, the Science of Genius.* New York, NY: Penguin Group, 2006.

Brettell, Sue (2007). *Communicating Your Personal Brand* 2007 Global TeleSummit Creating a Decade of Personal Branding (2007) *http://www.personalbrandingsummit.com*

Hoover, C. W., & Jones, J. B. (1991). *Improving Engineering Design, Designing for Competitive Advantage.* The National Research Council, Washington D.C.: National Academy Press.

Klinker, S. (2007). *Spinning Form: How to Tell Stories with Product Design.* Taken from Core 77 Design Student Guide: http://www.core77.com/hack2school/klinker.asp

Mitcham, C. (1998). *The Importance of Philosophy to Engineering.* University Park, PA, Penn State University: STS Program.

Mullet, K. (2003, November). *The Essence of Effective Rich Internet Applications.* Macromedia Experience Design Team.

Norman, D. A. (2003). Emotional *Design: Why We Love (Or Hate) Everyday Things.* New York, NY: Basic Books.

Nussbaum, B. (2006). *The Best Product Design of 2006.* Taken from BusinessWeek Magazine *http://www.businessweek.com/innovate/content/jun2006/* ©2008 The McGraw-Hill Companies Inc.

Pine, B. J., & Gilmore, J. H. (1999). *The Experience Economy; Work is Theater & Every Business a Stage.* Boston: Harvard Business School Press.

Pink, D. H. (2004). *A Whole New Mind: Moving from the Information Age to the Conceptual Age.* New York, NY Riverhead Hardcover.

Pool, R. (1999). *Beyond Engineering: How Societies Shape Technology.* New York, NY: Oxford University Press.

Stewart, V. (2006). *Math and Science Education in a Global Age: What the U.S. can Learn from China* Published By Asia Society. http://internationaled.org/mathsciencereport.pdf

Webster, M. (2007). *Miriam Webster Dictionary* Online: http://www.merriam-webster.com/dictionary/experience

KEY TERMS

Creative Divide: The division in human thinking between the analytical, engineering based approach to problem solving and the human centered approach to problem solving.

Creative Method: The somewhat abstract form of problem solving that relies heavily on the subconscious mind to draw connections and parallels between seemingly unrelated problems to form a unified, concrete design solution.

Engineering Method: The analytical and logical approach to problem solving that uses science, mathematics and physics to define, analyze and solve issues in order to ultimately improve the existence of life on earth.

Human Centered: A design philosophy where the functionality, purpose and aesthetic of a product or service is tailored to the needs, wants and desires of the user. Similar to the term *User Centered* coined by author Donald Norman in his book *The Design of Everyday Things.*

Poignant Dictator: A product development manager that oversees both engineers and designers helping them to communicate and facilitate efficient design solutions while still adhering to both the Engineering and the Creative Method of design.

Rational Designer: A human centered designer that works within the creative method but is still sensitive to the constraints of the engineering method and strives to find a unified balance by creating solutions that help the engineer do his job more efficiently.

Self Brand: The contrived image that an individual communicates to the world. The self brand can be anything from clothing and accessories to personalized web pages and social networking sites.

Sentimental Engineer: A logical engineer that works within the engineering method but is still sensitive to the importance of aesthetics, beauty and human interaction with the products that are being developed, striving to facilitate the designer to create objects that have meaning in the lives of people.

Chapter IV
Ambient Video, Slow–Motion, and Convergent Domains of Practice

Jim Bizzocchi
Simon Fraser University, Canada

Belgacem Ben Youssef
Simon Fraser University, Canada

ABSTRACT

The chapter describes the synergistic integration of distinct research and creation agendas, each firmly grounded in its own set of practices and methodologies. The authors participate in three separate domains of practice: humanities scholarship, scientific research, and artistic creation. They have continued to work within their respective specialties, but have also aligned their research and creation activities within a larger context that enriches their individual work. Humanities scholarship, artistic creation, and scientific research support each other at various critical junctures in the overall arc of the research. The chapter analyzes the details of each individual research or creation strand, and identifies the instances and the dynamics of mutual support and synergy. The mechanisms and attitudes that support the success of cross-disciplinary collaboration are identified and explicated.

INTRODUCTION

Digital convergence has transformed existing practices and driven new and emergent practices in all phases of our culture. In particular, it has transformed the details of practice in the three domains covered by our research agenda: humanities scholarship, scientific research, and artistic creation. Has the evolution of convergent perspectives within disciplines affected the process of collaboration across disciplines?

Fifty years ago, C.P. Snow argued that there was a growing gulf between the culture of the arts and the culture of the sciences (Snow, 1959).

Many have disagreed with the particulars and with the implications of his argument. However, most would agree there is necessarily a certain level of real difference in the general approach and the particular methodologies pursued within the two spheres. Nonetheless, they do share a common overarching goal: an increase in understanding of and insight into our selves and the world we share.

This chapter examines one attempt to work within separate disciplines with integrity, and at the same time to work across disciplines in order to maximize the cumulative impact of both individual and shared research agendas.

COLLABORATION AND CROSS-DISCIPLINARY PRACTICE

The topic of collaboration has been examined through various lenses. Bennis and Biederman examined collaboration in the business world (Bennis and Biederman, 1997). They found a number of factors which aided the development of successful collaborations. Their list included quality of participants, quality of leadership, shared purpose, ability to focus, optimistic stance, commitment to finishing, development of espirit-de-corps, and a sense of great work as its own reward. Schrage examined the same domain and found a similar list that included shared goals and a general level of competence, but added mutual respect, trust, effective communication, clear roles leavened by flexibility, and the examination of multiple representations of critical phenomena (Shrage, 1995).

Others have considered the question of collaboration across the two domains in question in this chapter (the arts and the sciences). Some have identified success factors that echo the findings drawn from the business world, but they also add other factors as well. Candy and Edmonds have the following list: shared vision, complementary interests, communication (including the develop-

ment of a shared language). Significantly, they also include "time" within their list of enablers, indicating the need for a sustained effort at cross-discipline collaboration (Candy and Edmonds, 2002). In separate studies, Vera John-Steiner (John-Steiner, 2000) and Oppenheimer (Oppenheimer, 2007) also stress the importance of common vision and shared values to dynamically harness distinct roles and traditional discipline-based approaches into a shared agenda. The Bridges Project brought together a group of artists and scientists to discuss their experience in arts-science collaborations. They stress the importance of language as a potential unifier, or disruptor, citing the need to maintain disciplinary language for purposes of precision, but also a concomitant need to clarify meaning across practices (Pearce, Diamond and Bean, 2003).

Finally, Wilson disagrees with Snow's basic assumption of fundamental differences. He maintains there is significant commonality in the two approaches: "...scientific and technological research should be viewed more broadly than in the past: not only as specialized technical inquiry, but as cultural creativity and commentary, much like art" (Wilson, 2002). We lean towards this interpretation, and although we each respect the perspectives and methodologies of our core domains, the history of our research program illustrates commonalities and intersections that join our interests and enquiries.

DISTINCT METHODOLOGIES

Divergent Methodology

On the face of it, the individual threads of our shared research practice are indeed rooted in separate domains and methodologies. Bizzocchi's methodologies includes classic humanities scholarship and his own artistic practice. Ben Youssef's methodologies derive from his background in both science and engineering.

Humanities Scholarship

The core of Bizzocchi's research is traditional Humanities-based scholarship. Humanities scholarship is ultimately based on the observation and analysis of creative works. The analysis will lead to understanding of these creative works within one or more of several overarching contexts: aesthetic, historical, cultural and social. Depending on the goal of the humanities scholar, the analysis will vary the emphasis between the viewer's perspective - the reception of the work, and the artist's perspective - the creation of the work.

Bizzocchi's Humanities scholarship concentrates on the moving image: cinema and video. His goal within this domain is to understand the design principles used by moving image artists to create experience and tell stories. This concern with the design, or the form, of media and media works is termed "poetics". The term and analytical perspective date back to Aristotle's seminal work, the *Poetics*. In the *Poetics*, Aristotle analyzed the structure and the elements of the dramatic works of the classic Greek playwrights. In the process, he described the design parameters and design guidelines for the classical Greek drama, including considerations of character, emotion, plot action, and narrative arc.

Bizzocchi's academic work is concerned with the poetics of the moving image, and in particular with the changes in poetics supported by the new digital technologies. He relies on several sources as the foundation of his analysis. The first is the scholarly and critical discourse around cinema and video. He is particularly interested in those scholarly works that are grounded in a consideration of the medium itself, that examine specific artistic works, and that analyze cinematic poetics, craft, and form. Scholarly works that are based in broader systems of cultural theory are of less utility for his work. His second source for analysis is an understanding of the new technologies. He and Ben Youssef must maintain an understanding of the current and future state of video production, post-production (editing and mixing), distribution and display technologies. The purpose is to use this knowledge to help in the understanding of current cinema and video works, and to anticipate the development of future directions in the art of the moving image.

The third source for his scholarly analysis is the works themselves. The core of his scholarship is the detailed examination of cinema and video works to reveal the artistic decisions that the creators made. This methodology is called "close reading". The practice of close reading involves the repeated viewing and reviewing of a work, accompanied by a detailed set of observational notes on both content and on media form and design. Bizzocchi uses these observations to develop deep understandings of the work, the creative decisions that went into the work, and of the medium itself. Close reading is an iterative process. During the process of performing a close reading, interim observations are reached with respect to the work and possible conclusions to be drawn from it. As these interim observations are made, earlier parts of the work are reviewed with the current observations and possible conclusions in mind. As this process repeats through several viewings of the work, the set of observations and conclusions gradually evolve together. In the end, both the work and the design decisions that form the work are revealed. When conducted with proper discipline and attention, close reading is an extremely reliable and effective methodology. A close reading of a well-designed film or video invariably yields interesting and useful insights. In the case of Bizzocchi's scholarship, the insights are concerned with the design of the work - the relationship between form, content, goal, and experience.

Artistic Practice

Bizzocchi's academic scholarship is complemented by his own artistic practice. He has worked in film and video for over forty years as director, cameraman, or both. His instincts in this area have been reinforced by his constant involvement in teaching these media over the same period.

Artists have many motivations, and many methodologies- the details of artistic motivation and artistic methods are unique to the individual artist. Bizzocchi's initial artistic motivation for his ambient video work was similar to the motivation of any competent academic researcher - it is curiosity. The curiosity here is the curiosity of creativity. It manifests in two sets of related questions. The first set of questions starts with "what is the nature of the emergent media form of ambient video"? It goes on to ask: "how will existing media practices inform this new genre, and how will media practice evolve in the process?". These questions grew directly out of his academic work. His writings about the effect of improving display technologies on the future of video practice included the predicted development of this form of "ambient video" art that was particularly suited to the new screens.

The second motivating question is intensely personal. It related to Bizzocchi's own arc as a filmmaker and video producer. As he considered his initial predictions for this new form, his own deeper creative instincts were sparked. He determined to work in this new medium as an avenue for his own artistic expression. He was excited at the prospect of his own growth as a director, cinematographer and editor. This has proven extremely rewarding to Bizzocchi as a video artist. His creative activity had been put aside in the pursuit of other interests. Now it has been revitalized and renewed with his ambient video work.

The dynamic between the scholarship and the artistic practice is interesting to review. As he worked in this area, he saw his own videos as demonstration "proof-of-concept" expressions of the future media developments he predicted in his academic writing. However, the purpose of the artistic practice is not to "prove" the scholarly writing. The purpose is rather to explore an artistic

space and form of expression that is interesting on its own terms as a creative direction. The art's aim is to feed the soul of the film artist, not the mind of the media researcher. However, the effect of the artistic practice is to do both. The video art does provide "proof-of-concept" to Bizzocchi (and to others who view his work) that ambient video is a creative form worthy of exploration. It also gives indication of some of the specific creative directions that will be useful for him and for other ambient video artists. In addition, the process of artistic experimentation inevitably reveals fresh creative directions not anticipated by the scholarly research. As these new creative directions are discovered in the production process, they become new research directions for the scholarly process.

In the context of this chapter, Bizzocchi's video art confirmed an observation from Bizzocchi's scholarly work - the importance of slow-motion as a creative direction. The video productions also uncovered a critical practical problem. Beyond specific limits, it is difficult to produce intenstive slow-motion effects in video platforms.

Scientific Research

Ben Youssef's methodology involves initially the identification of the research domain. This is followed by an extensive exploration of the state of the art of the relevant issues. From this investigation, he develops specific research problems. In the context of this collaboration, some of these research problems are related to Bizzocchi's work in both humanities scholarship and artistic practice. A given technical problem prompts ideas from which certain assumptions and hypotheses are formulated. The objective is then to test these hypotheses. Testing may be done by experimentation or by further collection of data. In Ben Youssef's work, this step typically includes modeling and simulation. Once the model is developed and verified, the results are analyzed and the hypothesis in question is then tested against these results. If the hypothesis is not supported by the evidence, it can be revised or rejected, and a new hypothesis is developed. If it is supported by the evidence, then it can be seen as contribution to knowledge in general and domain theory in particular.

In our current research agenda, Ben Youssef's objective is to develop a software tool to be used in video post-production to create interpolated frames for the purpose of slow-motion effects. He first tests the model with artificially created video sequences. This allows him to create a standardized situation for study, in which all variables are under his control, and in which the results of manipulating variables can be clearly identified and evaluated. If some correlation is found between variables, he can show that this is due to a causal relationship and not based on coincidence. A future step will be to test the model with actual live action video footage. The final test will be the ability of this tool to enable video artists to produce smooth and believable slow-motion. Successful completion of the research will enhance aethetic quality for both creators and viewers.

A SHARED RESEARCH AGENDA

Overview: Intertwined Research Threads

The overall arc of Bizzocchi's and Ben Youssef's research contains a series of intertwined strands: humanities scholarship, artistic creation, scientific and engineering research. Each strand focuses on its own set of intellectual and creative questions. Each stands on its own as a legitimate line of domain-specific enquiry with a unique context, methodology, and results. At the same time however, these separate strands are intimately intertwined with each other. At one time or another, each has supported the development of the others at critical moments. A detailed review of history of the different threads reveals that the joint effect of the shared strands is both cumulative and synergistic.

Humanities Scholarship: The Evolution of Video and the Ambient Video Form

Bizzocchi's core academic orientation is that of a media scholar interested in the evolution of film and video forms and genres. He is concerned with the identification and the understanding of current and future trends in the art and practice of the moving image. The development of these media is determined by complicated and interrelated dynamics of artistic practice, marketplace phenomena, academic and popular criticism and the effects of emergent technologies. The latter is the catalyst for much of Bizzocchi's research. The moving image is undergoing significant changes due to the effects of new technologies on the production, post-production, distribution, and display of moving images. As Gene Youngblood noted decades ago, "New tools generate new images" (Youngblood, 1970).

The combined impact of the new video tools is enormous. Sophisticated video cameras are becoming increasingly less expensive, higher in quality, and much more widely distributed. At the bottom end, consumer camcorders with HD capability are now available for less than a thousand dollars. Excellent "pro-sumer" cameras for beginning professionals and serious film students can be purchased for approximately $5,000. The sophisticated high-end "Red" camera retails for less than $18,000 (without lens and accessories).[1] Advances in editing and post-production technologies are even more striking. Software such as Apple's Final Cut Pro and Adobe's After-Effects is now widely available and relatively inexpensive. These software platforms combine the ability to handle high-definition imagery with an astounding array of choices for the sequencing, layering, manipulation, and transformation of moving images. Visual effects in the world of cinema were expensive to purchase and extremely cumbersome to implement. The film artist, unless she was addicted to the arcane technology of the optical printer, was forced to work second-hand through a specialized post-production effects shop that took days and dollars for each iteration of even a simple proposed cinematic effect. Today's digital moving image artists have on their desks a post-production platform far more versatile and less expensive, and have the further advantage of direct manipulation of the intended effect without the need, delay, and cost of an intermediary.

The effect of this explosion of sophisticated creative tools is amplified by the proliferation of multiple channels for the distribution of the moving images. Forty years ago, the primary venues for viewing the non-amateur moving image were the cinema house, broadcast television set (typically with three or four channels), and the 16mm school or club projector. Today, we have multiplex theatres, massive multi-channel cable and satellite services, videotape, DVD, legal and pirated digital video files, mobile video players, cell phones, game platforms, and the internet. The ascendancy of Blu-Ray will further increase the distribution and reach of high-definition moving images (Vancouver Sun, 2008).

The final piece to the technological puzzle is the even more compelling revolution in domestic display technology. What used to be a small fuzzy picture in a box in the corner of the room, is now a beautiful image that dominates the main wall in our living spaces. The combination of HD production standards with flat-panel display hardware has transformed video into a new medium. The aesthetic gulf between cinema and video has been eroded. The difference is still there, but it has been reduced dramatically, and the differential will continue to shrink as display technology improves further (at the end of this chapter, we will consider in more detail the inevitability and the specifics of these improvements).

Video is improving on every front: the capture of better imagery, more sophisticated manipula-

tion and processing of the image, multiple channels for the distribution of content, and increasingly impressive and visually stunning display environments. What is the effect of this set of changes? To answer such questions, humanities scholars often look back in order to anticipate the future. To understand the effects of current changes, it helps to look at the history of film and video. These two media are distinct, but they are intimately related. Video is the junior partner to cinema. Video has taken many forms - broadcast television, documentary video productions, video art - but the basic visual grammar and construction was derived from film and has been adapted to the realities of the newer medium. The foundation of both media is the sequencing of moving images and sound. As a consequence the underlying fundamentals are the same: continuity construction for the purpose of re-creating a "real" storyworld, or montage construction to create visual and narrative impact through the iteration of image. However, the treatment of composition and subject scale differed significantly in the two media. The single most significant design change in moving from film to video was the loss in visual quality. The beautiful cinematic wide shot of Sergei Eisenstein, Orson Welles and John Ford was compromised by the introduction of television. The landscape lost its visual impact, and the humans were rendered too small to carry sufficient emotional and narrative impact. Television was "a medium of close-ups". That is to say, visual storytelling in video relied upon the sequencing of a succession of tighter images, rather than the leisurely play of action and sound within a single wider image. This did have consequences for the editing process. The need to use smaller portions of the scene in video necessitated an increase in the cutting pace. Visual details and human reactions that would work in a cinematic wide shot, needed to be broken up and presented in temporal sequence in video.

A simple way to gauge some of the effects of improved image and larger screens on the grammar of television is to simply reverse the effects of the earlier poor imagery on cinematic storytelling. In today's large HD home screens, the wide shot is revitalized. Both its earlier cinematic functions have been restored. The landscape is once again a rich narrative environment with visceral impact. Story can be related in a sustained wide shot, yet retain narrative flow and emotional strength. As a result, cutting pace can slow down when a director or editor wishes to modulate pace.

In more general terms, the new video displays support a more pictorial aesthetic than the earlier television environments. The cinematographer's art is privileged: composition, textures and patterns, variations on scale, the play of light and shadow, the exquisite beauty of slow-motion - all are rendered more appealing in the new screen environments. In a different direction, the larger screens combine with the proliferation of post-production capabilities to support complex visual layering and transformations. The repeated playback capability of contemporary playback options support equally the simple visual pleasure of the pure image, the multi-mediated nature of more complex and layered visuals, and the multi-linear narrative outcomes of the split-screen or the complex plot structure.

New forms of moving image will develop within this complex of artistic creation, media development, and technological improvement. One form will be the development and maturation of a media form that has yet to find its voice. That direction is ambient video. These new, elegant frames on the wall can support a moving image art that is distinct from previous forms of cinema and television. Ambient video is video that does not require our attention, but rewards it whenever we wish. Ambient videos will play in the backgrounds of our lives, but will give us visual pleasure whenever we turn to them.

This new form has well-established roots. The archetypal example of ambient video is the Yule Log. The burning log has graced television sets in the winter holidays since its inception at WPIX New York in 1966. During the latest season, our local cable provider had five Yule Logs burning on separate channels at the same time. The video log, along with its close cousin, the video aquarium, are a well recognized commercial phenomenon, appearing in each new video format as it appears: broadcast, cable, VHS, Betamax, DVD, and internet. The video log and the video aquarium, like their real world antecedents, meet the functional tests for visual ambience:

- They never require our attention
- They reward our attention in any moment with visual pleasure
- We are free to turn away, or return, whenever we wish
- They work the first time we see them, and they sustain their interest over repeated viewing

In the words of Brian Eno, as he described ambient music: "…it must be as easy to ignore as it is to notice…" (Eno, 1978).

Ambient video takes, and will take, many forms. Some of it will be the expression of pure visual form - the play of abstract shape, line and color. This is an old tradition in cinema, including film artists from Oscar Fischinger and Hans Richter, through the Whitney Brothers, Jordan Belson, and the later hand-drawn work of Stan Brakhage. Some in this variation will be driven by the music track, instantiations of the ongoing search for "visual music" and the "color-organ" (Moritz, 2007). The visual interpretations of music embedded in Windows Media Player and iTunes are algorithmic and computational variations on this direction. Other ambient video works will also follow in the screen-saver tradition, but will combine representational imagery within an algorithmic and recombinant context. Still others will present ambient experience with a highly personalized and kitsch flavor, such as the digital photo frames that are appearing in the marketplace. Some ambient work will follow the logic of the web-cam, and we will see exotic or visually interesting locations playing on our own walls in real time.

The ambient form that occupies the bulk of Bizzocchi's research is yet another variation. He is most interested in ambient video works that are non-narrative, that use representational (not abstract) content, and are highly cinematic and pictorial in their visual treatment. These works will follow in the tradition of the video log and aquarium, but will transcend the kitsch limitations of those works into artistic representation. The natural landscape is a rich source for engaging ambient visuals: mountains, clouds, rivers, foliage, and fauna. Particularly useful for the ambient experience are those natural visuals that move slowly (such as clouds) or are rendered in slow-motion (such as water or fire). It is interesting to note that the initial bundles of HD services always include the channels that deliver this beautiful natural imagery: PBS, Discovery Channel and National Geographic. This programming is not fully ambient yet - it is cut with a pace and provided with a soundtrack that does not favor truly ambient experience. However, it can be used as a quasi-ambient background with the sound turned off. This is consistent with user studies of actual television attention, which show that much of the time the set is on, but viewers are not paying direct and focused attention to the programming (Kubey & Csikszentmihalyi, 1990; Fowles, 1992).

One commercial program in this area that is fully ambient in spirit and execution is the stunning series *Sunrise Earth* (Conover, Czuchra, & Caloyanis, 2004-2008). This work is built on beautiful locations, careful HD cinematography, very slow pacing, and the exploitation of the pictorial quality of high-definition visuals as best experienced on large scale flat-panel displays. There are a limited number of video artists who have been working in similar directions, although they have done so in the context of standard definition video. They include Simon King, William Kennedy, and Steve Lazur. Some of them have already made the transition to HD, and it is safe

to assume that if they continue to work in this genre, they all will do so. As this form becomes more recognized in the marketplace and in curatorial circles, more moving image artists will be drawn to the expressive possibilities of the ambient video form.

Artistic Creation

As Bizzocchi pursued his research into the prehistory, current state, and aesthetics of ambient video art, he became more and more interested in its creative capabilities. He was an experienced documentary filmmaker and video producer, with a variety of productions that had social and educational goals. His career growth in educational policy development and then in digital media research combined to slow his creative work to a virtual halt, but the draw of ambient video aesthetics revived it. He has produced three complete works in his *Ambient Video* series, and is in postproduction on a fourth work. His creative works are also collaborations. His cinematographer is Glen Crawford of Canmore, Alberta. Crawford combines a sure eye for composition and light with an intimate knowledge of the Canadian Rockies. The editing, music, and visual effects are the work of Bizzocchi's son - Christopher Bizzocchi. Each of the films is conceived by Bizzocchi, who then works closely with Crawford on location and with Christopher Bizzocchi in the editing and visual effects phase. Although the genre of ambient video is not generally understood by most critics, Bizzocchi's works have been well received, and have exhibited internationally in London, Los Angeles, Beijing, Melbourne, Perth, Vancouver, Victoria, Banff and Ann Arbor.

The three works are all shot in the mountains - primarily in the Canadian Rockies. The first was *Rockface* (2003). It consists of seven shots of mountain scenics, each one approximately a minute in length. The strength of the work is the sense of scale, composition and light. The natural content is subtly enriched - in each shot one or two human faces are visually embedded, at or just below the limit for recognition. These details would not be noticed by the casual viewer, but for the owner of an ambient video piece that plays repeatedly in a domestic living room, they will provide an unexpected visual addition. Continued viewing over time will eventually reveal all of the faces, and perhaps a causal connection for their appearances and disappearances in the course of the film. This liminal hint of narrative provides an opportunity for extension of the playing life of the piece on the owner's wall. In the same piece, the Bizzocchi's - father and son - discovered the visual elegance of gradual layered transitions from one shot to the next.

This use of layered transitions was fully exploited in the next two films of the series. *Streaming Video* (2004) is entirely based on the observation of water, starting with tiny rivulets and creeks, gradually growing into faster and more powerful flow, and finally culminating in an enormous waterfall. The third piece in the series is *Winterscape* (2007), a detailed visual essay on the many manifestations of snow, ice, clouds and water in the Canadian Rockies. A fourth piece, titled *Cycle* (currently in post-production), is similar in conception, but examines the change of the seasons. Technical quality is important to Bizzocchi's work, and each film has increased the resolution of the images. *Winterscape* was the first to incorporate some use of HD imagery, and *Cycle* has been shot completely on professional level HD equipment.

The works share three important creative directions. The first is an emphasis on the pictorial quality of the visuals. This grows out of Crawford's sure sense of composition and light exercised in the context of the stunning locations. The second creative direction is a reliance on gradual and layered visual transitions. These works differ from almost every other film in this regard. They have abandoned the standard cinematic device of the cut, and its occasional replacement - the dissolve. Instead, the transition from one shot to

a second shot occurs in stages. First, one element from a new shot is gradually layered over the existing shot, then a second element, and then a third, and so on until the second shot completely replaces the first. As soon as this is done and the second shot fills the screen, individual elements from a third shot are gradually and successively layered over the second, until the third shot fills the screen. This process occurs throughout each work, and is a substantive break with the long tradition of cinematic sequencing. The effect in Bizzocchi's work is the development of a sense of "magic realism" - a world that appears like ours on the surface, but follows its own rules of visual logic and flow.

The third creative direction of the series is the treatment of time as plastic, to be subtly slowed and speeded as the needs of the film dictate. Almost every shot has some manipulation of the time frame. Water flow is slowed down to give a slightly dreamy quality. Clouds are subtly sped up - they are still slow, but their own languid flow is more pronounced and dramatic with a slight increase in speed. Sometimes both effects are combined in the same shot, with some parts of the scene sped up while others are slowed down. [see Figure 1] Bizzocchi consistently relies on these time manipulation techniques, despite his frustration with the severe limits to the degree of slow-motion that can be applied within most video environments.

The water in the foreground will be slowed down in post-production, while the clouds in the background will be sped up. ©2007 Jim Bizzocchi. Used with permission.

The *Ambient Video* series of works has served several functions for Bizzocchi. First, working in conjunction with his two creative collaborators, these films have provided an avenue for a deeply satisfying artistic expression. Second, they have provided an informal validation and "proof-of-concept" for certain ideas from his academic scholarship. Third, he has discovered concepts in his creative work that point out new directions for

*Figure 1. Image from **Cycle** video [work in progress]*

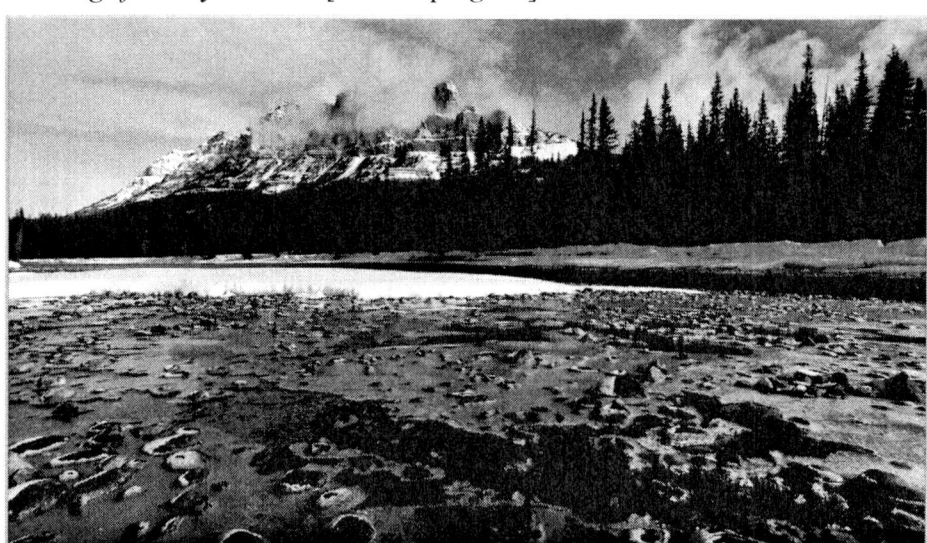

this scholarly writing. The power of visual layers and gradual transitions is one example of a concept that moved from creative production work into the realm of theory and scholarship. Finally, as we will see later, the difficulties in one aspect of production on his *Ambient Video* series provided impetus for his colleague Ben Youssef's work in scientific and engineering research.

Artistic Practice Integrated within Scholarly Context

At this point in the development of his art, it was important for Bizzocchi to situate his own creative work in a wider context. In the curatorial and exhibition world, it is critical for an artist to communicate a sense of their own work that includes its place within the broader artistic and curatorial discourse. It was also important for Bizzocchi as a scholar to understand the context that would illuminate not just his own work, but also that of other artists working in the cinematic and representational style of ambient video.

In the history of film and video art he found ample evidence of the three artistic interventions that characterized his own ambient video

work: the eye of the cinematographer, the deep manipulation of image in post-production, and the treatment of time as a malleable and plastic variable (Bizzocchi, 2008). Early film artists Germaine Dulac and Maya Deren both argued for the importance of image in the development of a film grammar that would fundamentally separate the moving image from other media forms (Dulac, 1978 and Deren, 1978). Many film artists, such as the avant-garde British landscape filmmakers, turned to nature and the scenic as sites for visual exploration that would reveal fundamental truths about both the subject and the medium itself (Dusinberre, 1975). In a more mainstream context, film artists from Eisenstein to Kurosawa have exploited the powerful relationship between composition and landscape as an expressive tool for cinematic expression.

The deep manipulation of image has just as strong roots in the history of cinematic and video art. This is the tradition of the trickster, that stretches from George Melies, through Harry Smith and Stan Brakhage, finding further expression in video artists such as Nam June Paik, Steina and Woody Vasulka, and most recently in Zbig Rybcsynski, Christian Boustani and Michael

Snow. The visual trickster's ability was increased when the moving image became electronic, and even more so in the latest digital post-production technologies. However, the spirit of the trickster is best expressed by the words of the cinema genius Stan Brakhage. He starts with a bow to Melies, crediting him with giving "the 'art of the film' its beginning in magic". Brakhage then embraces the essence of this direction as his own when he proclaims that the artist sees more powerfully because he "allow[s] so-called hallucination to enter the realm of perception" (Brakhage, 1978).

Finally, the creative treatment of time is an ongoing concern of film and video. Time has always been a protean element in the construction of moving images. Any treatment of motion is both based on time and a representation of time itself. Filmmakers concerned with the essential nature of the medium - structuralist filmmakers - often aggressively manipulate the time frame. Yoko Ono and her Fluxus colleagues radically slowed time down. Godfrey Reggio speeds it up in *Koyaanisqatsi*. Michael Snow (*Wavelength*) extended the duration of time, testing the limits of audience reception and patience. Andy Warhol broke those limits with his five hour static study of a subject sleeping (*Sleep,* 1963). Descriptions of the viewing experience of that piece sound like an archetypal ambient video experience, with the audience engaged as much in small talk, eating, smoking, and their own comings and goings as they are in the film itself (Koch, 1978). Curators have recognized the critical relationship of film and video to time in a variety of exhibitions. They have found their own touchstones to the manipulation of time. In the catalogue to one exhibition, Matthias Gaertner discusses a series of questions around slowness (Gaertner, 2000). In a second exhibition, the curator notes that the film and video artists seek to suspend or stop time (Cappellazzo, 2000). A third catalogue discussed the moving image representations of time with respect to location, mechanization, industry, globalization and the internet (Morgan and Muir, 2004).

Technical Background Research and the Identification of the Central Research Problem

Early on in our joint research program, we received a small amount of initial seed funding under the Research Development Initiative (RDI) program of the Social Sciences and Humanities Research Council of Canada (SSHRC). The aim of the RDI grant was to allow us to explore the state of the art in current video post-production technologies and tools. One of the goals of this project was to identify specific areas for future applied research into the development of more effective content creation tools for video artists and producers. The direction of our joint enquiry was guided in part by the aesthetic developments predicted in Bizzocchi's humanities scholarship.

The RDI research itself consisted of conducting interviews of video editors, special-effects technicians, video post-production service suppliers, and hardware and software manufacturers and vendors. In addition, our research assistants reviewed existing research that addresses creative and technical developments in film and video. For instance, we examined the technical capabilities of different video cameras for the consumer level to the professional. We also looked at the performance of the existing software tools for video editing in post-production.

We concluded that among the aesthetic trends we focused on initially (that is, slow-motion, complicated layers and transitions, split-screen storytelling), video slow-motion remains an area where further progress is much needed. Bizzocchi's research and practice had demonstrated that slow-motion can be an important tool for the new pictorial video enabled by high-definition home display units. It is one of the most dramatic and visually compelling phenomena associated with the moving image. The ability to easily achieve a visually pleasing slow-motion effect in a high-definition post-production environment will help video producers reach the full visual potential of the new high-definition display devices.

The problem is that most video cameras can't record in slow-motion in the same way film cameras can. Therefore, a constructed slow-motion must be applied during the post-production (editing and mixing) phase of a video project. With simple technologies, motion can typically be slowed to about one-half real speed with acceptable results. However, slowing beyond that usually results in unnatural and jerky renditions of motion. This is unacceptable for Bizzocchi's work, and for the work of other artists interested in exploiting the aesthetic possibilities of slow-motion in their work.

Producers need post-production technology that will make a number of "in-between" frames to lengthen the playback time of a given sequence. For intensive application of slow-motion (beyond a factor of ½ speed slow-motion) these "in-between" frames should not be mere duplications of existing frames. Simple duplication leads to the visual jerkiness that breaks the illusion of continuous motion. Instead, the "in-between" frames must be interpolated in post-production to blend the images from the live-action frames they extend. This interpolation should be elegant enough to yield an acceptable illusion of slow-motion at factors such as 1/4 or 1/8 of the original real time speed. This ability would allow the effective slow-motion rendering of a wide range of images such as moving water, fire, human, animal, and mechanical activities, without the expense and the restriction of shooting in film or very expensive specialized video cameras (the 1/4 to 1/8 time base extension is a useful and important base-level technical goal. More extreme extension of the slow-motion effect would lead to even more interesting results).

Our regular research meetings discussed and sometimes hotly contested the issues around this problem. In this process, we came to realize that we could use existing motion information across a number of video frames on either side of the location of where the interpolated frames are to be inserted. We decided to extend the information-gathering window beyond the two frames immediately adjacent to the interpolated gap. We called this extension process "widening the span" of the window of frames to be "sampled".

Scientific and Engineering Research: Slow-Motion

It is often necessary to create frames representing the input image at different times or locations. This is done by *interpolating* from the existing sample points or frame. It is a vital ingredient of time-base correction, video effects, audio pitch changing, and other tasks (Held and Marshall, 1991). Frame interpolation is important in standards converters that must convert frame rates, such as between 50 and 60 Hz. The approaches applicable to sample and line interpolation can be applied to frame interpolation by interpolating pixel values between adjacent frames using a frame memory. However, the result is unsatisfactory when motion occurs between the frames. This can be seen by considering a scene containing a small object that moves a distance of several pixels from one frame to the next. Interpolating pixels at the same screen location in each frame causes distortion at the edges of the moving object. This appears as a jumpy effect on moving objects known as *judder*. What is required is for the interpolation to process related pixels of the moving object, not just pixels at fixed locations on the screen. This can be accomplished by the use of motion compensation. Motion compensation is a very complex process, but has become common in standards converters, effects processors, and video compression processors.

Motion video has considerable redundancy from one frame to the next. Much of a frame may not be new information at all; some parts may be stationary and others may simply be parts of the previous frame that have just moved a little. In principle, anything that already exists in the previous frame need not be transmitted again; it can just be copied by the receiver from a stored copy of the previous frame. The task of figuring out what is new or old in a frame is motion compensation. It is based on processing the images in blocks. Given one complete frame as a starting point, blocks from the next frame are taken and compared with areas of the first frame to determine if there is a match at any position in the first frame. Keeping in mind that the block may have moved between frames, if a match is found, a *motion vector* is created for the receiver to use to predict that area of the second frame by copying it from the first frame. Recognizing that the amount of motion between frames is usually not too great, it is only necessary to search a small area surrounding the location of the block being tested. Even so, the amount of computation represented here is prodigious and it may be necessary to restrict the search range more than might be desired in the interest of practicality. Blocks that are not found in the previous frame must be coded by other means and transmitted in full (Tekalp, 1995). The generation of motion vectors is needed in the use of interpolated frames for the enhancement of slow-motion in post-production.

Ben Youssef set out to address the slow-motion problem by focusing on Motion Compensated Interpolation (MCI). The MCI process contains a first phase dealing with Motion Estimation (ME) and a second one handling Motion Interpolation (MI). These constitute fundamental problems in digital video processing (Wang et al., 2002). Typical ME tracks the movement of objects, or parts of objects, from one frame to the next. Chupeau and Salmon (1993) used a motion compensation algorithm where all pixels in the interpolated frames are assumed to have at least one motion vector from sets of adjacent frames {-1→0}, {0→1}, and {1→2}. This is known as AF-MCI for Adjacent Frame Motion Compensated Interpolation. In contrast, Ben Youssef's method always tracks motion across the crucial central time frame gap {0→1} into which interpolated frames will be placed. We call this the Interpolation Target Gap (ITG).

When performing ME, tracking a sum-of-luminance (SOL) value from an adjacent frame to another may lead to erroneous and erratic motion vectors due to the fact that there is not a 1:1 correspondence of similar SOL values from one frame to another. In other words, due to quantization and rasterization errors during filming, there is no guarantee that each pixel will map into any other frame. Aside from sampling errors tainting motion information, the AF-MCI method for ME is fallible because it assumes that each pixel has a good chance of being present in each adjacent frame. This assumption is perhaps unfounded because different parts of an object can be obscured, transformed, blurred, or occluded by another object; and therefore, each pixel will not necessarily appear in each adjacent frame.

A small amount of motion can result in large differences in the pixel values of a scene, especially near the edges of an object (Zheng et al., 2000). Also, a wider field of view and higher resolution (as in HDTV) makes it harder to fool the eye, and therefore more difficult to do motion compensation (Ben Youssef et al., 2005; Solari, 1997). The effectiveness of MCI will depend on the ME and MI phases as well as on source video scene complexity. We anticipate that pixel-wise MCI is an emerging viable alternative to conventional region-based techniques given the recent hardware advances and the availability of parallel computing on the desktop (Chupeau and Francois, 2000; Gorder, 2007).

Ben Youssef's method relies on using temporal motion compensation that analyzes a "Wide-Span" of multiple frames over time, thus the name WS-MCI. Performing ME on SOL values across the interpolation target gap for all motion vectors allows us to focus exploitation of available motion information in the time span that is most crucial. Additionally, WS-MCI can determine how an SOL value has moved over a wider span of time. Hence, it can better ignore movement 'noise' between adjacent frames because it does not assume that each pixel is present in every frame. This heuristic reduces quantization and rasterization error creeping into the motion vectors.

Motion Compensated Interpolation (MCI)

A video sequence is composed of discrete video frames where each frame is a 2D image. Motion estimation uses the source frames to deduce the motion vectors, and interpolation uses that information to produce the new interpolated frames. MCI can be iterated for each time gap in the source video sequence to produce a video sequence with multiple new frames. This results in slow-motion if the movie's frame rate is preserved when viewed. Multiple time gaps, into which interpolated frames are placed, represent multiple iterations of the MCI process. Also, the identities of the frame indices will shift forward in time for each MCI iteration such that the interpolation target gap spans different pairs of successive frames. In Figure 2a, only one new frame is inserted into the video sequence per source frame time gap; however, once motion information is gleaned from the source video, any number of interpolated frames can be produced between source frames.

The interpolation step uses motion information to deduce where pixels (SOL value positions) will migrate over time across the ITG. Figure 2b further illustrates that, for each MCI iteration, any number of interpolated frames can be produced within each ITG because the motion vectors pass through an infinite number of possible new frames within the time span.

Previous Work in Motion Estimation and Motion Compensated Interpolation

Motion estimation represents the major portion of motion compensation algorithm complexity (Solari, 1997), and many methods have been proposed. Region-based ME is ubiquitous throughout most current MCI methods because of its reduced resource requirements in processing time and memory (Andre et al., 2004; Chupeau and Francois, 2000; Lee et al., 2002; Solari, 1997). Some techniques use hierarchical block-based ME that provides better motion information over typical full search block-based ME (Choi et al., 2000). Because there is near-infinite variation in possible real-world video sequences, object-recognition-

Figure 2. Pixel-wise motion interpolation producing a) one, and b) many motion interpolated frames, shown in dotted outlines. ©2006 Belgacem Ben Youssef. Used with permission.

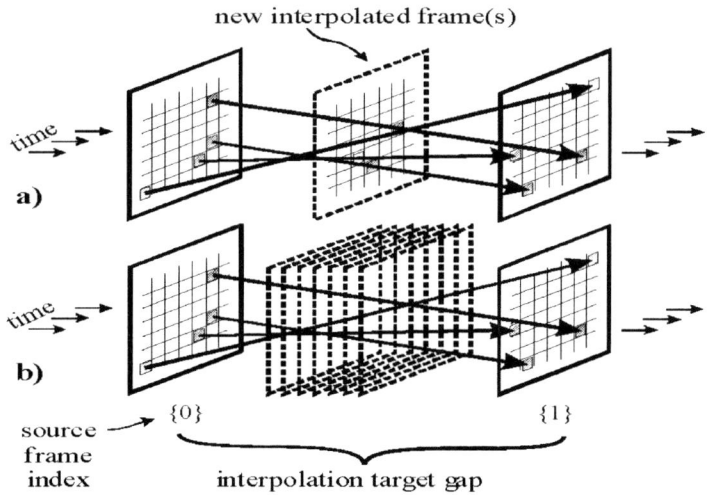

based ME is not plausible for generic MCI needs. Conversely, pixel-wise MCI may have distinct case-specific advantages that may become more evident as computers become more powerful.

MCI techniques provide the best solution in temporal rate up-conversion. Current techniques include frame repetition and linear interpolation by temporal filtering, but these methods may introduce motion noise and object blurring (Al-Mualla, 2003; Choi et al., 2000). Objectively comparing MCI techniques using various real-world video conditions is problematic because there is no control for experimental results. Consequently, there is a clear need for a test apparatus to objectively and quantitatively compare and contrast various video processing algorithms.

Proposed Method

The quality of video frames can be thought of as a measure of how close a set of interpolated frames approach a set of perfectly interpolated control frames in terms of the colors at each pixel in each frame. To rigorously analyze an MCI algorithm, we created an artificial video sequence to provide a reproducible and controlled sequence of data that could be directly compared to experimental results. For each test run, six artificial video frames, five control intermediate frames, and five experimental interpolated frames were created. The experimental frames were directly compared to the control frames in terms of the sum of their pixel color differences for all interpolated frames. This allowed us to quantitatively assess the quality of a set of experimentally derived interpolated video frames. The average error over all frames yields the metric *average-ECCE*. This denotes the average experimental-control-color-error.

All objects and scenes in the initial source frame are randomly created in a reproducible manner. This allows for parallel testing and analysis, thus yielding sets of results that can be directly compared to the control interpolated sequence. Ben Youssef's implementation of the

algorithms assumes a progressive-scan source video sequence. However, these algorithms could be applied to interlaced video if the frames are linearly de-interlaced first (Chupeau and Salmon, 1993). For each experiment, he performed paired t-tests using 40 test runs. The same source video frames were utilized for all experimental techniques.

Figure 3 displays the diversity of initial source scenes, and shows the chaotic nature of the "inkblot" object as compared to the simple circle object. Inkblots are created using a collection of dots based on random-walk theory. Because inkblots vary in size and are porous, a few blots can create a scene that is just as challenging to process as a scene with many simple objects.

To allow for SOL tracking that would mimic a natural scene, all objects are given 'texture' that doesn't change from frame to frame. This texture is produced by altering all the colors in each object by some random value that is as much as +/-20 for red, green, and blue. Texture was added so that the experimental MCI algorithms could distinguish between the areas within the same object that have the same SOL values. Moreover, a maximum speed of 40 pixels per frame was imposed. The translation's magnitude and direction are kept constant across the artificial video sequence.

Random objects are selected (inkblot or circle), assigned random sizes, translations, colors, and positions. Each one is then placed into an image layer which can be manipulated independently of other layers, thus allowing for control over experimental conditions. Object layers are used for constructing an artificial video sequence of frames. The number and type of objects in a scene can be altered to test the efficacy of Ben Youssef's three experimental MCI algorithms under different sample video complexities.

We assume that the 'best' interpolated frames should be as close as possible to the ideal control interpolated frames. The similarities between control and experimental images can be easily obtained for all the interpolated frames by com-

Figure 3. Representative source video frames showing a progression of source scene complexity. Each source frame is comprised of randomly positioned, textured, and scaled objects with random motion vectors. All these object-specific values are saved so controlled experimentation can be done. Frame a) contains 7 inkblots/circles, b) 7 inkblots, c) 14 inkblots/circles, d) 14 inkblots, and e) 21 inkblots. ©2006 Belgacem Ben Youssef. Used with permission.

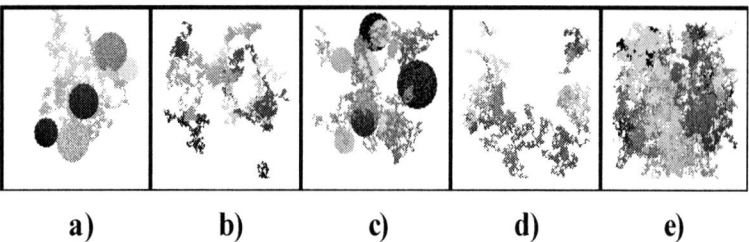

a)   b)   c)   d)   e)

puting the RGB-error-magnitude for all pixels in a frame and then averaging the frame color-error for all interpolated frames, thus obtaining the average ECCE (or Ave. ECCE).

Motion estimation is performed using pixel-wise sum-of-luminance (SOL) search that progressively funnels down through a hierarchy of down-sampled frames, using a decreasing search space at each level (see Figure 4). The SOL value at one pixel is searched for in other frames to find the end point for that pixel's motion vector. Pixel-wise ME is predictable in resource use, and produces dense and accurate ME information (Chupeau and Salmon, 1993). Multi-resolution representations of video frames have been shown to improve motion estimation (Wang et al., 2002). Each frame in the video sequence has its own hierarchy of 5 images. The lowest resolution is at the top, and the resolution progressively increases towards the bottom image, which is not blurred. As the search goes down the hierarchy of the stack of images, the search size gets smaller, known as funneling. Because we are using an artificially generated video sequence, the scene can be controlled to optimize several variables, such as funneling search sizes for each low-pass filtered hierarchy level. The funneling search size used along one side of the 2D search square is 41, 17, 13, 9, and 3 pixels, respectively.

Tracking a pixel's SOL value from one frame to another may not represent actual movement of a constant color pixel because pixel neighbors are used in the calculation. Searching based on sum-of-luminance decreases the dimensionality of the pixel search to 1D from an RGB value (3 dimensions). Because SOL values are fallible due to their color ignorance, using only adjacent source video frames for ME may be slightly more error prone than examining how SOL values move across the ITG. This time frame is where the interpolated frames are produced and it contains the most pertinent temporal/spatial information about how SOL values will move.

The three algorithms used in Ben Youssef's experiments are: AF-MCI, WS-MCI, and WS-MCI+SH. AF-MCI only creates motion fields across single time frame gaps, whereas WS-MCI always creates vectors across the interpolation target gap. WS-MCI+SH represents an enhancement to WS-MCI by adding spatial hinting to the generation of motion vectors. This extension uses motion vectors calculated for previous frames or nearby pixels as a first estimate to the correct motion vectors. That is, once a correct vector has been found for an area, it usually is maintained across that area. Whereas this approach works best in case of linear motion, which his generated test data is limited to, it could also be enhanced

Figure 4. Motion estimation using the funneling SOL search. ©2006 Belgacem Ben Youssef. Used with permission.

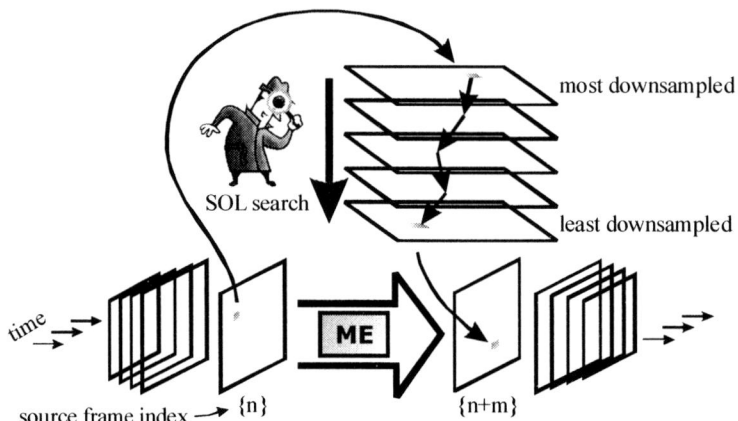

to use derivatives of velocity across the video frames and apply it for nonlinear motion.

Although the three MCI methods use the same source frames to get motion information, adjacent-frame and the two wide-span MCI techniques (collectively referred to as WSM for wide span methods) differ in their exploitation of motion information within the source video frames. AF-MCI creates only one motion vector that spans the ITG; however, both of WS-MCI and WS-MCI+SH motion vectors span this crucial time gap. Therefore, WSM more fully exploit the pertinent SOL-migration information across the time gap that is most motion-information rich.

Experimental Results

Experimental-control-color-error is concerned with pixel RGB values across all experimentally derived interpolated frames compared to control frames, and may be influenced by a number of factors such as: the object complexity and its size in a scene. In Ben Youssef's test runs, the scene complexity ranged from 1 circle to 21 inkblots per scene. Experimental and control interpolated frames were created and compared in terms of the sum of their pixel color differences for all interpolated frames and for each run. This allowed

us to quantitatively assess the quality of a set of experimentally derived interpolated video frames. The results, provided in Table I, show that scenes with several simple objects, random objects, or complex objects are all handled better by the two wide span methods. The column labeled "WS-MCI % Benefit Over AF-MCI" in this table represents the WS-MCI performance advantage compared to AF-MCI for all test runs at a particular scene complexity in terms of their respective average ECCE values. The same explanation applies to the last column in the table.

While there is a slight-but-definite benefit to the use of WS-MCI when there are many chaotic objects (e.g. scenes with a flock of birds or blowing leaves), this gain is further enhanced by WS-MCI+SH. This has the following implications: because the error rate will be lowest when vectors are correct, once a correct vector has been found for an area, it usually is maintained across that area. Therefore, by using adjacent vectors as starting vectors for each new search, Ben Youssef benefits from an improvement in the quality of his results. In addition, he was able to reduce the computational time requirements of WS-MCI+SH by 50%. The Ave. ECCE results are graphically depicted in Figure 5. An example of a control frame and an interpolated frame, the

Table 1. Quantitative comparison of AF-MCI with WS-MCI and Ws-MCI with WS-MCI+SH over a progression of artificial source video scene complexitY

Scene Complexity	# of Runs	Ave. ECCE for AF-MCI	Ave. ECCE for WS-MCI	Ave. ECCE for WS-MCI+SH, (CI= .99)	WS-MCI % Benefit Over AF-MCI	WS-MCI+SH % Benefit Over WS-MCI
1 Circle	40	1.86	1.58	1.09 +- 0.189	15.20	31.01
7 Circles	40	10.08	9.27	7.00 +- 0.471	8.09	24.49
7 Objects	40	18.53	17.57	10.33 +- 0.834	5.20	41.21
7 Inkblots	40	23.28	22.11	13.33 +- 0.936	5.02	39.71
14 Circles	40	19.47	18.78	14.51 +- 0.648	3.54	22.74
14 Objects	40	26.24	25.34	14.46 +- 0.648	3.43	42.94
14 Inkblots	40	34.40	33.51	25.83 +- 1.266	2.61	22.92
21 Inkblots	40	44.93	44.46	37.46 +- 1.318	1.05	15.74

Figure 5. The average ECCE values for AF-MCI, WS-MCI, and WS-MCI+SH. The lower the value, the better the performance in this case. ©2006 Belgacem Ben Youssef. Used with permission.

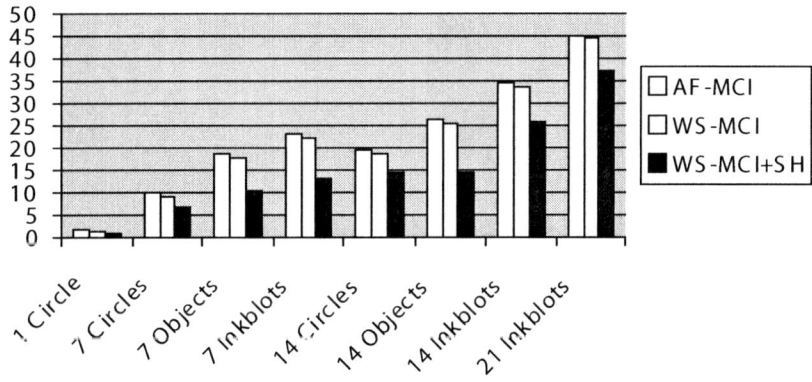

Figure 6. An example of a control frame (left) and an interpolated frame (right), the latter obtained via WS-MCI+SH, of an artificial video sequence used for testing Ben Youssef's method. ©2006 Belgacem Ben Youssef. Used with permission.

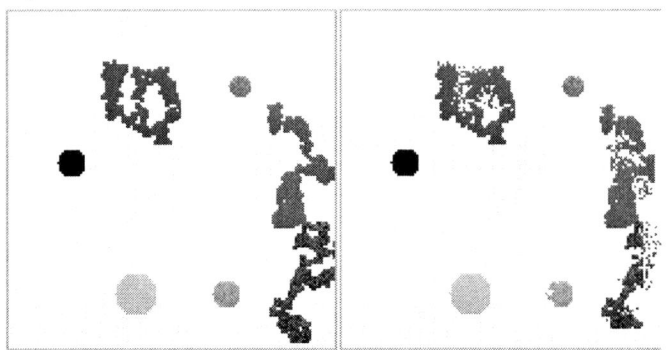

latter obtained by applying WS-MCI+SH, are are shown in Figure 6.

NEXT STEPS

Next Steps in Scientific Research

Future research will focus on making wide-span MCI more robust in terms of object tracking with regards to objects' colors, dealing with the problem of occlusion, and readying our method for testing using live-action video footage, encompassing both standard definition and high definition video. Other next steps will consider how WSM would handle nonlinear motion. For instance, Ben Youssef's preliminary work shows that using backward motion vectors produces interpolated frames that are more filled-in with sensible colors. This outcome is anticipated since backward motion vectors give more valid information about pixel movement across the ITG.

Ben Youssef will also study the impact of incorporating the size of the wide-span window, denoted by WS, as a user-controlled input parameter in future versions of his interpolation tool. WS refers to the number of input video frames used in the computation of interpolated frames. Hence, WS[2] and WS[4] would be equivalent to his current AF-MCI and WS-MCI implementations, respectively. Ben Youssef hypothesizes the existence of a "sweet spot" to the size of the WS window of input video frames surrounding the interpolation target gap. He anticipates the determination of this "sweet spot" for a particular input video sequence will prove to be a subjective process that is best left to the decision of the video creator.

Effective frame interpolation tools will allow both research partners to study the impact of varying the slow-motion frame rate to support more effective and sophisticated visual storytelling. Other techniques to be studied include experimenting with the basic frame rate for video production such as comparing "overcranked" 60 frames per second camera original footage with frame-interpolated post-production footage.

Our research collaboration extends to other related areas dealing with high-definition videoconferencing and high-dynamic range video. In the former, our work aims at enhancing the virtual interaction experience, overcoming many of the limitations of traditional group videoconferencing: the difficulty in making eye contact, the impact of one-on-one conversations, overall visual quality and the low-resolution imperative towards tighter composition for a higher magnification factor. Working with high-definition video presents the challenge of managing significantly greater amounts of bandwidth than standard-definition video. Our objectives include exploring advanced networking technology, determining how continuous streams of video can be maintained without latency issues, and developing the technical solutions to mitigate these issues. This research will also examine how the combination of HDTV and large video flat-screens changes the experience of videoconferencing; and potentially other shared visual environments.

The generation of high dynamic range (HDR) video is another emergent research area that will benefit from our combination of creative and scientific methodologies. Many practical applications would benefit greatly by providing more precise and possibly calibrated streams of temporally coherent data where the goal is to reproduce the appearance of images as perceived by the human observer in the real world. This assumption matches well to a diverse range of applications such as realistic image synthesis in computer graphics, digital cinema, documenting reality, tele-medicine, and some aspects of surveillance (Kang et al., 2003). The importance of HDR will continue to increase as displays covering a large luminance range become more readily available. Our work will involve the simulation of post-processing visual effects which rely on HDR pixel information such as glare, light and dark adaptation (night vision), and motion blur.

Next Steps in Creative Practice

The immediate next step in Bizzocchi's creative art practice is to finish his fourth film: *Cycle*. This film has been shot, and now he needs to finish the post-production. *Cycle* will continue to explore the expressive use of transition, layering, and time manipulation - including slow-motion. This work will differ significantly from Bizzocchi's other works in its increased visual quality. The first three pieces were shot primarily in standard definition. There was some prosumer HD footage shot for *Winterscape*, but the bulk of the film was shot in 16x9 standard definition digital video. The final version of the work is the sharpest of his first three films, but is nonetheless rendered in standard definition. *Cycle* is a co-production with the Banff New Media Institute (BNMI), and thus benefits from the creative support of the BNMI and the Banff Centre's excellent facilities and technical support. Crawford and Bizzocchi captured a series of strong visuals during the shoot, and the resulting HD rendition is very powerful. It translates exceedingly well to large image display - yielding a visual presence that is visceral in its impact.

When *Cycle* is completed, Bizzocchi will begin work on a different type of ambient video piece. His intention is to build an ambient piece based on urban imagery: transportation, industrial processes, and city scenes. The look and feel of this piece will therefore be quite different from his first four films, but the creative interventions will be the same: strong compositions, layered transitions, and the expressive manipulation of the time base. Bizzocchi anticipates that Ben Youssef's slow-motion algorithm will be ready for live action footage when he begins his urban piece. He looks forward to testing the technical limits of post-production slow-motion in order to maximize the creative impact of temporal manipulation.

Bizzocchi's work in nature scenics is not finished, however. He will return to the mountains to make more films. The visual interest of the environment, and the scope for creative treatment of scale, light and composition is compelling. There is also a compelling technical/creative problem to be addressed: the current limitations in video's contrast and dynamic range. The combination of dark foliage and earth with bright skies, clouds, water and snow is very difficult for current systems to render. At the same time, they are a creative opportunity for expressive imagery - if the technical capacity can be improved. Hence, the development of more effective systems for increased-contrast high dynamic range (HDR) videography is another technical challenge of great interest to both Bizzocchi and Ben Youssef.

Next Steps in Humanities Scholarship

Bizzocchi has two main future directions in his humanities scholarship. The first is to continue to track the roots and the current manifestations of ambient video. There still is much to do in understanding the history of film and video artists who experimented with the creation of large-scale, slow-moving pieces. At the same time, there is a steady acceleration in the number of contemporary artists and video producers who are exploring the creative implications of ambient video (or as some call it, "video painting"[3]). A deeper understanding of the past and the current nature of creation in this area will inform Bizzocchi's humanities scholarship, enrich his own artistic practice, and provide Ben Youssef with additional interesting technical challenges to research and solve.

At the same time, Bizzocchi will look beyond ambient video, and carry on with his broader research agenda into the future of the moving image. He will continue to examine the history and current practice in film and video, and ask how emergent technological trends will affect the ongoing evolution of the art of the moving image.

COLLABORATION ACROSS DOMAINS

What have we Learned?

We have benefited from our history of collaboration. Between us we have built complementary research projects that explore significant issues within their respective domains, but at critical points have also benefited from our shared perspective. Bizzocchi's humanities scholarship informs his art, and the practice of his art further informs his humanities scholarship. Our joint review of technical tools for video creation was a project that could only be done well by an artist and an engineer working together. In the process we discovered that the solution of a particular technical problem was at the same time critical for creative practice, and an interesting scientific research challenge. We look forward to the final step in the scientific process around this challenge, which is the testing of Ben Youssef's slow-motion algorithm in the context of Bizzocchi's creative video art. We also anticipate that the exploration of increased contrast high-dynamic range video systems will provide further opportunities for our cross-disciplinary collaboration. We therefore believe our history instantiates one model for developing a synergy between humanities scholarship, artistic creation, and scientific/engineering research.

What are the Characteristics of this Collaborative Model?

First, we are fortunate that our personal styles and values support our work. We are both driven by an intellectual curiosity that is based in our own specialties, but also extends to other domains, including each other's research. We are personally compatible - for example we both enjoy humor as an aid to sustained effort. We are both committed to working hard, and concerned about reaching intermediate goals and final deliverables. Early on,

we recognized this in each other, and took mutual pleasure in our steady achievement of outcomes in terms of planning processes, funding applications, core research activities, and publications. As a result we developed a strong sense of mutual respect and trust. This is critically important in a collaboration where each partner is genuinely interested in the other's research, depends on its ultimate success, but is not able to completely evaluate the ongoing details of his colleague's research process.

Second, our research agendas are discipline-based. Our shared program includes research questions that are important within our respective specialties. This is critical to the success of our program - it is a shared agenda that is thoroughly grounded in our primary domains of enquiry. At the same time, the combination of perspectives uncovers and then illuminates important shared questions - such as the importance of slow-motion in video, and the steps needed to increase its effectiveness. As Ben Youssef says: "Today's engineer needs to be both disciplinary and interdisciplinary, to have the breadth to see important challenges, and the depth to solve them." The same is true for artists, and for humanities scholars. A successful cross-disciplinary program is an effective way to support research breadth.

Third, we are fortunate to work in an atmosphere conducive to research and supportive of cross-disciplinary research. Our research has been supported by national and university research funding. Some of the funding has come from joint applications, other funds have been obtained by us as individuals, but earmarked towards components of our joint program. At the same time, we have benefited from a local climate that understands and values cross-disciplinary research. We are based in academic units that combine art and technology.

Fourth, we have stressed communication at every phase of our project. This is both more difficult and more important when the collaboration crosses disciplines. In this context, communication is critical, and is worth examining from several perspectives. A strong enabling factor is that each of us values clarity and precision in language. This holds for both our use of discipline-specific terminology and for our general stance towards speaking and writing. This commitment to clarity and precision, amplified by the need to discuss across disciplines, means that communication takes time and effort. This effort can be difficult in the short run, but a necessity in the long run. Ben Youssef once spent thirty minutes patiently explaining to Bizzocchi why the latter's use of "sampling rate" was inconsistent with standard engineering terminology. This correction led to a deeper understanding by Bizzocchi of critical factors in Ben Youssef's research.

Finally, the nature of Bizzocchi's humanities research served as a central unifier for all three core activities: humanities scholarship, artistic creation, and scientific research. This was due to his fundamental concern with "poetics" as the foundation of his scholarship. Poetics are "the creative principles informing any literary, social or cultural construction, or the theoretical study of these; a theory of form" (Oxford English Dictionary, 2007). The phrase goes back to Aristotle and his deconstruction of the form - the design principles - of Greek tragedy. It is an extremely practical and grounded humanities discipline. Aristotle's method was to review in detail the works of the classic playwrights and "reverse-engineer" their design decisions. Bizzocchi's methodology is the same - "close reading" of exemplary works to determine the key parameters and characteristics of cinematic and video creative decision-making.

An orientation to poetics as a preferred form of scholarship ensures a tight relationship to the other two domains. First, poetics is intimately connected to the fundamentals of artistic practice. Works of art are examined to determine the nature of the form, and to uncover the details of how to work successfully within that form. Further, once scholarship has clearly explicated an art form's

poetics, these principles for construction can be applied to the creation of other works of art in the same genre.

Because an orientation towards poetics articulates the fundamentals of artistic practice, it also helps to develop effective insights on the utility of existing tools, and the need for new ones. The application of this type of humanities scholarship (close reading and poetics) in the analysis of a technologically-based art practice has the potential to reveal the critically significant challenges for digital tool-makers: scientists and engineers. The combination of this type of scholarship with a parallel program in artistic content creation makes this potential even stronger. This process is exactly what happened in our research program. Bizzocchi's scholarship identified the importance of slow-motion in emergent forms of video. His art verified the effectiveness of this insight, and also revealed the inadequacies of existing production tools. This chain supported the need for Ben Youssef's scientific research. Without an orientation towards poetics, a different type of humanities scholarship, such as cultural studies for example, would not have played the same role in conjoining artistic creation needs with a significant engineering challenge.

Broader Considerations

Clarity of language and an orientation towards practical manifestations has other consequences. MIT's "Media in Transition" project takes a strong position on the responsibility of the academic to engage in public discourse. This necessitates a language and approach that is accessible and "… suspicious of specialized terminologies, a forum for humanists and social scientists who wish to speak not only across academic disciplines but also to policymakers, to media and corporate practitioners, and to their fellow citizens" (Thorburn, Barrett and Jenkins, 2003).

We have attempted to live up to these standards for orienting our research results towards both the media creation communities and to the general public. We regularly seek discussions with leading-edge practitioners in the video sector. We see these exchanges as directly relevant to our research. At the same time we have made ourselves available for a series of interviews with both print and online media outlets. In all of these encounters, we try to share insights that we have, and at the same time we gain from our focused discussions with knowledgeable print and electronic journalists (including the local daily newspaper, a national video "trade" journal, and local, nationally syndicated, and internationally syndicated online news services). We also reach out to those directly engaged in the areas of our research. Bizzocchi has engaged in email discussions with many artists and producers interested in the practice of ambient video. He maintains his own websites, which aim to make his scholarly research available and understandable for artists, academics and the general public.[2] He has also contributed to other websites such as Flickr, where ambient video and related forms are discussed. Henry Jenkins maintains that media scholars can accelerate the growth of an emergent medium through active engagement with both public discourse and with the medium's creative community (Jenkins, 2000). In the language of this chapter, discourse with the creative community, the press, and the public extends our shared research into even broader contexts.

FUTURE TRENDS

What is the future of this and related medium forms, and how will that affect research programs similar to ours? We believe the answer is: "more of the same". Video technologies will continue to improve, driven by the twin engines of Moore's Law and the digital consumer marketplace. Technological possibilities will increase (Brooks, 2004), and manufacturers will implement these possibilities in an attempt to maintain and increase market share. Specific examples already on the immediate horizon include the introduction of newer and better high-definition standards, extended-contrast high-dynamic range (HDR) video systems, larger screens and new screen technologies. Artists will continue to take advantage of these improved production and playback tools in the evolution of their own art and of the medium itself. Academics will have responsibilities in this state of ongoing emergence. There will always be a need for scholars to track the resultant artistic advances and for engineers to improve tools and processes. Convergent technological capabilities will continue to conjoin divergent professional domains. Because of this ongoing effect, communication and collaboration within and across these various sectors will become even more important in the future.

CONCLUSION

We believe that it is possible to build a substantive collaboration that crosses divergent domains, and that such a collaboration can strengthen the outcomes in these separate domains. In our case, a cross-disciplinary approach has increased the quality of Bizzocchi's humanities scholarship and artistic creation, and Ben Youssef's scientific research. The key success factor for our collaboration has been the identification of research questions that are significant in our own individual specialties, but also complementary in the context of our shared work. Other success factors include hard work, a sense of humor, personal compatibility, the development of mutual respect, and a constant commitment to communication and mutual understanding. External success factors that have strengthened our joint research are access to funding for collaborative ventures, and institutional recognition of cross-disciplinary collaboration. Finally, we note the central role that the analysis of poetics, in conjunction with a close reading methodology, has played in our overall research agenda. The orientation towards poetics and the details of video artistic practice has provided an initial impetus and an ongoing intellectual cohesion for our various threads of art creation, humanities scholarship, and scientific research.

ACKNOWLEDGMENT

The authors would like to acknowledge the support for their research and creative work received from the Social Sciences and Humanities Research Council (SSHRC) of Canada, the Natural Science and Engineering Research Council (NSERC) of Canada, the Banff New Media Institute, and the School of Interactive Arts & Technology and the TechOne Program at Simon Fraser University. Bizzocchi's creative work grows out of another set of collaborations - with Director of Photography Glen Crawford, and post-production specialist Christopher Bizzocchi. The authors thank our research colleague Dr. John Bowes for his involvement and ongoing support of our research agenda.

REFERENCES

Al-Mualla, M. (2003). Motion field interpolation for frame rate conversion. *ISCAS, 2,* 652-655.

Andre, T., Cagnazzo, M., Antonini, M., & Barland, M. (2004). Motion-compensated lifting-based wavelet transform. *IEEE International Acoustics, Speech, and Signal, 3,* 121-123.

Ben Youssef, B., Bizzocchi, J., & Bowes, J. (2005). The future of video: User experience in a large-scale, high-definition display environment. *ACM SIGCHI International Conference on Advances in Computer Entertainment Technology,* Valencia, Spain. (pp. 204-208).

Bennis, W., & Biederman, P. W. (1997). *Organizing genius: The secrets of creative collaboration.* Cambridge, MA: Perseus Books.

Bizzocchi, J. (2008). The Aesthetics of the Ambient Video Experience. *Fibreculture Journal, Issue, 11,* <http://journal.fibreculture.org/issue11/issue11_bizzocchi.html> - viewed July 1, 2008

Brakhage, S. (1978). Metaphors in vision. In P. A. Sitney (Ed.), *Avant-garde film.* New York, NY: New York University Press.

Brooks, R. (2004, November 2004). The other exponentials: Moore's law isn't alone. many technologies now improve so quickly it boggles the mind. [Electronic version]. *MIT Review,* Retrieved July 30, 2007 from http://www.technologyreview.com/Infotech/13863/

Candy, L., & Edmonds, E. (Eds.). (2002). *Explorations in art and technology.* New York, NY: Springer.

Capellazzo, A. (2000). Making time: Considering time as a material in contemporary video and film. *Palm Beach Institute of Contemporary Art,* Lake Worth, FL.

Choi, B., Lee, S., & Ko, S. (2000). New frame rate up-conversion using bi-directional motion estimation. *IEEE Transactions on Consumer Electronics, 46*(3), 603-609.

Chupeau, B., & François, C. (2000). Region-based motion estimation for content-based video coding and indexing. *SPIE Visual Communications and Image Processing, 4067,* 884-893.

Chupeau, B., & Salmon, P. (1993). Motion compensating interpolation for improved slow motion. In E. Dubois, & L. Chiariglione (Eds.), *Signal processing of HDTV* (pp. 717-724). Elsevier Science Publishers B.V.

Conover, D., Czuchra, D., & Caloyanis, N. (2004-2008). *Sunrise earth* [Television series]. United States: Compass Light Productions.

Consumers winners in HD wars. (2008, Feb. 19, 2008). *Vancouver Sun.* (pp. D3).

Deren, M. (1978). Cinematography: The creative use of reality. In P. A. Sitney (Ed.), *Avant-garde film* (pp. 72-73). New York, NY: New York University Press.

Dulac, G. (1978). Visual and anti-visual films. In P. A. Sitney (Ed.), *Avant-garde film* (pp. 33-35). New York, NY: New York University Press.

Dusinberre, D. (1975). *Avant-garde british landscape films (introduction to programme notes).* London: Tate Gallery.

Eno, B. (1978). *Music for airports (album liner notes)* (PVC 7908 (AMB001) ed.)

Fowles, J. (1992). *Why viewers watch: A reappraisal of television's effects* (Revised ed.). Newbury Park, CA: Sage Publications.

Gaertner, M. (2000). In H. Friedel, S. Gaensheimer & U. Wilmes (Eds.), *Moments in time: On narration and slowness* (). Stuttgart: Cantz Editions.

Gorder, P. F. (2007). Multicore processors for science and engineering. *IEEE Computing in Science & Engineering, 9*(2), 3-7.

Held, G., & Marshall, T. R. (1991). *Data compression: Techniques and applications: Hardware and software considerations*John Wiley & Sons, Inc.

Jenkins, H. (2000). Keynote address. Paper presented at the *Computers and Videogames Come of Age,* Cambridge, MA. Retrieved February 27, 2008 from http://web.mit.edu/cms/games/opening.html

John-Steiner, V. (2000). *Creative collaboration.* New York, NY: Oxford University Press.

Kang, S. B., Uyttendaele, M., Winder, S., & Szeliski, R. (2003). High dynamic range video. *ACM Transactions on Graphics, 22*(3), 319-325.

Koch, S. (1978). Andy warhol's silence. In P. A. Sitney (Ed.), *Avant-garde film* (p. 165). New York, NY: New York University Press.

Kubey, R., & Csikszentmihalyi, M. (1990). *Television and the quality of life: How viewing shapes everyday experience.* Hillsdale, NJ: Lawrence Erlbaum Associates.

Lee, S., Yang, S., Jung, Y., & Park, R. (2002). Adaptive motion-compensated interpolation for frame rate up-conversion. *IEEE Transactions on Consumer Electronics, 48*(3), 444-450.

Morgan, J., & Muir, G. (2004). *Time zones: Recent film and video.* London, UK: Tate Publishing.

Moritz, W. (2007). Retrieved 1/2007, 2007, from http://www.iotacenter.org/visualmusic/articles/moritz

Oppenheimer, R. (2007). The conversation continues: When artists and engineers first collaborated. In P. Jennings (Ed.), *Speculative data and the creative imaginary: Shared visions between art and technology* (). Washington DC: National Academy of Sciences.

Oxford english dictionary. Retrieved 2/2007, 2007, from http://dictionary.oed.com.proxy.lib.sfu.ca/cgi/entry/50182439?>

Pearce, C., Diamond, S., & Beam, M. (2003). BRIDGES I: Interdisiplinary collaboration as practice. *Leonardo, 36*(2), 123-128.

Schrage, M. (1995). *No more teams! mastering the dynamics of creative collaboration.* New York, NY: Currency Doubleday.

Shi, Y. Q., & Sun, H. (2000). *Image and video compression for multimedia engineering: Fundamentals, algorithms, and standards*CRC Press.

Snow, C. P. (1959). The two cultures. *Rede Annual Lecture,* Senate House, Cambridge, UK.

Solari, S. (1997). *Digital video and audio compression.* New York: McGraw-Hill.

Tekalp, A. M. (1995). *Digital video processing.* Upper Saddle River, NJ: Prentice Hall PTR.

Thorburn, D., Barrett, E., & Jenkins, H. (2004). Series Editors Media In Transition Book series. In L. Gittelman, and G. B. Pingree (Eds.), *New media* (pp. 1740-1915*)*. Cambridge, MA: MIT Press.

Wang, Y., Ostermann, J., & Zhang, Y. (2002). *Video processing and communications.* New Jersey: Prentice Hall, Inc.

Wilson, S. (2002). *Information arts: Intersections of art, science, and technology.* Cambridge, MA: MIT Press.

Youngblood, G. (1970). *Expanded cinema.* Cambridge and London: E. P. Dutton.

Zheng, W., Kanatsugu, Y., Itoh, S., & Tanaka, Y. (2000). Analysis of space-dependent characteristics of motion-compensated frame differences. *International Conference on Image Processing, 3,* 158-161.

KEY TERMS

Ambient Video: Video piece intended to play in the background, but to also give visual pleasure whenever it is looked at. Like Brian Eno's ambient music, ambient video "must be as easy to ignore as it is to notice". Also called "video painting".

Close Reading: Critical, highly detailed deconstruction and analysis of a text or a work of art.

Frame: In both film and video, a single image is referred to as a frame.

Frame Rate: Is the rate or frequency at which an imaging device records or displays a sequence of frames. As a rate, it is expressed in frames per second (fps), while as a frequency in hertz (Hz).

Frame Interpolation: Is a method of constructing and inserting new video frames between existing frames within a given video sequence.

High Dynamic Range (HDR): Means that the range of luminance (brightness) values has been increased.

Motion Estimation: Is the process of determining motion vectors that describe the transformation from one frame to an adjacent one in a video sequence.

Poetics: Originally (from Aristotle) the study of the form and structure of classic Greek drama. The term has since broadened, and now refers to "the creative principles informing any literary, social or cultural construction, or the theoretical study of these; a theory of form." (quote from Oxford English Dictionary, 2007)

ENDNOTES

[1] <http://www.red.com/cameras> - viewed Feb. 25, 2008.

[2] <www.dadaprocessing.com> gives an overview of all of his research, while <www.ambientvideo.com> concentrates on his ambient video work.

[3] The Montreal video collective Nomig has been working in this area for several years. They prefer the name "Video Painting", as do some other artists and critics. However, the practices of "ambient video" and of "video painting" are not yet clearly differentiated, and this point the two terms can be treated as interchangeable..

Section II
Creativity Unleashed

Chapter V
Randomness, Chance, & Art

Ethan Ham
The City College of New York, USA

ABSTRACT

Randomness is a slippery term that conveys different meanings in different disciplines. In mathematics, an individual number is random when there is an equal chance for it to be any number from a set of possible values. In computer science the term becomes more relative and numbers have varying degrees of pseudo-randomness. Information theory equates randomness with unpredictability and, at odds with other definitions, concludes that a higher level of randomness indicates a greater concentration of information; a message's probable denseness of information is highest when the message is partially surprising and partially expected. There is no fixed definition for what randomness means in art, but analogies can be drawn to how the term is used in other fields. For example, information theory's definition might suggest that artworks have the greatest impact when using a mixture of pattern and unpredictability.

INTRODUCTION

Randomness, if it exists at all, is a fragile state. This fragility isn't intuitive to us; our day-to-day lives seem filled with disorder and unconnected events. The precarious nature of perfect order is more easily understood. We know that the nature of the universe is for things to break down, clutter, and fall apart. We have scientific laws (the law of entropy) and folk laws (Murphy's law) to explain why order cannot be maintained for long. With that in mind, perhaps it is more understandable that order's opposite—randomness—is similarly

rare. Just as it is the nature of the universe for things to fall apart, it is also the nature of the universe for a cause to exist for every effect and for that effect to be determinable (at non-quantum levels[1]). But a truly random event has no relation to its trigger; the effect should not be deducible from the cause.

Look again at the very first sentence of this introduction and note the caveat of *if it exists at all*. The existence of randomness and the ability of humans to observe it is an ongoing debate. Knuth (1981) said, "People who think about this topic almost invariably get into philosophical discussions about what the word 'random' means. In a sense, there is no such thing as a random number; for example, is 2 a random number?" (p. 2). The goal of the chapter is to give a deeper understanding of randomness, how it is generated in computer science, and how it can be used in art.

BACKGROUND

Random is often used colloquially to indicate arbitrariness or things unrelated: random acts of violence, random thoughts, random encounters. A number of fields such as computer science, statistics, and informational theory have more rigorous definitions of randomness. But each of these fields uses the term in a way that is slightly at odds with the others.

As a starting point, let's establish what randomness means to a mathematician and, using that, build a working definition for what randomness might mean to an artist. In mathematics, an individual number is random when there is an equal chance for it to be any number from a set of possible values. When describing a sequence of numbers as random, we mean each number is statistically independent of the others; that the numbers in the series have no effect or relation to the others (Haahr, 2008). A random number or sequence is characterized as containing no meaningful information; if a number conveys some data (such as the result of a formula, a person's phone number, or the number of times the letter 'q' appears in this chapter[2]), then it is not random.

This trait of non-significance can be borrowed and used as a key characteristic of randomness in art. If an element in an artwork contains some meaningful information about the world around us, then the element isn't truly random. Consider this recipe by Tristan Tzara (one of Dada's founders) for writing poetry:

To Make A Dadist Poem
Take a newspaper.
Take some scissors.
Choose from this paper an article the length you want to make your poem.
Cut out the article.
Next carefully cut out each of the words that make up this article and put them all in a bag.
Shake gently.
Next take out each cutting one after the other.
Copy conscientiously in the order in which they left the bag.
The poem will resemble you.
And there you are--an infinitely original author of charming sensibility, even though unappreciated by the vulgar herd. (Brotchie, 1991, p. 36)

Would the resulting poem be random? Several aspects of this poetry generation process do seem analogous to our description of a random numerical sequence. However, the poem's recipe (or *algorithm*) is not rigorously random by mathematical standards. To improve the randomness of the process, we'd first want to remove any duplicate words so that common words (such as "the") wouldn't have a greater frequency in the poem. Second, we'd want to make sure that the slips of paper have identical sizes (otherwise, the larger slips would tend to float to the top upon being shaken and would bias our results). Finally, we'd need to question our basic ability to sufficiently randomize the slips of paper by shaking a bag. Several early attempts to generate random numbers (for use in scientific simulations) used slips of paper in bowls and bags, but were not able to generate sufficient randomness (Hayes, 2001).[3]

It isn't necessarily important to resolve the aforementioned issues for a work of art. Statistically rigorous randomness may be crucial (though elusive) in computer science and mathematics, but it is usually more than is required for stochastic artworks. In fact, giving common words a greater probability may even be desired. Even if we did wish to adjust the poetry-generating algorithm so that each word had a precisely equal chance of being drawn, we still wouldn't have a truly random poem. The source material of the article would largely determine the poem's content—a sports article would have an unusually large number of sports related words, a computer science article would be filled with computer jargon, and so on. The resulting poem would be unpredictable, yet through its vocabulary it would convey information.

The conveyance of meaning is an important distinction for the art-focused terminology proposed in this chapter: *random* refers to an unpredictability that communicates no information, whereas *chance* implies a basis in the real world, unpredictable yet meaningful. *Stochastic* describes randomness and chance either collectively or non-specifically. A painting in which colors were selected by rolling dice would have a random palette. A painting in which colors were selected based on the color of passing cars would have a chance-based palette. Both palettes could be described as stochastic.

CHANCE VS. RANDOM

The distinction between chance and randomness in art is a convenient taxonomy, but not every stochastic work cleanly fits into one category or the other. A painting's color choices being based on the movement of a ball in a sports game might convey some sense of the game or may simply seem arbitrary and better described as random, despite being determined by real-world events.

Tim Hawkinson's *Emoter* (2002) uses light sensors on a television screen to drive the facial expressions on a motorized photograph of the artist's face. If we were to categorize *Emoter*, we might describe it as chance-based—the sculpture's movements are triggered by whatever happens to be on television at a given moment. However, we might consider a fundamental characteristic of chance-based artwork to be the viewer having

insight into the details of the cause and effect relationship. Viewers of *Emoter* are able to deduce that the television screen's image determines the photograph's movement, but an understanding of what specific television image traits result in which specific motion remains elusive. The impression is one of random motion, even though the movement is *deterministic*; presumably if the television were playing a video repeatedly, each replay would result in the same facial performance.

What may classify *Emoter* as a chance-based, rather than random, artwork is how crucial the relationship between the television screen and the facial expressions is to the work's concept. The connection suggests that emotional reactions to television shows are as artificial as the medium itself. The ever-changing, yet repetitive nature of the eerie face evokes the ever-changing, yet repetitive nature of television shows.

Sabrina Raaf's[4] *Translator II: Grower* (2004-2005) is another work that is activated by chance

Figure 1.

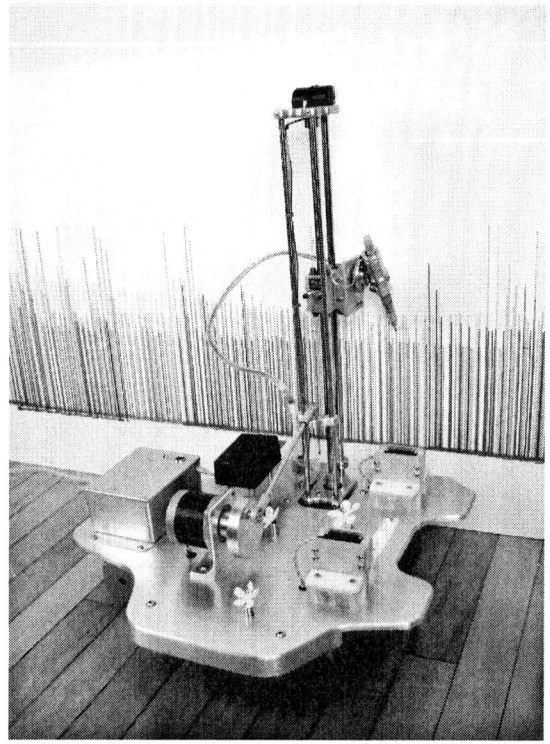

Image used with permission of Sabrina Raaf

factors. The artwork consists of a small robot that slowly works its way around a room, hugging the walls. A sensor near the ceiling detects the room's level of carbon dioxide and transmits the information to the robot. Every few seconds the robot draws a vertical green line on the wall—the higher the level of carbon dioxide, the taller the line. The lines become both a representation of grass and a bar graph tracking the carbon dioxide level (and consequently the presence of people) over time. The act of observing the artwork provides the chance stimulus that drives the art generation.

HOW RANDOM?

In the early and mid-1960s several researchers independently came to the conclusion that when looking at the randomness of a set of numbers, it makes little difference whether or not they were generated by a random process. What is more important than how the numbers were generated is how random the numbers appear to be (Chaitin, 1975). For example, using coin tosses to generate a series of ones and zeros (ones representing heads and zeros representing tails), the series 1010101010101010 is just as likely a result as the series 1011101001010111.[5] However, if we define "randomness" as the absence of a pattern, then the second set of numbers is random and the first set is not. Of course the first example's pattern is coincidental and one additional coin toss could have broken the pattern. But if we simply look at numbers in front of us and disregard our knowledge of how they were created, we would conclude that one is random and the other isn't.[6]

This data-centric approach views randomness in terms of complexity and lack of pattern. In 1965, while an undergraduate at The City College of New York, Gregory Chaitin proposed this definition and suggested that the randomness of a finite series of numbers could be measured based on the size of the smallest computer program that

can generate the series[7]. If we wanted to write a program to output 1010101010101010, the smallest set of instructions would be "print '10' eight times." For 1011101001010111, there is probably no algorithm shorter than "print 1011101001010111." Since the examples are relatively short series, there isn't a great difference in the size of the instructions needed to generate them. However, we could easily create a much larger series of numbers that follow the first example's pattern: "print '10' one thousand times." Now we have an instruction set that is 29 characters long that generates a number that is 2,000 characters long. We can establish that this series of 2,000 characters is not very random because the smallest possible program that can generate the series is so much smaller in size than the result. A very random number would require (as our 1011101001010111 example did) a program that is very close in size to the data it generates.[8] In essence, there are no shortcuts generating a random number because a random number has no patterns.[9]

Defining randomness in terms of the data's complexity was undoubtedly crucially influenced by computer science and the impossibility of having computers algorithmically generate true random numbers. Impossible because the very fact computers generate the "random" numbers using formulas means that the numbers have a pattern. Computer-generated "random" numbers are referred to as *pseudorandom* in acknowledgement of their algorithmic origin. Early in the history of computers it became apparent that not all *pseudorandom number generators* (PRNG) are of equal quality and, consequently, the numeric sequences they generate have varying degrees of randomness—some sequences are more random than others. This equating of "level of complexity" with "level of randomness" is in contrast with the idea that randomness is an absolute state; that something is either random or not.

The compressibility of data is affected by the data's level of randomness. Most electronic files (such as graphics, text, video, and sound) can be compressed; algorithmically processed into a file of smaller size. Generally speaking, the less varied the data, the greater the compression. To use an analogy, consider the text "See Jane run. Run Jane run. Run run run." This can be summed up as "Check out Jane running." That's a compression of 42.5%. We were able to do that by removing the repetitions and patterns. However, this is a "lossy" compression; the compressed text maintains the meaning of the original, but cannot be restored to the original verbatim. We could do a lossless compression of the data: "1=run:See Jane 1. 1 Jane 1. 1 1 1." Using this lossless compression we were able to reduce the size by 15% (lossless methods provide less compression than lossy methods).

A file of highly random data, however, cannot be compressed in a lossless manner—a compression algorithm cannot find any patterns to squeeze. This is not to say that programmers don't regularly claim to have created an algorithm that will compress random data—they do in a manner reminiscent of cranks claiming to have invented a perpetual motion machine. However, none of these claims prove to be valid.[10]

Assert(Random == Information)

In *Chaos Bound* (1990), Hayles explores a proposition[11] that a communication's level of randomness is an indicator of the amount of information it contains. Surprisingly, the greater a communication's randomness, the more information it is likely to contain. This is more understandable when we consider our data compression example from before. The "Dick and Jane" stories are highly repetitive and even a first grader might wish for something more complex and varied. Increasing the denseness of information results in a greater level of textual complexity, a lower compressibility, and (by Chaitin's definition) greater randomness.

We can increase the randomness/denseness of a communication (and consequently its components' informational probability) by compressing

it. An example of such compression is removing the vowels from text: "t wld gt vry ld vry qckly f ths ntr chptr hd n vwls." The compression results in an increase of informational content for each character in the message; the same amount of information is communicated in fewer characters. One cost of this denser information is an increase in decompression time; reading the information will take more time than reading an uncompressed version.

Does an increase of randomness always indicate a probable increase of information? If so, "xjblw9 fjmksdpgk kdo vnaie pxs fr" likely contains more information than "There is nothing like a dream to create the future.[12]" To avoid this kind of absurdity, the information theorists who developed the concept of information probability further surmised that every communication contains a mixture of information and noise (i.e., meaningless or garbled data). As a message's randomness/complexity increases, the probability that its components are informational increases. However, once a message reaches a halfway point in terms of randomness/complexity, the probability of information decreases and the probability of noise increases. So a message that contains very little complexity ("aaaaaaaaaaaaaaaaaaaaaaaaaaaa") and a message of extreme complexity ("WgAx;. UJ,B2Lf.WSI2;8FzRGeX") both have low probabilities of containing information.

These theories lead to the conclusion that a communication is likely to have the greatest concentration of information when the message is partially surprising and partially expected (Hayles, 1990). Likewise, artwork utilizing stochasticism is likely to have the greatest impact when the result is a mixture of pattern and unpredictability. That mixture is largely determined by how much control is maintained by the artist.

Control & Generative Art

Stochastic methods are often used as a way of relinquishing control. The Dadaists did so to emphasize the absurd. The Surrealists gave up conscious control as a way of tapping into the subconscious. Artists creating *generative art* give up control to stochastic processes to simulate the complexity of nature or the spark of creativity.

Generative art is art that was created according to an algorithm. Dadaist poems created using Tzara's directions (from earlier in this chapter) are generative artworks. Golan Levin (Zanni, 2004) argues that interactive and generative artworks are about "creating an *illusion of control*: the sense that the 'artist' has relinquished authorship to the user, or to some clever algorithm. In fact, this is a myth." In many cases Levin is correct—the truly defining characteristic of most generative artworks are the elements over which the artist maintains control.

It would seem that the percentage of decisions that are determined by stochastic data would directly correspond to the level of control abdicated by the artist, but this isn't the case. More significant than the number of random choices, is the breadth of variety that is manifested by those decisions. How the artist frames randomness can greatly throttle or expand an artwork's unpredictability.

For example, Jared Tarbell's *Node Garden* (2004)[13] uses a very large number of random choices every time it generates an image. However, those choices do not add up to works that look significantly different.

If the goal of generative art is the creation of a series of distinct artworks, then *Node Garden* would have to be considered ineffective. However, evaluating *Node Garden* using that criterion would be judging it against aims it did not have.

For generative artworks like *Node Garden,* randomness is a collaborator who does the grunt work—in *Node Garden*'s case, randomness takes care of the tedious work of placing all the picture's elements. As long as the artist's vision is maintained, the particulars of the execution do not matter too much. This is the approach Sol LeWitt took for his wall drawings. For those drawings, LeWitt limited his involvement to providing written directions for executing the drawings. The physical act of creating the artworks was relegated to teams of workers, whom LeWitt called draftsmen. In recognition of the collaborative aspect of the art, LeWitt always credited the draftsmen in the exhibition catalogs. One execution of LeWitt's drawing directions varies from another based upon the walls' dimensions, the workers' skill and care, and how freely the instructions[14] were interpreted.

Time-Based Art

When creating *Node Garden*, Tarbell's focus was developing an algorithm that gives a particular (and predictably) pleasing result, not an algo-

Figure 2.

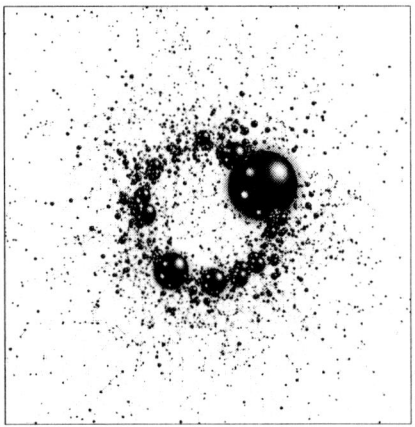 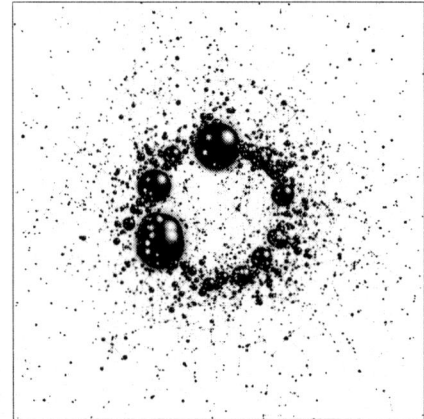

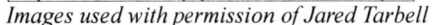

Images used with permission of Jared Tarbell

rithm whose output would continually surprise the viewer. Perhaps artworks like *Node Garden* are best thought of as a kind of performance. We wouldn't consider a particular performance of a play to be a unique work of art, even though it differs slightly from every other performance. Still, there is excitement and reward in seeing a live performance just as there can be real enjoyment in seeing *Node Garden* generate a new image in response to a computer mouse click.

A sense of time-based performance is present in many stochastic artworks, even those that are static and do not continually change. The heightened presence of time is due to its arrow being particularly straight in stochastic artworks. Part of relinquishing decisions to chance and randomness is that once the die is cast, the outcome is accepted without revision.

Erik Sommer[15] is a painter who uses mixtures of concrete and paint that peel off his canvases. Sommer regards the peeling as random and, unlike Jackson Pollock, does not "deny the accident" (Karmel, 1999, p. 22). Sommer does, however, rework the canvases until he is happy with the concrete's chance effect. A process that is random

Figure 3.

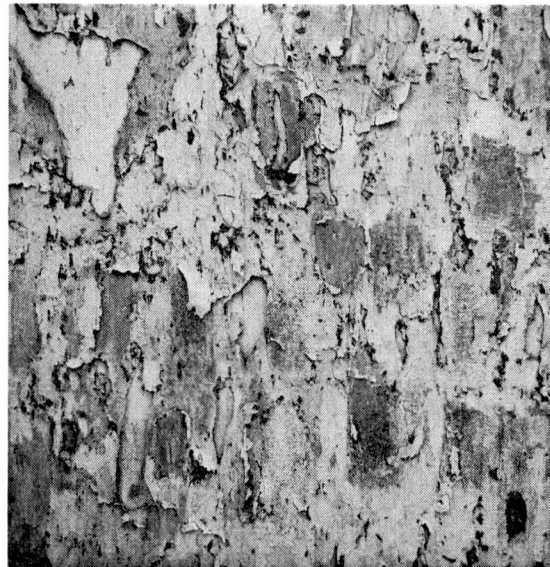

Erik J. Sommer, Taught, 2007, 48" x 48", mixed medium on canvas. Collection of the artist.

in the same way that repeatedly rolling a die until it comes up six is random.

By not fully committing to the chance outcome, Sommer's use of stochasticism is less constrained than Tarbell's. Sommer can comfortably allow a

Figure 4.

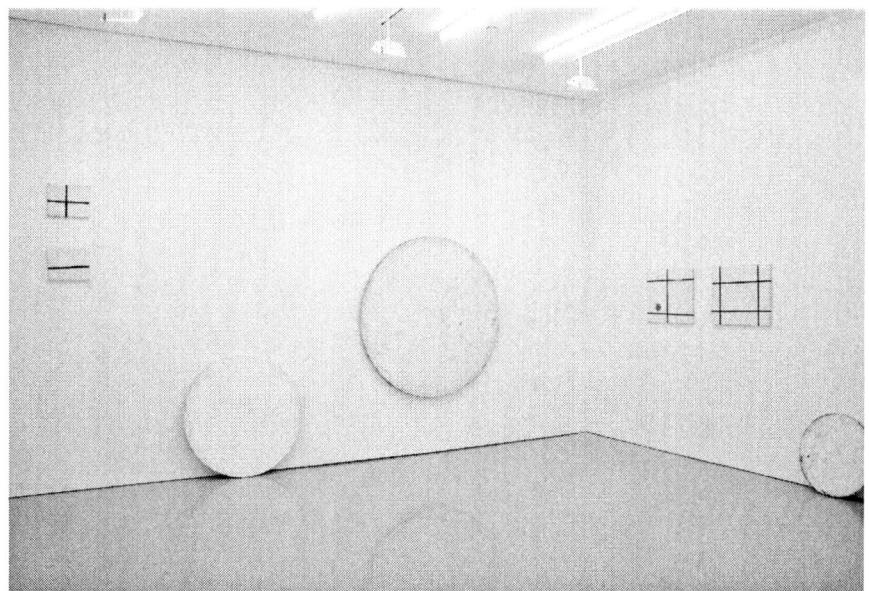

Mailed Paintings, Karin Sander. Photo © D'Amelio Terras Gallery, New York.

greater range of unpredictability because he can revise outcomes that go too far astray. In some ways chance is a more marginal element in Sommer's art—perhaps an unnamed assistant rather than a full-fledged collaborator. While Sommer's approach does harness chance's potential for unexpected outcomes, it also mitigates the excitement of a live, unedited performance.

Andy Goldsworthy's *Sheep Paintings* (1997-1998) achieve a greater sense of performance. The *Sheep Paintings* were created by laying out canvases in sheep pastures. Each canvas was "painted" by the marks of mud, feces, and urine surrounding a cleaner area that was protected by the placement of a salt lick. Goldsworthy (2007) says, "Whilst each painting is a result of chance, the choice of place, time, canvas size, food source and container radically affect its final appearance. The making of these decisions gives me the opportunity to work the canvases, albeit at a distance" (p. 153). Goldsworthy also notes that the work taught him "the importance of knowing when not to touch" (p. 153). Goldsworthy retains the artistic decision of when to remove a canvas from the field. If he chose to, he could also make an editorial selection of which canvases to exhibit without compromising the project's concept.

In contrast, the absence of the artist's hand is central to Karin Sander's[16] *Mailed Paintings* (2007). The paintings are pre-stretched, store-bought canvases that Sander mails unwrapped from various international locations to the art gallery. As the canvases are scuffed, stickered, and banded, they become a diary of their journeys. The viewer is very aware of the works' passage in time and place. If Sander were to intervene and adjust the aesthetic of any of these works, they would be wholly compromised—it would be as if a singer's live performance was lip-synced to another vocalist's voice.

Sascha Pohflepp's[17] *Buttons* (2006) is another artwork with a strong connection to time. *Buttons* is a camera without optical parts. When the camera's button is pressed, the camera does not record an image, instead it records the time. It then wirelessly searches the Internet for photographs that were taken by someone else at the very moment of the button press. Pohflepp (2006) explains, "After a few minutes or hours, depending on how soon someone else shares their photo on the web, an image will appear on the [camera's] screen." The photos are selected using a chance connection—two people happening to press a camera button at the same moment. Regarding the selected photograph, Pohflepp says, "In a way, it belongs half to the person who had pressed the button and still remembers that moment. Because of that connection, the photos are never dismissed as random, no matter how enigmatic they may be."

DSCN slide show (2002) by Philippe Blanc also makes connections based on photograph uploads. Each time a photo is taken using a digital camera, the image is given a default name. For

Figure 5.

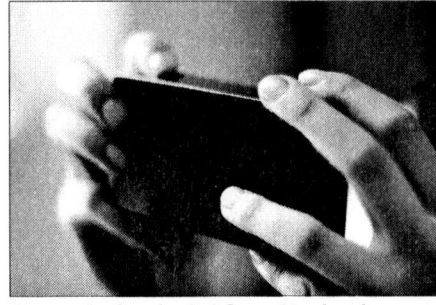 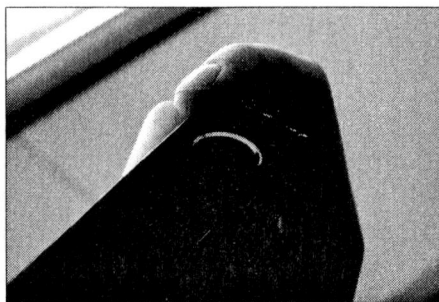

Images by Sascha Pohfleep. Used with permission.

example, the very first photo taken with a new Nikon™ camera is given the name DSCN001. Blanc created a program to search the Internet for photographs with the DSCN001 name. The result is a slide show of photographs connected by the shared experience of using a new camera for the first[18] time. Neither *Buttons* nor *DSCN slide show* generates new artwork. Instead the projects act as automated curators and select work based on a chance association.

Sometimes a work of art is completely deterministic yet is experienced by the viewer as being stochastic. This is the case for an on-going performance of John Cage's *Organ²/ASLSP* (1987), a musical composition in eight parts. *Organ²/ASLSP*'s sheet music comes with instructions that the performer should omit one of the parts, repeat one of the parts, and play the composition as slowly as possible[19]. The work's premiere lasted twenty-nine minutes. A performance that began on September 5, 2001 is intended to stretch out the music for 639 years (Wakin, 2006). The extended performance is located in a disused church in Halberstadt, Germany, the town where an organ with the first chromatic keyboard layout was built in 1361 (639 years prior to the originally planned start of the 639-year performance). The first twenty months of the performance were silent due to *Organ²/ASLSP* beginning with a rest. The first chord (two G sharps and a B in between) was struck on February 5, 2003 and lasted seventeen months. Small weights hold down the keys for the notes that are being played and the organ's pipes are changed to correspond.

Every note is predetermined, yet the audience has arbitrary musical experience. Whichever note happens to be playing is what the listener hears.

Control & Generative Music

Musikalisches Würfelspielen (musical dice games) were popular in Europe during the late 18th and early 19th centuries. Using published rules and a randomizer (such as dice or tops) players selected pre-composed musical phrases to create random musical compositions. Several such games were spuriously published under Mozart's name[20]. One such fraud, dating from 1787, used two six-sided dice to determine sixteen minuet measures and one six-sided die to determine six trio measures. This calculates out to $11^{16} * 6^{16}$ (or 1.23^{29}) possible compositions, though to the listener many of these variations would sound excruciatingly similar.

While the dice game has many possible variations, the musical phrases provided by the composer/game-designer maintain a high degree of control over the listener/player experience. In contrast John Cage typically relinquished much more control in his stochastic compositions. He viewed this as releasing himself "from what I had thought to be freedom, and which actually was only the accretion of habits and tastes" (Pritchett, 1988).

Beginning in the 1950s, Cage generated random music by tossing coins and using the *I Ching* fortune-telling process. Cage would start by randomly determining the broadest aspects of a composition (e.g., the key or time signature) and then proceed to the individual notes.[21] While Western music conventions may impose a general structure on these compositions, they are much more freeform than the Musikalisches Würfelspielen compositions. Cage's *Music of Changes* (1951), which was composed in this manner, does occasionally have the brief musical phrase that sounds conventional, but taken as a whole it is reminiscent of a cat walking on a keyboard. Comparing Musikalisches Würfelspielen and *Music of Changes* shows that the randomness of the input (which is the same for both) is only incidental to the resulting unpredictability. The greater factor is the process established by the artist.

Another Cage composition, *Imaginary Landscape No. 4* (1951) was composed in much the same way as *Music of Changes*. However, it introduced the added stochastic element of being written for twelve radio receivers instead of a piano. Each receiver has one performer to adjust volume and

Figure 6.

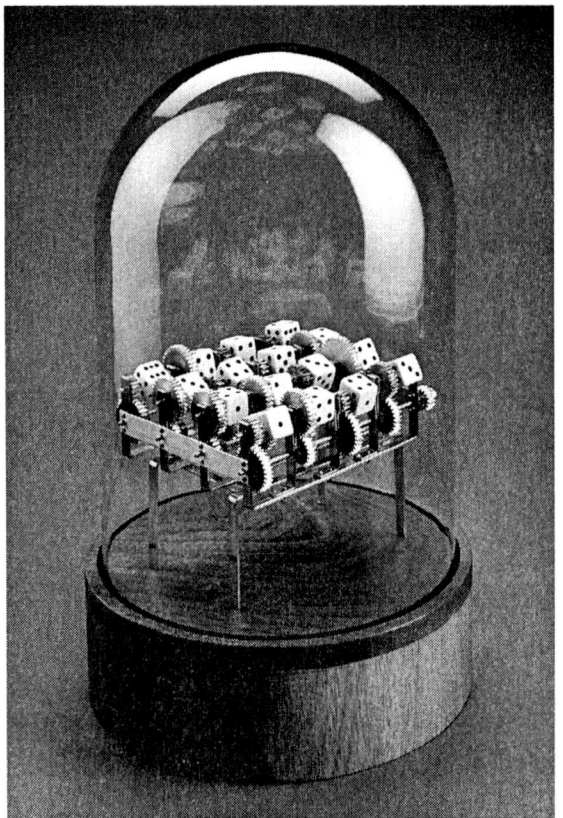

© *Marc Berghaus. Used with Permission. Photo by Doug Koch.*

another performer to adjust the frequency. *Music of Changes* incorporated a random composition technique, but does not change with each performance.[22] In contrast, *Imaginary Landscape No. 4* is tied to the time and locale of the concert so that each performance is entirely uniue.

Seeds & Pseudorandom Number Generators

Mandala #2 (2000), a sculpture by Marc Berghaus[23], consists of a grid of sixteen dice under a bell jar. Gears driven by a hidden motor turn the dice at different speeds. Berghaus (2007) explains, "Due to my use of unusual gear ratios (say, 1:1.7, rather than 1:2) in the gears that connect the drive shafts to the dice's axles, very few of the cycles

line up again at once, and it becomes impossible to predict the patterns of all the tumbling dice, despite the fact that all actual randomness has been stripped from them" (p. 49).

The rotation of *Mandala #2*'s dice is completely deterministic, so it may seem a poor simulation of randomness. And perhaps it is. But it is a very good, real world representation of how computers generate random numbers.

As mentioned earlier, computers cannot create true random numbers and instead they use formulas to generate pseudorandom numbers. Pseudorandom numbers are important for everything from shuffling the deck of a solitaire game program to negotiating communication on a computer network[24].

When preparing to use a pseudorandom number generator, programmers start by giving it a "seed" value. This seed is an arbitrary number used by the PRNG's algorithm to generate the series of numbers[25]. Without this seed number, a PRNG would generate the same series of numbers each time it was restarted. In a way, "the pseudorandom generator does not actually generate any randomness; it stretches or dilutes whatever randomness is in the seed, spreading it out over a longer series of numbers like a drop of pigment mixed into a gallon of paint[26]" (Hayes, 2001, p. 302).

Given this characterization of the seed as the source of the PRNG's randomness, one may wonder how a seed's value is determined since computers innate lack of randomness was the issue in the first place. One approach is to use input from outside the computer (such as user input) to establish the seed. But by far, the most common technique for creating a seed value is to use the current time (which on UNIX computers is expressed in the number of seconds that have elapsed since January 1, 1970).

Neglecting to seed a PRNG results in it giving the same default series of numbers each time it is restarted[27]. Likewise, a particular seed value will always return the same series of numbers. Years ago the author of this chapter programmed

Sanctum, a two-person online game. Every time a new game was started, the network would pass the same number to both players' computers to use as a seed value. That way, every time a random number was used in the game (for example, to determine whether an arrow hit a monster), both players' computers would have identical outcomes. Very occasionally, an error in the program would cause one player's computer to use a random number where the other player's didn't, with the result that one computer would be a step ahead in the series of random numbers. Once that occurred, every following use of a random number would cause the game state on the two computers to diverge further apart. Eventually the two players would be seeing completely different game states (an occurrence reminiscent of the science fiction genre of "alternative histories" where a slight change in history—in this universe or a parallel one—results in a very changed present).

Hardware Random Number Generator

Given the limited randomness provided by PRNGs, one might think that real world randomizers such as dice would provide better results. In 1965 statistician Frederick Mosteller had a unique opportunity to test this when Willard H. Longcor walked into his office and offered to record the results of a few million die tosses (Peterson, 1998). Mosteller compared the results to what would be predicted by distribution theory and found that Longcor's throws matched very closely (and the few places where it diverged pointed to errors in the theory). Coin flipping has also been extensively tested. During World War II an English mathematician spent his time in a prisoner of war camp tossing a coin—he came up with 5,067 heads in ten thousand tosses (Peterson, 1998).

But these accurate results seem to be the exception. The best pseudorandom number generators can outperform (in terms of tests of randomness) some physical number generators (Hayes, 2001).

Physical number generators' faults come from the same source as their virtue; the messiness of real world. That messiness can provide back doors for patterns to sneak in.

For example, British biometrician W. F. R. Weldon and his wife Florence spent a good deal of time rolling dice and recording the results to demonstrate the laws of probability. However, in 1900 the English mathematician Karl Pearson[28] analyzed the results of 26,306 of the Weldons' rolls and found that they broke the rules of probability—there were too many fives and sixes.

In 1977, Doyne Farmer fell in with a group of graduate students who were pioneering the field of chaos[29] at the University of Santa Cruz. Farmer was obsessed with using chaos theory to beat roulette[30] and spearheaded group forays into casinos with a computer hidden in a pair of shoes. While at a roulette table, information about the roulette tables spin, release of the ball, and so forth would be entered into the computer using toes. The computer would predict in which eighth of the wheel the ball was liable to stop and would transmit the results to a third shoe (worn by another member of the group) who would then place a bet. Apparently the system worked well enough for the group to make money (Kelly, 1995).

In 1955 the Rand Corporation published *A Million Random Digits with 100,000 Normal Deviates*, a 600-page book[31]. Rand used an electronic "roulette wheel" to generate the numbers. However, the machine proved to not be statistically random, despite repeated tinkering and modifications. Eventually Rand had to succumb to mixing and mathematically manipulating the numbers to have them pass statistical testing (Peterson, 1998 and Hayes, 2001). Rand's roulette table was in reality an electronic machine that generated a stream of bits (1's & 0's). In a twenty-five page introduction to the book, the method of number generation is outlined in order to assure the reader of the data's randomness. This introduction attributes the source of the bits to a "random frequency pulse source." Rand provided no further details, but the *Computer Handbook* (Huskey & Korn, eds., 1960) surmises that it was probably a Geiger counter and a low-grade radioactive material.

The same approach is taken by quite a few hardware random number generators—random number generators that don't use (or at least don't exclusively use) algorithmic methods for generating a number. John Walker's *HotBits* website[32] provides random numbers that are generated by measuring the nuclear decay of Cæsium-137. His method is to first detect the length of time between two electrons being given off by the radioactive material. A second set of electrons is timed and the two durations are compared. If there first set of decays has the shorter time interval a zero is generated, if it is longer then a one is generated. The resulting binary data is then converted to decimal numbers (e.g., 1001 would become a 9). This technique generates about 200 digits per second.

A less radioactive system is used at Random. org. The website provides random numbers generated using atmospheric noise (e.g., thunder). Yet another web-based random number provider is LavaRnd[33]. LavaRnd uses a web-cam with the lens covered. In the absence of light, the CCD chip on the video camera creates chaotic thermal "noise" which is put through a hash algorithm to remove unwanted, predictable patterns and is converted into random numbers. Atmospheric noise, video camera noise, and the fluid dynamics of LAVA LITE® lamps are all chaotic sources. A chaotic process is one in which minor variations in a process's initial conditions result in wildly different effects (i.e., the apocryphal "butterfly effect" where the flap of a butterfly's wing in a Brazil leads to a tornado in Texas). Chaotic systems often appear to be random even though their behaviors are entirely deterministic and based upon the initial conditions. So a random purist might argue that hardware random number generators using chaotic events are not truly random. This would leave just quantum event based (i.e., radioactive decay) random number generators as being truly random, however even that is in question. There is an ongoing debate as to whether even quantum events are truly random or simply chaotic. This is the kind of conundrum that led Knuth (1981)

Figure 7.

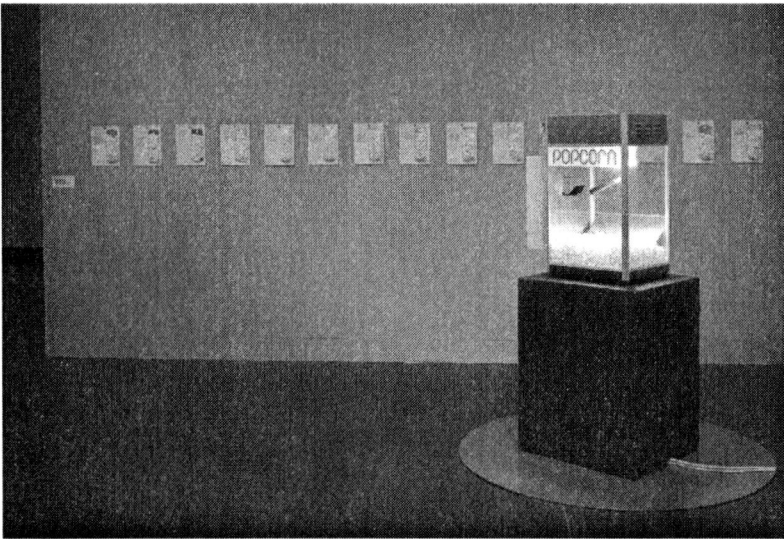

to pragmatically accept that "being 'apparently random' is perhaps all that can be said about any random sequence anyway" (p. 3). Even if one accepts quantum events as random, the numbers the generate may not be due to patterns sneaking from biases or flaws in the detection tools.

Nina Katchadourian's[34] *Talking Popcorn* (2001) translates the sounds of popcorn popping into a glossolalic babble. The sculpture consists of a microphone housed in a movie house popcorn machine. A hidden computer interprets the popping as Morse code and provides simultaneous spoken translation through a computer-generated voice. *Talking Popcorn* determines the Morse code by measuring the silences between popcorn pops in very much the same manner that radioactive-based hardware random number generators compare the durations between Geiger counter clicks. *Talking Popcorn* equates the longer silences as Morse code dashes and shorter silences as dots. These silences are measured relative to the running average speed of the popping so that as it speeds up, the pops don't become interpreted as an indistinguishable series of Morse code dots. The adjustment of popping speed smoothes out the particular, bell-curved popping cadence of a batch of popcorn and normalizes it into raw randomness. *Talking Popcorn* removes the real-world characteristic of popcorn building to a crescendo and then dropping off to the last few reluctant pops. Where some artworks, such as Hawkinson's *Emoter*, might be editorially classified as random (because the cause and effects are opaque to the viewer) even though it is actually based on chance events, *Talking Popcorn* is truly random through and through. There is no carryover of meaningful information because the triggering data's patterns have been smoothed out in much the same manner as LavaRnd does with its thermal noise events and the Rand Corporation did with its "frequency pulse" data.

Talking Popcorn's generation of information is reminiscent of a story that Hayles describes in her introduction to *Chaos Bound*. The story[35], which

Figure 8.

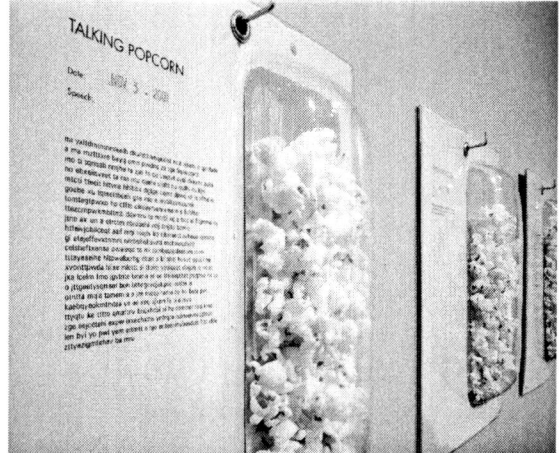

©2008 Nina Katchadourian. Used with permission. Courtesy of the artist and Sarah Meltzer Gallery, New York

comes from Stanislaw Lem's *The Cyberiad*, can be seen as a parable illustrating the relationship between chaos and information. In the tale, two constructors[36] are captured by a space pirate who pillages and hoards information. To gain their freedom, the constructors build a "Demon of the Second Kind"[37] for the pirate. The demon is designed to interpret the movement of stale air molecules as information. Whenever the motion of the molecules adds up to something intelligible, the Demon transcribes it onto paper tape using a tiny diamond-tipped pen. The pirate underestimates the amount of information contained within the chaotic motion, and he is soon buried in a mountain of paper filled with useless information: all the words that rhyme with spinach, why fan-tailed fleas won't eat moss, the sizes of bedroom slippers available on the continent of Cob, how Kipling would have written the beginning of *The Jungle Book 2*, et cetera.

The idea that we can be paralyzed by an overabundance of information seems even more relevant today (with the constant influx of information from the Internet, text messaging, emails, cel phones, and MP3 players) than when Lem wrote the story in 1967 or when Hayles discussed it in

1990 (a few years before the arrival of Mosaic, the first graphical web browser).

Unlike Lem's Demon of the Second Kind, Katchadourian's *Talking Popcorn* does not filter out the babble. In this regard, it is more like Borges's "The Library of Babel" which describes a universe composed of hexagonal, book-lined rooms. The narrator of the story posits that each book is unique and that every possible combination of text exists.[38] Since every possible book exists, the Library must contain the ultimate truth. There would also be many slight variations on the truth[39], and even more books filled with lies, and even more variations of those lies. But overwhelmingly the Library contains books of gibberish.

FUTURE TRENDS

The Internet provides a huge reservoir of data; a channel for receiving interactive stimulus; and a cheap and convenient platform for publishing art. These characteristics are attractive to artists exploring stochastic art, so the Internet is likely to continue to develop as an environment for random and chance-based art.

Physical computing—the use of microcontrollers and electromechanical devices—often uses chance and randomness (as seen in Raaf's *Translator II: Grower*). Recent innovations such as the Processing programming language and the Arduino microcontroller boards have made the electronics and programming required for physical computing much more approachable and popular. This trend is likely to continue. In the same way that creating a webpage has gone from a task for programmers to something pre-adolescents can do on social networking sites, we may see electronics development and the creation of electronics-based stochastic art become within reach of a more general public.

CONCLUSION

The motivations for using stochastic elements in art can range from a desire to free the creative process from conscious concerns to wanting to mirror the frenetic pace of our data-soaked lives. Both random and chance occurrences can be effectively used in art, but using chance may lead to richer, more interesting artworks because it brings an element of the world and a greater potential for resonance than the sterile isolation of true randomness.

REFERENCES

Berghaus, M. (2007). Simulated Chance and Staggered Gear Ratios. *Leonardo Music Journal, 17*, 1-90.

Borges, J. L. (1998). *Collected Fictions.* New York, NY: Penguin Putnam, Inc.

Brotchie, A., & Gooding, M. (Ed.). (1991). *A Book of Surrealist Games.* Boston, MA: Shambhala Publications, Inc.

Bryan-Wilson, J. (2003). Sol LeWitt. In Molesworth, H. (Ed.), *Work Ethic* (pp. 158-159). University Park, PA: The Pennsylvania State University Press.

Chaitin, G. J. (1975). Randomness and mathematical proof. *Scientific American 232*(5), 47-52.

Goldsworthy, A. (2007). *Enclosure.* New York, NY: Abrams.

Haahr, M. (1998-2008). *Introduction to Randomness and Random Numbers.* Retrieved January 15, 2008, from http://random.org/randomness/

Hayes, B. (2001). Randomness as a Resource. *American Scientist, 89*(4), 300-304. Research Triangle Park, NC: Sigma Xi.

Hayles, N.K. (1990). *Chaos Bound: Orderly Disorder in Contemporary Literature and Science*. Ithaca, NY: Cornell University Press.

Hayles, N.K. (1994). Chance Operations: Cagean Paradox and Contemporary Science. In M. Perloff, & C. Junkerman, (Eds.), *John Cage: Composed in America* (pp. 226-241). Chicago, IL: University of Chicago Press.

Hermida, A. (2005). *Sony shows off new PlayStation 3*. Last updated May 17, 2005 on BBC News website, http://news.bbc.co.uk/2/hi/technology/4554025.stm

Huskey, P., & Korn, G. (Eds.). (1962). *Computer Handbook*. New York, NY: McGraw-Hill Book Co.

Karmel, P. (Ed.). (1999). *Jackson Pollock: Interviews, Articles, and Reviews*. New York, NY: The Museum of Modern Art.

Kelly, K. (1995). *Out of Control*. New York, NY: Perseus Books.

Kelvin, L. (William Thomson). (1901). Nineteenth century clouds over the dynamical theory of heat and light. *The London, Edinburgh and Dublin Philosophical Magazine and Journal of Science, Series 6, 2, 1–40*.

Ketner, J., Herbert, L., & Volk, G. (2002). *Roxy Paine: Second Nature*. Houston, TX: Contemporary Arts Museum, & Waltham, MA: The Rose Art Museum, Brandeis University.

Knuth, D. (1981). *The Art of Computer Programming, Volume 2: Seminumerical Algorithms (2ⁿᵈ Edition)*. Reading, MA: Addison-Wesley Publishing Company.

Lem, S. (1974). *The Cyberiad: Fables for the Cybernetic Age* (M. Kandel, Trans). New York, NY: Harcourt Brace Jovanovich.

Noguchi, H. (1990). Mozart - Musical Game in C K. 516f . *Mitteilungen der International Stiftung Mozarteum, 38*(1-4), pp. 89-101.

Peterson, I. (1998). *The Jungles of Randomness: A Mathematical Safari*. New York, NY: John Wiley & Sons.

Pincus, S., & Singer, B.H. (1996). Randomness and Degrees of Irregularity. In *Proceedings of the National Academy of Sciences of the United States of America, 93*(5), 2083-2088. Washington, DC: National Academy of Sciences.

Pohflepp, S. (2006). Buttons. *Between Blinks & Buttons*. Retrieved February 22, 2008, from http://www.blinksandbuttons.net/buttons_en.html

Poundstone, W. (2005). *Fortune's Formula*. New York, NY: Hill and Wang.

Pritchett, J. (1988). *The use of chance techniques in the music of John Cage, 1950-1956*. Unpublished doctoral dissertation, New York University, New York.

Tippett, L.H.C. (1927). Random sampling numbers. *Tracts for Computers, 15*. London: Cambridge University Press.

Wakin, D. (2006, May 5). *An Organ Recital for the Very, Very Patient* [Electronic version]. *The New York Times*.

Zanni, C. (2004). Interview with Golan Levin. *CIAC Magazine*, 16 June 2004.

KEY TERMS

Algorithm: A set of well-defined instructions for completing a task.

Chance: In this chapter "chance" refers to unpredictable, but deterministic, events.

Chaotic: behaviors where minor changes in initial conditions can result in widely divergent results. Chaotic systems often appear random even though they are completely deterministic.

Deterministic: A situation where events are completely predictable based upon cause and effect.

Generative Art: Art that is created according to an algorithm. Generative art is typically intended to give the appearance of machine creativity.

Hardware Random Number Generator: A method for generating random numbers using a physical process, such as the nuclear decay of radioactive material. The generated numbers are often referred to as "true random" numbers in contrast with pseudorandom numbers generated by a pseudorandom number generator.

Pseudorandom Random Number: A number that was generated using an algorithmic process called a pseudorandom number generator (PRNG). Because the numbers are created deterministically they have the appearance of randomness, but are not truly random.

Quantum: Used in this chapter to refer to subatomic processes.

Random: used in this chapter to specifically refer to unpredictable events that are completely self-contained and communicate no information (in contrast to "chance").

Stochastic: having unpredictable characteristics. Used in this chapter to refer to both random and chance events.

ENDNOTES

[1] Later in this chapter we'll discuss random number generators that uses quantum events (i.e., the nuclear decay of radioactive materials).

[2] 52 times.

[3] Lord Kelvin (1901) tried generating random numbers by drawing cards and reported,

"The best mixing we could make in the bowl seemed to be quite insufficient to secure equal chances for all the billets [cards]." L. H. C. Tippet (1927) had a similar problem when he tried drawing a thousand cards from a bag: "It was concluded that the mixing between each draw had not been sufficient, and there was a tendency for neighbouring draws to be alike." The 1969 U.S. military draft lottery was flawed due to a strong reverse-correlation between the order in which the slips were put into the mixing bin (by calendar date) and the order in which they were drawn (Wetzel, 1998). In plainer English, potential draftees who were born later in the calendar year were placed into the mixing bin last and had a significantly higher chance of being selected for the draft.

[4] www.raaf.org

[5] Both series have a 1 in 65,536 (i.e., 2^{16}) chance of occurring.

[6] The idea that randomness comes from a number's method of generation rather than an inherent characteristic is what Knuth was driving at in his previously quoted statement: "In a sense, there is no such thing as a random number; for example, is 2 a random number?" (Knuth, 1981)

[7] A. N. Kolmogorov of the Academy of Science of the U.S.S.R. independently proposed a similar idea at about the same time as Chaitin. Unbeknownst to Chatin and Kolmogorov, Ray J. Solomonoff of the Zator Company made a similar proposal in 1960 as a method for measuring the simplicity of scientific theories. (Chaitin, 1975)

[8] The program we're describing is not a general-purpose number generator, but rather a program that is capable of generating a specific series of numbers. This is not the same as saying that the file size of a *pseudorandom number generator* (PRNG) is an indicator of its quality. It's also important

to realize that never generating a number that contains a pattern (e.g., 101010) is not a virtue in a PRNG. Coincidental patterns are commonplace and a PRNG that filtered out such patterns would be weaker than one that allows them.

[9] Chaitin's theory is related to earlier research on ciphers that was done by Claude Shannon, a pioneering information theorist who we'll reference several times in this chapter. Ciphers are algorithms that use an arbitrary piece of information (called a key) to encrypt data. Shannon determined that it is impossible to decrypt a cipher without the key if the key is truly random and is the same length (or longer) as the data that was encrypted (Hayes, 2001).

[10] Mark Nelson and Mark Goldman have issued separate challenges (with $100 and $5,000 prize monies) on the comp.compression newsgroup to anyone who achieves such compression. To date, no one has been able to legitimately claim the money, though programmer Patrick Craig attempted to take advantage of a loophole in Goldman's challenge.

[11] The idea originated in an article by Claude Shannon and was interpreted and expanded upon in a commentary by Warren Weaver. See *The Mathematic Theory of Communication* by Shannon and Weaver (1949).

[12] Victor Hugo, *Les Miserables*, 1862

[13] www.complexification.net

[14] Bryan-Wilson (2003) cites LeWitt giving directions such as "The lines should be made a close together as possible. They do not have to be regular but would differ with each person who does them." (p. 158)

[15] www.erikjsommer.com

[16] www.karinsander.de

[17] www.pohflepp.com

[18] Or 1,000[th] time since the default naming repeats after 999 photos.

[19] Hence the *ASLSP* in the title: "As SLow aS Possible."

[20] Mozart does appear to have created an unpublished dice game that used the letters of a friend's name to generate the composition (Noguchi, 1990).

[21] Once such a system of rules is established, it doesn't matter who executes them, human artist or machine. Cage eventually made extensive use of generative software, including a coin-tossing program, written by his assistant Andrew Culver. A list of programs used by Cage can be seen at www.anarchic-harmony.org/People/Culver/CagePrograms.html.

[22] Cage might argue that every musical performance is a unique work. This is the concept of his *4'33"*, a work in which the musicians play no notes, so the aural experience consists of the ambient noises of the audience coughing, et cetera.

[23] www.marcberghaus.com

[24] When two nodes on a network are attempting to send data simultaneously we don't want them to act like two cars trying to enter an intersection at the same time only to stop simultaneously, then try to enter the intersection at the same time again. To avoid such a scenario, the Ethernet protocol has both nodes pick a random number to determine how long to wait before trying to send data again. (Hayes, 2001)

[25] A series of random numbers created by a pseudorandom number generator will begin repeating within 2^n-1 results where n is the bit size of the seed number). A Sony PS3 console runs at 218 gigaFLOPS—i.e, 218 billion floating-point operations per second (Hermida, 2005). Assuming a 32-bit seed value and 100 floating-point operations for each generated pseudorandom number, a PS3 could run through the entire series of random numbers associated with a particular seed in 1.97 seconds. A 32-bit number has

4,294,967,295 possible seeds values (again, 2^n-1 results where n is the bit size), so it would take 268 years for a single PS3 to go through every possible series of random numbers.

[26] This idea of the PRNG's randomness stemming from the seed was expressed in the 1980s by Manuel Blum.

[27] A casino player once noticed that this was the case for a keno game. Each time the keno machine was powered off and back on, the number draw sequence repeated. The player won $600,000 as a result (Peterson, 1998).

[28] At about the same time he analyzed the Weldons' data, Pearson did twenty-four thousand coin tosses himself and came up 12,012 heads (Peterson, 1998)

[29] Chaos theory is more formally known as nonlinear dynamics. Later in this chapter we'll compare chaos and randomness.

[30] Claude Shannon (whose information theories are discussed earlier in the chapter), along with fellow Bell Labs researcher John L. Kelly, jr. and M.I.T. mathematician Ed Thorp, made a fortune in the early 1960s by successfully applying game theory to roulette and blackjack in Las Vegas. (Poundstone, 2005)

[31] The book had recently become available as a reprint. It can also be downloaded for free at www.rand.org/pubs/monograph_reports/ MR1418/index.html.

[32] www.fourmilab.ch/hotbits

[33] www.lavarnd.org. Landon Curt Noll, one of the minds behind LavaRnd, was also part of the team that created the lavarand (LavaRnd and lavarand are not the same, despite their confusingly similar names). Lavarand was a Silicon Graphics project in the 1990s. It involved using captured images of LAVA LITE® lamps to generate random, 140-byte seed values for feeding PRNGs.

[34] www.ninakatchadourian.com

[35] The short story has the burdensome title of "The Sixth Sally, or How Trurl and Klapaucius Created a Demon of the Second Kind to Defeat the Pirate Pugg."

[36] Constructors are magician-like sentient robots who can construct a contraptions (often artificially intelligent) for almost any purpose. *The Cyberiad* has the universe alternating between being populated by biological and robotic beings—each of whom eventually succumbs to tackling the challenge of creating the other (only to be overthrown by their creation).

[37] The first kind of demon is Maxwell's Demon, a creature described in a thought experiment that challenges the Second Law of Thermodynamics (also known as the Law of Entropy).

[38] It is further detailed that every book has 410 pages, each page has forty lines, and each line approximately eighty black characters. There are twenty-five "orthogonal symbols" consisting of a space, period, comma, and twenty-two letters. That calculates out to $25^{1,312,000}$ possible books (or, as the scientific calculator on my Macintosh puts it, "Infinity").

[39] There would be 1,312,000 books that vary by one character and $1.72*10^{12}$ books that vary by two characters.

Chapter VI
Holography:
Re-Defined

Martin Richardson
De Montfort University, UK

Paul Scattergood
De Montfort University, UK

ABSTRACT

When writing this chapter it became apparent that we were not only exponents of digital holography, but also the critics. This is a problem when it comes to new media. How can one begin to make objective critical theory on a subject when there are no historical or ideological structures that produce and constrain it? While other digital technologies prove well developed, semantic and expressive, digital holography has some way to go before any quantized analysis of the subject is possible. This paper explores the function of digital holography, seeking comparison from other media and explores holography's influence as a radical form of electronic digital three-dimensional image capture. Within this context we draw comparison with other forms of image making, from cave paintings in Lascaux (France), to Fox Talbot's early experiments to capture light, Corbusiers architectural designs of space, to early television transmission. They all have one unifying factor: the unfamiliar and the strange, emblematic to visual possibilities in our perception of space.

NEW VISIONS

The illusion of Three-Dimensional space may be traced back to a time in classical western history to the development of painting, and of its use to create visual likenesses, was through the utilization of mathematics by Italian architect, Filippo Brunelleschi, in the middle of the fifteenth cen- tury. His invention was used in the work of such famous artists as Piero Della Francesca, Albrecht Durer and others. In paintings made in accordance with the invention of perspective, it is possible to distinguish whether an object is situated in the foreground of in the background and to locate the point where all the lines describing the depth of the depicted space converge. Nevertheless, even this innovation did not solve the main problem, since the three-dimensional scene remained two-dimensional in the picture. Further perfection of the painting technique and the application of innovatory methods, such as the use of an improved camera obscure in the seventeenth century, were aimed at the inclusion of greater detail. But still the invention of perspective is a false, a man made calculation, which only serves to constrain the observers view rather than enlighten it.

A marked step in solving the complicated problem of detailed representation of our three-dimensional world was made by photography. In taking photographs, the lens is used to construct the image in the physical plane of the recording material. The lens is constructing a 'mini' three-dimensional form inside the camera and as we change focus we are selecting the physical plane, the area of interest we want others to see too. Since Fox Talbot's early experiments in the eighteenth century, photography has represented objects of the surrounding world in a far from perfect way. A vision we can take no further, a vision that, like language, has influenced society beyond all expectations. As we move into the digital epoch the prime illusion – digital holography – is patently waiting its turn.

As physics explores the quantum universe attempting the development of a holistic theory we can observe the holographic principle at work in almost every principle of life. Indeed, some even suggesting the universe may be compared to one giant hologram as the information of the whole exists in every constituent part; that a tiny little piece, a tiny particle, might contain all the information pertaining to everything that exists or has ever existed. As quantum holography reaches far into new theories to explain and 'un-lock' such baffling forces such as gravity, gravitons, gravitines - the very glue that binds our atoms together – even life itself, or our perception of it. A theory interestingly endorsed by many cults including the Institute of Noetic Sciences (IONS), founded by Edgar Mitchell who walked on the moon with Al Shepard during NASA's 14th Apollo mission. A note of caution however because as we shall see, the holographic principle itself is a spectacular tool of deception and arguably the most advanced form of technological visual deception to date, certainly one of the most intriguing.

Are holograms "Mere" illusions of objective visual reality or creative artefacts capable of expression, interpretation and deception? Certainly an emerging technology that offers the opportunity of usurping conventional two-dimensional with the third-dimension. Holographic Optical Elements (HOE's) usurp our perception of the 'real' in ways previously impossible. The alternative modern holography offers industry may be compared with the role of electronic circuits and microprocessors held at the beginning of the 60's as an alternative to the electronic valve. Mass-produced 'Holographic Optical Elements' are starting to replace micro-lens arrays, and 'Holographic Phase Memory' is poised-ready to replace today's standard magnetic hard-drives - we are about to start our journey into 'The Age of Photonics'. Modern holography is capable of integration within digital media exploding the limitations of its forbearers and launching its new development as hyper-media.

Like photography, digital holography is certainly more than a recording medium. It is a tool that can substantiate both fantasy and fact and comment subtly on the real world, blurring our grasp of that reality to the point where objectivity is submerged in a sea of imaginings. The extreme reality of modern holography challenges our understanding of what we mean by "real" yet it's essential ambiguity is as unsettling as their verisimilitude is reassuring in a post McLuhan age of "virtual reality" experiences, "reality TV" spectacles and "celebrities" because audiences aren't intrigued by technology, there're more concerned with fantasy. Those who take an interest in the science of modern holography will equally find interest in the explanation and theory of coherent light and the methods employed by those clever enough to capture its ability to record 'real' space via digital media. But there are other perspectives from which to view this progression in imaging, especially ones that emphasize the evolution of photonic applications as the rest of the world progresses along its speedy march into the future, discovering new human needs and new potentials for old ideas translated into digital. If you have read this far, you might by now share a sense that holographic imaging and its integration into useful every day life is not just a technological speculation, it is 21st century inevitability.

Early sculpture was essentially a method of creating a three-dimensional copy of a real object. As evidenced by examples such as the small statue known as the Venus of Willendorf (Austria) – dating back to the thirtieth millennium B.C and other objects of early Palaeolithic origin such as carvings on household articles of wood and mammoth tusk. The other, painting, transformed the three-dimensional world into its two-dimensional image. At Altamira, the artists utilised many naturally occurring dyes and materials such as charcoal and ochre to draw and build up images upon the walls of the caves, often following the natural patterns and formations of the rock to increase the three dimensional potential of the representational process and increase the level of verisimilitude in the work.

These very early forms of information storage and retrieval allowed for the development of communication both with other members of a communal group who were present at the time of the image making, and also to communicate ideas and concepts beyond the immediate moment. Historically this has allowed for the development of more complex societal structures. *Kathleen Burnett* describes the manner in which visual information was utilised in pre-literate societies as a means of communication.

Kathleen Burnett

Visual means of communication and information transfer have always existed from cave paintings to religious icons to Gothic cathedrals to paintings, sculpture, and other visual arts media. The information-poor, one might even argue, have historically relied on the visual media as their primary mode of reproducible information transfer. Certainly this was true in Western Europe before the growth of literacy, and even today scholars point to the democratizing effect of television. (Burnett, 1993)

Burnett's assertion that those with reliance upon visual media are likely to be *'information poor'* and that by default textual information is of a more developed nature is questionable. However it is true that historically visual media such as that which she describes has been utilised as an important tool for information transfer. In part this is due to the performative nature of the creation of images, whereby:

The performative elements of practice may be outputted through the artefact, the knowledge which is genuinely new may be "an 'understanding' rather than a mere explanation that is of central interest in research. (Sullivan, 2006)

When *Graeme Sullivan* talks of the artist being involved in undertaking a performance, he describes the nature of research investigation particular to artistic practice. The process-based nature involved in the production of artefacts opens up the possibilities for the development of ideas and concepts, which would be missed in a *"more streamlined process"* (Sullivan, 2006). The artefact within art practice existed and still exists both as a site of enquiry and a subject for investigation. With the development and expanding of the field of research through practice with the creative disciplines, the level to which the artefact exists as evidence of recording of ideas and concepts through the process of the creation remains a major focus of research and understanding.

Walter Benjamin

If the tasks which face the human apparatus of perception at the turning points of history cannot be solved by optical means, that is, by contemplation, alone. They are mastered gradually by habit, under the guidance of tactile appropriation. (Benjamin, 1935)

This is a description of how our knowledge and understanding of the universe we inhabit is shaped and developed by our depictions of it, it also serves as an explanation of what **Tim O'Riley** describes as an elliptic relationship between practice based acts of creation and the nature of thought relating to artistic creation, whereby developments in imaging are driven and re-contextualised by theoretical understanding, thus allowing opportunities for theory to re-inform the work (O'Riley, 2006). This is not an iterative approach as new ground is covered each time.

One similarity between the photographic and holographic process is the lens-based and optical nature that they share when capturing an image. The optical processes inherent in the recording imbue the hologram or photograph with certain amounts of unmediated visual information.

Susan Sontag

A photograph is an interpretation of the real; it is also a trace, something directly stencilled off the real, like a footprint or a death mask. (Sontag, 1977)

To a greater or lesser extent, all photographs and holograms bear a direct optical relationship to **'the real'** because both are made using light and light absorbent material. However, one of the key differences between a photographic camera and a holographic image capture device is the level of intervention into the **'real'** which must be made in order for the process of image capture to be successful. Current digital photographic technologies allow a simple point and click operation that may or may not involve any level of staging or elaborate sets. Holographic technology on the other hand involves a very significant level of setup. For example due to limitations of the chemical process the image capture must be made in a totally darkened room, similar to a camera obscure. It is also important for health and safety reasons relating to laser usage that the use of highly reflective surfaces must be tightly regulated.

The analogue or digital photographic camera will utilise the existing visible light sources (or possibly additional sources of the same type added by the photographer) and take a tracing of this light as it reflects back from the subject recording only its amplitude. What is depicted, in terms of light hitting the photographic negative or digital device is to a large extent, the same as what would have been visible had one been at the location of the camera at the moment the picture was taken. Somewhat differently, holography makes its readings by making an intervention into the constructed landscape of its set with a laser beam, which passes through the holographic plate (or film) and reflects back from the object to make an image on this plate that records the total wavefornt information, both phase and amplitude. While what we see when we view a hologram may appear to show the objects or people as they were, if one had been at the location of the image capture as it happened one could have seen nothing but a darkened black space in total stillness (because of health and safety implications it is usual, but not mandatory to use a visible laser beam for this process).

This of course has an enormous impact both on the nature of the subject matter and of the nature of the holographer's interaction with the **'real'**. A photograph may pass ***"for incontrovertible proof that a given thing happened"*** because, as described; a photograph bears direct visible optical relationship to the depicted (Sontag, 1977). In a sense it is also true that the hologram is a more accurate tracing of the real than a photograph. A hologram does not record the photons of light on a flat surface as a binary function, such as that performed by the optical/chemical or optical/digital processes of photography. In a different way altogether, a hologram re-creates the optical processes of the depicted space through a depth of substrate of the recording medium, rather than on the surface as an image. This means that because it records the optical reality not only of a single viewpoint, but also from a wide horizontal and vertical parallax, it is thus a 'tracing' in much more detail than a photograph. But as discussed briefly before, it is clear that the constructedness of the perceived reality in a hologram is not proof that any given action was genuine.

Victor Burgin

The intelligibility of the photograph is no simple thing; photographs are texts inscribed in terms of what we may call 'photographic discourse', but this discourse, like any other, engages discourses beyond itself, the 'photographic text', like any other, is the site of a complex intertextuality, an overlapping series of previous texts 'taken for granted' at a particular cultural and historical conjuncture. (Burgin, 1994)

The viewer of photographs must have a learnt ability to perceive and understand the visual information contained within the image, based on knowledge of discourses beyond the frame of the image in front of them. For the viewer to be able to comprehend an image they must have had experience of seeing reality in order to understand how that reality is represented in a photograph, they must also to a large extent need to have knowledge of both the long tradition of representation in art and also have a knowledge of the objects which the photograph is depicting before they can make an assumptions about any objects scale relative to others, or place objects in a position relative to other objects and the po-

sition of the viewer. The system of perspective in western art depicts the world in a similar way to the way that visual information reaches our eyes. Importantly though, this is not the way the world really is. (i.e. things that are far away are not really smaller.)

Many 'optical illusion' books, and textbooks on the subject of visual perception, display images depicting objects in a manner in which it is clearly observable that we are not seeing the real, but a representation. Once an optical illusion clicks into place in our minds, what we are looking at becomes more easily decipherable the next time we view the same image. As soon as a viewer knows what an image is depicting, they will be able to read from this; inferences of scale, distance, lighting, etc. In assessing the levels to which image types are intuitive we must pay attention to cultural differences of the observer.

Le Corbusier

If I hold up a primary cubic form, I release in each individual the same primary sensation of the cube; but if I place some black geometric spots on the cube, I immediately release in a civilised man an idea of dice to play with, and a whole series of associations that would follow. A Papuan would see only an ornament. (Le Corbusier & Ozenfant ,1920)

It is clear that a viewer who has never before seen a cow would have trouble deciphering the image, it is only when one knows the reality of what a cow is that one is able to draw associations from this, as Corbusier's 'civilised man' can draw associations from a cube embellished with particular arrangements of geometric dots. If one were to have seen similar animals, one could make suppositions or educated guesses; one could not however, make more defined assertions.

It is clear that a viewer who has never before experienced three dimensional space would have trouble deciphering spatial relationships in a photographic image, it is only when one has experienced spatial interaction that one is able to draw associations from this, as Corbusier's 'civilised man' can draw associations from a cube embellished with particular arrangements of geometric dots. If one has not experienced similar spatial relationships to those depicted, one could make suppositions or educated guesses; one could not however, make more defined assertions.

Holography is a method of constructing elaborate realities that appear hyper-real. They are at once accurate and indelible tracings of the real and also elaborate fakes. It is important not only to look at the process by which these images are manufactured, but also the differences involved in looking at or reading the images, clearly a full examination of this would lie outside the bounds of this essay but it is important to consider several of the factors involved.

In this regard, one can single out key differences. This is important because it makes a significant difference to the way in which the two mediums, photography and holography, depict the focused object, and the way that this information reaches is visually perceived. Because we perceive depth in holograms using range finding and stereoscopic vision. (For further information on this subject, read: Eye and Brain, The Psychology of Seeing by R. Gregory (1998)). We are un-reliant on '**perspective cues**' as a signifier of image depth. Holograms are intuitively comprehendible. They are also much more capable of creating the representation of depth in objects devoid of two-dimensional perspective cues. Stereoscopic vision is an ocular process that is not applicable to photography but remains usable when viewing holograms.

The way holographic information reaches our optical biological system to brain is dependent to reality whilst photographic is an interpretation, which, stripped of these perceptual capacities is much further removed from our perception of reality and closer to a language of representation, symbolism and language. Even if we remain

static and not take the opportunity to evaluate the image from multiple viewpoints, a hologram will appear, in terms of these optical processes, more like 3-dimensional reality and cognitively related.

NOT A 3-D ILLUSION

Holographic Optical Elements (**HOE's**) are a demonstration of holography's ability to reconstruct the optical realities of objects that are within the image. The most basic example of a HOE is a simple lens structure. *(For further information on this subject, read: Practical Holography by G, Saxby (2004)).* If a hologram is taken of a glass lens; when light is directed at the hologram, it will perform the same optical function as if the original glass lens was in the same position within the beam of light. This is a demonstration of a very fundamental difference between the nature of photography and holography. Photography does not, and is not capable of reconstructing the optical processes present either at the time of image capture or inherent in the object. Holography can be used to record this information and, by virtue of its ability to perform this optical function and under certain circumstances; recreate the actions of the original object in both space and time. It is important in this context to investigate how accurate either image type, or any other for that matter is at representing the world. Thus far this essay has discussed the level of accuracy to which the given methods of pictorial representation adhere in terms of how much they appear to be like the world. But is a system of representation where the signifier is indistinguishable from the signified really of importance or any value?

Perhaps a hologram is similar to a Lewis Carol creation, where he once produced a fictional representational of a map. This map was in fact, as large as Germany (the country that it represented). While every detail is shown with the highest level of accuracy, both the cartographic meaning and usefulness are made redundant by the world that it seeks to chart. In the end the people decide to use the entire country as a map of itself, as it is the most accurate possible map there is. (One also needs to remember that Carol himself had a deep fascination with, what was then, the new invention of photography later experimenting with trick photography alluding to fairy's at the bottom of the garden.)

This example of the uselessness of a map as a copy of reality relates not just to cartography, but also to all forms of image making. To a degree, one would be mapping or charting one's understanding and experience of reality if one were to make an image or representation of the world. In making this comparison we should clarify that the most accurate map may not be the most useful, and the representation, which appears most like the world visually may be a very poor 'tracing' of the real. On seeing images of the cave paintings at Chavaut, in the south of France, **Picasso** is said to have questioned whether we had made any progress at all (Morgan, 2003 p. 171).

It has never been spread out yet, said Mein Herr: the farmers objected; they said it would cover the whole country and shut out the sunlight! So now we use the country itself as its own map, and I assure you it does nearly as well. (Carroll, 1893)

In striving for a visual copy of the world that we seek to represent, we may actually be regressing rather than advancing from and building upon past ideologies. It is perhaps quite obvious, but also very important to highlight the fact that our sense of sight but one of a number of senses and therefore sight alone can never make a true representation of the world. Further to this it is important to remember that our sense of sight is actually a rather poor method of capturing and interpreting the information that is available. Among other limitations, of course, we do not see the entire spectrum of light; large swathes of possible ocular information are unavailable

to us. We do not see the extremities at either end of the spectrum, seeing neither infrared nor ultraviolet light. It can be argued that in seeking to show only the information that is available to us through our eyes alone, at a single static moment in time is a wholly insufficient method of picturing the real.

Wittgenstein

The picture is a model of reality, As a model it need not bear direct optical (or other) relationships to its subject in order to successfully represent it. Of course there are numerous possible methods for representing reality, many of which we may even be unaware of as of yet. (Wittgenstein, 1976)

In '*Slaughterhouse Five*', Kurt Vonnegut describes one alternative method of representation:

"Billy couldn't read Trulfamadorian, of course, but he could at least see how the books were laid out - in brief clumps of symbols separated by stars. Billy commented that the clumps might be telegrams.

"Exactly," said the voice.

"They are telegrams?"

"There are no telegrams on Trulfamadore. But you're right: each clump of symbols is a brief, urgent message-describing a situation, a scene. We Trulfamadorians read them all at once, not one after the other. There isn't any other particular relationship between all of the messages, except that the author has chosen them carefully, so that, when seen at once, they produce an image of life that is beautiful and surprising and deep. There is no beginning, no middle, no end, no suspense, no moral, no causes, and no effects. What we love in our books are the depths of many moments seen all at one time." (Vonnegut, 1996)

This is of course a fictional account, but highlights one interesting aspect of pictorial representation; while an image may be composed of innumerate distinct and disparate syntactic elements (such as pixels from a digital hologram), it is the interrelationships between these pixels and the structures that govern their distribution which we are able to read as a whole entire image. When we speak of reality we must question whether we are to include information that lies **'outside'** the boundaries of our perceptual framework. In the Tractarian ontology: the structure of reality is implicated in the structure of the world. For quite trivial reasons, the structure of the world is implicated in the structure of reality. Of course, it still remains a mistake to identify the world with reality.

Photography, particully through televisual events, is now part of the everyday; photographs are possessable objects, which are part of the fabric of everyday life. They are integrated into screens, print media, billboards and many other information dispersal platforms and technologies. Holography, on the other hand, is still very much set apart from the world that it depicts and inhabits, remaining sidelined, for use in hi-tech security devices and one off images. While holography is a medium that is still in the early stages of its development it is moving on apace, photographic imaging technologies are already highly developed. The tools for the production of photographic images have been democratised to such an extent that to be photographed is no longer an event in itself. There does remain a certain level of performativity to all photographs where the subject is aware that they are to be the focus of the image, but this has ceased to be to the extent to which it once was, and to the extent that within holography it still is.

Holography is utilised as a security device to confer genuineness and legitimacy to objects otherwise devoid of these properties. But the relationship between signifier and signified is not direct; the credit card or DVD will be made official by a small logo floating in space or an abstract pattern. The represented in the hologram need bear no relationship with the object to which it confers 'proof' of authenticity. Here it is the very limitations of the Holographic process that has been one of its key assets in becoming this signifier of authenticity. Holography is not a medium that is easily reproducible; it is possible to mass-manufacture the same hologram many times, though difficult and laborious to produce a single image or small multiples. It is in fact not what is depicted but the hologram itself that is a signifier of authenticity.

This notion itself brings up important questions of what is original and what is a copy or a fake. When a moment is captured in an image, the original reality will already have moved on in time, and while as discussed before with the help of Billy Pilgrim and the Trufalmadorians from **Slaughterhouse 5**, the perception of a depiction of reality as a single point in time rather than "**the depths of many moments seen all at one time**" is a wholly inadequate method of explaining or describing individual or collective reality (Vonnegut, 1996). The notion of image as a recording of a pinpoint in time is a distinctive characteristic of pictorial representation in photographic recording and post perspective western painting, but also to a lesser extent, historically, in holographic representations.

A representation may be considered exactly as that; re-presented, a copy or an interpretation, an explanation or description of the world. This is not to say that an image cannot be a reality in itself, exist as an object and neither be a transcendental message, nor a representation of reality. Simply that a representation or picture of the world cannot be wholly equivalent to the reality it seeks to represent. Holography can offer a four dimensional space, the medium is able to record images not only in three-dimensional space, but also in time. One of the issues historically related to this process within holography is that it has been tied to the recording of objects and actions.

This is significantly different to the ability to create and generate images that is provided by the new medium of digital holography. By drawing outlines or making analysis, whether qualitative or quantitative the new medium of digital holography is a tool uniquely placed for the rationalisation and expression of both the 'real' and imagined.

Creative applications of holographic imaging lie largely unexplored, as most practitioners in this area have focused to a large extent upon documentation and recording of pre-existing spaces and objects. The holographic output process has been utilised as a tool to embody pre-existing and imagined architectural spaces upon a planar surface, it will be possible to develop an enhanced resolution of spatial narrative structures (Pole, 1968).

The output of 3D and narrative digital spaces through an RGB dot matrix hologram has been developed by commercial organisations, as have camera systems for recording digital photography into holographic space. (Lane, 1982). Commercial organisations have also developed and utilised holographic imaging technologies as a "design visualization tool for engineers, designers and scientists", and have been able to produce full parallax, auto stereoscopic output for 3D digital imaging (Zebra Imaging, 2009).

FUTURE DEVELOPMENTS

As digital image capture, recording and editing devices become both more numerous and less costly, the possibilities are opened up for increased artistic investigation of the digital holographic process. Pioneering technological developments within the filed of holographic imaging will allow for the enhanced integration of technology and art – innovating the concept of viewer participation. Holographic technology will develop increasingly immersive experiences that are able to explore fact and fantasy in new and previously unforeseen directions.

This combination of interactive, immersive environments and their availability for creative applications will develop and expand 'real' spaces. Thus allowing for the investigation of non-hierarchical interactions, which subject the digitally recorded material to editing and interference, in order to produce work which utilises techniques "where expressiveness or tackiness come closer to the truth than a perfectly sharp, slick representation".

The holographer will be able to utilise materials and techniques, which investigate and broaden our understanding of spatial restrictions, in a manner that augments and relocates our relationship with depicted space and objective reality. Restrictions that are present in all other forms of image capture – from painting and sculpture to digital video, will be surmountable with the advent and increasingly widespread artistic use, of digital holographic technologies.

The acceptance of the holographic process by the art world has been sporadic, despite huge advances in technology over recent years. This is due, in part, to the focus of these developments upon commercial uses, which have made the holographic image ubiquitous, without increasing its quality or expressive possibilities to a substantial degree. The development of digital holography and its convergence with many other digital technologies will finally allow the holographer to transcend the desire to create a recording of 'real' space and investigate the intricate relationship between image surfaces and volatile, illusive truth.

The digital manipulation of recorded data, when output to digital holographic displays allows the holographer and artist to construct either lavish or minimal spaces. Thus, able to defy the conventions of previous holographic imaging, creative practitioners will move the exploration of the medium as being considered to be 'avant-garde' through its take up of new technology - toward creative practices that utilise technologies as a tool rather than an end in and of itself.

As increasingly expansive holographic environments become available, to be scaled to infinitely large sizes (through the utilisation of tiled 3d digital holography), we will see a move away from the reliance upon the frame as a portal or window into another reality. Holography now offers the possibility of constructing architectures and spaces, which wholly immerse the viewer to a level in which they become an active participant rather than an inert element of the artistic process.

Investing in this relationship between space, architecture and participant – the holographic output process will be utilised to give physical presence and materiality to ideas and creative explorations. The holographic artist will be capable of re-engineering thoughts, perceptions and experiences – both playfully and rigorously questioning concepts of interiority and exteriority, personal and universal - even the nature of lived experience itself.

REFERENCES

Benjamin. W. (1935) *The work of art in the age of mechanical reproduction.* Originally published in Zeitschrift für Sozialforschung.

Burgin, V. (1994). *Thinking photography.* Houndmills: Macmillan.

Burnett. K. (1993). *Toward a Theory of Hypertextual Design.* http://www.iath.virginia.edu/pmc/text-only/issue.193/burnett.193 accessed 26/02/2008

Carroll, L. (1893) Sylvie and Bruno Concluded . England. Macmillan & Co.

Gregory, R.L. (1998) Eye and Brain, The Psychology of Seeing. Oxford, England.

Lane, B. (1982) Stoscereopic Displays,"Processing and Display of Three-Dimensional Data". SPIE Proc. 367, 1982, pp. 20-32.

Le Corbusier & Ozenfant (1920). Purism. *L'Esprit Nouveau, 4,* 369-386.

Morgan, M. (2003). The Space Between Our Ears. Weidenfield & Nicholson.

O'Riley. T. (2006). Thinking Through Art, Reflections on Art as Research, (p. 94). Routledge.

Pole, R. (1968, January). 3-D Imagery and Holograms of Objects Illuminated in White Light. *Applied Physics Letters, 12*(1), 10 –12.

Sontag. S. (1977). *On Photography.* New York: Farrar, Straus and Giroux. P154 ISBN.

Sullivan, G. (2006). *Artifacts as evidence within changing contexts.* Working Papers in Art and Design 4 http://www.herts.ac.uk/artdes/research/papers/wpades/vol4/gsfull.html

Vonnegut, L (1976). Slaughter-House-Five. Laurel/Dell Books paperback.

Wittgenstein, L. (1994) Tractatus Logico-Philisophicus. London: Routledge.

Zebra Imaging (2009) Application. http://www.zebraimaging.com/html/industries.html

USEFUL WEB ADDRESSES

http://www.holograms.co.uk/exhibitions.html

http://www.geola.com/corporate.asp

http://www.zebraimaging.com/html/industries.html

Chapter VII
3D Sound Simulation over Headphones

Lorenzo Picinali
De Montfort Universtity, UK

ABSTRACT

What is the real potential of computer science when applied to music? It is possible to synthesize a "real" guitar using physical modelling software, yet it is also possible virtually to create a guitar with 40 strings, each 100 metres long. The potential can thus be seen both in the simulation of that which in nature already exists, and in the creation of that which in nature cannot exist. After a brief introduction to spatial hearing and the binaural spatialization technique, passing from principles of psychoacoustics to digital signal processing, the reader will be included on a voyage through multi-dimensional auditory worlds, first simulating what in nature already exists, starting from zero and arriving at three "soundscape dimensions", then trying to advance the idea of a fourth "auditory dimension", creating synthetically a four-dimensional soundscape.

INTRODUCTION

What is the real potential of computer science when applied to music?

Using physical modelling synthesis techniques it is possible to simulate as accurately as possible an acoustic guitar: of course, this is useful in terms of the opportunities made available to musicians to use an instrument they cannot in reality play, and in terms of acoustical studies of the instrument itself. But: Is that it? Once created, a guitar mathematical model can be altered as far as the imagination can extend: an acoustic guitar made of gold, with 40 strings, each of 100 metres in length, could virtually be created and played with a one-square-metre plectrum! Therefore, the potential can be seen both in the simulation of that which in nature already exists, and in the creation of that which in nature cannot exist.

Another kind of example will be discussed later in the chapter: instead of toying with the simulation of musical instruments, we shall try to create virtual acoustical environments through the simulation of three-dimensional (henceforth: 3D) soundscapes. The binaural spatialization technique will be used in order to achieve this goal; multi-dimensional soundscapes will be simulated not by placing real sound sources, such as loudspeakers, within the three dimensions, but by simulating the behaviour of our outer ear in terms of directional modifications brought to the sound input into the hearing system.

The mechanisms of spatial hearing will be investigated and analyzed, and three localization cues will be characterized and simulated. These are the Interaural Level Differences (ILDs), the Interaural Time Differences (ITDs), and the Direction-Dependent Filtering (DDF). Within the simulation of a real environment, these parameters would all be coherent with the position of the sound source: for example, for a sound source placed at 60° of azimuth, the sound would reach first the right ear and then the left (ITDs). Furthermore, it would be more intense at the right ear (ILDs) than at the left, and the sound would be filtered depending on the particular resonances of our outer hearing systems for that specific sound source location.

It must, however, be asked what could happen if the three localization cues were incoherent with the real position of the sound source. Of course, this is impossible in nature, and equally so in a standard soundscape simulation, when loudspeakers are placed in a 3D space. Still, achieving such incoherence is not impossible in a system based on headphones, where the signals sent to the hearing system are much more controllable, thus the whole reproduction system results may be much more flexible.

In this case, the binaural spatialization technique is useful not only to simulate a real 3D soundscape, but also to create new soundscapes, i.e., environments that are impossible to find in the real world. This seems to be one of the amazing new options offered by computer science:

while it could indeed be considered inessential to simulate a feature that already exists in nature, it is particularly interesting to create a feature that as yet has no existence in the real world.

To appreciate this new 'digital feature' fully, it may help to think about a voyage into multiple dimensions; the results may appear similar to Abbott's graphic achievements (*see* Abbott, 1999) when he wrote *Flatland* (frequent reference will be made to this book later in the chapter). A monophonic diotic signal (the same at both ears) could be perceived as a point sound source located in the middle of the head: zero dimensions, or the point. By introducing intensity and content differences between the two channels (ILDs) and creating a dichotic signal (different for each of the ears), it is possible to obtain a standard headphone stereo signal, with multiple sound sources located along a line between the ears (always inside the head): one dimension, or the line. Through introducing time differences between the two channels (ITDs), it is possible to obtain the sensation of the sound coming from out of the head, and with multiple sound sources located in a plane: two dimensions, or the square. Then, upon introducing a simulation of the DDF, with different frequency filtering for each virtual sound source, the perception reaches the third dimension, the cube. The auditory passage between these steps could be visualized as the graphic perception of a point that becomes a line, then a square and at the end a cube. What, then, about the fourth dimension?

The gain of a dimension can mean the coexistence of a concept that in a 'lower' dimension cannot be seen simultaneously. For example, when inspecting a square in a bi-dimensional world it will be possible to see just a line, one or two angles at the same time, and not more (in a bi-dimensional space, the viewed perception should be mono-dimensional, as in the three-dimensional the perception is bi-dimensional). Only when reaching the third dimension will it be possible to see all four angles of the square at the same time.

In a four-dimensional space, therefore, it may be possible to see, for example, the front and the back of a person at the same time… Yet how can this be rendered from an auditory perspective? The fourth dimension could be seen as the coexistence of entities that in a 'real' listening situation could not exist at the same time, such as a sound signal with an ILDs of a sound source placed at 60° of azimuth and at 0° elevation, an ITDs of a sound source placed at 300° of azimuth and at 0° elevation, and a DDF of a sound source placed at 0° of azimuth and at 90° elevation.

How on earth could all of this be created, if not with the help of computers?

Elements of Spatial Hearing

How can our hearing system determine the direction of the provenance of a sound, and therefore the position of a sound source, in a 3D soundscape?

In this chapter the mechanisms of spatial hearing will be described and analyzed, starting from a short overview of the external hearing system and continuing to a brief investigation of the three localization cues, the ILDs, the ITDs and the DDF, and of the binaural phenomena.

References for the topics discussed in this chapter can be found in Blauert (1996), Moore (2003) and Yost (2000).

Some Basic Notions

Before beginning this overview of the binaural phenomena and of the psychophysiology of the spatial hearing system, the definitions of a few terms will be attempted, in order better to understand that which is to follow:

- **Sound localization:** The judgement on the specific location of a sound source.
- **Sound lateralization:** It is feasible, most of all while listening to sound through a pair of headphones, to be unable to localize sound sources outside our head, but to perceive the sound as coming from inside the head, with sound sources placed along an imaginary line that starts at one ear and crosses to the other. This phenomenon is known as *sound lateralization.*
- **Localization cues:** Specific attributes of the sound event that are used by the hearing system in order to establish the position of a sound source in a 3D soundscape.
- **Monoaural:** Relating to or involving a sound stimulus presented to one ear only.
- **Binaural:** Relating to or involving a sound stimulus presented to both ears simultaneously.
- **Interaural:** Between one pair of ears.
- **Diotic:** relating to or involving a sound stimulus presented to both ears in exactly the same way.
- **Dichotic:** Relating to or involving a sound stimulus presented to one ear differently from the sound stimulus presented to the other ear.

Spatial Coordinates

In order correctly to localize a sound source in a 3D soundscape, a coordinate system needs to be established. Three planes need to be distinguished, each one with the origin placed at the centre of the head (*see* Figure 1):

- **Horizontal plane:** Placed at the superior margins of the two ear canals and on the inferior part of the ocular cavity.
- **Frontal or vertical plane:** Placed at an angle of 90° to the horizontal plane, it intersects with it at the superior margins of the two ear canals.
- **Median plane:** Placed at an angle of 90° to both the horizontal and the frontal planes, it constitutes the plane of symmetry of the head.

Figure 1. The spherical coordinate system (after Blauert, 1996)

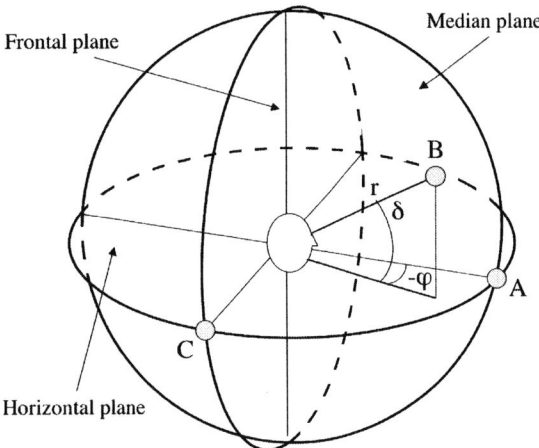

Using this system as a reference, the position of a sound source can be unequivocally defined by the *Azimuth* (φ, localization angle on the horizontal plane, calculated proceeding anti-clockwise), the *Elevation* (δ, localization angle on the median or frontal plane, calculated proceeding upwards) and the *Distance* (r, the distance between the sound source and the centre of the listener's head). In Figure 1, three sound sources are placed as an example; their spherical coordinates are:

- **A:** φ = 0°, δ = 0°, r = depending on the radius of the circles drawn
- **B:** φ = 345°, δ = 30°, r = depending on the radius of the circles drawn
- **C:** φ = 270°, δ = 0°, r = depending on the radius of the circles drawn

Outer Ear Overview

To introduce the human auditory system: the outer ear is its external part. It is composed of the pinna (the visible part), and the auditory canal or meatus. At a first sight, the role of the pinna seems to be quite simple: to convey the sounds that reach the head into the ear canal. However, if its particular shape is inspected carefully, its functions may be guessed as being far more complicated. The pinna in fact also significantly modifies the incoming sound, depending upon the angle of incidence of the sound itself and thus on the position of the sound source. This modification is mainly related to frequency filtering, especially for high frequencies (above 3000 Hz).

After it has been conveyed and modified by the pinna, the sound travels down the ear canal and causes the eardrum, also known as the timpanic membrane, to vibrate. After this point, the vibrations are transmitted through the middle ear by the ossicles, three small bones (the malleus, incus and stapes) that work as impedance converters and mechanic amplification devices through a complicated system of levers, and then to the cochlea, the last part of the peripheral auditory system and part of the inner ear.

The system as a whole does not, for our purposes, warrant scrutiny. The part of the peripheral auditory system involved in the mechanisms of sound modification linked to the source position, and therefore to the sound incidence angle, is in fact solely the external one, thus the outer ear.

The Mechanisms of Sound Source Localization

As defined previously, sound localization is the judgement made regarding the specific location of a sound source, performed through particular mechanisms fulfilled by our auditory system. In order to accomplish these mechanisms, the auditory system can work on certain particular attributes of the signal input into the ear canal: those are called the localization cues, and they can further be distinguished between interaural differences and monoaural attributes, as will be shown in the following sections.

The Interaural Differences

There are two kinds of interaural differences for the localization of sound sources at the left or at the right of our head:

- **ILDs:** Interaural Level Differences, the differences in terms of the pressure level of a sound stimulus between one ear and the other. They are generated by the presence of the head between the ears, the head that acts as an obstacle placed in the direct path between the sound source and the ear entrance (in this case, the ear is that opposite the position of the sound source).

- **ITDs:** Interaural Time Differences, the differences in terms of arrival times of a sound stimulus between one ear and the other. The differences are generated by the different paths that the sound wave needs to cover in order to arrive from the sound source to each of the ears: when the sound source is not located in the median plane, the distances between it and the two ears individually will differ, thus the sound wave will take a longer or a shorter time to reach the beginning of each of the ear canals.

Both perceptions are effective in order to perform the localization of a sound source placed on the left or on the right of our head; however, their importance varies according to the frequency bands covered by the sound source to be localized. The *duplex theory* (*see* Lord Rayleigh, 1907) is probably the most widely accepted hypothesis as to how the interaural mechanism works for sound source localization: the ILDs are more effective for the localization of high frequencies, while the ITDs are for low ones.

To try to explain this: as is known, low frequencies (between 20 Hz and 500 Hz, with wavelengths ranging from 16 m to 64 cm) have a wavelength that is much larger if compared to the diameter of our head (~17 cm), therefore they will not be scattered or absorbed by such a small obstacle. Thus, the ILDs result in being nearly irrelevant for low frequencies, and significantly larger for high frequencies: as an example, after Figure 2, it can be noticed that when a sound source is located at 90° of azimuth, the ILDs are at 1-2 dB for the 200 Hz, and at 20 dB for the 6000 Hz. Thus, the ILDs can be seen as a frequency-dependent parameter.

Figure 3, shows how the ITDs vary independently according to the frequency of the stimulus, only because of the speed of sound; therefore, the time taken by a sound wave to travel from the sound source to the two ears is dependent not upon

Figure 2. The interaural level differences divided by frequency and in function of the azimuth (after Feddersen, 1957). This diagram has been redrawn from Moore, 2003: due to this fact, it is not used as a reference, but only to give an idea about how the ILDs works.

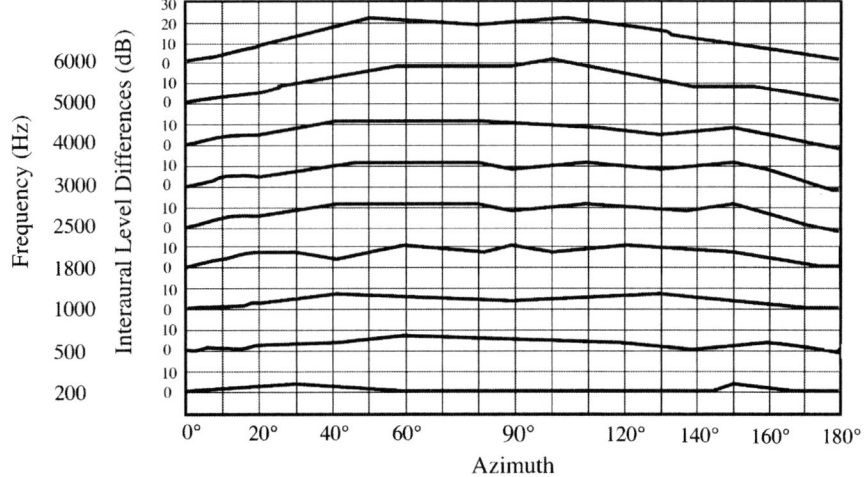

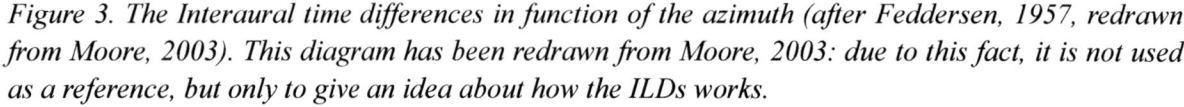

Figure 3. The Interaural time differences in function of the azimuth (after Feddersen, 1957, redrawn from Moore, 2003). This diagram has been redrawn from Moore, 2003: due to this fact, it is not used as a reference, but only to give an idea about how the ILDs works.

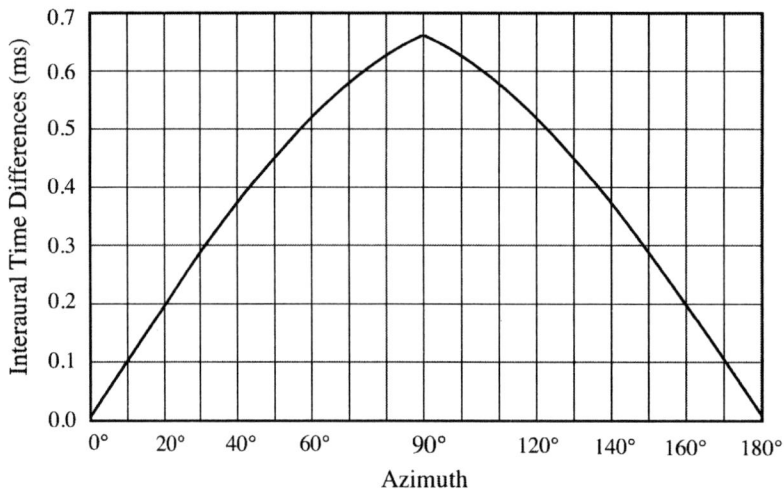

the frequency, but upon other physical parameters such as air temperature and humidity. Thus, the ITDs can be seen as a frequency-independent parameter.

However, the fact that the ITDs can be measured in terms of microseconds (a millionth of a second) creates some detection problems for frequencies whose period is comparable with that of the ITDs themselves. For example, if a 1000 Hz sound source is located at 60° of azimuth, the ITDs would be 0.5 milliseconds, exactly half of the period of the 1000 Hz frequency. In this case, it would be utterly impossible to establish the position of the sound source using only the ITDs as a determinant, because the sound waves at both ears would be in exactly the same phase, and would not be distinguishable except for the 0.5-millisecond difference in the onset of the oscillations. These problems occur also for higher frequencies and smaller periods.

In this scenario, the basis of the *duplex theory* may be understood: for certain frequencies, the ILDs seem to be the more reliable parameter for left-right localization, while for others the ITDs can be considered more effective. It has been calculated that for frequencies above 725 Hz (with

periods shorter than 1.38 ms, when the ITDs would start to create problems) the sound localization in terms of left-right detection is accomplished through considering mainly the ILDs, while for frequencies below 725 Hz (with wavelengths greater than 44 cm, when the ILDs would start to become irrelevant) the ITDs constitute the most important parameter.

The Cone of Confusion

Thus far, it has been explained how it is possible to determine the provenance of a sound from left and from right on the horizontal plane, but the question remains: how is it possible to differentiate sound sources placed in front of or behind the listener, or above or below? If two sound sources are located at 60° and 120° of azimuth (two positions that are specular, referring to the frontal plane), the interaural differences in the signals coming from them will be exactly the same. This problem is called the *Cone of Confusion* because plotting all of the positions of sound sources with the same interaural differences would generate the shape of a cone, with the head position as the apex (*see* Figure 4), and can be resolved merely with the help of a third localization cue, the direction-dependent filtering.

Figure 4. The cone of confusion. For each of the sound sources located in the circle (which can be considered as the base of a cone), r1 and r2 are respectively equal, therefore the interaural differences are exactly the same (von Hornbostel, 1920, redrawn from Blauert, 1996)

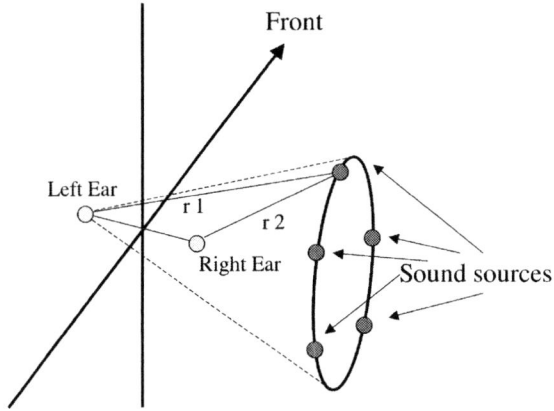

The Direction-Dependent Filtering and the Head Related Transfer Function

As said previously, the pinna has two main functions: while the first, sound gathering, can readily be understood, the second, direction-dependent filtering, appears more complicated. The dimensions of the pinna are far too small, if compared with the wavelengths of many audible frequencies, for it to function as a simple sound reflector: the dimensions of its cavities, instead, are comparable to $\lambda/4$ (where λ is the wavelength of a given frequency) of a large number of frequencies, and these can easily become sound resonators for sound waves coming from specific directions. Therefore, inside the pinna the sound is modified by reflections, refractions, interferences, and resonances that are activated for specific frequencies and, most significantly, for the incident angles of specific sound waves, hence the name *Direction-Dependent Filtering* (Batteau, 1967-1968; Shaw, 1968).

In order to abstract and simplify the principle, an empty bottle serves as an example: when blowing with the mouth close to the neck of the bottle, it is possible to generate a resonance whose frequency is determined only by the volume of the air inside the bottle and the dimensions of the bottle neck and not, for example, by the material of the bottle itself. However, in order to generate the resonance, the position of the mouth, the force of the air blown, and the inclination of the bottle need to be specifically selected (a choice that is usually achieved by trying different positions and speeds). It must be asked what would happen if we had a 'special' bottle, with more necks, more openings, and more cavities? There would then be many more resonances, and many more combinations of positions and blowing speeds in order to activate them.

This is what happens if the pinna is considered as a complex resonator: multiple resonances can be activated depending on the incidence of the sound wave. All of the resonances generated by the reflections and refractions on the shoulders and on the torso of the listener need to be considered, too. The sound input into the auditory canal is therefore modified through complex frequency filters that change their shapes depending on the position of the sound source. A couple of examples: for sound sources located above, there is usually a strong resonance around the 8000 Hz mark, while for the front-back detection of sound sources placed in the horizontal plane, the weight of resonances and absorptions at 3000 Hz and 5000 Hz are essential in order not to 'fall' into the cone of confusion effect. The combination of these filters, together with the interaural differences, is called the *Head Related Transfer Function*.

Individual and General Attributes of the HRTF

While some of the HRTF parameters can be considered constant for all human hearing systems, some others need to be considered individually, because they depend on idiosyncratic physiological differences between human beings, such as the shape of the pinna and the circumference of the head. When an HRTF is simulated, or binaural recordings are performed (see previous sections), both individual and general attributes should be considered, thus a simulation should be performed individually for each subject. All of the individual HRTF attributes mentioned in this chapter will be simplified and approximated, and for the current purposes it will be assumed that every single human being has the same identically shaped outer ear (and, equally, all other directional filtering elements such as the head, shoulders, and torso). For more information about this topic, *see* Møller (1996) and Katz (1996).

The Role of the Head Movements
Even if this is not directly relevant to what will be claimed in the following chapters, it is essential to underline that head movements are extremely important to sound source localization, and most of all for the front-back and up-down discriminations when the cone of confusion issue needs to be resolved. In fact, turning the head left-right or up-down causes important changes in the soundscape and in the relative positions of the sound sources, generating further information which can be considered as particularly relevant to correct sound source localization. As an example, if a sound source is located in the horizontal plane at 60° of azimuth, it could be easily confused with another located in the same plane at 120° of azimuth; on rotating the head to the left, if the sound source is really located at 60° of azimuth it will move towards the front (0° of azimuth); otherwise, if it is located at 120° of azimuth, it would move towards the left (90° of azimuth). Therefore, in critical situations, for example, with narrowband stimuli (sound stimuli with a narrow frequency extension, which can be filtered only with difficulty on the whole frequency scale) or in the case of a localization task in a particularly reverberant environment, when the localization cues cannot be precisely analyzed, the movements of the head are essential for a proper sound source localization.

Sound Source Localization on Three Planes

After this brief overview of the interaural differences and of direction-dependent filtering, it should now be understood how sound source localization is performed for a source placed in a three-dimensional soundscape. In order to simplify the problem, the mechanisms for the localization of the sound sources when these are placed in just one plane will now be addressed.

Sound Source Localization in the Horizontal Plane
In this specific case, the presence of dichotic stimuli is highly probable: in fact, solely for sound sources located exactly at 0° and 180° of azimuth (and, of course, 0° of elevation, simply because it is in the horizontal plane) a diotic stimulus can be input into the hearing system. Thus, for left-right discrimination (0°/180° to 180°/360°) the ITDs and ILDs are used, while for front-back judgement (90°/270° to 270°/90°) the DDF carries major importance. Due to the facts that all the three localization cues can be used and that more often a sound source would be located here (taking speech as an example), the horizontal is the plane where sound source localization performances are higher.

Sound Sources Localization in the Median Plane
While in the horizontal plane dichotic stimuli are far more common, whereas if a sound source is located in the median plane the sound will certainly reach the ear as a diotic stimulus: in fact, if the asymmetries of the head are ignored, the distances and the incidence angles between a sound source located in this plane and both ears are always the same. Therefore, the only parameter applicable by our hearing system is the DDF. For these reasons, the median one is the plane with which our hearing system has more difficulties in terms of sound source localization performances.

Sound Sources Localization in the Vertical or Frontal Plane

The frontal plane can be considered as the 'vertical version' of the horizontal plane: diotic stimuli are present only for sound sources located at 90° and 270° of elevation, whilst for all other positions a dichotic stimulus would reach the hearing system. Up-down discriminations are performed through analyzing the DDF. The localization accuracy on this plane is not as precise as it is in the horizontal plane, while certainly not as vague as for the median plane.

Simulation of the Spatial Hearing

A short yet comprehensive overview of the mechanisms of spatial hearing has been provided. How our hearing system can localize a sound source in a 3D soundscape has been examined, and the specific attributes of the sound input into our ear applied in order to achieve this have been noted. Therefore, phenomena that exist in nature have been described: every sound coming from every position is filtered by our body, our head, and our outer ear, and thanks to these it can be localized by our hearing system. The goal now becomes how to simulate this using a computer.

Spatial Hearing through Headphones

What does 'sound spatialization' mean? It could be considered as a synæsthesia, because the concept of space mainly refers to the sense of sight, while the word 'sound' is, of course, related to hearing; nevertheless, these terms may be associated, and such an association creates a new concept: the soundscape. An attempt to define the concept of soundscape could start with a simple question: what is the difference between a real listening to the sounds that we hear every day, for example, walking down the street, and the listening to a CD-DA (Compact Disc Digital Audio) of the same though recorded sounds from any stereo reproduction system? Independently of the origin of the stimuli, in everyday life we listen to sounds coming from sources located in a 3D space: we are in the middle of an immersive 3D soundscape, where for each sound we are more or less able to detect the position of its source. When we listen, instead, to a CD-DA, the sound is presented frontally. Using a standard stereo reproduction set-up (with the two loudspeakers placed in two of the angles of an equilateral triangle and the listener placed in the third), the sound that reaches our ear is not 3D, but mono-dimensional, in the sense that each sound source can be localized in one or the other loudspeaker, or on a imaginary line between the two. In fact, a sound that is played at an equal level from both loudspeakers would be localized exactly between the two, thanks to psycho-acoustic mechanisms that will not be discussed in this chapter (for more information, *see* Moore, 2003 and Blauert, 1996).

The main difference between the two listening situations has thus been defined: in the real one, a 3D soundscape is presented to our hearing system, while during a CD-DA playback the soundscape is mono-dimensional and frontal (of course, in this case the various interactions with the room where the CD-DA is played, interactions that can generate reflections coming from all directions and therefore stimulate the perception of a more spacious soundscape, are not now considered). Taking for a moment the playback of recorded sound: adding, for example, two loudspeakers behind the listener could help in coming closer to the experience of a 3D soundscape. If the four loudspeakers are placed at the corners of a square, with the listener located exactly in the centre, the sounds can be spatialized within a plane, thus a bi-dimensional soundscape can be created. In fact, changing the weights (the levels) and the sound contents of the signals sent to the four channels, sound sources can be virtually located within the square described by the loudspeakers (again, for more information *see* Moore, 2003 and Blauert, 1996). This is called *Quadraphonic* reproduction system (*Quad*), and it was the starting point for more famous and recent *surround* systems such as *Dolby Digital* (5.1, 7.1, etc.) or *THX Surround*: even if they more closely approach a proper 3D soundscape simulation, giving the impression of sound sources spatialized within a plane, they nevertheless lack one dimension.

What if four other loudspeakers are added above, generating a cube with eight loudspeakers at the apexes and the listener placed exactly in the middle? Using this specific system a third dimension (height) can be simulated. Now for some examples:

- If a sound is played at the same level from two frontal loudspeakers, the virtual sound source will be located in the middle of the line between the two loudspeakers. Through introducing differences in level between the two loudspeakers, the sound source can be moved into every position along that line.
- If a sound is played at the same level from four loudspeakers placed at the corners of a square, with the listener located in the centre, the virtual sound source will be located in the middle of the square described by the four loudspeakers (position of the listener). Through introducing differences in level among the four loudspeakers, the sound source can be moved into every position within that plane.
- If a sound is played at the same level from eight loudspeakers placed at the apexes of a cube, with the listener located in the centre, the virtual sound source will be located in the middle of the cube described by the eight loudspeakers (again, the position of the listener). Through introducing differences in level among the eight loudspeakers, the sound source can be moved into every position within that space.
- With multiple sounds played at different levels from the eight loudspeakers, a complex 3D soundscape can be generated.

Using this reproduction system with eight loudspeakers and, of course, a proper sound spatialization engine for the weighting of the respective signals in the eight channels, a 3D soundscape can be simulated (for more information on 3D spatialization techniques over loudspeakers' array, *see* Gerzon, 1973 and Pulkki, 1997). However, is this the only way to generate artificially a 3D soundscape? We used eight loudspeakers, therefore eight channels, in order to be able virtually to locate a sound source in a 3D space, but – don't we have only two ears each?

In the previous sections of this chapter, the mechanisms of the spatial hearing are well described: sound sources can be located in the three dimensions by using just two receivers. Thus, having sound reaching both ears by the use of a simple pair of headphones, it could be possible to eliminate complex and expensive multi-channel loudspeaker systems. Yet how can we simulate a three-dimensional soundscape when using only two channels? An answer to this question will be found in the following sections.

Note from the author:

It could be noticed that, in order to generate a bi-dimensional soundscape, it is not in fact essential to have four loudspeakers, as three placed at the corners of a triangle, with the listener in the centre, are sufficient; also, for 3D soundscape simulation, four loudspeakers placed at the apexes of a tetrahedron with the listener in the middle would be enough. This is absolutely true, yet I have tried to make the examples as simple as possible, and using an even number of loudspeakers seemed to aid clarity.

Dummy Head and in-Ear Microphones: Binaural Recordings

The first and easiest way to simulate a 3D soundscape through headphones is simply to perform a binaural recording using a dummy head microphone or a pair of in-ear microphones. A dummy head microphone can easily be made by taking a head mannequin with the dimensions of an average adult human head, with sufficiently precise pinna reproductions (in order to preserve the resonances, refractions and absorptions typical of a human HRTF), and placing two miniature omni-directional microphones at the entrances to each of the auditory canals. The recordings made through placing this device in the middle of a 3D soundscape, then played back through a pair of headphones, will give to the listener the impression of being exactly in the position of the dummy head, with sounds coming from every direction: left-right, front-back, and up-down. This result can be obtained even using the so-called 'in-ear microphones', which are simply two miniature omni-directional microphones placed inside the auditory canals of a subject, positioned at the entrance to the canal itself.

The fact that the microphones should be placed at the exact entrance of the auditory canal, and not at the position of the eardrum, may need some explanation: as happens with the pinna, even the auditory canal has its own resonances, while different studies (*see* Hammershoi, 1995) showed that these are not dependent on the angle of incidence of the input sound. Therefore, all of the localization cues are already present in the signal at the entrance to the ear canal, thus the microphone can be placed in that position.

A further observation needs to be made about the use of headphones for the reproduction of binaurally recorded sounds (and, as will be seen here, even for the reproduction of binaurally synthesized signals): when a stereo sound is played back through two frontal loudspeakers, the signal coming from the left loudspeaker will reach both the left and the right ears, exactly as will the signal coming from the right loudspeaker. This phenomena is called *crosstalk* and, in the case of binaural sound reproduction, it would generate many unwanted situations. In fact, when playing binaural sounds, it is really important that

the signal of the left channel should reach only the left ear, and that of the right channel only the right ear. Thus, the use of headphones is essential in this specific case: there exist systems that can be used to reproduce binaural sounds through stereo loudspeakers (*transaural* and *crosstalk cancellation* systems; as an example, *see* Tokuno, 1996), although they will not be discussed in this chapter.

The obvious problems linked to binaural recordings lie in the fact that the recorded 3D soundscape needs to be created in a real environment, using real sound sources or loudspeakers, and that the recorded scene cannot be modified after the recording. For these reasons, it cannot be considered a proper 3D sound simulation technique – simply a 3D sound recording technique.

Impulse Response and Digital Convolution

In order to simulate the directional mechanisms of the outer ear, this needs to be seen as a 'system', considering its mathematical definition. Given two families of signals, F1 and F2, a system is an apparatus that can transform each F1 signal into an F2 signal. It can be seen as a 'black box', the behaviour of which is described by the transform law S: F1 → F2.

In environmental acoustics, a system is a room or a hall; in a recording studio, a system is an outboard effect; in an orchestra, a system is a musical instrument. As regards spatial hearing, a system is the ensemble of all of those elements that participate in the modifications of the incoming sound depending on its incidence angle, thus the shoulders, the torso, the head, and the pinna. Without entering too deeply into the mathematical domain (if so, we would have to face other mathematical definitions such as *linearity* and *time-invariance*; for more information on these topics, *see* Rabiner, 1975), it may be stated that a system can be unequivocally described by its response to a specific signal: the impulse. This specific signal, known also as the *Dirac δ* or *δ(t)*, can be seen as a rectangle with an infinitesimal base and an infinite height: thinking about the sound, it is… an impulse, shorter than a clap, like a simple 'click'. A particular characteristic of this signal is that its frequency content is represented by a flat line running parallel to the X axis, related to frequency: it can therefore be gathered that the Dirac δ contains all the frequencies at the same amplitude.

If an impulse is reproduced within a system, the recording of the impulse itself, having passed through and been processed by the system, is called the Impulse Response, or IR, and it describes unequivocally the response of the system to all of the possible input signals. The system can then be simulated performing a simple mathematical operation between its IR and any signal that needs to be filtered. This operation is called *convolution*, and in the digital domain it can be seen as a series of multiplications between the samples of both the IR and the signal to be filtered. Here, it should be specified that the digital convolution is not a straightforward sample-by-sample multiplication of the two signals (*see* Picinali, 2006).

Therefore, knowing from example the IR of a music hall and convolving with it a musical signal, for example, a piano sonata, the recording can be listened to as it was played inside that music hall itself! This simulation becomes somewhat more complicated when multi-dimensional soundscapes are considered: as was said before, the spatial hearing system modifies the sounds differently according to the respective positions of their sound sources, thus the IR of this system needs to be recorded for all of the positions of the sound sources that need to be simulated. Assuming a sphere around the head of a listener or of a dummy head, equidistantly spaced positions can be sampled on its surface.

Reproducing an IR in each of the sampled positions, and recording them from the two microphones of the dummy head placed in the middle of the sphere, a database of IR would be created. Such a specific database is called a *Head Related Impulse Response (HRIR)* database (for an example, *see* Algazi, 2001); it represents the behaviour of the spatial hearing system for each sound source position around the head at a given distance (in this specific case, the distance is given by the diameter of the sphere where positions have been sampled).

Therefore, in order to spatialize a sound in a three-dimensional soundscape, an IR corresponding to the position of the sound source to be simulated (the azimuth and elevation angles) needs to be extracted from the HRIR and convolved with the signal to be spatialized. The resulting processed stereo signal, listened to through headphones, will give to the listener the impression of a sound coming from the desired position.

While this simulation technique seems relatively easy to realize, there are several problems linked to the IR extraction (again, *see* Picinali, 2006) and to the fact that performing convolution in real-time is a 'computationally heavy' operation, possible only with fast and powerful computers.

Isolation and Individual Simulation of the Three Localization Cues

Another approach to the 3D soundscape simulation over headphones could be to dissociate the three localization cues and to simulate them individually (similar approaches can be found in the literature discussing HRIR interpolation techniques for the simulation of moving sound sources; as an example *see* Hwang, 2006).

At first sight, the ILDs seem to be the easier to simulate. When sounds are mixed on a standard mixing desk, there is usually available a potentiometer, a *panpot*, to regulate the levels of the signals sent to the left or to the right output. It simply creates differences in level between the two channels, in order to localize the sound not in the centre, but in any position between the two speakers. However, as has already been seen in previous sections within this chapter, the ILDs form a frequency-dependent parameter, thus they vary throughout the audible frequency range. For this reason, ILDs cannot be simulated simply by reducing the level of a signal sent to the right or to the left channel; an equalization filter needs to be implemented in order to attenuate increasingly the high frequencies rather than the low ones (in fact, the slope of the filter needs to change according to the sound source position to be simulated), trying to follow the ILDs values given by the diagram shown in Figure 3. An equalization filter can be implemented even for the simulation of the DDF, both for the left and the right channels: the coefficients of the filter (the values of enhancement or reduction for each frequency band to be applied to the signal) need to change according to the position of the sound source to be simulated, in an attempt to follow the specific reflections, refractions, and resonances generated by the pinna and by the other directional filtering elements (head, shoulders, and torso).

The ITDs can easily be simulated using a simple delay: if the sound source is located in the right hemisphere, the delay would be placed on the signal sent to the left ear, while if the source is placed in the left hemisphere, the right ear signal is the one to be delayed.

Summing up those three individual simulations, calibrated specifically for the sound source to be simulated, binaural spatialization can be performed for each desired sound, yet two problems may be still generated by this binaural simulation method:

How is it possible to separate the ILDs from the DDF? If each is frequency dependent, through analysing the HRTF we would find that the ILDs are perfectly mixed with the DDF: we would be able to measure only the effect of the sum of the two, and not of one individually. In fact, referring to the psycho-acoustic properties of spatial hearing, there are no significant differences between these two parameters; they both modify the incoming signals, filtering them in frequency, thus they can be considered as non-divisible.

The implementation of equalization filters for the simulation of ILDs and DDF means that approximations need to be made in terms of the frequency response of the spatial hearing system. Even if studies have demonstrated how these can be considered as non-influential in terms of spatialization accuracy (*see* Kistler, 1992), binaural spatialization algorithms based on simple equalization filters are far from providing a high quality spatial sound perception.

Nevertheless, the flexibility of this approach makes it the more suitable for what will be performed in the following section: we thus have to approximate the ILDs as a frequency-independent parameter, and the DDF as that which carries all of the frequency content of the HRTF.

A Multi-Dimensional Trip

In the book *Flatland* (*see* Abbott, 1999), the author imagines the trip of a square, coming from a bi-dimensional word called Flatland, into other worlds with different numbers of dimensions: from the zero-dimensional world, a point with no dimensions and with all the inhabitants placed exactly in the same position, to the mono-dimensional world, a long line where segments lie one beside the other, and finally to the three-dimensional world, called *Spaceland*. As regards figures, the book obviously focuses on the visual stimulus: each added dimension is described as a change in terms of view, passing from single points to lines, and then to bi-dimensional figures. No reference is made to the hearing stimulus. Of course, Abbott cannot be blamed for having ignored multi-dimensional hearing. This was probably not among his intentions, and it is difficult not to admit that the visual sense is the most important for the human being in terms of environmental information gathering. Nevertheless, now that we have seen how to simulate three-dimensional soundscapes over headphones, an attempt at a 'multi-dimensional sound trip' could indeed be made.

Zero Dimensions: The Point

What is a zero-dimensional soundscape, and how can it be created?

Answering the first part of the question, it may be stated that a zero-dimensional soundscape can exist when the sound has absolutely no attributes providing us with information about the location of that sound in the space. From this, it is possible to assume that in a zero-dimensional soundscape the sound needs to be located in an undefined position or, even better, in a 'neutral' position, yet what about the middle of the head, in the centre of our auditory space?

Having no spatial attributes results in the situation that no localization cues can be simulated. There can be no interaural intensity or time differences, and no direction-dependent filtering. It is a purely diotic stimulus, with the same amplitude and arrival time at the two ears, and with no frequency filtering at all. To offer a practical example, a sinusoidal signal can be considered (a sinusoid is a pure tone, containing only one frequency and which, therefore, cannot be frequency filtered, but merely varied in its amplitude), presented through headphones at the same intensity and phase at both the ears.

No interaural differences, no direction-dependent filtering simulation, only a single pure tone: these result in zero dimensions, or the point.

One Dimension: The Line

Starting from the zero-dimensional soundfield and proceeding with an additive approach, we could create a mono-dimensional soundfield simply through introducing a spatial attribute within the selected signal, therefore modifying it and making it partially localizable.

The use of the expression 'partially localizable' has a precise meaning: visually talking, and referring to *Flatland*, a mono-dimensional world is composed of elements located along a line, therefore localizable in only left-right, front-back,

or up-down dimensions – but not in a combination of these. If we pass to the auditory domain, a better representation of that line is given by the phenomenon of sound lateralization. In that case, the sound source is not localized within a three-dimensional soundscape, but lateralized inside our head along an imaginary line drawn from one ear to the other.

Starting from the point, a single diotic pure tone, the first dimension can be added introducing Interaural Level Differences (ILDs) within that signal, transforming it into a dichotic stimulus. Two different pure tones could be created, both with different ILDs, therefore both lateralized along a line passing from the left ear to the right.

An objection could be made that the first dimension could be created through introducing ITDs instead of ILDs. The fact that, historically, the ILDs is the first localization cue to have been simulated and it is the most frequently used in terms of sound reproduction must be stated. Take again the example of the channel strip of a sound mixing console, with its *panpot* potentiometer: it moves the sounds from left to right simply changing the amplitude of the left or the right channel, simulating therefore the ILDs.

Only interaural level differences, no time differences between the signals at the two ears, nor direction-dependent filtering simulation, merely two pure tones: these result in one dimension, or the line.

Two Dimensions: The Plane

Proceeding with the method proposed before, the second dimension could easily be added introducing Interaural Time Differences (ITDs) within the stimuli sent to the two ears.

At this point, further objections might come from readers: the introduction of the ITDs has nothing to do with the addition of a second dimension and, most of all, follows not at all the visual notion of a line becoming a square.

These objections are all completely justified, yet... The important event to happen perceptually after adding the ITDs is that the sounds begin to be localized outside the head, and are no longer lateralized between the ears. When we listen to a standard stereo signal from a CD-DA or MP3 player, the sensation we have is that the sound sources are lateralized within our head: this effect is called *Inside the Head Locatedness* (IHL, *see* Blauert, 1996, and Moore, 2003), and is due to the fact that the only localization cue simulated in this specific case is the ILDs. Introducing differences in the arrival time of the signals at the two ears, the soundscape seems to expand outside our head, gaining a dimension which cannot precisely be defined as height or depth, but which obviously gives a more complex spatial characterization to the simulated soundscape.

This added impression could easily be compared with the gain of a new dimension: starting from the mono-dimensional soundscape, the two pure tones could then become more complex in terms of frequency content, for example, adding simple harmonic frequencies, in order to bring them closer to a 'real' sound – and of course introducing ILDs and ITDs.

Only interaural level and time differences, no direction-dependent filtering simulation, two 'slightly complex' tones: these result in two dimensions, or the plane.

For those of you conversant with computer music science and who wish to gain a clearer idea of the sensation described in the previous section, a simple experiment can be tried. Using audio editing software (a good, simple and cost-free one can be found in http://audacity.sourceforge.net), load the same mono audiofile into two different tracks, with the first track sent only to the left channel, and the second only to the right. Now, boost the second track with a +6dB gain (turning up the volume fader of that specific track until the +6dB level), and listen to the sound through a pair of headphones. It will be perceived that the sound source should clearly be lateralized to the right, remaining still inside our head. Next, delay the first track of 32 samples (this can be done using a VST plugin or by simply moving forward the audiofile within the track), which at 44.1 kHz corresponds to a delay of approximately 0.7 ms. The sound source would still stay on the right side of our head, but the sensation would be that the sound moves outside the head itself, passing from a simple sound lateralization sensation to a more complex, even if not precisely definable, sound localization one.

It needs to be said that this effect is highly subjective, and therefore it can differ according to the person who is listening to the spatialized audio. The only objective sensation is that through adding the ITDs, the sound gains in terms of spatial representation.

Three Dimensions: Space

The passage from a bi-dimensional to a three-dimensional soundscape should now be an easy task: after the introduction of a simulation of the last localization cue, the Direction-Dependent Filtering (DDF), the two 'slightly complex' tones can be substituted by various recordings of real sounds, and be freely spatialized within the three dimensions.

As learned from the preceding sections, this can easily be achieved performing a convolution between the signals to be spatialized and different HRIRs correspondent to the positions where the different sound sources want to be simulated, and performing what is called *binaural spatialization*.

Interaural level and time differences, direction-dependent filtering simulation, multiple 'real sounds' recordings spatialized on left-right, front-back and up-down positions: these result in three dimensions, or space.

The Fourth Dimension

Summarising what has been achieved in the previous sections, from a single diotic pure tone representing a zero-dimensional soundscape, a point, we introduced, firstly, ILDs in order to simulate a mono-dimensional soundscape, a line; then ITDs for a bi-dimensional one, the plane, and finally DDF for a three-dimensional one, space.

At this point, how is it possible to create an auditory fourth dimension within our simulation? How can it be presented in relation to a visual fourth dimension not precisely established? Could we proceed with an additive approach, therefore adding certain new parameters to the sound stimulus, or could something else be done?

An Idea for the Visual Fourth Dimension

In the first chapters of *Flatland*, the writer, a Square coming from Flatland, describes the life in its bi-dimensional world: the population is composed of triangles, squares and more complex multi-lateral figures, yet each individual is visually perceived as a line, changing its length depending on the orientation of the figure itself. Still, it remains a simple line. Within its explanation of its bi-dimensional world, the Square tries to make a comparison with our three-dimensional perception: sighting a island from a boat in the middle of the sea, it is not possible to see all of the bays and headlands along its coast, unless light gives us this information through different shadowing and reflecting effects. In a three-dimensional world, sight is only bi-dimensional, and the information that can be gathered about the third dimension are given by indirect factors, such as a stereographic view or light effects.

Similarly, in a bi-dimensional world, sight is mono-dimensional.

Therefore, in Flatland only half of the angles of an individual can be viewed in any one single moment: the chance of seeing all of the angles of the figure simultaneously seems absolutely impossible, as much as for us it is impossible to see both the face and the back of a person, or the external and the internal part of a house, in one single glimpse (of course, the mounting of multiple videos from camcorder recordings, the use of mirrors, or other stratagems are not contemplated in this explanation).

Only in the exact moment when the Square is conducted into a three-dimensional world, therefore is picked up from the plane where he lies and is lifted into the third dimension, is it possible for him to see the whole figures of its citizens, their angles, and their 'contents', all at the same time.

Given such terms, what could be our hypothetical visual impression if we are 'lifted' (there is no other verb to describe properly this transformation) into the fourth dimension?

An idea for a visual fourth dimension is that in gaining it, there may be the possibility of the coexistence of different elements that cannot coexist in our three-dimensional world: we might be able to see the façade of a building and at the same time its interior features, to invert a sphere as we would turn a sock inside out, and many other innovations.

From a mathematical point of view, these are possible and relatively simple processes while, visually speaking, it is difficult to gain a perception of what might constitute the fourth dimension. Nevertheless, even as for the Square it was impossible to conceive a third dimension, for us it is impossible to conceive a fourth one. We can merely attempt to guess what it might be – and this is exactly what we are going to do, from an auditory perspective, in the next section.

An Idea for the Auditory Fourth Dimension

A four-dimensional world may not only graphically, but also idealogically be seen as a place where there is the coexistence of stimuli that could not be present simultaneously in a three-dimensional world. From an auditory point of view the fourth dimension can be considered as the coexistence of auditory stimuli, as well as of single parameters within the same sound, which could not be present at the same time in nature.

To link it with the main topic of this chapter, 3D sound simulation, the fourth dimension could be created through introducing incoherence among the three localization cues. While adding a localization cue could, psycho-acoustically, mean the adding of one more dimension, creating incoherence between the mechanisms for the sound localization could signal the passage to a fourth dimension.

The use of a fully controllable reproduction system such as a pair of headphones (independent of the reproduction environment and its acoustical properties, and perfectly separated between the two channels) seems to be the perfect medium for our purposes. Repeated sound spatialization could be performed for the same sound source, using different azimuth and elevation degrees for each localization cue. The sound of a bee could, for example, be spatialized using azimuth degrees on the left of between 180° and 360° for the ILDs simulation, and azimuth degrees on the right of between 0° and 180° for the ITDs simulation, recreating then the DDF incoherently between the two ears (i.e., a DDF of a source above our head for the left ear, and below for the right one).

The perception of such a soundscape is certainly highly subjective; different individuals may experience completely different sensations. Nevertheless, the attainable goal of this kind of 'synthesis-simulation' is to create a soundscape that is spatially unnatural, completely new to our hearing system, and – most of all – which it is not possible to hear in a natural auditory environment.

This is as much as it is not possible to see both the total exterior of a closed box and what is inside it at the same time.

CONCLUSION

Simulation and alteration of nature: these are how nature can be simulated through the computer, and then altered, in order to create entities that do and possibly cannot exist. Returning to the question posed at the beginning, the example brought with this chapter wants to be an aleatory answer. If it is challenging to try to simulate precisely the spatial hearing of our everyday life, after having achieved this it is surely stimulating to take a step forward, and to try to modify the simulated soundscape in ways that are not possible within the 'real world' if we stick to the rules of nature.

Personally, my conviction is that the real potential of computer science when applied to music is to enjoy oneself, and this is basically what I have tried to convey in this chapter.

It needs be underlined that this does not an attempt to be a scientific interpretation of the fourth dimension. The auditory simulation of multiple dimensions through the use of one or more localization cues (the link between localization cue and auditory dimension is utterly aleatory) can be seen as a simple notion of how to use the potential of computers for the simulation and the alteration of nature.

REFERNCES

Algazi, V.R., Duda, R.O., Thompson, D.M. & Avedano, C. (2001). *The CIPIC HRTF Database*. Paper presented at the IEEE Workshop on Applications of Signal Processing to Audio and Acoustics, New York, USA.

Abbott, E. (1999, c1884). *Flatland: A romance of many dimensions*. London, UK: Penguin.

Blauert, J. (1996). *Spatial Hearing, the Psychophysic of Human Sound Localization*. Cambridge, Mass, USA: The MIT Press Cambridge.

Batteau, D.W. (1967). The role of the pinna in human localization. *Proc. Roy Soc, London, B168*, 158-180.

Batteau, D.W. (1968). Listening with the naked ear. In S.J. Freedman (Ed.), *The neuropsychology of spatially oriented behavior* (pp. 109-133). Homewood, IL: Dorse Press.

Feddersen, W.E., Sandel, T.T., Teas D.C., & Jeffress, L.A. (1957). Localization of high-frequency tones. *Journal of the Acoustical Society of America, 29*, 988-991.

Gerzon, M. (1974). Periphony: With-height sound reproduction. *Journal of the Audio Engineering Society, 21*(1/2), 2-10.

Hammershøi, D. (1995). *Binaural Technique: a Method of True 3D Sound Reproduction*. PhD thesis, Aalborg Universitetsforlag, Denmark.

Hwang, S., & Park, Y. (2006). *Time delay estimation from HRTFs and HRIRs*. Paper presented at the 8th International Conference of Motion and Vibration Control (MOVIC 2006), Kaist, Daejeon, Korea.

Katz, F.G.B. (1996). New approach for obtaining individualized head-related transfer functions. *Journal of the Acoustical Society of America, 100*, 2609.

Kistler, D., & Wightman, F. (1992). A model of head-related transfer functions based on principal components analysis and minimum-phase reconstruction. *Journal of the Acoustical Society of America, 91*, 1637-1647.

Møller, H., Sørensen, M.F., Jensen, C.B., & Hammershøi, D. (1996). Binaural Technique: Do We Need Individual Recordings? *Journal of the Audio Engineering Society, 44*(6), 451/469.

Moore, B.C.J. (2003). *An Introduction to the Psychology of Hearing*. London, UK: Academic Press.

Picinali, L. (2006). *Techniques for the extraction of the impulse response of a linear and time-invariant system*. Paper presented at the DMRN Doctoral Research Conference, University of London, UK.

Pulkki, V. (1997). Virtual sound source positioning using vector based amplitude panning. *Journal of the Audio Engineering Society, 45*(6), 456-466.

Rabiner, L.R., & Gold, B. (1975). *Theory and Aplication of Digital Signal Processing*. Englewood Cliffs, NJ, USA: Prentice Hall, INC.

Rayleigh, L. (1907). On our perception of sound direction. *Philosophical Magazine, 13*, 214-232

Shaw, E.A.G., & Teranishi, R. (1968). Sound pressure generated in an external-ear replica and real human ears by a nearby sound source. *Journal of the Acoustical Society of America, 44*, 240-249.

Tokuno, H., Hamada, H., Kirkeby, O., & Nelson, P. (1996). Binaural sound reproduction in a stereo dipole system. *Journal of the Acoustical Society of America, 100*, 2700.

von Hornbostel, E.M., & Wertheimer, M. (1920). Über die Vahrnehmung der Schallrichtung [On the perception of the direction of sound]. *Sitzungsber. Akad. Wiss. Berlin*, (pp. 388-396).

Yost, W.A. (2000). *Fundamentals of Hearing: An Introduction*. San Diego, California, USA: Academic Press.

KEY TERMS

Binaural: Relating or involving a sound stimulus presented to both ears. In a more general sense, the expression 'binaural spatialization' refers to the technique for the simulation of 3D soundfields over standard stereo headphones. When the binaural signals, previously modified by the mean of special processes called 'cross-talk' cancellation filters, are played through loudspeakers, the term used to define this situation is *transaural*.

Convolution: A fundamental mathematical function that involves sampling multiplications in multi-dimensional matrices. In the case of digital signal processing, convolution can be seen as a particular kind of multiplication between two vectors.

Fourth Dimension: There is no unequivocal definition of the fourth dimension. For most, scientists included, time is considered to be the fourth dimension, yet in this chapter an attempt at a different definition is executed.

Head Related Transfer Function (HRTF): The transfer function of the external hearing system and of all the other elements that contribute to the directional modifications of the signals input to the hearing system (torso and shoulders).

Impulse Response (IR): The recording of an impulse signal that has passed through and been processed by a given system. It unequivocally describes the behaviour of that specific system to all of the possible input signals.

Localization Cues: Specific attributes of the sound event that are used by the hearing system in order to establish the position of a sound source in a 3D soundscape.

Sound Localization: The judgement of the specific location of a sound source.

Soundscape: An acoustical environment or an environment created by sound.

Sound Spatialization: An ensemble of sound processing techniques oriented towards the simulation of multi-dimensional soundscapes.

Three-Dimensional Soundscape: The ensemble of the sounds heard in a particular location where the sound sources are placed in the three dimensions (x, y and z, or length, height and width).

Chapter VIII
Broken Cinema:
The Eye and Hand in a Time–Based Art

Raphael DiLuzio
University of Maine, USA

ABSTRACT

This is a guide for working with a visual art form using a digital time-based medium. This chapter will provide an overview of the necessary theories; processes, concepts and most important the "elements," needed to create expressive visual artworks through the technologies associated with this visual art form. It will examine in some detail how people can effectively visually communicate and express our artistic ideas and intentions through a digitally time-based medium. We have reached a point in time-based visual art where the tools and technology have matured enough to allow us to focus our attention more on process and concept rather than specific hardware or software tools. Please understand that more than a theorist, the author is a practitioner of the art form that he will discuss in the following pages. As such, at times, the author will be using his own empirical experience to support arguments in combination with or in place of the opinions of others. The ideas this chapter would like to address are rather complex and there is not enough space in a single chapter to put them forth in their entirety. Therefore, by necessity the atuhor will have to be brief, somewhat simplified, and a bit reductive. Nonetheless, he hopes it provides a basic understanding of the concepts and principles for creating visual art with a digital time-based medium.

"To See a World..."

*To see a World in a Grain of Sand
And a Heaven in a Wild Flower,
Hold Infinity in the palm of your hand
And Eternity in an hour.*

(Fragments from "Auguries of Innocence" William Blake)

INTRODUCTION

Visual art forms continue to evolve due to new innovations and the combination of traditional and emerging concepts, technologies, and processes creating new mediums and tools of artistic expression. How the artist works with these new mediums and how we, the viewer (or participant) understand meaning through visual perception is simultaneously evolving as well. The way in which the visual artists creates new works relative to changes in technology and the way we extract meaning from these new visual art forms and understand their inherent narratives in their new appearances is continually adapting.

During the Italian Renaissance, science and art would often intermingle, overlap and sometimes be engaged in by practitioners of both disciplines. Leonardo da Vinci was a perfect example of the type of creative artist who was able to work in both areas and combine them. Over time a divide occurred between these two disciplines that separated art from many of the technologies and innovations of science. In the ensuing five hundred years, from the Renaissance to the later part of the nineteenth century, the creative technology available for artists to use changed very little.

Driven by artistic urges and developing naturally out existing practices and technologies, innovations in printmaking and still photography, such as varnishing advanced the technique in oil painting of glazing and the camera obscura that was the predecessor of the photograph. Unlike prior artistic forms, the historical development of film, that is to say the motion picture, as a creative medium was unique. The medium did not arise from the needs or creative explorations of the artists. Instead, the invention of a new technology "gave rise to the discovery and gradual perfection of a new art" (Panfosky, 1997). No sooner had this art form been established then the reverse became true and much like other visual art forms the creative desires and explorations of art-makers spurred innovation and advancement of the art form itself.

Initially, the invention of the computer seemed to have little or nothing to do with art. Eventually, as we well know, the computer came to play a fundamental and influential role in evolution and creation of new creative art forms. Through its processing power and the ability to work with digital sound and image many pre-existing art forms including music, photography, drawing, and printmaking can now be created with the computer as a tool (please note, I specifically refer to the computer as a tool and not a medium). This has allowed the artist to creatively use and engage new technologies as they develop. It has also brought about a re-consideration of the relation between art and science and resulted in a tremendous amount of innovation and fusion in both areas.

The continued innovation of the digital process and its inclusion of motion picture technology created new opportunities of creative expression for the visual artists, very distinct from that of the traditional motion picture or contemporary cinematic mediums. Digital time-based visual art is the form that has resulted from the advancements in the technology and its accessibility as a creative medium for the artist. Used as a time-based tool it enables the artist to create visual work that embodies time as a formal element and engages it as an aspect of expression and narrative structure. Although it does share some of the formal qualities of the motion picture and cinema (as many art forms share qualities) it has a unique set of elements and principals all its own.

Working artistically and expressively with the available and evolving technologies for time-based creative practice requires consideration, understanding and application of the unique concepts, processes and principles that are fundamental to such a digital visual art form. This chapter will define a set of fundamental elements and formal principles specific to digital time-based art. It will also examine how the artist's approach to the creative process has changed relative to working with an art form that is distinguished by its unique and particular set of temporal qualities. The discussion will cover a definition of terms specific to a digital time-based art form; the structural aspects of a time-based art; its relation and similarities to other mediums such as drawing, poetry and music and how its narrative form may differ from the traditional Aristotelian idea of narrative or Poetics as well as the established and conventional form of cinema.

PART I: SEEING IN TIME

Background

For hundreds of years the tools and mediums available to visual artists were limited in their ability to express or "capture" time, as they were only able to create fixed images. Until the introduction of photography the artistic image could only mimetically reference a given particular space. It could not at all embody nor employ time as part of its expressive form. Although the photographic image was able to fix or record a fraction of a moment in time and show an actual space it too lacked the ability to truly convey time or utilize time as an expressive tool. In visual art prior to time-based imagery, the western system of instantaneous linear perspective would artificially fix the point of view of the onlooker in an imagined space. The only way the fixed image could remotely reference time was to be set in a series of sequential images. It is interesting to note that as soon as photography was invented, visual artists began to quickly explore motion though the capture of images in sequence (Muybridge, 1995).

The technology associated with time-based visual art rapidly matured from the last part of the 19th century to the beginning of our present century. Though, the ability to work in a very direct manner with this new art form similar to the "directness" of painting or music, for example, did not evolve as quickly. Nor had a set of formal principles and elements been defined and set forth as guides, for the time-based visual art maker. In their inception the tools for working in a time-based medium, unlike those of the painter, musician, and writer, made it hard for an artist to work individually on creative works. Nor were the tools as fluent in their ability to directly bring forth ideas like those available to the painter and artists in working in those other established mediums.

Initially the tools and processes were cumbersome limited by their linear analog form and "bulky" technology, requiring an "army" to produce a single work. As a result of continued inventions and advancements in digital and computer technologies, the time-based art form is now a powerful, fluid and compelling vehicle through which the visual artist can creatively express and communicate either individually or in collaboration with others. The contemporary visual artists can now create new art forms that employ time as a formal and expressive element.

Innovations in technology that allowed for the transition from analog processes to digital ones and from linear to non-linear editing are what brought about the current capabilities in working with digital tools in the creation of time-based artworks. This has made it a highly facile and expressive visual art form. The digital process has also "democratized" the time-based medium, making it a more common tool for individualized communication. This is evident in examples such as websites like YouTube and Google Video. The creative artist in working in this powerful visual form now has a seemingly limitless ability to manipulate the image not only in time but also down to the individual pixel. The development and evolution of the digital process has enabled the creative act done in a time-based medium to surpass that of other visual art forms.

Over the last century, advances in technology have provided artists with the creative and artistic means to express themselves in ways that were different from those of traditional mediums. Prior to the invention of cinematic technologies, in particular digital ones, the tools that were available to the visual artist were limited in their ability to the create only fixed image artworks. With the advancement creative digital tools artist have access to a medium that enables them to create visual artworks that are time-based in nature. This has altered how we perceive and understand meaning and narrative in visual art forms. The result is a shift in the viewer's perceptive eye from I what call the "silent-eye" to an "eye-in-time." This eye is more readily accustomed to recognize and accept time-based imagery and narratives that, in part or whole, contain combinations and instances of montage, superimposition, variability of frame rate, duration and non-linear sequence.

The shift in the eye's ability to perceive narrative from fixed imagery and forced perspectives, to its perception of narrative in a dynamic time-based imagery that may be entirely devoid of perspectival reference, is not unnatural. We exist in time, every moment of everyday we see the world around us moving in time and we in turn move with it. Our lives unfold as time progresses from past to present towards the future. The very narrative of our personal lives is directly related to the passage of time. Our natural inclination is to be drawn to and to be engaged by artistic forms that embody time either directly or mimetically. Historically, before the innovation of cinema, these art forms were restricted to the spoken word, written story and theatrical performance.

In effect, we've changed how we "see" and visually communicate through a time-based medium within the context of continually advancing technologies and complex visual structures. We've modified our eye, shifting away from the silent-eye, conditioned to the perception of narrative in fixed images and the artificial depth of forced perspectives (Baudrillard, 1998), to a more cinematic eye: the eye-in-time. This shift has affected the creative process of the visual artist. It has opened up new possibilities and means for artists to creatively express themselves and communicate through a time-based visual art form. It has also changed the manner in which the artists "hand," functions in the creative act of making a time-based artworks; how their hand is "seen" in the resulting artwork at the end of the creative process and how the viewer interprets and can engage this type of work. I will later refer to this change as the "hand in time." The artist's hand can manipulate the time-based visual work down to the individual frame, even the very pixel and employ time as formal and conceptual part in the act of artistic expression. The facility that

the technology allows, makes it the same as that the eye-hand connection in drawing. That links the mind's eye to the vehicle of expression be it a pencil or brush; or the physical relation of the musician's hand to the instrument.

Historically the term "visual arts" refers to the creation of those works that are, of course, primarily visual in nature. Until the end of the twentieth century drawing, painting, printmaking and photography were the principal forms of creative expression in the visual arts. As the twentieth century transitioned into the twenty-first, invention and innovation led to the development of the technology for capturing pictures in sequence on film; aptly termed the motion picture. Through the end of the twentieth and into the twenty-first century film completed the transition from its initial analog format to a digital one. It became ready to be used in a way separate from traditional cinema and be re-defined as a new means of individual artistic expression.

It is clear that in its initial analog form, the medium was too cumbersome to allow it to readily match the type of personalized or individualized explorations of creative expression, process and artistic immediacy that other visual art mediums shared. Continued innovation in the technology and its shift to a digital format addressed these issues. Its birth out of technological innovation is not the single factor that makes this visual art form unique among those more traditional forms. It is distinguished by its connection to technology and how that connection enables the medium to do what prior visual art forms could not; employ, express and embody time by real means. Thus, it is a true *time-based* visual art form.

Cinema as a visual form has become codified, decadent and has even reached the limits of its narrative abilities due to its dependence on a commercial market as well as the costs associated with and level of production involved in making a cinematic work. In any medium how the viewer engages or encounters the creative form establishes a context that defines how the work is received, viewed and understood. Film quickly established a set of conventions and traditions in direct connection to its distribution and presentation. Whether it is seen in a theater, DVD, broadcast or online, how the individual accepts the work, views it and understand its narrative; and in the end individually interprets it is predetermined.

Cinema, even when digitally produced, is indeed a type of time-based media in terms of its formal structure. Although based on my prior description and conditions I set forth, it is not a true time-based visual art form. Which is not to deny that film cannot be, or is not in part, a kind of art form all to its own. In general, unlike the work of art, it is intended as entertainment; while the intention of the artist and resulting artwork may not necessarily meant to be entertaining at all. In fact, this is sometimes quite the opposite. A visual artwork is the singular clear intention of the artist and is, directly or indirectly, their imaginative or creative self-expression. It visually reveals descriptively or through metaphor the inner narratives of the individual who is responsible for its creation.

The conventions and context of cinema have been defined through its commercialization. In turn the context limits the acceptance of its narrative structure as well as the possibility of expanding the concept of what that structure can be. An artwork, unlike the established cinematic form, can be presented in any way the artists chooses from the public setting, gallery and museum to online, performance and interactive installation. In this chapter I am describing a digital time-based visual form that is an artwork. In so being the context becomes open and changes how we accept and view its structure and interpret and understand its narrative form.

Time-based media, as an art form, is not necessarily dependant on large production budgets, huge crews or costly means of distribution. The artist can work alone and choose their venue of presentation. By nature of its dependency on con-text of presentation, convention of purpose and multiple levels of collaboration, it is difficult for cinema to express a clear singular intention. This is why cinema falls short in its ability to expand its narrative beyond its context, conventions and manufacture.

The combination of the broad variety of the types of presentation formats and the recent development of live visualization coming from the VJ culture along with the possibilities of the technology have resulted in narrative forms that break away from those established and maintained by cinema. The Aristotelian model of Poetics, a narrative with a beginning middle and end is no longer a determinant. In digital time-based art narratives can be open, with seemingly no beginning or end, although they may move in time they share in a paintings (or fixed images) ability to express a type of narrative that neither starts nor stops but continues in time. This results in the possibility of time-based visual art forms that can be described as having a *"broken or open narrative structure"* or being *"open cinema"* and further for the creation of a type of artwork that can even be termed a *"time-based painting."*

As human's we have a unique ability to communicate through variety of means from the spoken to the written word. We are compelled to transmit and record our experience, thoughts, reflections and dreams to one another through as many means as possible. Aside from voice and hand, think of the variety of external devices that we use daily for communication, from a newspaper or book to a cell-phone, radio and computers. In my opinion our desire to convey our internal thought is almost as strong as our survival instinct. No matter what the tools, mediums or methods of communicating we are engaged in a process of creating, recording, sending, receiving and decoding narratives.

Our narrative inclination is a natural one and has been with us since the origins of language. In its earliest forms it was expressed through oral traditions and storytelling. Theater has been a long-standing part of human civilization beginning with primitive dramas that most likely developed from ritual, dancing and play-acting. I would assert that the more developed narrative practices of historical accounting; storytelling and theater were the first instances of a kind of time-based art. Unlike other earlier fixed visual arts (painting, sculpture etc.,) the aforementioned creative forms use of time was both malleable in its accounting and essential as an element of creative expression and intention.

In our daily life we experience time as it passes from one moment to the next; the past is a memory experienced, while the future is what can be imagined. In oral and written time-based narrative art forms, time can expand or compressed; moments can move from present time to past or future with only the turn of a word. The ways in which we use time in storytelling enables us to create compelling descriptions of persons, places or events that capture our attention and can allow us to imagine into the narrative itself.

The Seeing Eye

Our discussion must begin with the definition of a simple term the "seeing eye." As visual art makers the eye we refer does more than simply see as it plays a fundamental role in the creative process. It allows the hand of the maker to recreate pictorially what they see before them or imagine (in their mind's eye) through a variety of mediums using tools from charcoal to paint. In traditional terms, we call the connection between the eye and the hand kinesthetic coordination.

The eye and the hand work together in a physically harmonized process so that the hand in essence draws what the eye sees. When learning to draw, the visual artist builds muscle memory through such practices as "blind-contour" drawing. This develops kinesthetic coordination strengthening the direct connection between the eye and the hand. As the eye moves along the contour of a form the hand follows and records

this movement on paper, canvass or stone moving in harmony with the eye.

The Silent Eye

The tradition of image making or mimetic representation, through the visual artists imitation of the world, serves as a means to encode narrative information into fixed images or structures, such as paintings, architectural elements or sculptures. Although the function of encoding narrative into a fixed form has been an important and long-standing method of communication, it is somewhat contradictory to the nature of narrative itself, for narrative by its very nature includes a sense or description of time. The contradiction lies in attempting to encode a transient element, a thing that is "fluid" or moving, into a fixed form. The fixed image cannot accurately convey the complex, temporal experience of a narrative that can move us back and forth between past, present and future.

Before the invention of the cinematic medium we had only analog technologies that allowed for the creation of fixed image visual artworks. Our eye became conditioned to extract narrative from such fixed images by looking at the complex relations of icons, symbols and indexes, using the intellectual tools to decode meanings (Minor, 1994). As I stated earlier I use the term the silent-eye to describe the way in which we perceive narrative and extract meaning from fixed imagery.

Although we can decode the narrative content in such work, we do not experience a sense of time from it. Our only temporal engagement is in the amount of time it takes to understand or extract meaning from the work and the duration of our gaze. Either of which is quite different from the sense of time we experience in a written, oral, theatrical and cinematic work; or the in the narrative experience of our daily lives; moving through time from moment to moment. For the longest time, the closest approximation to this in visual art was sequential imagery, which has been present from Egyptian hieroglyphs to modern comics and graphic novels.

Until the invention of new technologies that allowed for the creation of the motion picture or cinematic medium, fixed imagery could only barely allude to or approximate, through a sequence of images, the experience of time. The viewer could not overcome the inherent contradiction discussed earlier. Only the spoken, performed and written word were those art forms that came closest to resolving this contradiction. The early technologies that gave us the motion picture and later, the digital technologies that gave us non-linear video and composite effects easily overcame this problem and brought word and image together into a full fledged time-based art form that could express the complex and subtle qualities of time.

The Eye In Time

As mentioned before, early motion picture technology began to resolve the conflict between the silent-eye and what I'll refer to as the "eye in time." The early analog format of film not only allowed for the creation of a medium that moved the viewer through time; it also gave way to new perceptual and visual theories; new means for creating and perceiving complex alternate narrative structures. Through this newly evolving eye-in-time, in contrast to the silent-eye prevalent until the early twentieth century, we developed the ability to extract meaning from complex narrative structures and non-linear sequences in visual time-based artistic works.

The eye-in-time can perceive and recognize subtler more abstract constructions of narratives, even when those structures are seemingly, "broken or opened." Meaning there is not the obvious traditional Aristotelian narrative plot construction of beginning, middle and end. The broken or open time-based art forms can take advantage of any combination of nonlinear sequence, montage, superimposition, frame-rate and duration. These

works can be experienced through a variety of visual means, from digital time-based installation, live performance and VJ-ing to broadcast, net-cast, gaming, WIFI and digital signage. The ability to work in a time-based visual art form, augmented by digital tools, processes and technologies, allows the artist to create new expressive and artistic works that are more akin to our natural way of experiencing the world and our own individual personal narratives.

PART II: THE HAND IN TIME

Cinematic Contributions and the Principles of a Time-Based Art Form

Over the last hundred years a cinematic "language" has been established that contains two important distinctions in how the cinematic narrative is constructed through, either through *mise en scène* or *montage*. In the former, the visual narrative is straightforward. One shoots and shows the narrative time in a scene, as it would be in life, with no cutting or editing away from the scene or a change in the angle of the camera, the point of view. The later, montage, allows for a more complex type of visual narrative to be established. One can shift the scenes point of view by cutting form one camera angle to another. Further, the scenes narrative structure can be altered by instances editing back and forth from present to past and or future; even cutting away to other scenes outside of the context of the primary one.

The cinematic concept of *mise en scène* is very specific to cinema as it deals traditionally with how a narrative is seen and shown in real time, as such I will not include it among the principles of time-based art. Montage, on the other hand is quite different. We will use part of the cinematic definition but extrapolate and add to it as well. Montage is one of the five major *principles*, along with superimposition, frame-rate, duration and non-linear sequences of the digital time-based art form. I make the distinction between the *principles*, mentioned immediately before, and the *elements*, frame, clip, sequence and time-form or time-composition, which I'll address later. The distinction is made because the principles address larger themes. They can operate independently or in combination while the elements by necessity are interdependent. They exist in sum in any given digital time-based visual artwork.

The principles, montage, superimposition, frame-rate, duration and non-linear sequence are best understood as fundamental qualities that have numerous and wide application across the breadth of the medium. They operate both structurally and descriptively giving voice to narrative through expression, meaning and concept. Either of the five can be used individually as overarching independent structures of a work without loss or compromise. The combination of these increases the dynamic potential of the work. A time-based work can be formally centered round the single use of superimposition (or any one of the principles) to the exclusion of the others and still be a strong and vital representation of the time-based art form. The artist can also use the principles more expressively and interpretatively as they have a strong relation to semiotics and metaphor. While the elements, frame, clip, sequence and time-form are rudimentary and rely upon one another in forming the whole of a given time-based artwork's overall formal structure.

Let's begin with montage. The Soviet concept of it shaped the way in which modern time-based works are conceived. Filmmakers such as Lev Kuleshov, Dziga Vertov and Vsevolod Pudovkin developed the early theories on the montage effect. Sergei Eisenstein, who was among the Russian filmmaking pioneers, believed that the montage effect could convey powerful ideas and concepts. He proposed a more radical and powerful use of montage, unlike the straightforward French idea of montage as simply cutting clips or sequences together into linear scenes or sequences.

Eisenstein realized that when two individual shots or clips, each with a discrete meaning, collided (were cut together, i.e., montaged); from this collision of independent clips (Eisenstein, 1969) a third meaning, separate from the two could be inferred. The idea is similar in concepts to semiotics and the formation of meaning in language (Ortny, 1993) through the relationship of signs, symbols, icons and indexes (Minor, 1994).

Montage, enables us to create complex levels of meaning and it also allows for time to be shown in a non-linear way, i.e., running back and forth, skipping from present to past and future. Early examples of flash back and forward were expertly explored in the classic films, "*A Christmas Carol*" (Dickens, 1910) and "*It's a Wonderful Life*" (Capra, 1947).

Contemporary examples of the use of montage to create alternate narrative structures that utilize more complex shits in temporality are the movies "*Pulp Fiction*" (Quentin, 1994) (in which Quentin Tarantino was perhaps one of the first filmmakers to really try to push the temporal shifts in a given work), "*Run Lola Run*," (Tykwer, 1998) "*Time Code*," (Figgis, 2000) and "*Memento*" (Nolan, 2000). A creative use of montage can present the viewer with types of experiences that exist out of time and space, such as dream states, empowering the creator of the work with an incredible temporal and spatial malleability.

American filmmaker Maya Deren pushed the creative use of montage early on in her highly stylized and alternative form of cinematic storytelling. She used montage to juxtapose different narrative storylines and edit together dramatically different scenes to create extraordinary shifts in time, space and place. Her experiments with temporal shifts and spatial jumps influenced future styles in editing, adding to the time-based language that formed in the later part of last century. I would even venture that her films, which were far and away from the commercial or the conventional were very early manifestations of a time-based visual art form. This is especially true given that she often shot (or set up the shots) stared in and edited the works on her own, making them creative expressions of personal intention.

Superimposition is the ability to make a single composite image from two or more frames, clips or shots. In effect a composite image is a blending of two or more images, or better still, layers. The composite is made through any one of a number of means including; levels of transparency and opacity; chromakeying, using a blue, green or other solid colored screen to drop out part of the image and replace and it with something else and other layering as described in the preceding paragraph. It is similar to collage in pictorial art, in that you are indeed layering the imagery. The difference is that it moves in time and each layer can also move with its own independent time. In a time-based media superimposition can also "bend" time, bringing past, present and future into the same frame. Much like montage, it too can present the viewer with types of experiences that exist out of time, such as dream states. It can also change spatial relations by bringing any number of images together in the same frame.

Superimposition can even be used in a manner similar to the glaze effects in oil painting. The digital tools used for working in this time-based form allow for an image to be broken down into layers. Each layer can have independent attributes from transparency to blending effects and more. Blending refers to the under-layer's influence on the layer above it, quite the opposite from oil paint where the top layer is the greater influence. As the digital image deals with luminous or emitted light versus reflective light in pigment (the pixel as opposed to pigment) you can determine what types of actions a blending mode, or layer, can have. This can range for the simplest affect as in the amount of transparency to more complex relations such as additive or subtractive levels of a colors influence.

Frame-rate refers to the ability to vary the speed of the film in adjusting the rate of the moving frame by showing more or less frames per second.

The minimum amount of frames needed to show relatively smooth motion is, arguably, between twelve to fourteen frames a second and is referred to as persistence of vision. In the early years of analog filmmaking to the camera had to be hand cranked in order to advance the film. By varying the speed at which one cranked the camera a cinematographer could achieve slow motion; over cranking, more frames per second or fast motion; under-cranking, less frames per second.

Using the digital tool this can now be done either while shooting or in postproduction. When done in shooting it requires a particular camera (like the Panasonic Vari-Frame or newer cameras now being produced for the prosumer) that allows you to shoot in an adjustable frame-rate. By increasing the frames per second you can create slow motion and the opposite effect, time-lapse, or fast-motion results from decreasing the number of frames per second just like the older method of hand cranking.

In postproduction you can use the computer to easily speed up or slow down an image. It must be pointed out that when using a computer to slow down an image digitally at a certain point it becomes a bit "choppy," This is due to the fact that you are not really adding "true" frames with new pictorial information. The computer is creating extra frames and blends of frames based on mathematical algorithms that interpret and generate an inter-frame based on the visual information between two individual frames. To further slow down the image digitally requires even more inter-frames to be created that, again, causes the motion to become much less smooth. Although artistically, this effect is sometimes desirable.

The computers power imparts a temporal malleability to the time-based form that allows for the exploration of time through its direct and visceral expansion, contraction and even to bring it to a complete stop. Apart from the mere analysis of motion, that slow-motion and its opposite reveals, variability in frame-rate is a highly expressive tool. The temporal qualities of a digital time-based visual artwork, the adjustments and shifts in time that can be created in a sequence, have strong similarities to the temporal shifts in music and even to the rhythm and pacing of the recited poetic verse.

Duration is independent of frame-rate as it is the amount of time for which something is shown or continues not necessarily the speed of the thing shown. The first use of the "freeze-frame" is a perfect example of duration as it was quite radical when first used and has now become almost a standard visual device. Duration is highly important in a time-based art form as it can be used to create great rhythmic structures. In one sense it provides a kind of musical structuring. You can use duration to create visual beat even matching the tempo of a musical score.

You can achieve counterpoint through the frequency and contrasting amounts of duration in a work. Unlike superimposition that relies on the computers power, duration was achievable in an analog form but the computer allows for it to be pushed beyond what could be physically achieved by hand. Although much simpler in concept then montage, and the other time-based principles, it is singularly effective in creating dynamic visual structure.

A young filmmaker Lars von Trier, in the film *De Fem benspænd* (Leth, 2003), (*The Five Obstructions*), challenges an elder filmmaker, Jørgen Leth, to remake his film, *Det Perfekte menneske, (The Perfect Human)* (Trier, 1967), five times, each one with a different set of obstructions. One of these is to limit edits to no more then twelve frames. The resulting work is perhaps one of the more excellent examples of duration used in a time-based art form. I refer to the piece as an artwork as Leth's original work *Det Perfekte menneske*, and Trier's, *De Fem benspænd* are not created in a traditional cinematic context. Like the work of Maya Deren they are experimental and artistic works that are highly divergent from mainstream commercial filmmaking.

Non-linear sequence is a very difficult device to employ in a traditional narrative film as it can easily distract and confuse the viewer. It is typically used to go from present to past and future and back again to establish complex relations and differing parts of the narrative structure. Or it can be utilized to obscure the overall structure and force the viewer to become a participant and work to unravel or decipher the meaning of the narrative structure of the film. The classic films, "*A Christmas Carol*" (Dickens, 1910) and "*It's a Wonderful Life*" *(Capra, 1947)*, used non-linear sequencing to break the temporality of the narrative and make the overall structure more complex. In more recent films, such as, Quentin Tarantino's, "*Pulp Fiction*" (Quentin, 1994) and Christopher Nolan's, "*Memento*" (Nolan, 2000), non-linear structuring was used to either obscure the overall narrative or develop a deep sense of mystery.

Elements of a Time-Based Art

In every art there is finite set of elements that are the basic building blocks of the particular art form and as such are unique to each individual art form (Kandinsky, 1979). In drawing and painting, it is the point, line, plane and shape (as well as tint, shade and color but for our purposes I will not include these). In a time-based art form it is the frame, clip, sequence and time-form. How the artists understands and uses the formal elements in the creation of a time-based medium are important to define and consider.

First lets examine point, line, plane and shape in terms of geometric theory. The Greek mathematician Euclid (*c.* 300 BC), in his *Elements of Geometry (Heath, 1956)* describes the basic following postulates:

A point is that which has no part.
A line is breathless length.
The extremities of a line are points.
A plane is breathless length and width.
The extremities of a plane are lines.

A form is breathless length, width and depth.
The extremities of a form are planes.

A comparison can be made with the elements of a time-based art form. In so doing it will help the artist working in this form to more deeply see the individual parts relation to the whole and understand in the end the Euclidean notion that the whole is greater then the sum of its parts. The comparison I am referring to is as follows:

A frame is that which has no part.
A clip is breathless length.
The extremities of a clip are frames.
A sequence is breathless length and duration.
The extremities of a sequence are clips.
A time-form is breathless length, duration and time.
The extremities of a time-form are sequences.

If we accept the assertion that the basic element of a time-based work is the frame then it becomes readily apparent how the formal structure for an entire work can be built around this tiniest of elements. The digital tool allows for us to easily separate out this basic element and for the artist to work individually and facilely with it. By understanding that the frame is, in essence, a manifestation of the original archaic fixed-image that historically preceded time-based work the artist can understand several things.

These include, how to visually compose the pictorial aspects of the visual composition and how to begin and end a work. It is also important for the time-based artist to take care that the individual frame is considered, full of intention and charged with the same creative integrity that will run throughout the entire work.

The clip in a time-based work is simply any number of frames beyond one. It is a very unique concept quite unlike the cinematic idea of a shot where *mise en scène* is the goal. In a time-based art work the clip does not have to contain an entirety of action, purpose or even narrative. The clip is a brief independent visual "mark" that moves in time. It is very much akin to the mark in drawing or the individual word in a poem; a singular aspect that in itself does not describe nor complete the work. Alone by itself it may, in a very general sense, refer to something but it has no extended meaning or narrative. It can show action event, but unlike the cinematic shot he clip only begins to act as a part of the language of the medium, it is not language itself and must be seen as almost "mute" to a degree unable to communicate anything beyond the most minimal visual information.

The sequence is where the language of this medium begins to reveal itself. A sequence contains a series of clips; the relation from one clip to one another is what begins to generate meaning. In principle it is related to the concept of semiotics as the individual clips can act as a kind of sign and signifier. In a sequence the contrast of the clips is what leads to expression in line with the thinking of Sergei Eisenstein. It is from the continued succession of collisions of different clips in a sequence that meaning is generated. In understanding this, the time-based artist has a very powerful tool for expressing content, developing thematic ideas and working towards creating narrative structures.

I used the term "time-form" to refer to the overall structure of the completed time-based artwork. I specifically chose not to refer to it as a movie or a video as the time-based artwork does not (and in my opinion should not) have that convention and contextualization applied to it. The time-form has a structure that can be broken or open not necessarily closed. It is a structure that contains a "new push-pull between static/fixed and motion/moving. It is important to recognize this is not a simple relation of opposites though, for it only fully functions when they happen simultaneously, or are both present and absent, so one is never left with just one of the dualities, but is held in tension between them" (Smith, 2008). This holding of the tensions is what allows the time-based art form to become dynamic and break the conventions of those other kinds of forms which have time-based qualities but whose narratives are not open or broken.

As I stated earlier, narratives written, spoken or performed are forms of time-based media and have similar qualities, such as the compression or expansion of time and the ability to shift between temporal states from present to past or future. What they cannot do is duplicate superimposition, or approach the effects of montage, frame-rate or duration. Further they lack the ability to allow duration to create rhythmic contrasts. Again, unlike the cinematic as this medium is not site or presentation specific there is no predetermination on the part of the viewer as to how they should accept or interpret the work based on the context in which they encounter it.

CONCULSION

The cultural effect of creative expression through time-based art is profound. As the technology and the associated "language" of the medium (time-based versus the fixed picture) rapidly evolves, so too does our acceptance of the experience of the medium, our ability to perceive it and extract meaning from it. Further, because the digital time-based art form includes montage, superimposition, frame-rate, duration, and non-linear sequence the media have affected how we contextualize and understand our own personal and cultural narratives relative to these concepts.

It is important to realize that the digital tools we have for creating time-based artworks are merely only tools. As in any medium the tools themselves do not create nor possess an aesthetic and I would assert the in contrast to Marshall Mcluhan's belief that "the medium is the message" (Mcluhan, 1994), they do not impart or alter the message or intention of the creative artist. The digital tools are inert, lifeless without capacity or capability. Their potential can only be achieved through a kinesthetic relation to the human creative ability. Without a sound understanding of the fundamental principles and elements of any creative art form the tools can most often only produce the most mundane and coarse results.

Through experience application and understanding the artist creating digital time-based artwork can achieve the highest level of creative work and make art that is fully expressive and wholly embodies their intentions. The tools, digital or other, become transparent when the artist understands the fundamental aspects of their medium. The tools are not arbitrators that mitigate the aesthetic or message of the artist. In fact they augment the artist's ability and allow for the creation of more profound artworks.

As creative individuals we have struggled through the ages to find a means of expressing ideas and imaginings in a way that is both direct in expression and akin to how we experience the world. Digital tools have given us the ability to create works that more closely approach actual types of expression of what we can imagine in our minds eye and in our dreams. Digital processes have augmented and advanced our physical bodies and the limitations of fixed media, in essence changing how we see and communicate within the context of continually advancing technologies and complex visual structures.

REFERENCES

Arnheim, R. (2006). *Film as Art*. Berkeley, CA: University of California Press.

Arnheim, R. (1974). *Art and Visual Perception*. Berkeley, CA: University of California Press.

Baudrillard, J. (1998). *The Consumer Society: Myths and Structures*. (p. 116). London, England: Sage Publications Limited.

Bazin, A. (2004). *What Is Cinema?* Berkeley, CA: University of California Press.

Capra, F. (1947). *It's a Wonderful Life.* Writers: Van Doren Stern, P (story). Goodrich, F. (writer). Liberty Films II. United States.

Dickens, C. (1910). *A Christmas Carol.* Edison Manufacturing Company. (short story). United States.

Douglas, K. (Ed.). (1990). *From Marxism to Postmodernism and Beyond.* Baudrillard, Jean. *The Consumer Society,* (p.109, p. 33). Stanford, CA: Stanford University Press.

Eisenstein, S. (1969). *Film Form: Essays in Film Theory.* Washington, PA: Harvest Books.

Eisenstein, S. (1969). *The Film Sense.* New York, NY: Harcourt Books.

Eisenstein, S. (2006). *The Eisenstein Collection.* Oxford, England. Seagull Books.

Figgis, M. (2000). *Timecode.* Red Mullet Productions. United States.

Heath, T. L. (Trans.). (1956). *The Thirteen Books of Euclid's Elements, Books 1 and 2.* Euclid. Mineola, NY: Dover Publications.

Kandinsky, W. (1979) *Point and Line To Plane.* Mineola, NY: Dover Publications.

Kazin, A. (Ed.). (1977). Blake, William. *The Portable William Blake.* (p.150). New York, NY: Penguin Books.

Kracauer, S. (1997). *Theory of Film.* Princeton, NJ: Princeton University Press.

Leth, J. (1967). *Perfekte menneske, Det* (The Perfect Human). Demark. Laterna Film.

Levaco, R., (Ed.), (1975). *Kuleshov on Film: Writings by Lev Kuleshov.* Kuleshov, L. Berkley, CA: University of California Press.

Mcluhan, M. (1994). *Understanding Media: The Extensions of Man.* Boston, Mass: MIT Press.

Metz, C. (1990). *Film Language: A Semiotics of the Cinema.* Chicago, Il: University of Chicago Press.

Minor, V.H. (1994). *Art History's History.* Saddle River, NJ: Prentice Hall.

Mitry, J. (2000). *The Aesthetics and Psychology of the Cinema.* London, England: Athlone Press.

Muybridge, E. (1955). *The Human Figure in Motion.* Mineola, Dover Books.

Nolan, C. (2000). *Memento.* Nolan, J. (short story Memento Mori) Nolan, C. (screenplay). France: Newmarket Capital Group.

Ortny, A. (1993). *Metaphor and Thought.* Essay, Reddy, M. J. New York, NY: Cambridge University Press.

Panfosky, E. (1997). *Three Essays on style; Style and the medium in motion pictures,* (p. 93). Boston, Mass: MIT Press.

Smith, O. (2008). *Raphael Di Luzio and The Concept of Time-Based Painting,* (p.16). Wynwood, Florida: Wynwood The Magazine of Art.

Tarantino, Q. (1994). *Pulp Fiction.* A Band Apart. United States.

Taylor, R. (Ed.). (2006). *Vsevolod Pudovkin: Selected Essays.* Oxford, England: Pudovkin, V.Seagull Books.

Tykwer, T. Director. (1998). *Lola Rennt,* (Run Lola Run). Tykwer, T. (writer). 1999. X-Filme Creative Pool. Germany.

Vertov, D. (1985). *Kino-Eye: The Writings of Dziga Vertov.* Berkley, CA: University of California Press.

von Trier, L. Director. (2003). *De Fem benspænd,* ("The Five Obstuctions"). Denmark. Almaz Film Productions S.A.

KEY TERMS

Clip: Any number of frames beyond one. A clip can show action event, but unlike the cinematic shot he clip only begins to act as a part of the language of the medium, it is not language itself and must be seen as mute to a degree unable to communicate anything beyond the most minimal.

Duration: Duration is independent of frame-rate as it is the amount of time for which something is shown or continues not the speed of the thing shown. Duration can be used to create rhythmic structures. It can provide a kind of musical structuring. You can use duration to create visual beat even matching the tempo of a musical score.

Eye in Time: How we perceived and interpret time-based visual imagery after the invention of motion picture technology.

Frame: The basic element of a time-based work that cannot be divided or separated or edited any further.

Frame-Rate: The amount of frames per second (fps) at which the work is either, recorded, edited in postproduction or played back. Frame rate refers to the ability to vary the speed of the film in adjusting the rate of the moving frame by showing more or less frames per second. The minimum amount of frames needed for relatively smooth motion is, arguably, twelve frames. This is minimum is referred to as persistence of vision.

Montage: The cutting together of two separate clips or sequences each with an independent meaning and the result of which is to produce a third meaning independent from the two.

Sequence: A series of clips in which the relation from one clip to one another is what begins to generate meaning. In principle it is related to the concept of semiotics as the individual clips can act as a kind of sign and signifier. In a sequence the contrast of the clips is what leads to expression and meaning.

Silent Eye: How we perceived and interpreted fixed visual imagery, such as paintings and photography prior to the introduction of motion pictures.

Superimposition: The ability to make a single composite image from two or more frames, clips or shots by laying one image over another by one of several means including chromakeying, transparency and blending effects.

Time-Form: The overall structure of the completed work. The time-form has a structure that is open not closed. It is a structure that contains a push-pull between static/fixed and motion/moving.

Chapter IX
Ambivalent Interplay

Heejoo Kim
Columbia College Chicago, USA

ABSTRACT

The human vision, the most ubiquitous receptor of the human senses, has been the prevailing sensory organ for a noticeable manifestation of visual arts. Nevertheless, in the aspect of new technology art, the embodied experience through senses dismantled and amalgamated in hybrid aspects. Explicitly, new media artists perceive that interactive technology is evolving rapidly in such a short period of time. Rather than engaging in technology more interactively, however, it seems they are scrutinizing the subsequent progression of the phenomenon in interactive art. Artistic experiments have predominantly been transferred through the human sensorium in interlaced approaches: touch, sight, smell and hearing have synesthetic qualities in their interactive connections in between works and viewers. Recently digital art performs in multi sensory forms of knowing and communicating. There are investigating perceptual and emotional mechanisms of involuntary synesthetic experiences. This artistic phenomenon is not only historically intriguing, but may also contribute to present synesthesia research. The functions and interrelations of the synesthetic approaches in new media arts and neurological researches are discussed separately.

INTRODUCTION

The synesthetic experimentation by artists is arguable. Many different types of art works are based on deliberate contrivances or interfaces of sensory fusion and not on involuntary senses of cross wired association. Therefore, most artistic approaches with hybrid sensory fusion are not inside of the domain of biological synesthesia research. On the other side, in contrast, some researchers assert that synaesthesia[1] is social, cultural, but not a biological phenomenon: it is cultivated and formed by trained exercises. As a matter of fact, we recognize the 'synesthetic experience of being' in all forms of art throughout the history—in poetry, painting, sculpture, music and noticeably new media art, such as: interactive cinema(installations), artificial reality, net art, wearable art, telematic art, game art and even mobile art. The emerging conjunction between new technology and new media art links intimately to the way humans appreciate and understand the ecology of art. If we need to mark the distinction between old and new media, inevitably, the leading distinction is interactivity. Nevertheless, the issue of "interactivity" is controversial. How about the reciprocal influences between a painting and a viewer? Even though any traditional form of art stimulates only one or two sensory organs, it is an unavoidable consequence that viewers respond to the pieces of old forms of art in different ways. Then, what is the core perception of art media these days? What redefines the definition of interactivity in new media art history?

Although discussing how traditional and emerging art media have inherent connections in the way of interaction with viewers, we all cannot deny the radical differentiation between passively watching and dynamically participating. New technologies reconstruct media and psychological influences the senses. If the traditional painters, filmmakers, or performance artists present the circumstance for speculation, the contemporary media artists transform the viewer to participant, and incite more senses to engage their works. Furthermore, more than ever before, many different types of recent media art educes and coalesces sensations that are normally experienced separately. The synesthetic experience has made an impact on human interaction culturally, mainly through technology oriented interactive art forms. This discussion is focused on multi sensory Interplay; ambivalent senses in new media art, particularly interactive projection, net art, game art and mobile art.

EARLY HISTORY OF SYNESTHETIC EXPERIMENTATION IN ART

In the mid-nineteenth century synesthetic experiments had been placed in art movements. Those movements were frequently shown in the writings of composers and visual artists. The device, such

as "clavecin oculaire" which produced color of lighting based on music tone, was invented by French scientist, Jesuit Caste in the eighteenth century. Other inventors like Jameson, Kastner, Bainbridge Bishop and Rimington researched those devices. In 1893, Rimington developed and named his device "color-organ", and had a successful concert playing Wagner, Chopin, Bach and Dvorak with colors corresponding to their music (Peacock et al., 1998, p. 397-406).

Alexander Scriabin, a Russian composer and pianist started experiments in synesthesia in the first decades of the twentieth century with Wassily Kandinsky, a Russian painter, printmaker and art theorist. At that time, concerts with color lightings were sensational events. Scriabin was intrigued by the development of idiosyncratic tonal language and a poetic, philosophical, and aesthetic vision. On the other side, Kandinsky developed his paintings based on hearing tones and chords. Both, Scriabin and Kandinsky discovered and built up colored hearing in their early years. According to Scriabin, colors are often associated with tonality, and his quality and intensity of synesthesia did not always consistently exist. Whenever, he emotionally engaged with music more than usual, the synesthetic sensations of color would become intense. Not every single pieces of music would elicit synesthetic sensation. Some of classical music did not evoke his synesthesia. Therefore, attention has become drawn about his synesthesia by other researchers and became controversial. The major aim of his experiment with the auditory and visual perceptions explored the artistic potentials of the simultaneous playing of sound and colors.

In the case of Kandinsky, he theorized colors with tones and geometric figures and their relationships (Brougher, 2005, p. 89-179). During his youth in Moscow, he discovered his fascination with color symbolism and psychology. In his early age, he recognized that he was unusually stimulated by color. When the main art stream moved towards Impressionism and cubism, he started geometric abstract painting. Biomorphic forms and non-geometric outlines appeared in his paintings. His early abstract paintings named with musical titles such as 'Composition' and 'Improvisation' were experienced by synesthetes[2]. Those elements, such as lines and spirals, were distinguished from other artists like Mondrian and Malevitch. His main concern was the experiments with multi cross sensation and sensory perception of movement. His intension was not solely focused on sensory fusion like Scriabin (Vanechkina, 1980, p. 107-115). While being already familiar with the multi sensory system of the consonances of color and sound, he got more interested in experimentation with dissonances of color, shapes and tones. He seriously started exploring synesthetic experiences with other synesthetes. He researched the perceptual interaction of the simultaneous presence of color, sound and movement to evoke deeper sensations. The synesthetic associations of Scriabin and Kandinsky were not only external experiences, but also internal mental experiences. They explicitly sensed the color while they were listening to music internally and often externally. The existence of synesthesia was slightly different from Scriabin to Kandinsky. Scriabin rarely experienced any synesthesia phenomenon at all without some emotional attachment. And Kandinsky had one ultimate ambition for his synesthetic challenges: to create compositions that touch the strings of the soul (Kandinsky, 1912/1982). Even though they seemed to have little distinction in this same realm, both tried to explore the emotional dynamics of synesthesia in their own way.

SYNESTHESIA AND SYNESTHETS IN NEUROLOGICAL TERMS

Dr. Richard E. Cytowic who was known for rediscovering synesthesia in 1980 sharply distinguished neurological term of synesthesia from multi sensory joining experimentations. He defines synesthesia as the involuntary physical experience of the stimulation of one sensory modality reliably caused a perception in one or more different senses. Also, he asserts the term 'synestheia' is often employed by people who described 'multi across brain sensation'. In other words, 'artistic experiments with sensory fusion' is not in the realm of the neurological research. Then, what is the neurological definition of 'synesthsia'?

The word 'synesthesia', meaning "joined multisensory perception," shares a root with 'anesthesia' meaning "no sensation." It originates from two Greek words, syn (together) and aisthesis (perception). Synesthetic phenomenon engages any combination of senses and these aspects are specific to every individual person. The characteristics of Synesthesia are defined by the phenomena of cross-modal associations that are durable, discrete, generic, memorable, emotional, noetic and involuntary, but elicited. It refers to a reciprocal sensation: 'Auditio colorata', assigns to seeing specific colors when hearing certain musical tones. 'Phonism', is an esoteric sensation, for example, hearing sounds when staring at letters or colors. 'Photism', is an abnormal sensation of light or forms, that occurs upon hearing loud sounds. Generally, two or more senses are automatically and involuntarily joined such as a voice. For example, the voice is not only heard, but it possibly can be felt, seen, and/or tasted. Synesthesia is a physical phenomenon of the brain, not the result of imagination, cultivating or learning. It differs from metaphor and deliberate artistic intension such as visual sound. Some multi associations are much more common than others: colored hearing is more often the case than the combination of taste and smell. The most common synesthesia joins color to letters and numbers. Some form of synesthesia may occur in 1 in 23 people. It runs mostly in families, and appears to be inherited. Synesthesia is a genetic feature, not a disease (Cytowic, 1989, p. 849-850). As a matter of fact, Dr. Cytowic considers synesthesia as a normal brain process that occurs in families but, is not part of every family member. So to speak, everybody has a synesthetic quality of tendency; however most of people just cannot attain it.

Consequently, synaesthesia is the involuntary physical experience of a cross sensory association. The stimulation of one of the senses can cause more than one or more senses to activate. This neuropsychological interpretation attempts to locate the phenomenon of synaesthesia besides the borders of metaphor, symbolism of sound and artistic multi sensorial experimental approaches. Considered from this perspective, it is an exceptional function of the brain, occurring outside of the cortex in the left hemisphere. Often synethesia functions as an assist to the sub consciousness. It shows a remarkable development of the sense of order, balance, and accuracy. From the perspective of the receptor cells, in biological terms the sensory organs involved different kinds of sensations; vision, hearing, smell, taste, and touch. For instance, our color perception, the palette of visible colors, is smaller than the range of audible sounds.

Therefore, the perspectives people get from 'synesthetic experience' can be variant. The abstraction of the correspondence of sense impressions occurs naturally or is created primarily in the multidisciplinary context of the art form. So to speak, there are two contradistinctive approaches: synesthetic experiences by synesthetes and art as a synesthetic experience. Synesthetes, people who have, or had, the medical condition synesthesia, commonly associate numbers, music, and/or letters, with the experience of colors, or may have a three-dimensional view of a year, month, and a day as a map. They apply their eminent experiences to support their creative and inventive practices, and many non-synesthetes have attempted to conceive forms of art that would perceive what synesthetes feel. Synesthesia is not only known for its inherent attention, but also for the intuitiveness it may give into cognitive and visceral processes that happen in synesthetes and non-synesthetes alike. Definitely those two aspects are grouped

into discrete categories, whereas they share assertive essential concepts that allocate the new junction between the senses.

Cytowic concludes that artistic trials of synesthesia would be historically intriguing; nevertheless, they are not relevant to the neurological or biological study of synesthesia. He puts various deliberate multi-sensory art works under the term "sensory fusion", and places them outside the history of involuntary synesthesia (Cytowic, 1993).

SENSORY FUSION EXPERIMENTS IN MODERN ART

As we discussed previously, the definition of 'synesthesia' in art needs to be differentiated from 'synesthesia' in biological and **neurological** research. The meaning of synesthetic art conveys the experience of some kind of cross wiring in the brain and the abstract connections between unrelated inputs through projection, interaction and sound. New media artists stimulate arbitrary relations between seemingly distinct perceptual entities such as: visual sound, wearable projection, virtual data navigation, and traveling cinema. As a consequence of the new interactive media, our consciousness, senses and body will emerge into new experiences with unlimited multi-sensory synesthetic qualities. Through the new technology, the synesthetic experience is enhanced in a more communicable and accessible way. The origins of the synesthetic experience are to be found in paintings, sculptures, performances and films. Paradoxically, various forms of art are considered synesthetic. Although, there is a significant difference between *'personal synesthesia'* and *'created synesthesia': personal synesthesia can be a natural presentation of cross sensory perception. Nevertheless, created synesthesia is the consequence of artificial synesthesia.*

Hybrid sensory associations challenge the traditional view of the five senses. In artificial synesthetic experiences many attempts have been made to simulate this evidence through the use of aesthetic techniques, such as metaphor. This experience is based on the fact that our perception is created by our consciousness from the very instance of its inception. Our perception is never equally present as observation, but it can evolve into projection. People have a radical desire to pursue something that they cannot have. People want to sense something beyond their ordinary sensation. As such, the synesthetic approach in art which presents the fragmented pieces of desire; it seems as if, the senses of feeling, vision, smell, taste and hearing are mixed into certain forms of art.

In modern arts, often senses are reconstructed into the visual surface. Sound, smell, and tactile environment can integrate with vision. What we hear becomes observed corporeality. Through esoteric experiences of sound and image, we sense a realistic imitation of senses. Sound and vision stimulate sensory organs in ambivalent ways. The relation between conflicted brain attempts can be described in the synesthetic procedure. A vision can emerge into silent voice. 'The Scream', by Münch, for example, even though this is a two dimensional painting, it conveys silence through vision. We see the person is screaming. We feel and hear his sound with our eyes. It produces the image of moments, a metaphor of extreme silence.

Recently, the word, Synaesthesia, is used in art as a metaphor of imagination or feeling, not as a co-sensation. Specifically it is an inter sensory association with a psychological and emotional nature. In visual art forms, it can be either passive or active, having various edges of emotional eidetic happenings. It is known as, "the multi mingle of contiguity". Correspondingly, metaphors have pretty obvious motivation for evoking feeling. Considering this phenomenon of features revealed in the gestalt of the visual and acoustic images. Emotional influences are inherited in art. In this case, synesthetic artistic approaches

would be considered as a participation of the operation of feeling through the subconscious level. In connection with this, synesthesia and sensory fusion would be related to the realm of non verbal and sensuous cognition, side by side with the visual and acoustic process of emotions. For example, the visual sound is a force which offers the interrelation between vision and hearing brain mechanisms.

This is a multi across brain function that manifests itself in human interaction. It is not an abnormal, but a norm. Moreover, it would be characterized as simultaneous accomplishment, sensuous in its manifestation. Synesthetic art practices social interplay. It is cultivated and reflected. The nature of this practice occurs in the conclusions of anomalousness. In other words, the fundamental meaning of synesthesia has a clear distinction from so called 'sensory fusion'; this word has a somewhat reciprocal interrelation.

INVISIBLE INTERFACE: THE OLFACTORY VISION

The olfactory organ is an intuitive retroaction. In most situations, smell is the first and predominant reaction to stimuli. It warns you that something is burning before you look at the blackened, smoking frying pan. It makes you relax before you taste a cup of jasmine tea. It makes you recoil before you touch stinky socks. Even though smell is a rudimentary sense, it is also at the prominence of neurological research. Then, how do we obtain odorants, and convert them and define them as smells?

Smell is one of the chemical senses, like taste. When an odorant actuates the sensory cells inside of the nose that identify smell, they go through mechanical incitement to the brain. The brain then analyzes diagrams in an electrical mechanism as explicit odors and olfactory sensation turns into cognition. We are continuously breathing air and collecting relevant information about the surrounding environment. Then, the chemicals are defined by our sensory systems. For instance, odor molecules are tiny enough to be vapor so that they reach the nose and then deliquesce in the mucus. At this point, smell, unlike taste, passes through a long signal path. It seems that we appear to have an innate capability to ascertain aversive odor and therapeutic scent. More than any other senses smell is the most intuitively connected to the parts of the brain system that affects emotion. The olfactory chemicals in the brain interpret sensation into perception: smell links our emotional behavior, and also, mood and even memory.

Animals, like dogs and horses, have stronger olfactory system than humans. Animals can detect emotional behavior with their sense of smell. For example, dogs can smell fear in humans. Fear is an internal nervous system response when animals face a dangerous situation. In humans, the autonomic nervous system controls stress repercussion. This repercussion causes the body to prepare the physical reaction for an immediate self protective mechanism. When humans are in danger or in an intensive stress situation, the body reacts with a fight or flight reaction. For example, in fight or flight there is an increased breathing rate, higher heart beat, and an increased sweat gland activity. Particularly, the human sweat glands respond to emotional stress and stimulation. Stress or pleasure causes an intensification of natural odor which makes animals smell fear or aggression. Therefore, if an animal is in front of a person, that animal is able to recognize the odor of the human as fearful or aggressive, if the person is feeling those emotions. The role of olfactory is to communicate alarm in between animals and humans. The smell of sweat reveals emotion.

Smell does not link only emotion, but it also links to memory. Can you recall the smell of your grandmother's sweater? We usually remember the smell before we define the past as cognition. Smell is linked to memory intimately. Smell elicits sub consciousness. We all have our unique

smell, even though most of the times this is not recognizable or is hidden under deodorants and perfumes. In daily life, we deal with the impact of smell without either understanding or knowing it. Smell influences our motivation and behavior. We are interacting with smell: this is invisible mind interface.

The Berlin-based, Norwegian artist, Sissel Tolaas created odors. For her show in 2006, "The FEAR of Smell — the Smell of FEAR" at MIT's List Visual Arts Center, she let viewers rub the nearly vacant white wall to release the smell that is produced with scent-reproduction technologies. She synthetically collected each man's odor. Then, through a micro encapsulation[3] process, she eventually ended up with special paint: scratch and sniff. When the viewer touched and rubbed the surface of the walls, microscopic capsules let the smell out into the air.

The sweat donors for this show were from all over the country. Since men sweat more, she explains how she collected only male sweat with the special cell phone size device that she invented. It captured a man's armpit sweat when they felt likely fear. Later, the packages of collections arrived in the mail at her office in Berlin. Then she used technology to analyze her collection of sweat to the aromatic scents of the perspiration.

So called, odor artist Tolass, who has degrees in math, chemistry, art and language, started working with odor. She said "I realized we have only two words to communicate about smells: bad or good," "I thought, something is wrong with that." She was collecting odors from all around the world, and building a library of 6,763 scents. Due to her collections, the perfume industry, IFF (IFF has developed scents for Prada, Calvin Klein, and Ralph Lauren) allows her to collaborate with them on scents. She has also worked with companies like IKEA and Volvo to add a scent to their brands. This would include adding scents to paints that IKEA uses in the paint that goes on the furniture. For Volvo she developed scents that were added to car leather. She is the most controversial figure in the fragrance industry.

Her intention of creating those kinds of unusual odors is not repulsive, but she tried to enhance our olfactory centers to be more definitive. We can smell up to 15,000 different smells, however, recent technology has invented fake aromatic scents that disguise all natural odors. We cannot imagine America without antiperspirants and deodorant advertisements. The smell of sweat is something that hides under synthetic fragrances. Our olfactory centers are overloaded with fake scents. What she says about the West is that they are, "the smell-blinded." She explored the real odors from the world that we exist in right now. Her simulations not only revealed the reality of the human instinct but, it also mutates cultural bias. Tolaass describes, "I tried to find out what has been done with smell, and I realized that this is a field where not much has changed. I thought: Something is wrong here. We have a nose, but it's suppressed compared to what we know of eyes and ears." For her MIT(Massachusetts Institute of Technology) exhibition, she set up the visual metaphor of smell. Viewers touched the smell and it stimulated their other sensory organs and emotions. This combination of experiences tweaks the brain mechanism which evolves from passive interaction between work and viewer to a more closed relation.

VISUAL AND ACOUSTIC EMBODIED EXPERIENCES

Digital art, such as video and new media art, originates from the coalition of human vision to support enhancing our experiences in our culture. Synesthesia characterized visionary aspects in every field of digital art. Visual perception is the capability to interpret facts from visible light reaching the eyes in psychological terms. Everything you look at is sent to your brain for processing and storage like a video tape. It sounds very simple, yet the sense of sight is actually considered the most intricate among the five senses.

How do your eyes work? Basically, what you are actually seeing are beams of the light bouncing off of the surfaces. The light rays enter the eye through the transparent protective layer on the surface of your eye, known as the cornea. Then the light rays pass through the dark circle in the center of your eye, the pupil, and into the lens part. Depending on the amount of lights, the iris makes the size of the pupil change.

Therefore, if you are in a dark environment, your pupil would enlarge to let in as many light rays as it can. Behind the pupil is the lens and it focuses the images through the back surface of the eyeball, called the retina. The retina is filled with rods and cones. Rods identify shapes and cones identify colors. Consequently, both of these types of cells send the information to the brain through the optic nerve. The brain turns images upside down which stay in the retina. The brain does this in a specific place called the visual cortex. Visual attention seems closely connected to "consciousness", this is very suggestive. Our subjective experience of the yellow of a lemon may be identical with the activation of yellow sensitive neurons in the brain. Our vision is one of the main control systems in sensory reaction with other senses.

Then what about the hearing? Ears are extraordinary sensory organs that pick up sounds, translate them, and send them to your brain. This is an exceptional process that it is completely mechanical. The sense of smell, taste and vision relate to chemical reactions, but the hearing procedure is completely based on physical operation. First of all, to understand how the ears work, we need to understand what sound is. An object generates sound when it vibrates in the air. That object could be anything, such as a solid, a liquid, and an air. Mainly, we hear sound flowing through the air in our surroundings. When something is oscillating in the environment, it quivers the air particles around it. Those air particles shake up the other air particles around them, and carry this vibration through the air. This is called compression. This depends on the variation of the sound frequency; we hear different sounds from different vibrating objects. A high pitch sound means a high wave frequency. That means the air pressure fluctuation switches back and forth more quickly. The level of air pressure, the wave's amplitude, defines the volume of the sound. The outer part of the ear catches the sound waves. This structure which has a number of curves, determines the direction of the sounds. The brain system analyzes distinctive patterns and determines where the sound comes from. Also, it determines the horizontal position of the sound by comparing the wave frequencies coming from two ears. Once the sound waves pass through the ear canal, they vibrate the eardrum which is rigid, and very sensitive. The eardrum can be easily fluctuated by the slightest air pressure. Therefore, the compressions and rarefactions of sound waves push the eardrum back and forth. High pitch sounds make the drum shake more rapidly. The eardrum is the absolute sensory element in your ear. This pressure amplification passes the sound information into the inner ear and it is translated into the nerve system. Finally, the brain understands the information of the sound waves. The visual artist perceives and experiments with this connection of the brain, vision, and hearing which combines and creates different environments.

The work of Bill Viola, The Crossing, is a sophisticatedly combined film, digital video, and audio with a large scaled dual video projection. In a completely dark space, from each side of a double screen, a human figure is walking from a deep distance towards the viewers in slow motion. As soon as the synchronized figure sequences fill up each screen eventually, on one side, a tiny flame starts to appear at the bottom part of the body, and on the other side, a small drop of water begins to drip and gradually is poured on his head. Subsequently, the fire and water amplify and devour the figure. The overwhelming slow motion and high intensity stereo sound engulf the viewer. Then, the fire and water stop, the figure

gradually disappears and the screen goes dark. A few moments later, the cycle begins. The crossing is a transcendent conjunction between traditional painting and sculpture, and moving image and sound. Nevertheless, this is not a physically interactive piece; it is immersive visually and remains an indirect interaction with the viewer. They seem to listen to this image and become overpowered by sensation. Watching a firing flame and pouring water induces mixed sensory systems.

IMMERSIVE VISUAL AND AUDIO MOTIONS

New media art has deeper connections to the human-computer interaction. There is an invisible or visible interface that viewers interact with in this new surface art. Cognitive psychology has been the most significant impact on our understanding of the process and importance of human-computer interaction. The definition of 'cognition' is 'the mental process, such as perception, reasoning, problem-solving, etc, which enable humans to experience and process knowledge and information.' Psychologists that have researched this field of study underlie work in artificial intelligence and cognitive science(Bruandet, 2004, p. 288-298). However, the non-computational aspects of cognition, such as sensation, support much of the real human cognition.

Most of the computer artifacts now in interactive aspects are not 'cognitive' devices. Modern human-computer interaction has sensual and intuitive enhancers, and these aspects will grow and develop. The computers, as tools, do improve perceptual and speculative abilities. New technology cannot make viewers think more intelligently, but it can allow them to experience more in sensual ways. Thus, human-computer interaction interface is mostly an issue of doing sensuous ergonomics. The meaning of 'sensuous' is 'appealing to the senses aesthetically'(Bruandet, 2004, p. 288-298). So to speak, it evokes the rich stimulation of sensory systems through body sensation and mental constructions of reality.

Computer technology expands in multiple ways of art presentations that can be from a text interface to an immersive virtual interactive environment. Media art is transformed and awakened by the new interface and environment. Therefore, most interactive media art demonstrates this process to people who play with the interface. This is a meditative augmented reality environment displayed with sound, light, and some other coded, accumulated visual elements. On the other hand, many new media artists use sensors, cameras, or even mobile phones as their tools with codes. Another aspect of trends required the particular computer skill of the user through user interface. The developers of media worlds, such as games, challenge the user to learn and explore the strategies. The approach to this progress is that different media could be evolving within the same interface. The most essential aspect of interacting with computer systems is intuitiveness which examines the importance of computer functions as perceptual rather than conceptual tools. With enhanced synesthetic qualities, not as 'cognitive artifacts,' new technology, including coding, motion detecting, data base, etc., empower the human creativity.

Expanded film formats, such as virtual reality and interactive video installations, more than any other mediums, have enhanced the intercorrelation of the multidisciplinary human reactions. They evokes intuition, emotion, sensation, and even scrutiny. Intimate combinations of technology and multimedia arts offer subtle, but complex expressions. The interactivity of digital video provides unlimited spatial facts: it creates real-time experience and interconnectivity. Camille Utterback's work engages a connection, which participants themselves unconsciously perceive. She invents the system based on a motion tracking video camera to manipulate the motion or image of participants. One of her interactive video installation pieces, Untitled 5, creates or-

ganic and painterly strokes by the movement of people. This intension was achieved by meaning of capturing in motion the flowing trajectories of movement in lines, made visible by an algorithm. It was one of the earliest interactive video capturing installations. Seeking a more significant meaning of capturing the essence and sensation of speed and motion, the speed applied to action renders them immaterial and invisible. The appearance of people replaced paint stokes and lines. This visualization is fluid. With the system of 'Untitled 5', Utterback designed a distinctive connection between art, technology and human interaction. Seeing is touching at a distance. In her work, people literally touch invisible tactile interfaces and see how they move from left and right and up and down. This interface becomes a living canvas; the image appears, morphs, and accumulates. She invents multiple ways of seeing and touching an environment simultaneously. The impossible becomes possible. This motion painting emerges from an invisible surface and visible atmosphere.

'Come Closer: Encouraging Collaborative Behavior in Multimedia Environment', a project by Squidsoup, uses wearable technology and collaborative interactions to vindicate and investigate our notion of personal space and proximity to others. If 'Untitled 5' is defined by unintended movements, 'Come Close' relies on a more physical existence of human beings. Technically, it has a very similar composition to Utterback's work. Even though it shares an intimate concept about the relationship between the installation and the participants, the core way of presentation is different. Three dimensional graphics projected in space seduce people with attractive floating colors, shapes and motions which are visually and physically enticing. In front of this interface, participants recognize their presence in both real and virtual space. The floating pieces of shapes react by a participant's optic breathing energy. The collaborator of Squidsoup, Cliff Randell and Anthony Rowe, put speed and force into a three dimensional graphic environment, as a search for 'simultaneity', and for a 'synthesis' between what is remembered and what is seen. This interactive installation implies not only that its tightly compressed subject is so dynamic, but that it is a fragment of a constantly changing visual experience. It reflects a desire to go beyond the virtual reality and, perhaps beyond ordinary senses.

The people, Squidsoup, are a London based group of artists, musicians and interactive designers that create works within an interactive environment. They were originally started by Anthony Rowe and James Lane, after meeting at Middlesex University in 1996. Now, they separate Squidsoup into three different areas, squidsoup.com, squidsoup.org and the game. Members of the group are from various fields: photography, education, sound art, illustration and digital media art: Squidsoup.com is a commercial site. What they are doing in Squidsoup.com is creating sound, film, video, animation, photography, typography, graphic and web design. They focus on user experiences in a working atmosphere by using an innovative and immersive interface within the design process throughout interactive multimedia. In squidsoup.org, they experiment with arts. Their work is intuitive but, unexpectedly interesting. The works combine sound, interactivity, and the joining of physical and virtual space to generate an immersive experience. The most fascinating thing in their works is that the viewers are not in a passive role anymore. The works give participants full active control of their interactive experiences. They provide a fertile space which allows the viewers to go beyond the edge of experiments technically and creatively.

'Untitled 5' or 'Come Closer' has similar interactive atmospheres which do not include sound interaction. Other works by squidsoup.org, such as Driftnet and Freq2, are constructed musical environments that viewers can navigate. Driftnet has a subtitle called, 'flying like a bird.' Participants need to stretch their arms and mimic a bird in front of the screen so they can "fly"

through a musical composition. A screen, showing three dimensional environments, reacts to the participant's presence. By flapping and tilting their arms and bodies without any other device, 3-D animation[4] on the screen tries to create simulated movements allowing the navigation through the virtual space. This is a beautiful and retrospective act letting the viewer play like a child on the playground. Sound becomes a major element in 'Driftnet.' By navigating through the immersive space, participants are interacting with the musical composition which involves the user in an intuitive, highly sensory way. The main purpose of their work is to allow participants to take active control of their experiences and movements. Driftnet I was exhibited at Shunt (London Bridge, UK) from June 13-22, 2007 as a public trial. The first prototypes were shown at Future of Sound events at SAGE (Gateshead UK) and Goldsmiths (London UK) in early 2007, as part of the Cybersonica artists' showcase.

Driftnet is a collaboration by Gaz Bushell, Anthony Rowe and Ollie Bown. Compared to other works that Squidsoup did, Freq2 is very close to the concept of 'Driftnet', but, emphasizes the sound part more. In this experimentation, participants use their body as a tool through which to create and control sound. They literally draw the sound with their whole range of body movement. Lines are the essential part of the audio wave form. They use webcam technology to capture the forms of the user's shadow outline. These waveforms are instantly visible as they change their movement. The playback of this piece is dynamic and real-time. What you see is what you hear. Freq 2 lets participants recognize their spatial and temporal moments through the relation to the wave graphics of sound. The visual part, an abstract three-dimensional landscape, extracts in real-time into the depth of space, leaving a touch of the interactions that happened. This is the memory of what has gone before. It is reflected motion through sound with a long journey of interplay. The sounds generated from live waveforms. This sound has been built into the compositional soundscape[5] with a range of pitches and rhythm components. Freq2 reacts to the passing people as well as the direct participant's interaction, manipulating an unreal reality soundscape for the space where it exists. This work was commissioned for Cybersonica 06.

ART IN THE NET SPACE

As new technology continuously evolves, internet space is growing as a new field of art. Surfing the internet brings you totally different experiences: it enable us to explore the distinctive information which is the telemetric nature of communication. Through linking the concept 'tele' and 'synesthesia', we sense that the transmission of the vast data information generates a synesthetic connection. Recently society has become digitalized and remarkably accelerated. A telemetric culture is a rapidly emerging and continuously envisioning perception and conception. Synesthesia is a blurring the border of different senses: the intermingle of vision and sound, and vision and smell. It is possible that all five sensorial interrelations are happening. Consciousness, body and senses confront with unusual experiences with synesthetic qualities. It is a result of the new way of communication, and is the new media. The concept of synesthesia is connected with the notion of the era. Our stimulation of the senses automatically draws to the meaning of association or interrelation. Therefore, synesthesia is an exceptional fact in creative movement and aesthetic interpretation. 'Cybermedia', merges and crosses the edge of internal and external areas: our senses emerge into tele-senses[6]. Virtual spaces have already expanded in diverse formats. As of yet, virtual reality does not only make the unreality quite an accessible reality, but, equally makes it practical. Marshall McLuhan, the media theorist, defines and analyzes the space as visual and acoustic. The view of the world dominated by western culture shows the perception of the world as a linear and logical thought. He explained this phenomenon in his publication: Global Village: "If man is able to transpose the workings of his central nervous system into electronic circuits, he will be on the brink of externalizing his consciousness in the computer. One could conceive of consciousness as a projection of internal synaesthesia towards the outside world, which in general coincides with the traditional description of common sense. Common sense is this specific human ability to translate one particular kind of experience towards all other senses and to present the result of this process as one global mental image."

Digital art reveals itself as a brand new aesthetic challenge. It digs into many potential dimensions, such as the World Wide Web. Browsing through cyberspace, virtual space is not unreachable anymore. The meaning of computer technology experiences is sensing and interacting with others without necessary physical contacts. Through navigating virtual environments, we share and communicate different thoughts and feelings. Images, sounds, motions and corporeal experiences are intermingling and becoming a fusion of mediums, resulting in creating audible images and visible sounds. Being on-line, an individual is connected with others in real-time without distance limitation. Cyberspace is, indeed, an unlimited, ambient space for artists and people who appreciate their works. With the manifestation of new media, the concept of art itself has become tele-transmittable, and responsive. In other words, cybernetic space promotes a new genre of perspective: tele-connection and interaction. It is apparent that tele-synesthetic experiences provide both natural senses and digitally enhanced senses.

New media art has been evolving in various forms. In "Future Shock", written by the sociologist and futurologist Alvin Toffler, he mentions a personal perception of "too much change in too short a period of time." Indeed, in such a short period of time, media art encompasses the art works that are created by new media technology, including internet art. Internet art is often called net art, and it uses web interfaces as its primary medium. *The Dumpster*: a visualization of romantic breakups for 2005, an interactive online visualization created by Golan Levin with Kamal Nigam and Jonathan Feinberg, is a collection of data-based artworks based on "blogs." This project is navigable through virtual space. Using the mouse the user can drag and click images. That allows people to take a look at the intimate detail of personal relationships and experiences. Information visualization has formed as hybrid graphics, scientific information analysis, and human-computer interaction. This used to be a traditional tool for scientists and engineers, and now it is an innovative medium for artists. By artists, it reuses new narratives which were hidden under the databases. Also, information visualization used for artistic material generates new perspectives of our society and culture as well.

The graphical interface of 'The Dumpster', categorizes the similarities and differences between patterns of the romantic pain. The break up data for 'The Dumpster' is real which extracts from millions of online blogs, provided by Intelliseek. Data is searched and analyzed for phrases such as "break up" or "dumped me." Normally, data gathering and analysis are not open to the public. These unrevealed private feelings are posted on the blogs so that other people read them. The new media theorist, Lev Manovich describes 'The Dumster' project as a "group portrait." However, this project might be reflecting us as individuals in a self portrait that media technology allows to assemble. Certain social groups' sensibilities, such as American Teenagers, are simultaneously real and abstract in this project. It expresses the intimate emotions associated with linguistic and image-based interfaces. Using very personal diaries we can navigate, compare and contrast certain social groups of people's experiences, and respond to similar emotional traumas. Re-used data based by artists create a tele- synesthetic connection from heartbroken individuals all over the world. This is a new way of presenting our social and cultural issues. The 'Cyberspace[7]' that we are interacting with now is already a popular multimedia component. Due to the development of internet technology, we have social interaction on a world-wide scale at an enormous speed. What happens in this communication is shared experiences in time-sensitive situations. In this situation, rapid communication is expanded sensations and perception of that sharing. This evolvement of new technologies, in particular multimedia art, such as net space, is emphasized on sensation and communication more than facts, thoughts, knowledge, and understanding. It could cause the issue of lacking knowledge so that it causes a critical intellectual deficiency. However, it also opens a new opportunity to explore this phenomenon in rich ways. In this trend, normal people receive the experience of synesthesia.

CYBER TOUCHING: INTERACTIVE INTERFACE

The sense of touch is human perceptions and dynamics. Touching could be thrilling, exciting, bizarre, and eccentric, or a taboo: it might be the most underappreciated of our five senses. Nevertheless, touching is how people connect with people, objects, and all different kinds of our surroundings. The sense of touch starts from the epidermis: the top layer of the skin that contains numerous nerve endings all over your body. The skin which covers all of the body protects from critical organisms that cause infection. At the same time, it provides us information about surrounded environments of outside of body. When we touch any kind of surface our skin informs us if that a specific texture is odd or familiar, sticky or smooth, hot or cold, hard or soft. All information we receive from our senses like sight, hearing, smell, and taste, go to our brain nerves through nerve endings, such as touching. The epidermis sends messages to our brain informing us what kinds of surfaces we are feeling. Then the brain decides immediately what we need to do according to what we touch. Moreover, the nerve endings feel the pressure of surfaces as well. This is how we get bruises and red marks of the skin depending on the pressure outside of our body. If we feel severe pressure, it is, we can say 'pain.' Pain is often combined with the feeling of itchiness, burning, stabbing, sharpness, ache, moving, tearing, throbbing, and numbness. It can be telling us something urgent: and is part of the automatic response of our body, regardless of the circumstance. Thus, the sense of touch is operated by intuitive reflex, and also touch is very much linked with emotions. This is related to muscular and nerve contractions which contains the emotional response. It is an inevitable phenomenon that physical reaction causes emotional response.

From the human interaction point of view, new levels of technology lead people closer to their intuition in order to use tools and interact with media arts. Intuition is an ability which we learn in the early age of our life before we recognize consciously: immediate instinctive understanding. For example, the Apple iPhone works based on human intuition by the basic physics of the body. As users rotate the phone itself, they change the view of the phone from portrait to landscape. Also, using touch screen technology, people navigate interfaces more instinctively. Those new styles of interaction draw more senses working simultaneously with greater extents than ever before. The iPhone isn't the only device to involve the human computer interface. The touch screen makes the relationship between human and machines a more natural and intuitive interaction. This is a computer display screen that is sensitive to human touch, allowing a user to navigate the system by dragging or hitting the pictures or words. It can be a substitute user interface with applications generally requiring a mouse or keyboard. It also makes the wide variety of applications, such as a public display system, retail system, customer

self-service system, computer based training, and assistive program more accesible. It can help give information more easily and rapidly by simply touching the display screen. Especially, in the fast paced environment like a retail business, touchscreen systems make work faster and more efficient. How about the automated bank teller (ATM) or airline e-ticket booth? Those kinds of system save waiting time and even labor.

Those technologies include virtual reality, tangible integration and even the game industry. In the art game genre, synestheic experience is amplified and shared by single and/or multiple players. Many of art games borrow their graphical sources and play strategies from traditional commercial games like Packman, Pong, or Super Mario Brothers. Compared to classic games, art games tend to question player's psychological approach and simultaneously figure out their conceptual messages. For example, *Electroplankton,* is an interactive music video game created by Toshio Iwai and published by Nintendo for the Nintendo DS portable gaming system. Basically, this unique software allows a player to interact with animated "plankton" and create music through ten different kinds of plankton themes. This unique visual and audio experience engages users without traditional game elements, such as scores or subjectivity. Toshio Iwai has developed the new way of approaching the game as a musical instrument combining visual interfaces. There are ten different themes with ten different styles of touch screen inputs to manipulate by players using speakers or headphones. Each the creature, Electroplanktons, produce their own unique and various sounds. Using random touch screen taps, players can make their own composition of sound pieces. Some sounds are created by the path based on the motion of the Electroplanktons. The direction of motion affects the sound of the pitch, and the path effects the length of the note. Most of the planktons are controlled by the touch screen. The microphone is another important element in this play. The players can record real-time audio along to a series of rhythm tracks, and their own voices or their environmental sounds through the microphone as well as the touch-screen. This project has a definite entertainment quality. However, can we say this is a certain type of game? Obviously, it borrows NintendoDS hardware and the way of play is very similar to other games. But, ironically, there is no goal set up, no storytelling and no score. This offers players to manipulate and respond to the multiple senses, such as: touching, speaking, seeing and hearing. It is a completely unusual experience not just in the NintendoDS platforms, but in the entire game market. It remains a huge potential in the gaming genre as an interactive toy.

Later, Toshio Iwai enhanced this game as an innovative visual instrument. With Yamaha, he developed Tenori-on, which is attractive for attention to its unique features and refined design. This is a device that enables anyone to compose sound and enjoy dots, lines, and lights. Literally, the meaning of TENORI-ON is "sound in your palm" in Japanese. Even though, the origin of this project from Toshio Iwai's former project, Electroplakton, TENORI-ON has more aspects as an instrument. Although, it contains some limited samples, it is not a sampler or a synthesizer. It has a tactile x and y matrix interface plus other buttons that adjust many different kinds of elements, such as notes, tones, rhythms, tempos and etc. I mentioned this is close to an instrument rather than a game, but this is a stylized piece of interactive visual sound art rather than a simply 'instrument.' In other words, this is a fully self contained musical instrument that allows people to produce music like sounds. Specifically, the interplay between lighting, line, and sound grabs your attention immediately. It is incredible and mesmerizing. Also, players can play a synchronized sound or send and receive sounds between other Tenori-on. The function of this piece moves between figuration and abstraction, visualization and conceptualization. This is a brand new way of musical and optical communication using a touch-screen device.

FUTURE TRENDS: BEYOND THE SPATIAL AND TEMPORAL LIMITATION

These days, mobile phones are being used widely. Artistic expression is spread out through digital video projections, games, web browsers, and mobile phones. Even though, this is an initiative movement in the new media art field, it is a very efficient and an ingenious way of spreading art works. Also, compared to game art or internet art, there is less spatial, temporal or material limitation; a mobile phone is a much more flexible and convenient device. Mobile art can take a concrete form in mobile interactive cinema, games, Bluetooth or wireless Ethernet based controllers and streaming video devices. These culturally linked forms of expression will use mobile devices as a new method for communication. Another point of interest is a movable screen or new form of remote controller to bring cross wire perceptions of realities. So it replaces other recent devices, such as Pocket PC's, I-pods and MP3 players. Although, we haven't seen much "mobile art" on our cell phones yet, this is definitely an initiating and emerging movement. It becomes more possible to search specifically designed web pages using recent mobile devices. It could also connect and transform other hybrid new media art, such as games, videos, and web art. This new and different style of communication and interaction is developing and becoming a new genre of tele-synesthesia[8] media art. Tele-synesthesia is based on the nature of the synesthesia, but it is the emergence of a new type of digital tele-contact. The recent digital revolution results from the fusion of telecommunication within hybrid multimedia. Beyond the limitation of time and space, tele-synesthesia is expanded and extended from the synesthetic aesthetic, which are the traveling senses or tele-culture. This digital era optimizes the quality of our life and future. We are overcoming the limitation of our temporal and special boundaries by using innovative technology.

CONCLUSION

Through our senses, we feel, see, smell, and hear our surroundings. Since we always use our senses, we hardly pay attention to them. However, each of these senses is irreplaceable and valuable. What we perceive through the nervous system defines our experiences. As a specific sense of interaction in an integrated perception, sensory fusion is an essential human ability. The term, 'synesthesia', has been displayed since about a century ago. Nowadays, this is quite common in aesthetics, although there are some controversial issues that remain in between neurological science and art boundaries. In order to comprehend the aesthetic of art, we need to understand the human sensory mechanism. Synesthesia is a critical characteristic of new media art. New media art contains a synesthetic complex, particularly in interactive art forms. The level of interplay between views and arts has been developed by the emergence of multi sensory functions. Synesthetic interplay often occurs with trained artistic individuals. It is quite an important feature in art theory and mono sensory arts. The function of synesthesia evolves considerably in the new situation with new technological tools, such as electronics, sensors, computers, and internet spaces. Recent developments in hypermedia, multimedia and virtual reality has challenged the traditional perspective of human computer interaction. This brought us many other possibilities of our various perceptions. There are differences between 'sensation' and 'perception.' Sensation is what stimulates our sensory organ, perception is what is experienced mainly as consequences. In synesthetic art, a sensation that normally transmits one perception actually produces another perception or more, so that the sensation stimulates other multiple sensory systems. Synesthetic media has been already appearing in various forms. This phenomenon alters how sensations are perceived. For example, recent information technology conveys rich experiences that were previously not available to most

other people. The purpose of new technology is to broaden our realm of sensation to feel reality more fully. This is a crucial point in the stage of developmental technology and its impact on our culture. In the psychological approach, understanding new media art corresponds to analyzing synesthesia experiences in human interaction. In terms of creativity, specifically, hybrid, cross wired sensory experience stimulates aspects of the human sub consciousness. In the emerging media arts, synesthetic interactions are increasing and becoming something to be inspired by. Based on new empowered technology and highly developed research, this interplay will become more sophisticated, intricate, and immersive. Ambivalent senses will be obtained as well.

REFERENCES

Aderson, Eh. (2000). Lightness perception and lightness illusions. In MGazzaniga (Ed.), *The New Cognitive Neurosciences*. Cambridge: MIT Press.

Blais, J., & Ippolibo, J. (2006). *At the Edge of Art*. New York, NY: Thames & Hudson.

Bowlt, J.E., & Long, R-C.W. (1984). *The Life of Vasilii Kandinsky in Russian art: a study of "On the spiritual in art" by Wassily Kandinsky*. Newtonville, MA: Oriental Research Partners.

Brougher, K. (2005). Visual-Music Culture. In K. Brougher, J. Strick, A. Wiseman, & J. Zilczer (Eds.), *Visual Music: Synaesthesia in Art and Music Since 1900*. New York: NY, Thames & Hudson.

Caroline, A. J. (2005). *Sensorium: Emboded experience, technology and contemporary art*. London, England: MIT Press.

Cytowic, R.E. (1989). *Synesthesia: A union of the senses*. New York: Springer.

Cytowic, R.E. (1996). *Synesthesia: Phenomenology and neuropsychology*. New York: Psyche.

Cytowic, R.E. (1997). Synaesthesia: Phenomenology and neuropsychology. In S. Baron-Cohen, & J. E. Harrison (Eds.), *Synaesthesia: Classic and contemporary readings* (pp. 17–39). Massachusetts: Blackwell.

Gibson J.J. (1966). *The Senses Considered as Perceptual Systems*. Boston: Houghton Mifflin Company.

Grau, O. (2006). *MEDIA ART HISTORIES*. Cambridge: The MIT Press.

Grice, H.P. (1962). Some Remarks About the Senses. In R. J. Butler (Ed.), *Analytical Philosophy, First Series*. Oxford: Basil Blackwell

Hajo, D. (2000). *Wassily Kandinsky 1866–1944: A Revolution in Painting*. New York: Taschen.

Jütte, R. (2005). *A History of the Senses: From Antiquity to Cyberspace*. Cambridge: Polity Press.

Kahn, D. (2001). The Sound of Music. In D. Kahn (Ed.), *Noise Water Meat: A History of Sound in the Arts*. New York, London: The MIT Press.

Kandinsky, W. (1963). *Concerning the Spiritual in Art: And Painting in Particular*. New York: G. Wittenborn.

Maeda, J. (2007). *Aesthetics and Computation Group*. http://acg.media.mit.edu (accessed May 2007).

Marks, U. L. (2000). *The Skin of the Film: Intercultural Cinema, Embodiment, and the Senses*. Durham, NC: Duke University Press.

Manovich, L. (2001). *The Language of New Media*. Cambridge: The MIT Press.

McLuhan, M. (2007). Visual and Acoustic Space. In C. Cox & D. Warner (Eds.), *Audio Culture: Readings in Modern Music* (pp. 67-72). New York, London: Continuum.

Peacock, K. (1988). Instruments to Perform Color-Music: Two Centuries of Technological Experimentation. *LEONARDO, 21*(4), 397-406.

Rius, M., Parramón, J.M., & Puig, J.J. (1985). *The five senses: Sight*. Hauppauge, NY: Barron's Educational Series.

Vanechkina, I. (1980). Complex approach to the research on A.N.Scriabin's light-music conception. In *Problem of complex research of art creativity.* - Kazan: Izd. KGU, (p. 107-115).

Varese, E. (2007). The Liberation of Sound. In C. Cox & D. Warner (Eds.*), Audio Culture: Readings in Modern Music* (pp. 17-21). New York, London: Continuum.

ENDNOTES

[1] Synesthesia means a condition in which one type of stimulation evokes another sensation.

[2] Synaesthete means a person who experiences synaesthesia.

[3] Micro encapsulation means a process by which very tiny droplets or particles of liquid or solid material are surrounded or coated with a continuous film of polymeric material.

[4] 3D animation means animating objects that appear in a three-dimensional space.

[5] Soundscape means an atmosphere or environment created by or with sound.

[6] Tele-sense means sensing computers technologies and telepresence capabilities.

[7] Cyber space means the electronic medium of computer networks.

[8] Tele-synesthesia means sensation evoking by a telematic use of new media such as the travelling senses.

Chapter X
The Aesthetics of Net dot Art

Yueh Hsiu Giffen Cheng
Yuan Ze University, Taiwan

ABSTRACT

The development of net art originates from the rising of net media generally. During the past two decades, Net art has overthrown the standards of traditional aesthetics, just like the conclusion given by Walter Benjamin that: "Technology and techniques restructure the human sensory apparatus (Esther, 2000, p. 42)." Even though we can admire the unique aesthetics of net art as they appear in some postmodern art movements, Net Art is in such a multi-polar form that one cannot easily find a single point from which to admire it. Based on the research of literature reviews and case studies in net art, this article tries to discuss the characteristics of net art, and classify net art into eight catalogs (Email Art, Non-linear Narrative, Online Performance, Information Art, Game art, Collaborative creation, Internet Community, Physical Interaction) in order to emerge the aesthesis of net art from the general aspect of contemporary arts.

INTRODUCTION

This chapter discusses the characteristics of net art according to different categories, attempting to emerge a breakthrough of net art aesthetics from the phenomenon of its chaos. However, there are some difficulties that I have encountered during the course of cataloguing my findings. First of all, the form of net art depends on technical media; as technology moves on, the forms of net art are getting more difficult to neatly organize. This is why David Ross describes net art as purely ephemeral (Ross, 1999, p. 37). Net art is a historical factor which is proceeding, and it is indeed difficult to determine the category of net art in the progressive tense. Yet, how can an uncategorized art be discussed as to its standard of value? I believe this question also bothers critics of post modern net art.

Rudolf Arnheim argues that in a period of transition from a "post-modern" era to an "information" era, aesthetics is no more regarded as a question of form or style, but of philosophy. It is not about disorder, structure or fracture, but to sense everything, reaching a status so that every thought connects to all human beings (Arnheim, 1996, pp. 117-120). When aesthetics meets philosophy, the discussion is not just limited to the reaction of vision and psychology, but reaches a higher level of concept and creation of concepts. Philosophy has components of logic and thought, and therefore includes a relatively rational argument. Tilman Baumgartel writes in the essay "Net Art. On the history of Artistic Work with Telecommunications Media" that "Net Art is almost a type of new Jerusalem where that which is impossible in the "real world" should happen: global "herrschaft" – or free communication for all, consumers who become producers, social networking over and through geographical and social borders, direct information exchange beyond economic constraints and without filtration through the mass media...... From an art-historical perspective, an important aspect of net art is that in the meantime not only texts but also the most diverse media (film, sound, graphics, animation sequence, photographs, 3-D simulations, etc.) are there in the Internet, next to each other, and can be transmitted. Everything that can be translated into bits and bytes can be brought online (Baumgartel, 2001, pp. 152-159)". Therefore, net art is so broadly metaphysical that it is inappropriate to simply divine the movement according to its materials.

George Fifield, a curator at the Decordova Museum and Director of Boston Cyberarts, writes that, "Interactivity is the great question of this newest art form. What form of interactivity will most engage the audience and provide a lasting aesthetic experience - emotional, rich, and satisfying? After centuries of linear narrative and the painted square, artists are looking at ways that the art itself can engage with the viewer and modify the artistic experience (Foote, 2003)". Dr. Amy Dempsey in turn says that: "Internet Art is democratic, and interactivity is its key feature. Images, text, motion and sound, assembled by artists, can be navigated by viewers to creative their own multimedia montages of which the ultimate

'authorship' is open to question. Viewers become users (Dempsey, 2002, p. 286)". In a kind of summary David Ross simply states that, "interaction is the nature of Net art (Ross, 1999)".

It is useful to turn to a practical industrial means of judging values by focusing on how awards are handed out. In Austria in 1997, the judges of the Global Electric Art Award stipulated eight rules for judging Net art works submitted in competition. These were: A). Use of technology: the technology application should be creative. B). Grammar: creativity in using language links. C). Structure: structural creativity in the texts. D). Public service: creativity about online public service in the work's concept. E). Net-awareness or self-reflectiveness. F). Co-operation: coordination between text and components, and between human and network. G). Community and identity: creativity in the degree of Net communities' interaction. H). Openness: the creativity of open texts (Spaink, 1997, pp. 4-6). This annotated list covers most forms of Net art works, and so I made use of these eight rules and the aforementioned scholarship to divide the categories of works I found by the interaction modes between the audience and works.

BACKGROUND (DEFINITIONS OF NET ART)

Net art was added to the list of Prix Ars Electronica in 1995, and since that time juries of the prize have never stopped discussing the question: What exactly is art on the net? There is a perpetual debate about net art: about its definitions, its boundaries, its function, its aims and its means; about accessibility, politics, status; about high culture versus popular culture; about inclusion versus exclusion; about funding and money; about art and anti-art (Spaink., 1997, pp. 1-2). These art topics are always being discussed; net art is not an exception either. In order to the make an impartial judgment on the Net Version of Prix Ars Electronica, the juries of the prize announced a statement about Net art:

Net art does not equal taking whatever is on your gallery's walls, converting it into something the computer can digest and then making that accessible via a homepage. Net art deals with the consequences of what it means to be on the net and the implications of the choice of making that particular technology one's medium (Spaink., 1997, p. 2). This simple definition implies that net art is not the same thing as arts online.

In 2000, Andreas Brogger published an article called "Net art, web art, online art, net art?" In this article, he proposed a tentative definition of net art. He argued that "net art is art that cannot be experienced in any other medium or in any other way than by means of the network (Brogger, 2000)". The rise of the net started only ten or so years ago. Many people are still ambivalent about the definition of net art. However, it needs a clear setting of its boundaries. Ben Davis proposes the following view: searching for art via network is an art activity of networking. Art on the net is not material, but a kind of circumstance. The net is a thinking field which can present different aesthetic ideas. In this field, different ideas can have mutual communication (G. Chen, 1998, p. 81).

The former curator of the Walker Art Centre, Steve Dietz, has offered a useful definition of net art: "Internet art projects are art projects for which the Net is both a sufficient and necessary condition of viewing/expressing/participating. Internet art can also happen outside the purely technical structure of the internet, when artists use specific social or cultural traditions from the internet in a project outside of it. Internet art is often, but not always, interactive, participatory and based on multimedia in the broadest sense(Wijers, 2005)." In addition, Chiara Alinovi published an article titled "NET.ART The art of our times" in the Stedelijk Museum Bulletin in year 2003. Alinovi points out that net art as an art that is especially made for, and published on, the Internet (Alinovi, 2003).

"Documenta X" of 1997 is a turning point for net art, the short history of net art makes it seem like a baby that is still a toddler. Net art may be crowned with a halo of modern technology, but since it is still undergoing rapid transformation, not only is it hard for experts and scholars to make a definition, but also audiences become confused when looking at it. No wonder, in the end, people use "new media art" to name the art form combination with technological industry.

As David Ross has noted (in a text attempting to establish "21 Distinctive Qualities of Net. Art"), there is often a high degree of intimacy between the user and the art on the web. We can view it, and use it, in our home. The interaction is individualized (though not always) (Ross, 1999). In the almost ten years since David Ross's published article, although the form of net art is changeable, it hasn't developed within the scope of these rules so far? Does it mean that a miniature form of net art development has gradually been fixed? Or, is net art out of fashion? I think that it is hard to answer these questions -- the situation is the same as the fact that it is difficult to define new media art.

CATEGORY OF NET ART

Email Art

Mail Art is an art movement that uses mail delivery as its medium. It proposes that art is an expression of dialogue and emphasizes that there are no standard regulations for participation in this art practice other than use of the mail service. Everyone can be a mail artist as long as one has ideas and sends mail. Obviously, this boundless mode of Mail art incorporates a spirit of Fluxus and Dadaism. The Fluxus movement, which developed during the 1960s, pursues artistic creation outside pure art and emphasizes that any action or object can be an artistic creation. This concept has made mail art, which communicates through mail and letters, a new member of the art history for a good reason. Fluxus' influence on Mail art includes the construction of techniques and ideas, and, more accurately, it restores Friedrich Schiller's and Immanuel Kant's "Play impulse" theory regarding the origins of art.

Ray Johnson, the Father of Mail Art, established The Correspondence School of Art in New York in 1962, making efforts to represent and categorize art through the mail. In order to invent an exchange mode as a form of art, and to make communications pass from artist to artist and cross from country to country, Ray Johnson tried out many playful experiments in the art of message exchange. The most classic one was that Ray Johnson tore off pictures, attached them to different postcards and sent them to many people. Among the numerous receivers, some people sent back the postcards and added their opinions and drawings. Thereby, the postcards delivered by postmen became an interesting form of mail art with a lot of people's contribution. Collage is most common form of mail art construction, which is similar to the ready-made art of Dadaism.

The rubber stamp is another common component of mail art. Some artists like to create their own stamps while some like ready-made stamps or various types of stamps to decorate artistic mail. The repeating action of a single pattern matches the repeating creation of Dadaism. Writing design is another basic component of mail art. Some people use ready-made printed writing to do a collage, while others like to draw special handwriting with pen-and-ink. This expression of words in art is just like the conceptual artists' metaphor of words, but the uniqueness of mail art is its' postmark.

In the 1980s, a huge number of mail art exhibitions declared the peak of this art movement. Later, in the 1990s, following the introduction of Net media, artists competed using their creative talents on the Net, so, alongside snail mail postmen, Email becomes another deliverer, which also represents a special trope of net art creation. The form of Email Art is very extensive, including picture delivery, re-editing, language communi-cation, ASCII expression, even attachments of music media, etc. – such a colourful art.

The great difference between Email art and traditional Mail art is "immaterial", but tangible material is the essence of traditional Mail art, hence there is a great gap between Email art and traditional Mail art; a gap waiting to be overcome. According to Tilman Baumgartel: "That mail art never really accomplished the cross-over into the art industry lies in its particular qualities, and in that it is also similar to contemporary net art. From its nature, mail art was a network matter and like the Internet, the Mail Art network had no central point and theoretically was open to all - which, paradoxically, did not increase its visibility. On the contrary, it was the network character that made mail art artists a closed group to which one either belonged or did not. The most important reason that the art industry never really became interested in mail art may be mainly due to the lack of works able to be exhibited or sold - similar to net art. The works were not more important than the communication process itself" (Baumgartel, 2001, pp. 154-155). Hence, even in this convenient net era, there are still many mail artists who prefer to send traditional analog mail to preserve its handmade beautiful touch.

"Landscape Exchange" (http://98.to/pcd/) is a piece of mail art that combines electronics and tradition analog modes of production. The composer Mingda Xie invites artists all over the world through his website to participate in the creation of this Email Art work. First he held in his hand a postcard ready to send, standing at a scenic spot, and took a photo. He emailed this photo to a participant who also received the paper postcard. When the participant received the postcard, he could do the same thing: "taking a photo and posting the same postcard" on to someone else. With the idea of "Landscape Exchange" and the cooperation of participants, the same postcard will show up in two photos with different scenery in the background. This gives a record of the interlacing of time and space and also the wonderful application of conceptual art to mail art. The method of digital photos in addition to emails made this "Landscape Exchange" mail art flow in both virtual and real worlds, thus successfully overcoming the absence of touch in Email Art. Furthermore, it increased the multipolarity of traditional mail art.

Walter Benjamin points out in relation to the mechanically produced representation of an "original", context is central to the meaning of a work of art. The difficulty in making art for the net is foremost a problem of context (Walter, 1955, p. 221). Artistic works produced by technology may have lost the "uniqueness" of traditional art because it is duplicable; therefore their "aura" has disappeared. Nevertheless, Walter Benjamin argues that the characteristics of technological duplication have increased the value of exhibition. When one is watching or admiring art, without the desire to own the piece or any part of it, one can be purely close to the pleasure and inspiration of art (Hu, 2005, p. 37), perhaps this is a factor in the new aesthetics of Email Art.

Non-Linear Narrative

From the beginning of human civilization, narrative works have doubtless been a basic mode of artistic expression across its' eight great fields (painting, sculpture, architecture, music, literature, dance, theatre and film), or, to narrow down to the fine arts: painting, sculpture and architecture. However, "non-linear" characteristics have now overthrown the faith people have had about art for nearly twenty centuries, and the aesthetics of art has entered the period of Post-Modernism.

Post-Modernism, which began in the 1960s, does not have a traditional fixed form of expression. It can be called an anti-rationalistic aesthetic, emphasizing the "openness" of art works. From the aspect of technology, the birth of the "non-linear" mode of practice must be attributed to the rapid development of computer technology. The multimedia interactive CD title developed during the 1990's could be seen as a test version for non-linear works. Following the maturation of net technology, a collective creation of internet works has become the latest version, which comprises interactive characteristics with readers/audience and represents a breakthrough in the concept of "open works". The two complements of interaction and openness realized the concept of non-rationality in Post-modernism and the spirit of regarding the reader/audience as the subject. Roland Barthes, a French writer and theorist, proposed the theory of "the death of the author" in 1967. Barthes argues that although the structure of literature is given by the author, the author's influence no longer exists when he has stopped writing; instead, it is replaced by the structural symbols of the text itself and the symbols from the reader's personal experience while reading. Roland Barthes stated that the immortality of a work is not because the work gives different meanings to different people, but because the work implies different meanings to everyone – the time of reading is the time of writing. Thereby, the author and the text are born at the same time, and the work becomes an open text; readers become authors, and the birth of the reader must be at the cost of the death of the author (Barthes, 1977, pp. 142-148).

The theory of "the death of the author" by Roland Barthes corresponds with the notion of "collaborative creation on the Internet". The notion of an author in a non-linear narrative falls apart as new meanings of the text are re-transmitted after collaboratively constantly being recreated or rewritten by its users in real time on the Internet. Roland Barthes proposed another philosophical statement, "From Work to Text", in 1971: texts have extending meanings, which can be extended boundlessly. Texts have numerous meanings, which form an *intertextuality* of the text itself. A text does not make reading a passive process of receiving, but is a script for active play (Barthes, 1977, pp. 155-164).

George P. Landow in his Hypertext 2.0: Convergence of Contemporary Critical Theory and Technology mentions that "the Hypertext Theory has much in common with the Writerly Text Theory of Roland Barthes. They both have the textuality and intertextuality that dismiss hierarchy; they are the direct expression of post-structuralists' textual theory (Landow, 1997, pp. 4-5, 25)." In "From Work to Text", Roland Barthes foresaw the direction of development for the future, and outlined some of the implications of the aesthetic of non-linear works that would have an impact on postmodernism: Artistic works imply that the trend of the commercial market and the transmission of meanings is too narrow and closed. Yet the openness and multi-vocality of texts have surpassed the works – this is indeed what is envisaged in post-structuralism and net art culture specifically.

The non-linear fiction "Blindspot" (http://adaweb.walkerart.org/project/blindspot), combining words, voice, pictures and interaction with readers, by the New York writer Darcey Steinke, became an early classic net fiction work in 1999. The main structure of this fiction describes a housewife's feeling of waiting for her husband to come home. The background colour of the pictures is mostly grey, like the elegance and tenderness of the language, which resembles the romantic but gloomy characteristics of noir novels. On each page of this work, there are many links to paragraphs or pictures pointing to branches of other episodes to describe the housewife's mood, and some pages even allow readers to leave personal opinions or comments. So the readers become indirect authors who transform the story of this novel cooperatively. "Blindspot" joined the Ada exhibition project in 1990 as a non-linear fiction at an early stage. The work simply used the basic Hyperlink function of the internet, obviously so as to lower the influence of technology to a minimal level and preserve the writing style of Darcey Steinke in a logical way.

Postmodern art digested the diversity of art while at the same time interlinked different categories of art. In the field of net art, there is no single art factor which exists alone, and those factors become the "text," as it was called by Roland Barthes. Darcey Steinke presented a net text, "Blindspot", that spoke across media with voice, animation, film and words. Each artistic factor had its own multi-meaning language, which intersected and formed intertextual episodes – the quintessence of non-linear narrative.

The young net artist Santiago Ortiz from Colombia is fond of visual creations that use word play. His recent "Spheres" (http://moebio.com/spheres/), posted on rhizome.org in February 2007, represents a new form of word game and a dialogue of possibilities. Santiago Ortiz used 122 words as the base, allowing readers to link any two words and leave personal opinions about the words in the Pop-ups. With an increase in participants, the connections between these two words become more complex or elaborate. After a careful calculation, 122 words bring out 7381 linking lines, and each linking line may have several responses. After those users' participation, a sort of narrative story is born from the text. However, all of the sentences are fragments, and each reader receives different messages, so the process of reading is like a jumping melody, which assembles poetry that implies the beauty of uncertainty often at work within post-modern literature. "Spheres" realized the concept insisted on by post-structuralism: "Post-structuralists hold that the concept of "self" as a singular and coherent entity is a fictional construct. Instead, an individual comprises conflicting tensions and knowledge claims (e.g. gender, class, profession, etc.). Therefore, to properly study a text the reader must understand how the work is related to his own personal concept of self. This self-perception plays a critical role in one's interpretation of meaning (Lash, 1991, p. 175; Mizrach)."

Online Performance Art

Performance art, which began in the 1960s, is a public performance that individuals or groups perform for the public at specific times and locations. Due to the influence of Dadaism and Happenings, Performance art emphasizes creation of occasional and ad-lib acts; it is an avant-garde performance and an invention of an event. Performance art can be any situation that involves four basic elements: time, space, the performer's body and a relationship between performer and audience. There is no other restriction on the definition of Performance Art (Goldberg, 1988). However, when Performance Art encounters Net Media, four simple factors will have complicated the interaction and transformation that results.

First of all, as for the factor of time, there is only a clock on the computer screen but no central standard time in the world of the net. For example, when a piece of work is exhibited online, from the beginning to the end is simply the change of link passage from new page to saved page. Time plays no more than the role of virtual setting in the pages. As for space, the concept of distance is zero in the net world, or one can say the only distance is from the computer to the plugs. In a net world, a gallery can be the balcony of one's own house, or the highest building in the world (Taipei 101). This space is just the arbitrary actual place of the user. However, physical performances by performance artists on the net and participation of the audiences are factors difficult to assess. Because interaction is the basic route of communication with the computer, net performance artists and audience have to produce a connection by interacting with a computer first. In general, when Performance art encounters the Internet, the artists have to take control of the game rules at both ends, then they can make good use of the Internet media.

Prof. Jack Bowman from Kentucky in the USA, as a performance artist, poet, writer and performer, based on his research into and love for performance art, defined ten precise rules for performance art in the 1990s; some of them are worth mentioning here for a comparison with Internet media. Bowman notes that, "performance art is usually immediate, which means the time of its being". The 'time being' on the Internet refers to the time when a user is facing the software system, not the time for users at the two ends behind computers. So, the definition of "immediate" on the Internet has a gap with its definition in the actual world. Bowman also argues that, "performance art can involve the audience with taste, smell and sounds not available with electronic media and not practical with conventional theatre. This is due to the usually small audience" in attendance (Bowman, 1990, p. 1). The Internet on the other hand delivers a small performance for thousands of users, but it also makes the haptic barriers of taste and smell impossible to solve with current technology. Hence, when we discuss the characteristics of Internet media, Jack Bowman's definitions are still worth considering.

Traditional Performance art includes Body Art, Fluxus, Happenings and action poetry, but now that the net has introduced a new platform media, how will these traditional art forms change? The best person to answer this question must be Arcangel Constantini who began to discuss the art value of net competitions in 2000. Constantini constructed an online competition in 2002, called "Infomera" (http://www.museotamayo.org/infomera). The motivation of this project comes from the aim of creating an ephemeral net art work. "Infomera" was an online competition of two artist teams, each of seven members. Each game took 120 minutes. Within the time limit, two artists separately used webpage components, such as videos, voice, Java software, animation or hypertext, to kill the other one's webpage. The screen was updated every minute, so every audience in the world could watch the progress online. In order to increase the participation level of the audience, Arcangel Constantini allowed every registered member of the audience to throw tomatoes or hiss at participants, and they could even join the judgment-making at the end of the game.

From the creative expression of Arcangel Constantini, we can understand his concept that says that in new media art as in performance art, "the artwork has been transformed into a structure that relies on a constant flux of information and engages the viewer/collaborator the way a performance might (Morris, 2001, p. 9)." In addition, Mark Napier also thinks the definition of New Media is similar to that of performance art, since the software is something you basically perform on your machine (Morris, 2001, p. 10). It doesn't matter if the online contest is one of Arcangel Constantini, or "Auto Portrait" of David Bouchard, or even "Tendril" of Ben Fry, they are all performance within the mode of programming language.

Data Visualization/Information Art

Data Visualization can be traced to the development of electronic art or computer art, or more precisely, another form of art brought about by science. The composition of Data Visualization comes from a huge amount of resource materials, to categorize various sorts of data using the methods of science, mathematics, and logic, then to visualize the patterns of the statistical contents. Data Visualization of net art explores the subjective experience composed by net data, and expresses it in abstract and reflective visual forms. Generally speaking, Data Visualization works have a component factor reminiscent of Op Art that involves carefully calculated patterns and strongly contrasting colors. From the aspect of design, both Data Visualization and Op Art purely use visual stimulation and sensation to challenge traditional aesthetics of sense and sensibility.

The Op Art popular in the 1960's emphasized creating an illusive space by non-physical invention, and uses meticulous methods of science to stimulate the visual nerves to produce a systematic art work. From the viewpoint of the contemporary world, there is neither weight nor shape to computer information – it is pure pattern. "The

term 'immaterial' refers to a somewhat daring neologism," said Jean Francois Lyotard in an interview about the Centre Pompidou, "It merely expresses that today - and this has been carried through in all areas - material can no longer be seen as something that, like an object, is set against a subject" (Baumgartel, 2001, p. 153).

So-called Data Visualization discusses how to make those "immaterial" things visual through precise science. Data Visualization uses elegant scientific lines to declare another form of new aesthetics, and realizes the theory of Paul Brown: "Science is evolving into a new science called art (Wilson, 2002, p. 28)." I regard Data Visualization as the "New Aesthetic of Op Art" under the rhetoric of post-structuralism.

In the category of net art, a large part of Data Visualization explores the large amount of information exchange, such as users' moving tracks on the screen, or searching the clues of hyperlinks to other websites. Visualization of a moving track which was originally invisible expresses the art of delivering information; this is what I will introduce in this section: trace and collect moving tracks of net users, then show this large amount of information data through the visual patterns of mapping.

In 2006, Lisa Jevbratt was chosen by The Swedish Public Art Council (SKR) to work on "The Voice". This project will last for three years and uses the website of SKR (http://www.statenskonstrad.se) to collect source material from 2006 to 2009. When users use the search engine of this website, "The Voice" will record the page data which were connected successfully by users, including the connecting time and total time of being read. The data collected by "The Voice" will visualize the read record on the website: word size indicates the frequency of being searched for, colours and frames demonstrate the frequency of being read. The first and last words being searched are listed in the left-upward and right-downward corners of the screen. "*The Voice*" explores the moving tracks that people leave in the public space and collects the "quest" clues left by the populace through the action of net searching. It forms a collective consensus, collective identity and collective voice.

Mapping Data Visualization, coming from modern business of information, aims to present the context of complicated information and to present a visible pattern of originally invisible tracks through a mode of mapping. J.B.Harley-Who is credited with bring a cross- disciplinary approach to cartography- argued that maps are social construction of the world: "Far from holding up a simple mirror of nature that is true or false, maps redescribe the world, like any other document, in terms of relation of power and of cultural practices, preference, and priorities (Abrams & Hall, 2006, p. 12)." If we admire the beauty of Data Visualization from an aesthetic angle, large and complex ever changing order in chaos flux is beautiful.

Game Art

Regarding art theories, we have discussed Friedrich Schiller's and Immanuel Kant's "Play Impulse" theory that art originates from the behavior of art creation while playing; for example, painting came from daubing, dance from hop-scotch, and music from nonsense sounds. However, their original potential has been forgotten gradually by civilization, or it is said this essence as a hidden play characteristic has been forgotten by traditional art education. Among the series of art movements of the 1960's, we see a play characteristic in art again. It is an interesting and amusing form of art that enters the populace again; art is not so lofty any more. The play concept of art shows the internal spirit of logical phenomena, the play characteristic representing a reflection of the time, expressing the crises and confrontations in the cultures of prosperous societies (Chen, 2004, p. 434).

Following the development of technology, games become another invention, even a cul-

ture of beauty in another new media art. Lev Manovich in his book The Language of New Media mentioned discussions about the aesthetics of electronic games. Ars Electronica proposed to regard electronic games as a form of art in 2001. The movement of combining games and art was brought to birth silently, and in the waves of dispute of course it was blown into the field of net art.

An American artist, curator, writer and gamer Anne-Marie Schleiner, described these possibilities in an interview in 1999: "As a curator I am interested in the notion of art as culture hacking, art with a critical agenda that seeps outside the boundaries of prescribed art audiences and engages itself with a broader public (ie. the gaming public). Art that finds cracks in the code and hacks into foreign systems. I also want to invite a cross-pollination of gaming and art strategies by providing artists with tools and techniques developed by game hackers and exhibiting game patches created by gamers as art (Schleiner, 1999, p. 2)." No matter whether it is net strategy or pure net art, interaction effects are the real motivation for net art (Chen, 2002, p. 109).

Natalie Bookchin's "The Intruder" (http://www.calarts.edu/~bookchin/intruder/) is a standard net art work that regards games as [providing direction for] invention; this is a representative work discussing play art. Natalie Bookchin uses narrative stories as a structure and simple games, such as shooting, table tennis and tracing, to create play interaction with users. After users finish an action, such as shooting or picking a ball, the narrative voice-over will give an imperative for the next action. In order to see the whole story, users will try hard to meet the rules and requirements of the game. The playing process of this work consists of game history in one part, and another, larger, part of the experiment in applying physical relationships to words.

"Contemporary art had already begun to move toward fun and entertainment, with the tacit consent of the art community (Cornwell, 1996), " which is especially true in an age of new media art. Artists before were afraid of being labeled as "playing" or "game-like", but now we can answer loudly that "playing art is another form of aesthetics of new media art". It is like Lev Manovich's opinions about the two games of Doom and Myst; he argues that games bring people back to a primary world of fairy tales (Manovich, 2000, p. 245). Compared with the complicated real world, legends and fairy tales are indeed more attractive. The ultimate goal of art is to feed people with spiritual food, and playing art suits modern people of high technology, doesn't it? When Anne-Marie Schleiner was asked her opinion about artists adding game-like aspects into art work, she answered: "Artists can bring a critical and perhaps more diverse agenda in terms of age, gender, and politics to computer games. Artists are adept in approaching cultural artefacts in a manner that merges form and content with an attuned awareness to cultural belief systems that are embedded in aesthetics (Schleiner, 1999, p. 4)". I remember when I was studying at Fushing Art School in Taiwan, one of my teachers often mentioned his philosophy that: "art creation is first simple then becomes complex, and then moves from complex to become simple. Yet the latter "simple" is no more simple in the end." Perhaps Game Art will return to this principle form of creation, but the difference is that it applies high-tech tools to stimulate a visual reaction. The concept of playing looks simple, but its cultural meaning is quite complicated and never simple.

Collaborative Creation on the Internet

Following the coming of the Web 2.0 Age, sharing and collaborative creation has become the developing mode of net resources; "in twenty-first-century culture, collaboration seems the order of the day (Inge, 2001)." As the relationship between users and net applications moves from dissemination to participation, from the personal website of a

single path to the blog of mutual feedback, and from the online Encyclopaedia Britannica to the Wikipedia co-edited by everyone, Web2.0 has become the name for collaborative wisdom and collaborative contribution.

The Web 2.0 Age emphasizes the development of de-centralization, collaborative creation, re-mixability, emergent systems and other attributes of users' experience (Lin, 2006), so users play the central role. The concept of Web 2.0 seems to match Roland Barthes' theories of *the* "Writerly Text" and "The Death of the Author". The so-called "writerly text" refers to the decentralisation of textuality and intertexuality. When readers/audience are reading /watching the works, they can add their opinions to the works, such as open texts which involve co-editing or collaborative creation. Following

Based on case studies of collaboration art projects and literature review studies, I have analyzed four characteristics of collaborative creation within internet media: Playing participation; the growth of the art form; The Verbality of Art, and The Transferring of Authorship. According to the outcome of this research, a new way of appreciating the new form of net art has emerged.

PLAYING THE GAME: IMPORTANT NET ART FACTORS ATTRACT USER PARTICIPATION

Interaction between works and users is a key factor in net art. The integral exploration of a work demands the default path of a creator and also the complete participation of users. Net works, which require interaction, ask for a certain period of time for the users to finish browsing and operating the work. Unfortunately, most people have limited patience toward art.

According to America's Harper's Bazaar magazine, audience members at an exhibition only stayed for from five seconds to three minutes in front of each work (Yeh, 2002, p. 109). So, the most important consideration for creators to think about is how to attract users to participate in interaction with net art works. I found an interesting fact from examining numerous net art works of collaborative creation, that many works consisted of playing factors; it seems that the creators hope to attract participants through the inducement of games. In traditional art education, we learned how to admire a painting, how to see a sculpture or how to listen to a melody; this kind of education made people a passive audience. By contrast, in the field of interacting net art, the audience has to be the active agent, otherwise the admiration of art works cannot proceed. The question is: how to make a passive audience become positive participants? I think this is the reason why many net art works have the attributes of games. Through the inducement of games, passive audiences become participants in art creation voluntarily. In the process of the game, users spontaneously explore all messages delivered, so net art works can be displayed integrally.

The group of Sulake's "Habbo Hotel" (http://www.habbo.com/)is an online friend-making website exclusively for youth, and it hopes that youth can learn appropriate social life through this virtual social field. This website takes the game concept as its structure to create a big global hotel chain. Anyone can live in the "Habbo Hotel" after a simple registration online, and free membership is the reason why it is so popular. In this hotel, user can use fake names, false sex, or even shape a perfect person for themselves to meet. Because there is no identity validation, there is not the personality burden of real life, which is another reason why the website is very successful. The game structure and vivid virtual motions in "Habbo Hotel" drive users to spontaneously make a contribution to the website, and to accept the social experience acquired from this hotel; the integrity of this work has been achieved perfectly.

THE GROWTH OF THE ART FORM: THE SHAPE OF WORKS CHANGE WITH USERS' PARTICIPATION AND CONTRIBUTION

The net is an easy-to-use medium, and art works that take the net as their preferred medium must allow users to enter the work easily. Hence, the net becomes a public space, and art accomplished on the net also becomes a type of open public art. According to the definition from Britannica Encyclopaedia, so-called public art is art work exhibited in a public space, allowing the public to participate or touch the works. On the other hand, for net art works, interaction and audience participation are the main factors. If the works can reflect the users' interaction as a contribution to the works, can they accomplish the ideal of public art itself? Or it should be called another kind of exhibition of online public art?

Jeffrey Shaw, a famous Australian new media artist states that now with the mechanism of the new digital technology, the artwork can become itself a simulation of reality – an immaterial digital structure encompassing synthetic spaces which we can literally enter. Here, the viewer is no longer a consumer in a mausoleum of objects; rather he/she is a traveler and discoverer in a latent space of sensual information, whose aesthetics are embodied both in the coordination of its immaterial form and in the scenarios of its interactivity manifest form. In this temporal dimension, the interactive artwork, in each time is restructured and re-embodied by the activity of its viewers (Foote, 2003). In other words, for an integral work of collaborative creation, the performance of its art form must change with users' participation and contribution; the art form is not controlled by artists only, but is constructed as well by contributors to the work. If looking at the status of net collaborative creation from a psychological aspect, reflecting the footprint of participants directly on the actual works, it not only encourages users to visit the work again to find their own footprint, but also elevates positive emotions from participants having made a contribution.

The comic website "Renga" (http://www.renga.com) by Japanese artists Rieko Nakamura and Toshihiro Anzai, applied the principles of growth and change to their net collaborative creations. In Japanese, "Ren"= Linked and "Ga " = images. As the name shows, it is a work using picture links as their primary mechanism for change. The interesting thing is that all picture links have associated metatabs with specific symbols within them. The "Renga," considering users' different personal experiences, allow participants to upload pictures in accordance with individual cognition, and to link to extant pictures on the website. The pattern of the whole page is like a climbing vine changing continuously so that no one can predict the final situation of the display.

THE VERBALITY OF ART. ART DOES NOT BRING MYSTERIOUS COLORS ANYMORE BUT EXPERIENCE SHARING AND DIALOGUE INSTEAD. ART BECOMES A VERB

Due to development of modern digital technology, art now shows a multi-polarity, especially in those net art works relying on technology as a disseminating platform. There are many types of easily operated software available in the market, for users to create personal image works, animation, and even websites. Extremely intelligent creation can be produced by the fool-proof operation of this software – this is the biggest contribution of digital technology to art. Hence, art creation is no more the privilege of a small group in society, but an opportunity for everyone to create. As well as the interactive characteristics of the net, the definition of art creation becomes worth discussing. As Ben Davis said: "In a certain sense, the act of finding

Figure 1. Interface of the CyberFortune.info (Images by Yueh Hsiu Giffen Cheng)

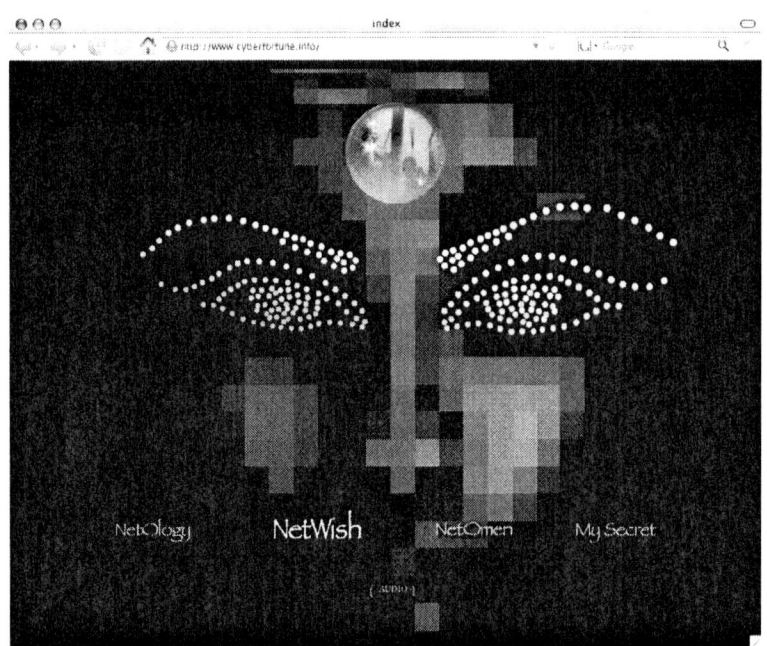

Figure 2. NetWish section from the CyberFortune.info (Images by Yueh Hsiu Giffen Cheng)

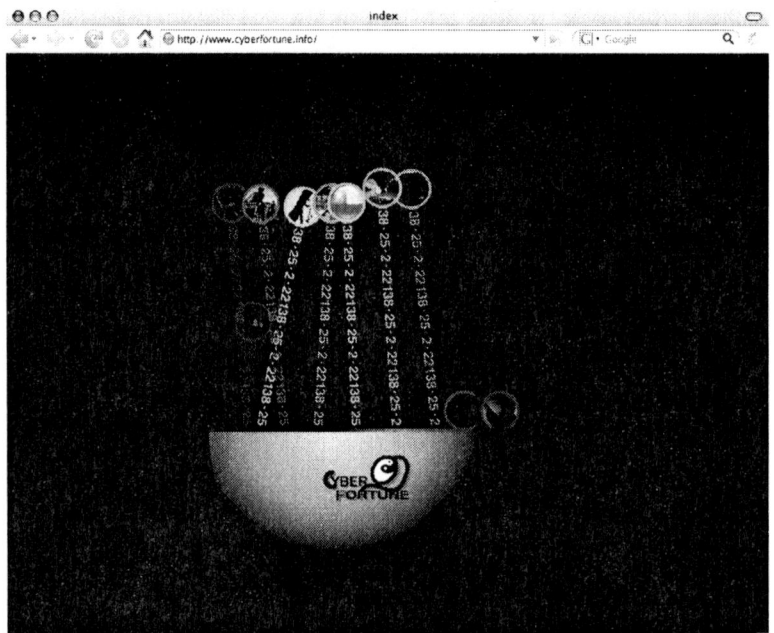

art on net is a net art activity itself...Net art is not something, but an environment...In the field of net, a thought field that different aesthetics can be proposed, different concepts can communicate with each other (Chen, 1998, p. 181)."

Szyhalski Ding, a net artist and Professor at the Minneapolis College of Art and Design, claims that, "the Internet is a public space; it's just a much more populated and busier public space. It has its own rhythm and logic" (Foote, 2003). From the viewpoint of these scholars, the aesthetics of net art becomes an expression of conceptual art. The integrity depends on users traversing the art work to explore it. Hence, when users visit the work, both the work and the participants are conducting an art performance. In terms of collaborative creation, art creation brings no mysterious colours anymore, but experience sharing and dialogue instead. Art becomes a verb.

"CyberFortune" (http://www.cyberfortune.info) an experimental project created by Yueh Hsiu Giffen Cheng incorporates the theme of Taiwanese traditional fortune-telling practices, with the function of collaborative creation based on the internet medium. It focuses on an assumption about fortune-telling: The Internet and computers may have a mysterious power to predict something that people do not know yet. To fully realize the collaborative execution of this creation project, it will require public involvement.

"CyberFortune" responds to the form of the viewer/participant's engagement, thus stimulating and enticing users to participate and contribute to the project. Accordingly, the practice of fortune telling has become an indispensable basic element for constructing "interactive" net behavior with users. While appreciating the traditional lore, one is also experiencing and creating at the same time an alternative digital form that combines ancient and technological fortune-telling.

TRANSFERRING AUTHORSHIP: ARTISTS OF COLLABORATIVE CREATIONS BECOME THE EDITORS OF PROJECTS, AND USERS PARTICIPATING IN THE PROJECT BECOME ARTISTS AT THE SAME TIME

When art works are not physical objects any more, the boundary of authorship becomes more blurred (Foote, 2003). When works involve participants as creators beside or instead of authors, the relationship between authors and works starts to break down. Who created the work? Who finished the work? Who are the contributors behind the screen? Looking from the mode of net collaborative creation, we can see clearly that the works have been contributed by net users. In fact, the artists themselves made the least contribution to the works.

Julian H. Scaff argues, "For now to have to capacity to view the digital artwork means also to have the capacity to (re)produce it infinitely, and to change it endlessly. Not only is authenticity in question, but the idea of authorship is almost obsolete (Scaff, p. 2)." As Shu Lea Cheng mentioned in an interview: "In the net art projects I have been doing, the characteristic of 'mass participation/involvement' has been emphasized a lot. The net is a media through which the world can 'enter' the artworks easily, and the artworks are completely within the 'public domain'. Under this concept, I think so-called 'authors rights' are to some degree overthrown (Cheng, 2000)." Hence, during the process of collaborative creation, artists become the editors of projects, and users participating in the project become artists at the same time.

"Screening circle" (http://artcontext.org/wire/art/2006/screeningCircle.html) by Andy Deck exhibited in the Whitney Artport in March 2006, is one of the net art work concerned with the transferring of authorship. Users can draw personal images on "Screening circle", or change other people's images. After users draw, then the images are displayed in whirling images around the screen. Therefore every participant who makes a contribution to this website becomes the artist-creator of "Screening circle". Besides, "Screening circle" is fun – from the bright colors to adorable images, all conform to the aesthetics of computer games, which corresponds to the "playing participation" principle mentioned before.

The process of collaborative creation on the Internet is what actionists are pursuing: the value of an artwork is created simply in a short time (hour, minute, second) not in long-term preparation (a month, year or century). Andy Deck's unpredictable action to invite net users to join in the process of creation realized the immediate creation style of actionists and realizes the aesthetics of net collaborative creation.

Internet/Digital Community

A digital community is a social group convened by the dissemination of Net media; just as in the interaction modes of other communities in daily life, a collective goal gives birth to the communal relationship between the individuals. Peter Kollock in his book "The Economies of Online Cooperation" suggests that there are three motivations for digital communities to participate with each other: Anticipated Reciprocity, Increased Recognition and Sense of Efficacy.

The so-called "anticipated reciprocity" refers to the reciprocal relationship between communities. People do not always get equal feedback when contributing to a digital community, but will receive respect from the community. For example, the engineer who publicizes the open source code always receives thanks and regards from the beneficiary. The idea of "increased recognition" refers to the fact that community members hope to be honored for making contributions, so as to gain identity or self-appreciation.

Following the development of technology and changes in human society, net communities soon became a new form which is virtual but influential; even the field of net art has a record of being influenced. Digital communities have become a rising trend on the net and appear to be shaping the political and cultural nature of the web as an interpersonal and collective forum and meeting place. The British Publisher Penguin publicized the creative project "a million penguins" (http://www.amillionpenguins.com) in February 1, 2007 to call net communities together to compose an experimental novel by a million people. This six-week project takes the mode of Wikipedia and allows community members to edit any part of the novel freely. It is obvious that this project of "*a million penguins*" showed the great size and extent of the net community; as for the result of this collaborative creation, it is "humorous" at best, but has a long way to go to be a mature creation.

Blogs are perhaps the best representative today of the digital community at work and play. Since the establishment of the first blog, simply called Blogger (Blogger.com) in 1999, within a few years, blogs multiplied into many millions. The power of Internet to rapidly adopt new technology is best seen with the uptake of blogs. Easily-operated blogs give people the possibility of both creativity and communication. Anyone can be the owner of a blog to write, post photos or compose any form of artwork in the virtual world of Net community.

Community has considerable power, but the community's power via the virtual Net is even more amazing. Because users concealed in the net do not suffer from the social pressures of normal life, interaction between communities and the factor of being carefree are the source of the net community's power, which also becomes a route to inspire art creation.

Interaction with a Real Installation Combined with Internet

Gerfried Stocker, the Managing director of Ars Electronica, in an essay on the art of future writes:

"The art of tomorrow is the art of the media". Gerfried Stocker thinks that the possibilities of being able to communicate at any time and everywhere give rise to concomitant new prospects for social, and thus artistic, interaction. It is no longer the technological possibilities but rather the socio- cultural structures of the information society that are decisive in the context of an art of tomorrow.

The internet as a cultural and commercial sphere is the basis and breeding ground of innovative, inspired modes of doing artistic work, which have an impact over an enormously expanded area due to the network linkages that are immanent in the medium (Stocker, 2005, pp. 27-28). Some artists regard the net's virtual character as the new domain of Utopia, but some others still prefer analog artworks that are touchable. The latter combine physical installations with the net serving as a component to create a hybrid medium, and in so doing, demonstrate the net's character of having no temporal and spatial gaps; they exhibit actual effects of the virtual net on the actual physical world.

The highlight of 2006 in Taiwanese in new media circles was the "Baby Love" project by Shu Lea Cheng. The real installation used in the "Baby Love" is duplicated babies made of silicone gel, sitting in rotating cups of different colors. Net users can upload their favorite music to the database of "Baby Love" and interact with the live show. The audience in the museum can choose any song uploaded by net users and enjoy the fun of sitting in a rotating cup with a duplicated baby. Each rotation and bumping with other cups will cause the music to be mixed, reconstructed or restarted. All new musical combinations produced by exchanged data while duplicated babies bump into each other in the rotating cups will be returned to the net, so net users who uploaded the music can listen to their contributions again.

Artists have made good use of net cultural characters combined with real installations for interaction. Thereby, net users and gallery audiences, non-related initially, can cross the temporal and spatial barriers to produce an interesting connection on different levels of interacting relation. Shu Lea Cheng said, in my email interview about real installation works combining with the net: "The concrete/tangible display is the body of the work. I would not call them display. As installation, these works transcend gallery space with a net connection. Via the web, public wharf, public intervention, the work is not a self-enclosed, contained entity, but is ever-expandable in the part of software and human interaction (Cheang, 2006)".

Hans-Georg Gadamer, a German philosopher, in his text "Truth and Method", writes that: "The work of art is not an object that stands over against a subject for itself. Instead the work of art has its true being in the fact that it becomes an experience that changes the person who experiences it. The "subject" of the experience of art, that which remains and endures, is not the subjectivity of the person who experiences it but the work itself (Gadamer, 1993, p. 102)." From the previous example, we realize that in the field of new media art, the interactive relationship established between artworks and audience is indeed communal and coexisting. It is especially true of installations involving the net, because a work without an audience's interaction cannot awake the conceptual metaphor in depth, and in depth experiences can't be expected to be felt by the audience.

CONCLUSION

As technology moves on, the category of new media art is getting broader. Many people have almost lost their ability to appreciate art and to recognize points of value in the surging waves of art. After the Impressionists broke down the traditional forms of art, impression and colour became the mainstream of aesthetics. Soon afterwards, Fauvism and Cubism joined the battle of colour, with fervent expression of shape and composition. After WWII, following the concept of ready-made art from Dadaism, suddenly the rules of art appreciation became illusory. Many people could not accept Dada's 'revolution simply for the sake of revolution'. But looking down the long river of history, starting with Dadaism as an incubation period, the revolutionary movements later on, such as those of event art, conceptual art and Fluxus, could be developed. During the transformation of art forms in the past half century, people have gradually acquired a more tolerant art appreciation. Net art comes naturally from those schools of post-modern art mentioned previously. Considering the wide acceptance and understanding of conceptual art, art happenings and Fluxus art, together with the classification of each net art form mentioned in the earlier article, the aesthetics of net art can be better appreciated now as a continuum of these flows of signs and their meanings in a social and cultural context.

REFERENCES

Abrams, J., & Hall, P. (2006). *Else/where: mapping new cartographies of networks and territories.* Minneapolis, MN: University of Minnesota Design Institute.

Alinovi, C. (2003). *NET.ART The art of our times-4/2003 Stedelijk Museum Bulletin* [Electronic Version]. Retrieved 27 Jun 2006 from http://www.stedelijk.nl/oc2/page.asp?PageID=390.

Arnheim, R. (1996). From chaos to wholeness. *The journal of aesthetics and art criticism, 54*(2), 117-120.

Barthes, R. (1977). *Image, music, text/ Roland Barthes; essays selected and translated by Stephen Heath.* New York: Hill and Wang.

Baumgartel, T. (2001). Net Art. On the history of Artistic Work with Telecommunications Media. In W. Peter, D. Timothy, & Zentrum fur Kunst und Medientechnologie Karlsruhe (Eds.),

Net-condition : art and global media (pp. 398). Cambridge, Mass: London: MIT Press.

Bowman, J. (1990). *Performance Art* (Publication.: http://www.bright.net/~dapoets/performa.htm

Brogger, A. (2000). *Net art, web art, online art, net art?* (Publication. Retrieved Jun 2006: http://www.afsnitp.dk/onoff/texts.html

Cheang, S. (2006). *E-mail interview with Shu Lea Chang by Yueh Hsiu Giffen Cheng.*

Chen, G. (1998). Art on the net, is not equal to net art. *The journalist.*

Chen, Y. (2002). The boundary of virtual wisdom and stratagem. *Artist Magazine.*

Chen, Y. (2004). Art relaxed. *Artist Magazine.*

Cheng, S. (2000). Traveling between virtual and reality world- Shu Lea Cheng (Publication. Retrieved Apr 2006: http://goya.bluecircus.net/archives/004358.html

Cornwell, R. (1996). Artists and interactivity: Fun or funambulist? In *Serious Games*. London: Barbican Art Gallery/Tyne and Wear Museums.

Dempsey, A. (2002). *Styles, schools and movements: an encyclopaedic guide to modern art.* London: Thames & Hudson.

Esther, L. (2000). *Walter Benjamin electronic resource: overpowering conformism.*

Foote, J. (2003). Net Art: A New Voice in Art. Challenging Perceptions of the Virtual and Physical (Publication., from History & Philosophy of Mass Media Final Paper: http://babel.massart.edu/~jfoote/netartpaper.html

Gadamer, H. (1993). *Truth and method* (2nd rev. ed.). London: Sheed and Ward.

Goldberg, R. (1988). *Performance art : from futurism to the present* (Rev. and enl. ed.). London: Thames and Hudson.

Hu, S. (2005). Experiencing the Climax of communication. In *Climax- The Highlight of Ars Electroniza* (pp. 204). Taipei: National Taiwan Museum of Fine Arts.

Inge, T. M. (2001). Theories and Methodologies Collaboration and Concepts of Authorship. *PMLA, 116*(3), 623-630.

Landow, G. P. (1997). *Hypertext 2.0* (Rev., Amplified, ed.). Baltimore: Johns Hopkins University Press.

Lash, S. (1991). *Post-structuralist and post-modernist sociology.* Aldershot, England ; Brookfield, Vt., USA: E.Elgar Pub.

Lin, x. (2006). 2006 Web 100. *Next Publishing Corp.*

Manovich, L. (2000). *The language of new media.* Cambridge, Mass.; London: MIT Press.

Mizrach, S. Talking pomo: An analysis of the post-modern movement (Publication. Retrieved 6 Feb 2007, from Florida International University: http://www.fiu.edu/~mizrachs/academentia.html

Morris, S. (2001). Museums& New Media Art- A research report commissioned by The Rockefeller Foundation. [Electronic Version] from http://www.cs.vu.nl/~eliens/onderwijs/multimedia/mma1/college/@archive/refs/Museums_and_New_Media_Art.pdf

Ross, D. (1999). Net.art in the Age of Digital Reproduction (21 Distinctive Qualities of Net.Art) [Electronic Version]. Retrieved 20 May 2006 from http://switch.sjsu.edu/web/v5n1/ross/index.html

Scaff, J. H. Art and Authenticity in the Age of Digital Reproduction [Electronic Version]. *Digital arts institute.* Retrieved 03 Mar 2007 from http://www.digitalartsinstitute.org/scaff/index.html

Schleiner, A. M. (1999). E-mail interview discussing artists and the computer game industry by Jim

McClellan (Publication.: http://www.opensorcery.net/jiminterview.html

Spaink, K. (1997). Prix Ars Electronica 1997- net Jury statement, July 1997 [Electronic Version] from http://www.spaink.net/english/ArsPrix97.html

Spaink., K. (1997). Prix Ars Electronica 1997- net Jury statement, July 1997 [Electronic Version] from http://www.spaink.net/english/ArsPrix97.html

Stocker, G. (2005). The Art of Tomorrow. In *Climax- The Highlight of Ars Electroniza* (pp. 204). Taipei: National Taiwan Museum of Fine Arts.

Walter, B. (1955). *Art in the Age of Mechanical Reproduction, Illuminations.* New York: Schocken Books.

Wijers, G. (2005). Preservation and/or Documentation: The Conservation of Media Art (Publication. Retrieved 04 Sep 2006, from The Netherlands Media Art Institute: http://www.nimk.nl/en/

Wilson, S. (2002). *Information arts: intersections of art, science, and technology.* Cambridge, Mass.; London: MIT.

Yeh, J. (2002). Net. Art - Exhibition. *Artist Magazine.*

KEY TERMS

ASCII: (American Standard Code for Information Interchange) ASCII is a code for representing English characters as numbers, with each letter assigned a number from 0 to 127. Computers often use ASCII codes to represent text, which makes it possible to transfer data from one computer to another.

Blog: A noun and a verb, the term "blog" is a contraction of "web log" and is a web-based publicly accessible personal journal.

Cyberspace: Coined by William Gibson in the 1984 sci-fi novel *Neuromancer*, cyberspace describes the non-physical terrain created by computer systems.

Generative Art: According to Adrian Ward, a term given to work which stems from concentrating on the processes involved in producing an artwork, usually (although not strictly) automated by the use of a machine pragmatic instructions to define the rules by which such artworks are executed.

Conceptual Art: Also known as "idea art", the conceptual art movement began in the 1960s and extended to the late 1970s. Its tenet embodies the notion that the primary purpose or function of an artwork is to convey its idea or concept; to that end, the physical creation is almost superfluous to the work. Some conceptual artworks are complete when the artist has detailed the piece through text, never fabricating it in a tangible form.

DADA: A nihilistic art movement that flourished in Europe early in the 20th century; based on irrationality and negation of the accepted laws of beauty. Dada artists were later involved with Surrealism. Dada was influential to several art movements, including Fluxus.

Email List: Or list server A shard distribution list wherein a message sent to a designed email address redistributed the message to all list subscribers. A mailing list that is a administered automatically is called a list server.

Fluxus: Is an international network of artists, composers and designers noted for blending different artistic media and disciplines in the 1960s. Fluxus is often described as intermedia, a term coined by Fluxus artist Dick Higgins in a famous 1966 essay. There are hundreds of types of Fluxus. Dada is often used to induce coagulation in the art, although some Fluxus is curdled with ideas from Situationalism or Neoism or with extracts of various species of Pop-Art.

HTML: Hypertext Markup Language is a scripting language that makes it possible to establish links between documents and arbitrary nodes.

Hypertext: Linked segments of text that can be navigated by a user. Hypertext originated in Theodor Nelson's concept of the "docuverse", a space of writing and reading where texts could be electronically interconnected by anyone contributing to the networked text. While the World Wild Web essentially is a hypertext environment, hypertext software existed before HTML.

Hypermedia: An extension that supports linking graphics, sound and video files in addition to text elements. The web is partially a hypermedia system since it supports graphical hyperlinks and links to sound and video files.

Note: Above information cited from *Digital Art* by Christiane Paul, *Internet Art* by Rachel Greene, and *Art of the digital age* by Bruce Wands.

Chapter XI
A Graphics Tablet as a Fine Art Tool

Nicola Quinn
University of Limerick, Ireland

Annette Aboulafia
University of Limerick, Ireland

ABSTRACT

People have used tools for artistic expression for millennia. Relatively recent is the use of digital technology to afford the creation of art. However, many draw into question digital technologies conduciveness to creativity during the artistic process. Recent developments of digital technology for artists have lead to the creation of a graphics tablet from Wacom Technologies. It is claimed that the graphics tablet is more favorable to creativity than other existing digital technologies. This chapter addresses this issue through a qualitative study of five artists using the Wacom graphics tablet, in particular the artist's own experience using the graphics tablet is explored. The outcome of this study indicates that the graphics tablet is a useful tool. However, there are still several improvements required to advance the graphics tablet to a stage suitable for fine artists.

INTRODUCTION

Technology and art are no longer distinct entities (McCullough, 1998). There are two different although related perspectives when discussing technology and art. One is where the technology becomes embedded in the art itself. The other perspective is where the technology acts as a tool for creating art. The first perspective will be briefly outlined followed by a discussion on the second.

Technology based art allows users to have a more intimate and captivating artistic experience as their actions depict what happens in the piece. Pier & Goldberg (2005) explain how works of art that use technology as a means of joining the artist and the audience give life to the artist's ideas. When technology becomes part of the art, the user may enter the art piece so to speak, as the technology is the art. This is in contrast to the more traditional way of creating and presenting art, usually as static images viewed passively by people. Technology based art is increasingly geared towards interactivity and innovation in user interfaces (Edmonds et al, 2004). It is art that lives in a space; it appropriates the space to its own artistic ends. As argued by Bester (2003) the space itself and the artifacts that are enclosed within it are the work of art.

As such, computer technology has extended the concept of art and has provided new ways of knowing and judging art (Binkley et al, 1994). This topic will not be explored further here, where the focus is on information technology as a means to create the art, rather than applying technology as a delivery method.

The second perspective is where the technology acts as a tool. Tait (1998) comments on how artists since the beginning of time have struggled with tools, for example which colored mud would stick to the cave wall and keep its color. Fine artists spend countless time and energy making their tools an extension of themselves. Fine artists are those that have a need to express their vision or opinion through the form of a painting, a photograph, a print or sculpture (Tait, 1998). For instance, when learning to paint, artists will spend unlimited time adjusting to the brush, the feel of the paint, the techniques of painting, the feel of the canvas and in turn will master the art of painting. An important feature of any tool for artists is that it should provide the artist with the freedom of expression to create their works of art.

In conjunction with being a tool to support the activity of creating art, the tool can also influence the creative process itself. It is likely that information technology has a different influence on the artist's activity than simple hand tools. Whale (2002) argues that computers and technology are not only enabling artists and designers to extend the scope of drawing, but to help the understanding of the aspects of drawing itself. For instance,

he argues that "[...] there are more compelling reasons for using computers, arising from their ability to spawn genuinely new approaches to drawing and to contribute to our understanding of this most central of human activities." (Whale, 2002, p.65)

COMPUTER SUPPORTED TOOLS FOR ART CREATION

The graphics tablet (by Wacom Technology Inc. n.d., Figure 1) is at the forefront of the movement towards computer supported fine art creation. It is one of the leading technological tools that enable the artist to create art. The graphics tablet has been designed with the artist in mind, and is mainly used by graphic designers, cartoonists, architects and artists. The tablet has a flat surface that allows the user to draw an image using a stylus pen similar to drawing on paper.

Through using this stylus pen (Figure 2) the artist can apply the effects of paint through the techniques of drawing. The stylus pen resembles a normal writing pen in shape, but is slightly

Figure 1. An artist using a Wacom graphics tablet

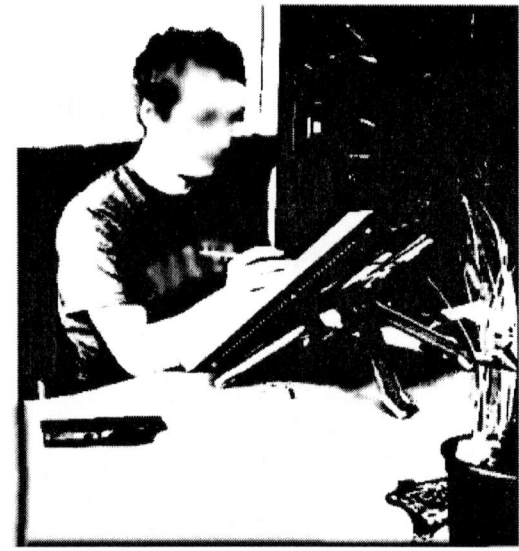

longer than an average pen. It is equipped with a pressure sensitive eraser, a long grip area, and a sensitive pen tip that allows the artist to have more control over line width, color saturation and opacity. Near the pen tip there is a switch which acts like a right click button on a mouse. The pen also incorporates a 'flip it over mechanism' to switch from paintbrush to eraser (Figure 2). This is particularly useful, as the artist is not required to change implement in order to switch to the eraser. It is also mimics the real life mapping of a pencil with the eraser on the end, that when you switch to eraser you simply flip the pencil. Using this real world mapping will ensure that the artists use the eraser, and know where to find it.

Recent developments allow the display screen to be incorporated into the tablet, whereas previously the image was only displayed on a computer monitor. This allows the artist to draw directly on the screen, and therefore have a more seamless experience. The angle of the screen can also be adjusted to suit the artist. This is similar to how an artist would typically use an easel and adjust the angle of the canvas while painting.

Graphics tablets are commonly used with a computer as an alternative to a mouse, keyboard and standard monitor. It is assumed that using a pen as an alternative to a mouse draws the user closer to activities such as drawing, writing, and usual pen interactions, as opposed to computer commands (Buxton 1997). Global Ergonomic Technologies (1998) compared user's posture when using a pen and a mouse. They describe

the neutral position as one where your body parts reside when completely relaxed. The results from their study demonstrated less deviation of posture from the neutral position when using the pen rather than the mouse. The mouse results have shown excessive deviation from the neutral position. This excessive deviation can lead to severe body aches and repetitive strain injury (RSI). Pen use results in a posture more natural than during mouse use and thus appears to be a bio-mechanically superior input device.

The stylus pen is pressure sensitive, which is a form of force or haptic feedback that provides haptic perception to the user. The usage of haptic devices is a recent trend and has been applied in the medical area, especially. Haptic devices allow users to 'feel' their interfaces and interactions. It refers to technology, which interfaces the user via the sense of touch by applying forces, vibrations and or motions to the user. Haptic perception is one of the perceptual systems we use to obtain information about our environment. It is the process of recognizing objects through touch. It involves a combination of somatosensory perception of patterns on the skin surface and proprioception of hand position and conformation. Without the appropriate integration of proprioception input, an artist would not be able to brush paint onto a canvas without looking at the hand as it moved the brush over the canvas.

Haptic devices allow the artist to have direct contact with a virtual instrument, which is able to produce real-time images. Here we will list a few

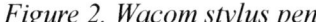
Figure 2. Wacom stylus pen

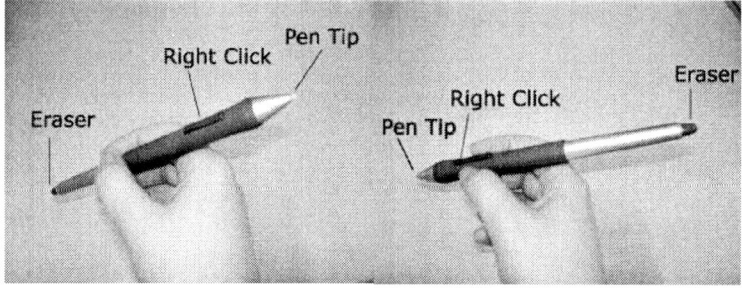

tools that have been developed for artists, which use haptic devices one way or another.

Gregory, Ehmann & Lin (2000) have a project inTouch, that is a 3D interface that allows the user to paint and edit a polygonal mesh using a PHANToM from (Sensable Technologies 1993). The PHANToM is a piece of hardware that allows a user to touch and manipulate virtual objects through a stylus pen that is attached to a mechanical arm. Gregory, Ehmann & Lin state that by giving the user a sense of touch, haptic feedback overcomes the restrictions of 2D interfaces such as a mouse and keyboard.

Baxter et al (2001) have developed a tool named DAB that uses a deformable 3D brush model. This tool also incorporates force feedback allowing the user to receive more realistic control of their brush. DAB empowers the user to paint in a natural form with all the benefits of using a computer. Allowing the user to paint in natural form and yielding haptic feedback is helping to bridge the gap between technology and traditional methods. The 3D brush model is used in conjunction with a PHANToM. They have developed a bi-directional paint model that enables easy loading of complex blends to the brush. Baxter et al's multi-layered paint model incorporates real life features such as drying, blending, bi-directional flows and complex brush loading. Their haptic system allows the user to paint directly on the canvas displayed on screen. The space-bar toggles the virtual palette, used for paint choosing and mixing.

Baxter, Wendt & Lin (2004) have developed a system IMPaSTo, which captures a large range of styles similar to oils and acrylics. IMPaSTo is based on their previous system *DAB*. Baxter et al.'s interactive model captures the full range of physical behavior for oil and acrylic paint, for example viscosity. This system can be used in conjunction with either a PHANToM or an Intuos pen tablet. The Intous pen tablet developed by Wacom Technology Inc. (n.d.), is where the display is separate to the monitor. The user draws or writes on the tablet while viewing their standard monitor. Both devices serve as physical metaphors for the virtual brush. Their system allows users to dry paint instantly, keep paint wet indefinitely and adjust the opacity of the paint at any time. This allows more creativity in the manner of which the paintings are created, as you are not restricted by the paint's physical qualities.

Yoshida et al (2004) have developed a system, Sumi-Nagashi based on a Japanese art technique of the same name. This system also gives haptic feedback to the artist by allowing them to hold a stylus type paintbrush over a digital canvas. The haptic feedback is produced through a device called the Proactive Desk. The Proactive Desk is a digital desk with force feedback. It allows users to handle both digital and virtual objects on the desk with realistic feeling. Sumi-Nagashi literally means ink-floating. The user paints on a digital canvas using a device shaped like a paintbrush. Sumi-Nagashi is attempting to let the painter feel the viscosity and adhesion of paints on the canvas through touch.

Flagg & Rehg (2005) have developed a system that allows the artist to paint one layer at a time. The tool is targeted towards educating novice artists in the techniques of painting. Their aim with this tool is to allow artists to create art using traditional media in a computer-assisted manner. This system employs multiple projectors to create an interactive display on the artist's canvas. The aim of this display is to aid the artist with the details of the brush stroke orientation, position and texture. For selecting colors the artist uses a Wacom tablet with a pen attached to the opposite end of the paintbrush. Flagg & Rehg argue that the use of multiple projectors helps to guide the artist toward a final desired painting. In order to apply this tool, the artists have to plan their work beforehand. The artist's painting will also have to be created from back to front in a series of layers.

Pier & Goldberg (2005) describe an interactive art piece called Fluids. In Fluids the artist aimed to transmit fluidity. The technologists used water and wind to give form to the piece. They designed a method to obtain data from the water and air in order to relate it to the user and events in the virtual world.

Henzen, et al (2005) introduces an interactive drawing tablet, Electronic Ink, which they hope will support the drawing experience closer to that of pencil and paper. Electronic Ink is made of a thin layer over switching electrodes whose color can be switched electronically from one stable state to another e.g. white to black. They are concerned with animators and therefore mainly black, white and grayscale drawings. They envisage their system being used from the beginning, therefore giving a digitized artwork throughout, unlike presently when the artwork is created on paper and then subsequently digitized for post-production. This is different from the Wacom graphics tablet as it is designed mainly for animation and not painting or drawing in the traditional sense.

Each of these technological tools as discussed above have tried to support the artistic process of drawing and painting by applying computer-supported tools based on haptic feedback. However, they each require the artist to use technology that is quite different to what they are accustomed to, for instance using a PHANToM, a Proactive Desk, or projectors. These technologies may not appeal to the wider audience of artists, as they are unrecognizable from the traditional methods. The graphics tablet uses implements that at least resemble and have similar usage as the pen, pencil or paintbrush that artists have used for millennia.

To date there exists very little literature assessing how well the graphics tablet is actually used and perceived by artists. The next section reports on a qualitative study that was conducted involving five fine artists that used the graphics tablet to create an art piece.

EXPLORING THE USAGE OF A GRAPHICS TABLET BY ARTISTS AIM, DESIGN AND SET-UP OF STUDY

The aim of the study was to explore fine artist's experience of using a graphics tablet to create an art piece and how this tool may facilitate fine art. Rather than using large samples we applied a case study method or research strategy. Case study research is an empirical inquiry that investigates a phenomenon within its real life context. It involves an in-depth examination of a single instance or event (a case). The case plays a supportive role to aid in the understanding of the issues. The study was conducted with five fine artists (cases). The case study method is but one of the methods used in the qualitative approach. A qualitative approach is an inquiry process of understanding a social or human problem, based on a number of methods, including the case study. A qualitative approach was chosen for this study, as we were concerned with the artist's experience and attitudes.

To ensure a broad collection of data through the case studies, a triangulation of data collection methods was implemented applying two or more methods of data collection (Mason, 1996). The data was collected using interviews, note taking, video and audio recordings.

The procedure of the study was as follows. Each participant was asked in advance to bring an item to draw or they could choose to draw the still life, which was set up in the test room. A briefing was given to each participant, explaining the procedure and aim of the study. This was followed by a short interview inquiring about their current artistic process and their use (or lack of use) of technology.

The interview was conducted in a coffee room in Plassey House at the University of Limerick. It is a very friendly and homely place, which exhibits art works (Figure 3). As intended, this setting affected the participants in the sense that they all started talking about the room and its paintings,

Figure 3. Plassey house

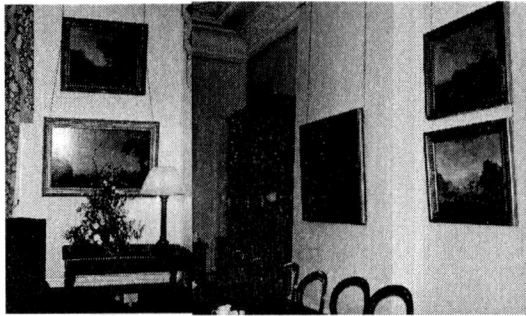

and this in turn created a 'creative mood'.

Following the interview the artists were introduced to a 21 inch Wacom Cintiq tablet. The tablet was surrounded by paper, sketchpads, pens, pencils and pastels to make the artist feel more comfortable in an environment attempting to act as a replica of their own creative space. They were asked to draw or paint a picture in order to get familiar with the tool. The artists were allowed to use as much time as they wanted and they were encouraged to verbalize their thoughts while using the tool. Asking the participants to air their thoughts uses the think-aloud method. The think-aloud method can give insights into how participants reason and their intentions while using a system (Helander, 2005). For all participants it lasted three to four hours, which included short breaks. The session ended with an interview, which addressed their experience with the new graphics tablet. The goal of this last interview is to understand the artist's experience of the graphics tablet. A secondary goal is to see if the artist's normal behavior changes due to the introduction of the graphics tablet.

A video camera was placed in front of the artist to catch their behavior and facial expressions. An mp3 recorder was placed on the table to the right of the artist's position. Both the pre interview and post interview were mp3 recorded. The data gathered from these interviews and observations was transcribed and categorized.

In total three males and two females participated in the study. They all practice fine art (in their cases painting). Their ages were from the early twenties to the early sixties. Only one person had no experience with computers. The others had from two to twenty years of experience with e-mails, Internet browsing, editing and adjusting photographs. Only one of them (the youngest) had used a graphics tablet before (the older model of graphics tablet) as he found that the end product effect looked more natural than when he used a mouse.

The participant's computer experience is an important factor as it may be an indication of whether or not they are open to using new technology. Three of the participants were very open to using this new technology, whereas two were a little wary. The artists that were a little wary used mostly traditional methods in their creation of art. They also had limited interactions with computer technology and therefore were anxious about attempting to create fine art with a tool as foreign to them as a graphics tablet. The three artists that were more open about the new technology were in general more adventurous in their creation of art. They also had more extensive computer experience, and were used to trying new things and learning new technologies.

Each case study followed a study plan, which outlined steps to follow during the study. This enabled the replication of methods in each case study. This was not intended to reveal identical results but aided in giving structure to each study. The study sample was chosen as a range of contrasting cases to compare findings. The studies did not explore the learning period required to become efficient in the use of a graphics tablet.

RESULTS

The results from the case studies were categorized into five topics. Each of the topics seem to have a particular relevance for the participating artists

Figure 4. Successful facilitation of the creation of fine art

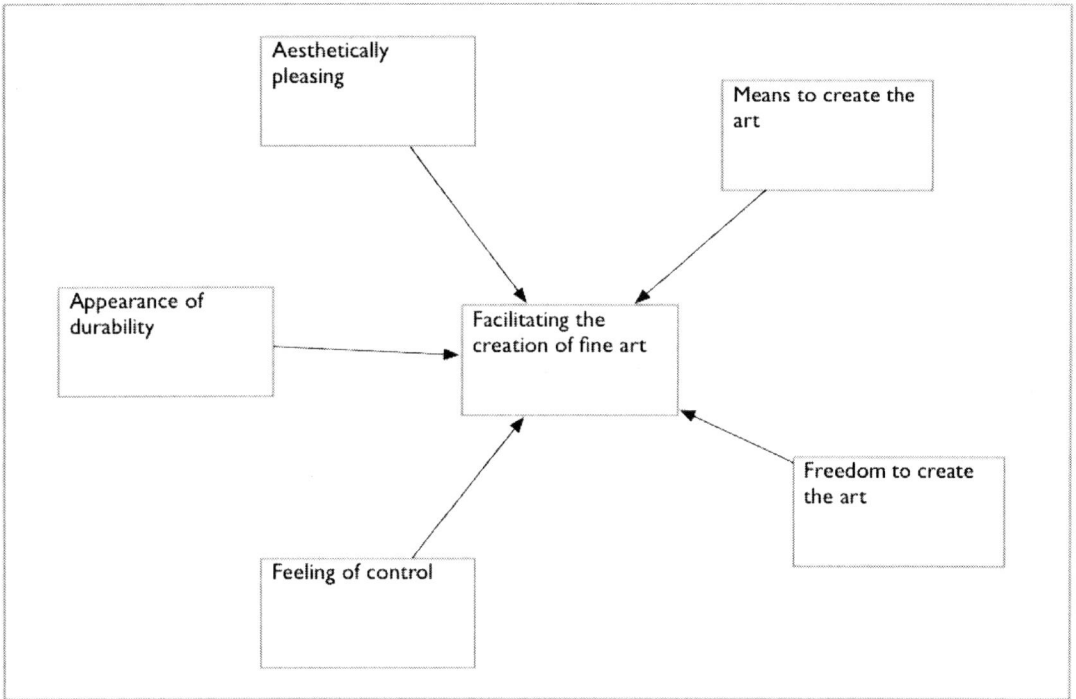

in the process of applying the graphic tablet to create drawings or paintings (Figure 4).

In no specific order we here discuss each of the topics:

a. **Means to create art:** Technological tools for art should provide the means to create art. A graphics tablet allows the artist to create art through the techniques of drawing and painting. The graphics tablet allows the user to hold a stylus pen to draw on the surface of the graphics tablet similar to drawing on paper or painting on canvas. However, quite a few procedural changes are needed compared to the traditional way of creating a drawing or a painting. The artist cannot use the techniques crafted over years of manipulating the paint on a brush as the paint is now digital with a graphics tablet. Using a graphics tablet, the artists were applying paints to the digital canvas, yet applying the techniques of drawing. For instance, the graphics tablet provides the drawing techniques of shading and cross-hatching, but using the graphics tablet the artists had to apply 'paint' giving the effects of drawing. A graphics tablet is one of the few effective ways of allowing the artist to create a painting through the techniques of drawing.

Vygotsky (1971) studied the differences between painting and drawing in the tra-ditional sense. Essentially the difference is with space. Painting disregards the flat two-dimensional drawn image and com-pels the perception of a three dimensional image. A drawing on the other hand can represent a three dimensional space but still remains two-dimensional. We perceive drawn images as three-dimensional but we still appreciate the play of lines on the two dimensional plane. When painting the most common tools are the brush, and knife (to sharpen edges.) For drawing the tools vary from pencil and pen to charcoal, pastels and crayon. However, all of these traditional tools can be used in paint, drawings, prints or sculptures but for different ends. In sculpture the artist can use clay, but may also use a pencil or pen to draw certain aspects on the clay and in the end may paint their sculpture. The artist is not restricted to using just one tool; they can use any tool that will produce the effect they seek. Technology can also be used as it offers a new perspective or way of combining these tools in digital art.

Other procedural changes were discovered and explored by the artists. One artist started with the object and worked back to the background, which is the opposite of his usual work process. Another artist used minimal layers when usually he would have layered everything. And one of the artists found it difficult to lighten its image. He started out very dark as intended, but then went to lighten the image and found the only suitable tool to use was the eraser, this entailed inserting a solid white color in certain areas to lighten them. Whereas with real paint these lightened areas would be a pale mix of colors. His technique carried the intention of lifting the paint from those areas in order to lighten them.

Holding the stylus pen also differed from the traditional way of painting. One of the artists found the graphic tablet similar to drawing on an easel, except for the way the pen was held. Usually, artists would hold a pen or paintbrush at a slant, whereas with the stylus, the pen was held straight to the screen. Another procedural change – although perhaps of minor importance - was viewing their piece during the process. When evaluating their painting, the artist would usually stand to view the picture from different angles and distances. With the graphics tablet the artists sat, as this was more convenient for working with the graphics tablet. The artist did not view their painting from different angles or distances in order to see areas where improvement is needed. This would most certainly affect the overall painting, as the artist may be too involved in the painting to see the bigger picture, so to speak. By standing back and looking at the painting from a distance, the artist gains a certain objectivity that may be lost when using the graphics tablet.

All of the artist's procedure changed due to using the graphics tablet, one way or the other. This change is likely to affect the acceptance of applying such a tool. In addition change of procedure is likely to change the end product and thus also have an impact on whether or not they decide to use a graphics tablet in the future. It was mentioned, however, that they did not believe that it was possible for other people to say whether or not their product was made with a computer – which was an initial concern for most of them. This implies that there is a social and cultural aspect with regard to the value of procuring digital art. The artists commented on how they attempted to make their paintings more 'natural' and less like the product of a computer. This suggests that if it looked like the product of a computer it would be less valuable to the art world.

b. **Freedom to create the art:** Technological tools for art should provide the freedom to create art. The graphics tablet provided a certain freedom to scribble and tidy lines without worrying about the effect. It was felt that they did not have to put in as much preparation as compared to painting on canvas. As such, it allowed them to draw more freely as they could erase or start again without any hassle. This is in contrast to painting on canvas where each line you paint has a consequence. Three of the artists also mentioned how they thought the graphics tablet was very useful for loose, freehand drawing. Another advantage was mentioned by one of the artists, who thought that using the graphics tablet was a 'space saver', as one did not have to set up the paints and working area.

The added value when using the graphic tablet of more 'freedom to create art' is likely to support the artist's procedure in a positive way, and thus reinforce a positive experience of using the tool.

c. **Feelings of control:** Technological tools for art should make the artist feel in control of all events pertaining to the creation of art. All five artists experienced difficulties in controlling certain functionalities while they were painting, such as right clicking, reaching to the top of the graphics tablet to change tools or going into menus by accident. They all had difficulty using the stylus pen to change their tool, e.g. brush size. The reason for this is that the menus are placed at the top of the graphics tablet and it seems to be an awkward position to reach when one is in the process of painting. Double clicking was found to be nearly impossible as the artists had to tap the tablet twice in quick succession. The position of the right click button on the pen also posed a problem. All five artists hit this button by accident at some point in their process, which immediately broke their concentration and "rustled their feathers." It was argued that lack of controlling certain functionalities inadvertently restricted them in their work process.

A few times the artists were looking for functionalities that that were not available. For instance, two of the artists found it difficult or almost impossible to get the exact shade or tone that they were searching for. If a traditional palette had been available the exact shade or tone would have been achieved, as argued by one of the artists. Acquiring their desired shade or tone with digital paint proved difficult, as the skill of mixing physical paints is less useful in the digital realm. Digital applications use color space co-ordinates to describe the different combinations of colors. However, users do not think in terms of color space co-ordinates, and therefore have difficulty finding the color or shade that they would like (Baxter et al. 2001).

Yet again, not all of the artists took advantage of all the functionalities that were available, for example the different methods of painting (water color, paintbrush, pastels, and pencils). The reason for this, as explained by one of the artists, was that many of the functionalities were not visible without exploring the menus. When painting in the traditional sense all of the artist's tools are on display, whereas with the graphics tablet most of the tools are in menus and/or hidden from view. This organization of tools may be optimal when using a mouse as the mode of interaction. However, it becomes difficult when using a stylus pen and graphics tablet as the mode of interaction.

These problems of functionality and lack of certain functions resulted in some degree of frustration, confusion, and feelings of losing control over the entire creative process. If users do not feel in control they associate this with a negative experience (Schrammel & Tscheligi, 2006). Negative experiences are likely to influence the artists work process in a negative direction. Picard & Klein (2002) assert that even a small change in emotional state can significantly impact creativity and problem solving. Frustration can lead to many unpleasant side effects, for example the decreased ability to concentrate, and the decreased ability to think creatively.

d. **Appearance of durability:** Technological tools for art should convey an appearance of durability. All of the three male participants worried about damaging the screen, particularly when they were using the shading technique. They were worried about scratching the screen permanently. The impression that the screen was not durable is associated with them comparing the graphics tablet display to that of a typical computer monitor. This in turn could affect their end product and dissuade them from using a graphics tablet in the future.

e. **Aesthetically pleasing:** Technological tools for art should appeal to the artist's aesthetic preferences. The overall appearance of the graphics tablet was enticing to the artists for the creation of art. It was argued that it was visually very similar to using an ordinary pen or pencil, as it resembled a canvas on an easel (a freestanding upright support for a painter's canvas or a blackboard).

In summary the participating artists had both positive and negative experiences of using the graphics tablet. Their positive experiences were related to the added value of the tool, compared to their usual way of painting such as the freedom to scribble and tidy lines without worrying about the effect. This also means less effort in order to produce their work.

The artists' negative experiences were related to problems with functionality such as awkward positions for changing tools, and lack of - or not knowing all of the functionalities. Although this often led to feelings of frustration and lack of concentration it did not seem to impact their overall positive impression of the graphics tablet.

Explaining experiences is a very complex matter. Many known and unknown factors are likely to be involved in the way people feel and think. Here we list a few of such possible explanations.

One explanation could be that the problems of functionality were either minor or not frequent enough to influence their overall positive experience to a negative one. This, however, may change if they were to use the graphics tablet often in their work. Such minor problems could become major due to frequency and subsequently affect their experience. On the other hand, the more they use a graphics tablet these problems of functionality may be eliminated.

Their negative experience could therefore be related to lack of practice. Any kind of tool-use affects not only our practical activity, but also our way of thinking. Next to change in techniques, Holloway (2005) argues that fluid use of a computer requires the artist to employ a linear understanding of the tool in contrast to the curvilinear understanding needed for analog tools. When using analog tools all of the functionalities can be understood from the outset, whereas when using a technological tool the user is required to explore the tool in order to understand the functionalities available. Following this, one could argue that the artists in this study were not accustomed to 'exploring their tool' that is necessary in order to discover all the functionalities.

Another possible explanation could be that their overall positive experience is due to the fact that the artists were exploring a new tool. According to Schrammel & Tscheligi (2006) exploration is usually felt to be satisfying to the user even if they do not discover or produce anything. Having a challenge that matches the ability of the user is also argued to be a potential prerequisite for a positive experience (ibid). Each participant was challenged to produce an artwork either based on a picture they had brought with them or on a still life.

In relation to how the graphics tablet successfully facilitates the creation of fine art, the graphics tablet faired adequately. It satisfied two of the five concerns completely.

It facilitated the means to create art, and it appealed to their aesthetic preferences. Although it created more freedom to scribble and tidy lines it did not provide complete freedom in terms of the ability to entirely rotate the Wacom tablet. And it did not allow the artist to control all events pertaining to the creation of art or convey an appearance of durability.

Surprisingly for the artists, the end product was so similar to their 'real' creations, that they did not believe it was possible for viewers to see the difference. This was in stark contrast to their impressions before they used the graphics tablet. Many of the artists had admitted that normally they would not even attempt to create fine art with a computer related tool. The reasons for this were that they presumed it would be difficult to use and it would not give them the ability to create good art that rivaled what they could create with traditional methods. Not being able to see much difference between art created using traditional methods and a graphics tablet was another positive experience and it may be the most important criterion for the acceptance of the tool.

FUTURE TRENDS

The value of exploring user experiences of new tools and identifying its usefulness over existing methods cannot be disputed. However, this in itself does not change the actual design of the tool. Designing tools for others is a challenging task as it also means designing activities, techniques and procedures for the users. We strongly advise user centered design approaches when designing tools for artists. User centered approaches explore users' needs and activities, and involve users throughout the design process. If artists had been involved in the design process of the graphic tablet, many functionality problems - like the ones identified by the artists in the reported study – might have been solved before the final product was released.

Other design considerations should also be mentioned. The intention with designing new tools is usually the improvement of practice, one way or other. For the user this entails new working techniques and procedures. New tools do not simply replace old tools - new tools produce a reorganization of hand and mind work. Bolter & Grusin (1999) state that tools and applications designed for artists often mimic traditional art and craft methods, i.e. representing one medium in another. They also argue, that this is not always the best solution, as new tools can be used to produce new techniques of art or crafts, which could improve their practice or become more pleasing. A few points should be mentioned here. Even if designers try to reconstruct the same techniques and procedures, it will always produce new work processes, no matter how minor the changes may seem. Computer supported tools for artists entail a set of pre-defined tools; the application itself, the various toolboxes from which the application is composed and the computer system. Whether or not the new techniques will improve their practice is a matter for investigation.

The case studies that we have conducted have shown a few 'added values' using the graphic tablet. At the same time they found that some of the techniques were frustrating. In any kind of tool application there is a learning period, and it is not possible from this investigation to predict how a prolonged usage may change their experience and acceptance of the tool.

Other design considerations are related to the issues of usability, including training time and need for special skills. More general guidelines for design include simplifying tasks, making things visible, achieving the correct mapping and designing for error (Norman, 2002). In addition designers need to take into account individual differences. In his attempt at creating artistic tools, Buxton (1997) discovered that each artist has very particular skills and tools, and that it was precisely these skills and tools that are poorly portrayed in computer systems. He suggests that tools created for artists should reflect the diversity and complexity in other media. This could be achieved by broadening the interaction with technology beyond standard input devices (mouse and keyboard).

One avenue to broaden the interaction with technology would be to incorporate multimodal interaction, which provides the user with multiple modes of interfacing with a system beyond the traditional keyboard and mouse input/output. Pen based input and haptic feedback have been applied in the graphic tablet. MacKenzie, Sellen & Buxton (1991) compared the performance of a standard mouse with a stylus pen and tablet for the tasks of dragging and pointing. They found the highest index of performance for the tablet and stylus pen during pointing and for the mouse during dragging. More interesting is that they also note that a stylus pen and tablet have the potential to perform as well as a mouse in direct manipulation systems (for example, Microsoft Windows' drag and drop system), and may out perform the mouse when the user activities include drawing or gesture recognition.

This would suggest that future paint software should be designed for interaction with a stylus pen rather than with a mouse when used with a graphics tablet. However, keeping in mind the comments provided by the artists, a few usability issues need to be addressed, such as ensuring that all menus, buttons and items of interest are placed within easy reach of the user and have a certain level of visibility of the functions.

Fitzmaurice et al (1999) present an exploratory study comparing paper to a Wacom digitizing tablet. Their results indicated a requirement for a freely rotatable drawing surface analogous to the artists in the study reported here. The Wacom tablet used in this study was an older type of tablet than the one discussed here. However, even now the graphics tablet is still not freely rotatable. It can be rotated slightly to the right or left, but the problem arises when a user tries to rotate it 180 degrees as it has a wire protruding from the back of the tablet which stops it from rotating completely. Our participants also found this to be a problem when using the Wacom tablet.

The future of digital art is dependent on the advancement of technological tools such as the graphics tablet. When creating tools for art the artist should be involved throughout the process, and the five parts of the facilitation of art model should be adhered to. In saying this, the evidence from this chapter would lead us to believe that it may be possible for traditional fine artists to make the transition from physical to digital art effortlessly in the future.

CONCLUSION

Clearly there are promising new applications to afford the creation of art. However, the acceptance of new tools is dependent on the advancement and usefulness of such tools and the willingness of artists to use these tools. The Technology Acceptance Model (TAM) introduced by Davis, Fred D. et al. (1989) explains user acceptance of a technology based on user perceptions. The model distinguishes between perceived ease of use and perceived usefulness. Perceived ease of use is the extent to which a person believes that using the technology will be free of effort. Perceived usefulness determine a user's intention to use the technology. Perceived usefulness is likely to be influenced by the perceived ease of use. Or as more simply expressed by Venkatesh (1999), the easier the technology is to use, the more useful it can be.

The above study has explored the experience or as expressed above 'the perceived ease of use' by artists to create a painting through digital means. The five artists participated in the study for three to four hours. This is obviously not sufficient to make any substantial conclusion about the usability or 'perceived usefulness' of this tool. Further studies could be conducted for an extended period of time, for example a comparative study where the artists produce the same picture with and without a graphics tablet, thus comparing the graphics tablet with traditional painting techniques. The participant's errors and time taken could be recorded.

The types of software that lend themselves for use with a tablet, such as Adobe Photoshop, Paint Shop Pro, Corel Painter, MS Paint could also be explored and possibly improved for usage with a graphics tablet. Another important investigation could be to explore the haptic feedback embedded in the stylus pen, such as investigating whether or not haptic feedback in conjunction with the graphics tablet is of benefit.

A fundamental issue to explore is the relationship between the use of artistic tools and their impact on the final art piece. Holloway (2005) comments on how with any work of art, the quality of the image is dependent on the artist's skill in manipulating the tool set. Based on our studies, it can be argued that working with a computer provides many options as well as limitations but the art produced is still dependent on the artist's skill.

The following conclusions can be drawn from these studies. Fine artists to create art can use a graphics tablet. It should be thought of as another way to create fine art. Each of the artists enjoyed the experience. They did however have problems using the graphics tablet. Future paint software should be designed for interaction with a stylus pen rather than a mouse. The easier the graphics tablet is to use for artists, the more useful it will become. The graphics tablet does supply an avenue for traditional fine artists to pursue their art through digital means. Using a stylus pen as an alternative to the mouse immediately sends a different message to the artist. They will not be abandoning their skill to try and master creating art with the mouse. They can still use their

drawing and painting skills to create art in the digital realm. An important point to make is that the graphics tablet allows the artist to create art that does not look computer made, and that may not have been possible using traditional fine art methods. However, the attitude towards art that is created using a computer needs to be addressed, as this is not encouraging for fine artists to migrate from traditional means. This is still the largest stumbling block stopping digital art becoming successful as an art form. The graphics tablet is helping to alter the art world's perception of fine art created using technology. This alteration can only be to the betterment of fine artists, both using traditional and digital means.

REFERENCES

Baxter, W., Scheib, V., Lin, M.C., & Manocha, D. (2001). Dab: interactive haptic painting with 3d virtual brushes. In *'SIGGRAPH '01: Proceedings of the 28th annual conference on Computer graphics and interactive techniques'* (pp. 461–468). ACM Press, New York, NY, USA.

Baxter, W., Wendt, J., & Lin, M. (2004). Impasto: a realistic, interactive model for paint. In *'NPAR '04: Proceedings of the 3rd international symposium on Non-photorealistic animation and rendering'* (pp. 45–148). ACM Press, New York, NY, USA.

Bestor, C. (2003). Installation art: image and reality. *SIGGRAPH Comput. Graph., 37*(1), 16–18.

Binkley, T., Entis, G., Maxwell, D., & Smith, A. R. (1994). Computer technology and the artistic process: how the computer industry changes the form and function of art. In *'SIGGRAPH '94: Proceedings of the 21st annual conference on Computer graphics and interactive techniques'* (pp. 494–495). ACM Press, New York, NY, USA: Chairman-Jane Flint DeKoven.

Bolter, J., & Grusin, R. (1999). *Remediation: understanding new media*, London; Cambridge, Mass: MIT Press.

Buxton, B. (1997). Artists and the art of the luthier. *SIGGRAPH Comput. Graph., 31*(1), 10–11.

Davis, F.D., Bagozzi, R.P., & Warshaw, P.R. (1989). User acceptance of computer technology: A comparison of two theoretical models. *Management Science, 35*(8), 982–1003.

Edmonds, E., Turner, G., & Candy, L. (2004). Approaches to interactive art systems. In *'GRAPHITE '04: Proceedings of the 2nd international conference on Computer graphics and interactive techniques in Australasia and South East Asia'* (pp. 113–117). ACM Press, New York, NY, USA.

Fitzmaurice, G.W., Balakrishnan, R., Kurtenbach, G., & Buxton, B. (1999). An exploration into supporting artwork orientation in the user interface. In *'CHI '99: Proceedings of the SIGCHI conference on Human factors in computing systems'* (pp. 167–174). New York, NY, USA: ACM Press

Flagg, M., & Rehg, J.M. (2005). *Oil painting assistance using projected light: Bridging the gap between digital and physical art.* Gvu technical report git-gvu-05-35. Georgia Institute of Technology.

Global Ergonomic Technologies (1998). *Comparison of Postures from Pen and Mouse Use.* Guerneville, CA, U.S.A. URL: http://www.wacom.com

Gregory, A., Ehmann, S. & Lin, M. (2000). inTouch: interactive multiresolution modeling and 3d painting with a haptic interface. In *Virtual Reality* (pp. 45–52). IEEE, New Brunswick, NJ.

Helander, M. (2005). *A guide to human factors and ergonomics.* Taylor and Francis.

Henzen, A., Ailenei, N., Fiore, F.D., Reeth, F.V., & Patterson, J. (2005). Sketching with a low-latency electronic ink drawing tablet. In *GRAPHITE '05:*

Proceedings of the 3rd international conference on Computer graphics and interactive techniques in Australasia and South East Asia' (pp. 51–60). New York, NY, USA: ACM Press.

Holloway, J. (2005). *Promise, paradox and opportunity.* URL: http://moca.virtual.museum/editorial/holloway.htm

MacKenzie, I.S., Sellen, A., & Buxton, W.A.S. (1991). A comparison of input devices in element pointing and dragging tasks. In *CHI '91: Proceedings of the SIGCHI conference on Human factors in computing systems* (pp. 161–166). New York, NY, USA: ACM Press.

McCullough, M. (1998). *Abstracting craft : the practiced digital hand / Malcolm Mc-Cullough.* London, Cambridge, Mass: MIT.

Norman, D.A. (2002). *The Design of Everyday Things.* New York: Basic books.

Picard, R.W., & Klein, J. (2002). Computers that recognize and respond to user emotion: theoretical and practical implications. *Interacting with Computers, 14*, 141–169. URL: http://www.sciencedirect.com/science/article/B6V0D-459BFXM-3/2/1aea6019fe1bb3835dd6e2480658a68e

Pier, M.D., & Goldberg, I.R. (2005). Designing interfaces for art applications. In *CW '05: Proceedings of the 2005 International Conference on Cyberworlds* (pp. 172–178). Washington, DC, USA: IEEE Computer Society.

Schrammel, J., & Tscheligi, M. (2006). Experiences evoked by today's technology - results from a qualitative empirical study. In *20th International Symposium on Human Factors in Telecommunication.*

Sensable Technologies (1993). *The phantom.* URL: http://www.sensable.com

Tait, W. (1998). The space between: fine art and technology. *SIGGRAPH Comput. Graph. 32*(1), 17–19.

Vygotsky, L. (1989). *Psychology of art*. MIT Press.

Venkatesh, V. (1999). Creation of favorable user perceptions: Exploring the role of intrinsic motivation. *MIS Quarterly, 23*(2), 239–260.

Wacom Technology Inc. (n.d.). *Wacom cintiq 21ux*. URL: http://www.wacom-europe.com/int/products/cintiq/whatto.asp?lang=en

Whale, G. (2002). Why use computers to make drawings? In *C&C '02: Proceedings of the fourth conference on Creativity & cognition* (pp. 65–71). ACM Press.

Yoshida, S., Kurumisawa, J., Noma, H., Tetsutani, N., & Hosaka, K. (2004). Suminagashi: creation of new style media art with haptic digital colors. In *MULTIMEDIA '04: Proceedings of the 12th annual ACM international conference on Multimedia* (pp. 636–643). New York, NY, USA: ACM Press.

KEY TERMS

Cross-Hatching: Marking or shading with two or more sets of intersecting parallel lines.

Fine Art: Fine art is any art produced or intended primarily for beauty rather than utility (Collins English Dictionary).

Graphics Tablet: Computer peripheral that allows hand drawn input using a stylus pen rather than a mouse.

Haptic: Relates to the sense of touch.

Multimodal: Having two or more modes of interaction

Proprioception: The sense of relative position of neighbouring parts of the body. Unlike sight, taste, smell, touch, hearing and balance by which we perceive the outside world. Proprioception is a distinct sensory modality that provides feedback solely on the status of the body internally.

Somatosensory: Related to the sensations that involve parts of the body not associated with primary organs, for example, light touch, pressure, pain, and temperature.

Stylus: A pointed instrument used as an input device on a pressure sensitive screen.

Technology: Any computer related digital tool.

Chapter XII
Information Visualization and Interface Culture

Greg J. Smith
serialconsign.com, Canada

ABSTRACT

This text seeks to contextualize the history of and discourse surrounding information visualization. It positions visualization in relation to broader 20th century visual culture and addresses the evolution of the interface as a ubiquitous tool and the aesthetics for understanding the organization of information. A timeline of precursors to the Graphical User Interface (GUI) is developed and a survey of recent related history and theory is conducted to deliver additional perspectives on information aesthetics. The text concludes with a brief survey of several recent visualization projects to illustrate the variety of fields being engaged and enriched by contemporary information design.

INTRODUCTION

In recent years, there has been a scramble to delineate what may be a new frontier in visual culture. Artists, theorists and designers have rushed headlong into an ambiguous realm that has been simultaneously described as data art, information architecture, infographics and most frequently, information visualization. What do these terms mean? Furthermore, what range of creative and communicative activities do they encompass?

This evolving initiative to define information visualization can be largely attributed to the influence of designer Edward Tufte, who taught statistics at Princeton in the mid-1970s. His dissatisfaction with existing literature on the "graphic standards" of information design inspired his seminal text *The Visual Display of Quantitative Information,* which he self-published in 1982. According to Tufte (2001), information visualization is about creating "instruments for reasoning about quantitative information" (p. 9) with the utmost importance placed on clarity and precision.

Information has become an aestheticized commodity, one driven by an increasing visual literacy. Contemporary media allows users to cross-reference and interact with complex flows of data and statistics while registering multiple world-views. Once limited to scientific and industrial research, quantitative data analysis is now accessible "in our living rooms and at our breakfast tables" (Danziger, 2008, p. 11). This chapter will position recent interest in information visualization as resulting from the widespread proliferation of interface culture over the last thirty years. Before delving into the nuances of the interface it is worthwhile to pause and schematize a working definition of information visualization that will act as a foundation for subsequent discussion.

What constitutes information visualization? This is a complicated question to answer as identifying the boundaries of new disciplines is always a contentious affair. Using Tufte's "instrumental" definition as a springboard, this discussion will consider information visualization as the *distillation of a body of data into a meaningful graphic representation*. Problematizing this notion of "informative" representations is of particular interest to media theorist Mitchell Whitelaw (2008) who has identified data as a "substrate" of information, a body of "raw material" that can be curated and contextualized into legible forms or celebrated as abstraction. For the purpose of this discussion we will consider charts, graphs, maps, and time-based interactive pieces as potential candidates for inclusion in the domain of information visualization. In many ways, visualization can be considered a "cartographic" enterprise—as the map delineates territory, the visualization renders data, connections, time and space.

DEFINING THE INTERFACE

An interface defines the boundary between two entities. It abstracts the interior language of a system and serves as an operable membrane through which this system can be manipulated. Although we tend to associate this relationship with the control of technology (i.e. software directing hardware), we can abstract the notion of the interface to read the practice of information visualization. An information "map" is an abstraction of a dataset into a more accessible, legible form that can be quickly scanned, comprehended and potentially even reconfigured by a user.

As a paradigm, interface culture is most clearly exemplified by the Graphical User Interface (GUI). The GUI is a virtual environment that has become so ubiquitous that we have become blind as to how much it colours our perception. This "hidden in plain sight" perspective on the interface as a paradigm was the subject of Steven Johnson's text *Interface Culture* (published at the height of Microsoft's power, before the dot-com bubble burst), which presciently employed the GUI as a cipher to read the sweeping economic and technological changes of the mid 1990s.

Many of the organizational qualities of the GUI have now been deployed in other forms of media. The influence of interface aesthetics can be seen in print, motion graphics and gaming and has radically altered the nature of broadcast design – one only need compare archival television news broadcasts from the 1980s to the info-blitzkrieg of present-day cable news. There has been much speculation about the wide-ranging implications of pervasive computing, a technological revolution that will shortly transform the world around us into a networked system of "intelligent" objects. This research will posit the history of information visualization as being directly tied to the proliferation of interface culture, an ideological and technological shift that has already occurred. It will also assume there is some truth to the notion that increasing bandwidth devoted to the transfer of data will demand more sophisticated frameworks for interpreting this deluge of information (Johnson, 1997).

In *The Language of New Media* (2001), Lev Manovich updates the thought of art historian

Erwin Panofsky by referring to the database as replacing perspective projection as the "symbolic form" of contemporary culture (p. 219). In his seminal writing on the Renaissance (1972), Panofsky identifies perspective as "picture space" and describes the transition of painting surface from a simple material underlay into a "window through which we look out into a section of the visible world" (p. 120). According to Manovich, the non-hierarchical digital inventory is now an equally important paradigm. However, does it not follow that the means through which we interact with this information is more meaningful than the structure of the backend? Secondly, can we draw any connections between the "standardized displays" and experiences associated with GUI-based operating systems and contemporary information visualization?

Once visualization is considered as emerging from interface culture, an entirely new historical vector is activated, one which can be used to trace the roots of visualization back more than 50 years. This enables our analysis to move beyond standard conversations of graphic clarity and timelines of recent representational techniques and software developments to frame the discipline of information visualization as emerging out of a broader sampling of 20th century visual culture.

This text will explore the relationship between information visualization and interface culture through three investigations: a historical examination of significant early technologies, prototypes and imaging techniques that anticipated the GUI, a summary of discourse and research that cuts through, informs and problematizes visualization, and a survey of contemporary work.

SIGNIFICANT RELATED TECHNOLOGIES

In tracing the genealogy of information visualization there are a number of potential historical discourses to draw from. The study of information design usually employs statistics, demographics or cartography as choice vantage points from which to consider the discipline. Recent interest in the work of William Playfair (1759-1823) and Charles Minard (1781-1870) is proof positive of the legitimacy of this backstory in the eyes of many design historians. Given that the goal of this conversation is to delineate a connection between visualization and interface culture, this analysis will instead look to early and mid-20th century imaging technology and the roots of the GUI as a starting point. What follows is a list of several key developments in imaging and interface technology that have had significant influence on information design.

It is worth noting that one of the most famous images associated with Charles Minard is his graphic detailing the ill-fated March of Napoleon into Russia in 1812-1813. It is no accident that one of the first complex information graphics schematized a military campaign, considering the longstanding tradition of technological and informational innovation being advanced by the gears of war. Instead of looking to the techniques of Minard, what can we learn from related mapping and interface endeavors from the 20th century?

This portion of the text will examine the history of radar, Vannevar Bush's Memex, the development of the Head-Up Display in military aviation and the birth of the GUI. In tracking these developments we may better understand the intersection of 20th century informatics and imaging technologies as well as the roots of pervasive interface culture.

THE BATTLE OF THE BEAMS

Of the many battles that took place between the United Kingdom and Germany during this war, the "battle of the beams" was one of the most decisive. This conflict pitted nascent British and German radar technology against one another with aerial dominance of the skies over England hanging in the balance.

Radar was first developed by the German inventor Christian Huelsmeyer for the purpose of collision avoidance in nautical navigation. Huelsmeyer publicly demonstrated his system in 1904—it operated by firing radio waves at targets and detecting their reflections. Over the next two decades, European and North American scientists would further develop this research and the range of radar systems extended from several to 25 miles. By the onset of the war, radar was emerging as a viable tactical tool. The crux of British-German radar warfare emerged from the German air force's utilization of the "Knickebein" and "X-Gerät" signal transmission systems to enable nighttime bombing runs over Britain. The Luftwaffe bombing raids were executed with surgical precision and this presented a sea change in aerial warfare to which the British military had to respond. Fortunately for Britain, a rudimentary radar network had been implemented before the onset of the war and it was able to serve as the cornerstone in a comprehensive British defense strategy that would ultimately "out-visualize" their German opponents.

In 1937, a prototype radar network was set up along the perimeters of Great Britain. Dubbed "Chain Home", the system consisted of a line of transmitter stations positioned at 50 mile intervals around the perimeter of the United Kingdom. Led by scientist Robert Watson-Watt, the British military capitalized on this system to develop state-of-the-art methods for enemy detection and fire control. This advanced mapping of the airspace over the United Kingdom acted as a "force-multiplier" allowing the British defenses to concentrate the aircraft where they were needed most and coordinate supporting anti-aircraft fire. Chain Home was monitored by oscilloscopes and operated as follows:

When a pulse was sent out into the broadcast towers, the scope was triggered to start its beam moving horizontally across the screen very rapidly. The output from the receiver was amplified

and fed into the vertical axis of the scope, so a return from an aircraft would deflect the beam upward. This formed a spike on the display, and the distance from the left side—measured with a small scale on the bottom of the screen— would give the distance to the target. By rotating the receiver goniometer [a tool for measuring angles] connected to the antennas to make the display disappear, the operator could determine the direction to the target... while the size of the vertical displacement indicated something of the number of aircraft involved. By comparing the strengths returned from the various antennas up the tower, the altitude could be determined. (Wikipedia, 2008)

This imaging technology provided the British forces with an early warning system by providing realtime data tracking German aerial activity over, or approaching, the United Kingdom. Radar-based defense networks have been described as "electromagnetic curtains", an upgrade to the medieval notion of fortification in which brick and mortar are bolstered and extended by telecommunication infrastructure (De Landa, 2003, p. 5). This technological development provided Britain with the strategic edge it required to turn the tide in the ongoing air battle against Germany. The oscilloscope based radar system would eventually give way to the Plan Position Indicator (PPI) display, which is now universally associated with radar technology.

A DESKTOP FOR THE AGES: VANNEVAR BUSH AND THE MEMEX

In July 1945, the engineer Vannevar Bush published an essay entitled "As We May Think" in *Atlantic Monthly*. This visionary text hypothesized a device called the Memex, and his description of this apparatus revolutionized thought about and the storage and manipulation of information. In outlining the Memex, Bush anticipated hypertext, the notion of personal "desktop" computing and foreshadowed the development of the GUI. The Memex capitalized on emerging magnetic tape technology and utilized it as a super storage medium capable of archiving vast amounts of information. The system could be loaded with thematized collections of texts and images and was operated through multiple screens that facilitated navigating and annotating this body of information. The best way to understand the operation of the text is to refer back to Bush's own words that explain a scenario in which the system might be used:

...the owner of the memex, let us say, is interested in the origin and properties of the bow and arrow. Specifically he is studying why the short Turkish bow was apparently superior to the English long bow in the skirmishes of the Crusades. He has dozens of possibly pertinent books and articles in his memex. First he runs through an encyclopedia, finds an interesting but sketchy article, leaves it projected. Next, in a history, he finds another pertinent item; he ties the two together. Thus he goes, building a trail of many items. Occasionally he inserts a comment of his own either linking it into the main trail or joining it, by a side trail, to a particular item. When it becomes evident to him that the elastic properties of available materials had a great deal to do with the superiority of the Turkish bow, he branches off on a side trail which takes him through text books on elasticity and tables of physical constants. He inserts a page of longhand analysis of his own. Thus he builds a trail of interest through the maze of materials available to him. (Bush, 1945)

The Memex facilitated the creation of connections between texts, a system for user annotation and, most importantly, a non-linear means of navigating "writing space". Implicit in Bush's contextualization of the device was the idea that the computer could be a tool for personal research,

rather than simply facilitating institutional or commercial data analysis. More important to this discussion is the fact that the Memex provided a comprehensive interface for managing multimedia content as well as an early inluential prototype for desktop computing.

A GIANT STEP FOR MILITARY IMAGING: THE BIRTH OF HUD

As stated in the discussion on radar technology, the importance of military research in helping advance 20th century imaging technology cannot be understated. The aerospace industry has been a continuous driving force in the evolution of graphic interfaces due to the complexity of aerial navigation and warfare. Another important benchmark in graphic display technology was the development of the Head-Up Display (HUD) in military aviation in the late 1960s. Originally developed by Specto Avionics, the HUD was originally proposed as a navigation aid to assist pilots with nighttime or rugged terrain landings. An article appearing in *Aircraft Engineering and Aerospace Technology* announced HUD to the world in 1968. This new display technology was described to be:

focused at infinity to appear superimposed upon the pilot's view ahead of the aircraft. The pilot need not remove his gaze from the view ahead of the aircraft in order to obtain data about the aircraft's performance. With conventional instruments there is a significant delay after the pilot realizes he needs data before he can re-adjust his eyes to obtain it. Obviously a display which enables the pilot to absorb data continuously without any conscious effort must be more efficient than a display that requires constant activity. ("Specto Avionics – Head Up Display", 1968)

Ready access to information about velocity, altitude and heading made pilots more in tune with their aircraft and kept their eyes glued to the sky rather than the various display surfaces that line the cockpit. There is an almost cybernetic subtext to the HUD, as it provides the operator of an aircraft with a mainline to the vital signs of their vehicle, imposing this data on top of their visual field, blurring the line between pilot and aircraft.

HUD information was originally projected via cathode ray tube (CRT) and has since been deployed through a range of delivery systems ranging from the commonplace (LCD) to those straight from the pages of speculative fiction (direct retinal projection).

THE GUI AND PERVASIVE INTERFACE CULTURE

The interface was previously described as the boundary between two entities. In terms of computing, it represents a software overlay that facilitates the user interacting with lower-level functionality. It is easy to overlook the GUI, as it has become the primary means through which people interact with information on a day–to–day basis. It is not unusual for a contemporary knowledge worker to spend the day administrating commercial logistics in a spreadsheet application and follow that with an evening navigating the interlocked architectures of social networks and content management systems. The GUI is of interest to the study of information visualization because in having these systems at our fingertips, we have developed a comfort in manipulating information while simultaneously multitasking across numerous applications. These developments have led to increasingly refined expectations of visual clarity in interface design.

The research underpinning the GUI was conducted at the Augmentation Research Center (ARC) at the Stanford Research Institute, by a team led by Douglas Englebart. In the 1960s this team developed the "oN-Line System" (NLS), which featured an early implementation of hyperlinked text. This work would serve as the basis of the first graphical user interface for the Xerox Alto computer developed at Xerox's Palo Alto Research Center (PARC) in 1973. The Alto featured the first deployment of now-standard features such as windows, buttons, icons and widgets and also made use of a pointing device, the mouse – also the brainchild of Douglas Englebart.

The Xerox Alto (1973) and its successor the Star (1981) were the first computer systems to embrace the now ubiquitous notion of the "digital desktop". These systems took the idea of the desktop as a work surface and translated the administrative and research activities associated with this space, abstracting them into metaphor. This interface minimized the user's contact with hardware and replaced esoteric methods of communication (punchcards, cumbersome "terminal" displays, etc.) by making the picture plane of the screen the site of interaction. Through the combination of fledgling visual and physical interface technologies the act of computing had become a graphical experience. Nowhere was the emergence of this visual paradigm driving the interface of personal computing more clear than in the runaway success of the Apple Macintosh in 1984.

A stripped-down version of Apple's prototype system Lisa, the Macintosh introduced the general public to a wide variety of GUI functionality that we now consider commonplace. These included fixed-height scrollbars, the trash can, the drag-and-drop process and a file system where all content was represented graphically. The system also made significant advances in type design, visual consistency and user experience, all areas that Apple would begin to build its brand around. The Macintosh was a runaway success and while Microsoft Windows would become the dominant operating system and GUI, it was most certainly Apple who went on to set the standard in computer interface design over the next 25 years.

With the crystallization of the GUI we can identify a connection between several of the technologies that comprise this timeline. Douglas Englebart, one of the key architects of the personal computing jump, started his engineering career by serving in the US Navy from 1944-1946 with the explicit purpose of enrolling in a year-long training program dedicated to radar and other emerging tele-imaging technologies (Eklund, 1994). The genesis of informational "screen culture" within a military test bed is easy to overlook given the much-heralded corporate research endeavors and subsequent commercial applications that we generally associate with the history of personal computing.

Why is acknowledging this group of technologies and platforms as underpinning information visualization important? Visualization is very much tied to the history of the screen, or more specifically the picture plane. Track this lineage and it leads you to related representational strategies and paradigms. The GUI, and the technologies that preceded it, have fundamentally changed the way we perceive information, this applies not just to computing but other media as well. We now exist in an era of pervasive interface culture, and with it is the implicit understanding that information is modular and, more importantly, a site for interaction.

INFORMATION VISUALIZATION PRECEDENTS AND DISCOURSE

In their recent paper "Towards a Model of Information Visualization" (2007), design scholars Andrea Lau and Andrew Vande Moere identified the availability of software and datasets, wide-ranging interdisciplinary thought and increased bandwidth as factors in helping to "facilitate the growth and importance of information visualization" (p. 6). While these factors speak to the abundance of cartographic mashups and social maps we've seen in recent years, this text takes the position that these factors are "effects" rather than the cause of an increased interest in information aesthetics.

In looking to develop a nuanced reading of causes and the nature of information visualization and interface culture Lev Manovich's scholarship on recent media history is a good point of departure. Manovich's reading of the database is useful precisely because it searches for the roots of contemporary modes of informational organization in cinema – the dominant medium of the 20th century. In tracing recent developments in visualization this discussion will also consider the sizable influence of John Maeda, Ben Fry and provide counterpoint by way of Alan Liu's articulate critique of information aesthetics.

LEV MANOVICH'S CALL FOR AN INFO-AESTHETICS

Lev Manovich was one of the first contemporary media theorists to critically engage information visualization and it is a subject that he has returned to often over the last decade. To Manovich, visualization represents an aesthetic project that speaks the language of the contemporary era. He identifies it as a subset of mapping and acknowledges the practice as deriving from the "new priorities" of networked culture, a society concerned with "making sense" of and "producing knowledge from information" (2005). This is similar territory to that traversed by Mitchell Whitelaw as cited in the introduction. To Whitelaw (2008), there is a world of difference between information and data, and while Manovich is speaking directly to the potential of information to illuminate the world around us, Whitelaw is invested in articulating the possibilities of data as a material. This "fork in the road" in discussions of information-fueled design or expression is worth keeping in mind as it highlights quite different destinations for any "data practice" or related aesthetics.

In *The Language of New Media* (2002), Manovich turned to proto-cinema directors such as Sergei Eisenstein (1898-1948) and Dziga Vertov (1896-1954) to contextualize new media. Eisenstein's development of montage editing and the index of experimental techniques employed by Vertov in *Man With a Movie Camera* were used to question the role of narrative in the 20th century. Manovich pointed to the production of film—the collection, storage and editing of footage—as database logic incarnate. He turns to the shooting schedule (often asynchronous to the narrative of a film), multiple takes, and material left on the cutting room floor to highlight that "production logistics" rather than the story arc yield a database of "possible films" (p. 237). This notion of an "expanded cinema" becomes a touchstone for evaluating new media.

Manovich employs a similar technique in reading the connection between certain moments in art history and visualization. To Manovich, a key by-product of the practice of visualization is an anti-sublime reading of the world. Where a romantic painter like Joseph Mallord William Turner (1775-1851) would set out to capture the incomprehensible majesty of light, space and atmosphere, a contemporary visualization would consolidate an incredible amount of data and dispassionately communicate it in a single frame or through a custom interface. Both of these creative pursuits speak to what could be described as representation and summarization, but in romantic painting there is inclination to revel where visualization merely renders. While many visualization projects celebrate hypercomplexity, this is most certainly a novel aesthetic that sits well outside the domain of classical and romantic art. The fact that Manovich considers visualization in relation to the broader history of representation is important because it allows for more engaged readings of the discipline than simply weighing the practice in relation to the history of graphic communication.

Manovich repeatedly gestures towards the widespread deployment of digital production,

editing and special effects technologies in the image arts (film effects, graphic design, etc.) in approximately 1993 as a definitive event that signaled not so much the arrival of "new media" but new thinking about media. It was at the possibilities of digital compositing that led to the contemporary image, which he identifies (2007) as being characterized by hybridity and remixability and describes as "meta-media". In using the term meta-media, Manovich describes images comprised of some combination of live footage, animation, typography, motion graphics and information overlays. This consolidation of media and techniques lends itself quite readily to a comparison with the multiscreen/multitask experience of the GUI.

The conclusion to Manovich's essay "Data Visualization as New Abstraction and Anti-Sublime" (2002) identifies an overt connection between visualization and science in that it can "help explain the patterns that surround us." He points out that the utility of the medium of visualization is in its ability to highlight connections and linkages that might otherwise go undetected. Manovich ends this text with a speculation that the true potential of visualization lies in communicating our personal connection to the flows of information that surround us:

If daily interaction with volumes of data and numerous messages is part of our new "data-subjectivity," how can we represent this experience in new ways? How new media can represent the ambiguity, the otherness, the multi-dimensionality of our experience, going beyond already familiar and "normalized" modernist techniques of montage, surrealism, absurd, etc.? In short, rather than trying hard to pursue the anti-sublime ideal, data visualization artists should also not forget that art has the unique license to portray human subjectivity – including its fundamental new dimension of being "immersed in data." (Manovich, 2002, p. 12)

In using Manovich to read visualization a first instinct might be to rely only on his exploration of the database or moving image. Ultimately Manovich sees the true value of visualization in its capacity to reflect the manner in which we as individuals engage the information around us.

JOHN MAEDA, BEN FRY AND THE ERA OF POST-VISUAL ARTS

Another key figure with a decidedly unique personal response to the information around us is John Maeda. Having reaped the benefits of a traditional design education and bolstered this with a playful engagement with technology, Maeda has emerged as one of the preeminent design educators of this generation. In the foreword to his autobiographical monograph *Maeda@Media*, Maeda identifies computation as having pushed us into an era of "post-visual arts." To add fuel to the fire, he titled a 2001 show at Tokyo's NTT InterCommunication Center "Towards Post Digital." These provocative descriptions are not the words of a contrarian, rather the suggestion that the information revolution has already happened and that the creative class is charged with making sense of it all after the fact.

Several themes are evident across Maeda's body of work. His software applications and multimedia design are generally process-oriented with a nuanced sensitivity towards materiality and an emphasis on engaging the means of production. Maeda is acutely aware how much the trappings of personal computing affect the way we perceive the world around us. Many of his projects revolve around developing custom, experimental tools rather than mastering proprietary software platforms. Beyond this, Maeda has repeatedly explored everyday means of interaction (mouse, keyboard, application window, desktop tower, monitor, etc.) as creative opportunities in and of their own right. For Maeda, the endgame in working with technology is to work through it, to be liberated:

There needs to be a concrete set of core advancements in the tools we use, not just incremental updates. To realize such a future, more artists must be unafraid to peer deep inside the machine and directly affect a deconstruction of the software systems that imprison all digital expressions. (Maeda, 2000, p. 225)

This statement could very well serve as a synopsis of Maeda's wildly successful tenure at the Massachusetts Institute of Technology (MIT) Media Lab from 1996 to 2008. While at MIT, Maeda led the Aesthetics Computation Group (ACG), a design and computation studio that set out to prototype "advanced architectures and processes to enable the exploration of unimagined spaces and forms." (Maeda, n.d.) A remarkable group of designers contributed to this fertile project, one of whom was Ben Fry, a programmer and visualization scholar who has indeed "peered deep inside the machine."

Fry's ACG Masters thesis *Organic Information Design* (2000) set out to consider how interactive visualization systems could embody organic qualities in response to flows of information (p. 13). Fry's basic hypothesis was that while the representation of static data had been researched extensively, the relevance of this body of knowledge started to waver when computation permitted interactive time-based representations. To this end, Fry set discussed a variety of organic phenomena (e.g. adaptation, metabolism, homeostasis and reproduction) and deployed them in his Valence and Anemone projects, two incredible visualizations that consituted a large part of his thesis work. Employing a methodology similar to that of Tufte, Fry created his own archaeology of visualizations and turned to forward-thinking software projects from the early 1990s. Discussed work included Mitchel Resnick's *Starlogo* (1994), Ramana Rao and Stuart Card's *Table Lens* (1994), Lisa Strausfeld's *Financial Viewpoints* (1995), and Martin Wattenberg's *MarketMap* (1998). He also penned an extended homage to John Conway's experiments with simulating cellular autonoma through the Game of Life (1970). Fry ultimately concluded that organicism in information design could allow users to "engage in an active deconstruction of a data set" and to "pull apart complexity" via the realtime manipulation of the rules driving the representation (p. 16).

Fry further developed this body of research through his dissertation *Computational Information Design* in an attempt to reconcile data mining with interface design. He chose the field of biology, specifically the complex realm of genetic classification, as a proving ground in which to develop reactive systems for visualization. The bulk of Fry's thesis work was coded in Processing, an open-source programming language co-developed with Casey Reas, whom Fry quoted (2004) describing their project as moving "graphics and concepts of interaction closer to the surface" (p. 126). Processing has emerged as one of the platforms of choice in visualization and the programming environment is clearly the heir apparent to John Maeda's earlier Design by Numbers software-for-artists initiative.

CHARTJUNK, ADMINISTRATIVE DEBRIS AND THE LAWS OF COOL

Since this discussion started with Edward Tufte, it is useful to return to him in attempting to further parse the discourse surrounding visualization. Tufte's vocabulary for discussing information design is packed with phrases like "chartjunk", "wasted ink" and "administrative debris" (2008). Clearly a reductionist, Tufte values clarity and an accurate representation of the data being communicated above all else. While this reading of information design certainly makes sense considering established protocols in statistics, science and modern cartography, are there other ways that we can read Tufte's zeal for skeletal minimalism in broader information design? The digital humanities scholar Alan Liu has developed

a very useful critique of contemporary information aesthetics that can enrich and problematize Tufte's work.

Alan Liu's text *The Laws of Cool: Knowledge Work and the Culture of Information* (2004) presents a comprehensive dissection of interface, ideology and the workplace in the information economy. The book excavates a large body of 20th century management theory while tracking the transition of the dominant "work" paradigm from the assembly line to the cubicle farm. Nestled in the heart of this text is "Information Is Style", a chapter that speaks directly to this exploration of visualization and the interface.

According to Liu, by the time the web had arrived we were already living in an era defined by a "single, great canvas", one that had consolidated all of the posters, typographic experiments and, most importantly, the ethos of modernist design (p. 207). Liu identifies this monocanvas of screen, browser and application as a generalized information interface and points out that we have "imported" our aesthetics for reading this sprawling construct directly from Modernist design ideology. To Liu, there is a direct connection between best practices and usability in interface and software design and the whitewashed universality of the International Style. One noteworthy distinction between these points of reference is that where Modernism celebrated a fusion of form and function, the contemporary info-consumer is completely attuned to the separation of style and content – what can be described as the expansive divide between raw data and how we experience it on screen (Johnson, 1997). In this new milieu, design becomes a delivery device for data, and the "thirst" for stylized representations of information is the primary reason that a web page is more like a "glossy consumer magazine than cargo off a truck" (Liu, p. 215).

To Liu, information design (and the appreciation thereof) is an ideologically driven process. Liu would identify the recent widespread interest in visualization as further evidence of a general complicity with contemporary modes of production, proof positive of the triumph of the information economy over creative domains. While this perspective is quite cynical, it is interesting to weigh against Tufte's purist "function over form" tendencies. Viewed in this light, the practice of visualization can be read as a love affair with bureaucracy, design for "efficiency obsessed" technocrats. Can we politicize visualization? It is outside the scope of this discussion to answer that question but we should acknowledge that visualization can (and no doubt will) be subject to broader cultural discussions in the near future.

Perhaps the key paradigm in contemporary information visualization is shifting from discussions that orbit around clarity towards the possibility of depth in and the creative exploration of the interface. Seen in this light, the "means of interaction" becomes a medium, one which yields potential as an educational and communicative tool that allows the user to scan and interpret large bodies of data and be an active agent in reconfiguring the representation of this content.

The GUI has profoundly shaped our relationship with information. It has enabled rich and complex, multivalent connections to and interpretations of the "flowing data" that now permeates all facets of life. While information visualization certainly can be read in relation to statistical or cartographic traditions, it is crucial to recognize the intrinsic connection between this medium, and interface culture as a way of life and means of perception.

Recent Developments & Contemporary Work

Considering the degree to which computation drives contemporary visualization, it is of little surprise that the field is evolving rapidly. Some key recent developments include the wildly successful *Deisgn and the Elastic Mind*, a spring 2008 show at the Museum of Modern Art (MoMA) in New York City which included a diverse ray of interactive visualization projects. A few months later, MoMA added one of these pieces (a version of *Cabspotting* by Stamen Design) to their permanent collection. Also of note is the significant initiative at the New York Times to deliver increasingly sophisticated multimedia information graphics as part of their current rebranding as a "21st century newspaper". These developments speak to the gradual shift of visualization as a hermetic tool for specialists to a meaningful visual language ready for application across mass culture. Mike Danziger's 2008 thesis, *Information Visualization for the People*, does an excellent job of exploring the increasingly social and accessibility-oriented nature of visualization. One only need look as far as the projects making the rounds in the art and technology networks on the web to sense that we are in the midst of a significant cultural shift in terms of how we represent and communicate information.

Given the investment this text has made in a historical reading of information visualization, a suitable ending point would be to examine some choice examples of contemporary work. This selection of projects illustrates an exciting spectrum of representational techniques and points to a range of communicative, geo-locative, archival and polemical possibilities for information visualization.

Completed as a part of the aforementioned *Computational Information Design* (2004), Isometric Blocks (Figure 1) is an application for scanning patterns in clusters of genes that tend to occur together. When comparing the genomes of two individuals, single letters will vary over every few thousand units of genetic code. These changes are called single nucleotide polymorphisms (SNPs) and tend to occur in distinct patterns that are referred to as halotype blocks. To visualize a range of halotype blocks mapping the SNPs of a sample group of approximately 500 people, Fry created an interactive application that possessed multiple modes of representing this data set. A user can seamlessly switch between these

Figure 1.

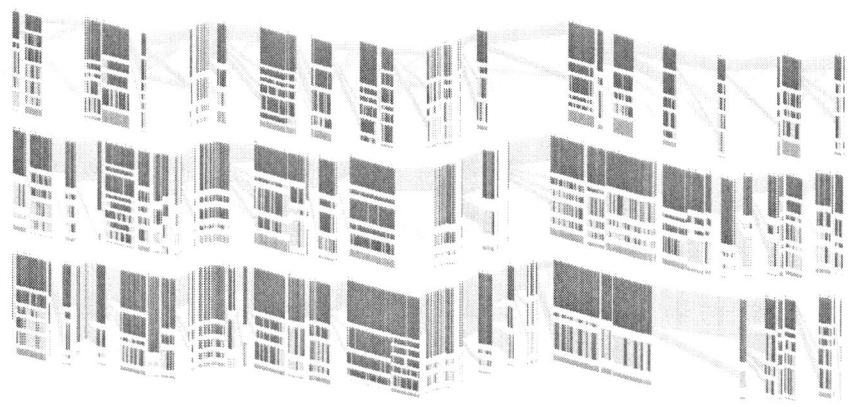

Ben Fry – Isometric Blocks
http://benfry.com/isometricblocks/

views in an effort to understand and interpret this data and the system also possesses the ability to display additional "advanced" information for expert users.

In reading this visualization, the relative thickness of each column indicates the percentage of the survey group with a given halotype block configuration. Moving along the x-axis, representations for increasingly rare halotype blocks are visible. The colors in each row depict possible variations for each SNP, with the darker tones representing more common SNPs and the lighter tones rarer SNPs. Figure 1 is a 3D view of the data set rendered as an isometric projections where each block is offset in the z-axis. The bottom row includes lines noting the transitions between blocks. In the application Fry has created for presenting this data the visualization can be viewed in several other 2D and 3D modes permitting a range of scaled and quantitative means of interpreting the information. Beyond this, the visualization can be "tuned" to alter the range of information represented in a given view. All of

this functionality delivers a customizable viewport for exploring this complex body of genetic information.

Ben Fry approached this project with the zeal of a cartographer rather than that of a scientist. He has noted that in dealing with extremely large, exponential bodies of data, a *Powers of Ten* style of representation is inadequate as genetic data yields "plateaus" of interest which need to be explored and developed rather than zoomed-through (quoted in Abrams, 2006.) Beyond this, the complexity of genetic information necessitates multiple viewports which allow the exaggeration, reconfiguration or muting of certain parameters. Isometric Blocks provides a window into the world of contemporary genetics that can speak to the specialist and educate and inform the lay person.

Oakland Crimespotting (Figure 2) is a data mashup that spatializes crime statistics from the city of Oakland's Crimewatch service across maps culled from the Microsoft Virtual Earth database. Launched in 2007, the project is updated with fresh

Figure 2.

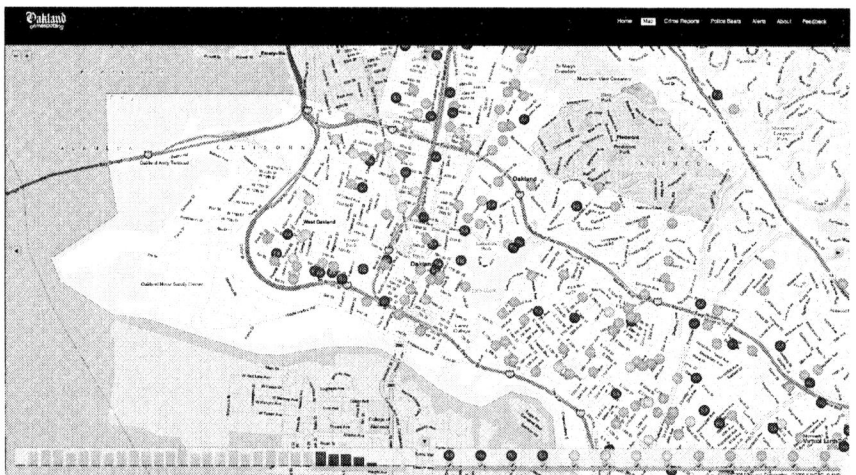

Stamen Design – Oakland Crimespotting
http://oakland.crimespotting.org/
© 2007 Stamen Design. Used with permission.

crime reports daily and provides a web interface for viewing the crime statistics for the last 30 days in the greater Oakland area. Crime types are categorized and color-coded to group violent offenses, property-related crime, and criminal activities that affect quality of life and the exact location of each crime is identified. The system also allows users to isolate blocks of time, types of crimes and specific police beats to provide a very flexible means of understanding how crimes play out across the city.

Oakland Crimespotting is controlled through a simple menu that spans the bottom of the screen. The left side of this interface allows you to select a timeframe and the right side allows the selection of which offenses are being displayed. Users can zoom into specific regions or beats to see detailed information for a few city blocks and zoom out to see patterns of crime distribution emerge across the entire city. Navigating this system is a purely visual exercise that is certainly much more intuitive than reading police press releases or tracking crimes covered in the traditional media. The project expands upon Adrian Holovaty's Chicago Crimespotting project (2005-2008) with an added

emphasis on user experience and an interface that is graphical rather than text-based.

Most importantly, Oakland Crimespotting allows users to cut through the rhetoric and stereotypes that can cripple neighborhoods. The project gives citizens the tools to develop an understanding of what asocial activities are taking place throughout the city and in doing so through this project they no longer have to rely on archaic civic infrastructure. Systems such as this can serve as research tools for communities to foreground issues that may not be getting adequate media coverage or political attention.

Atlas of Electromagnetic Space (Figure 3) is an interactive visualization, completed in 2008, is both an index of the radio spectrum of electromagnetic (EM) frequencies and a database of art projects that utilize this bandwidth. In reading the visualization, a viewer can switch back and forth between two views, with each privileging a specific body of data (EM frequencies or projects). The x-axis utilizes a logarithmic scale to spatialize a very broad range of frequencies (1KHz-100GHz) transforming an enormous dataset into a manageable range of information that

Figure 3.

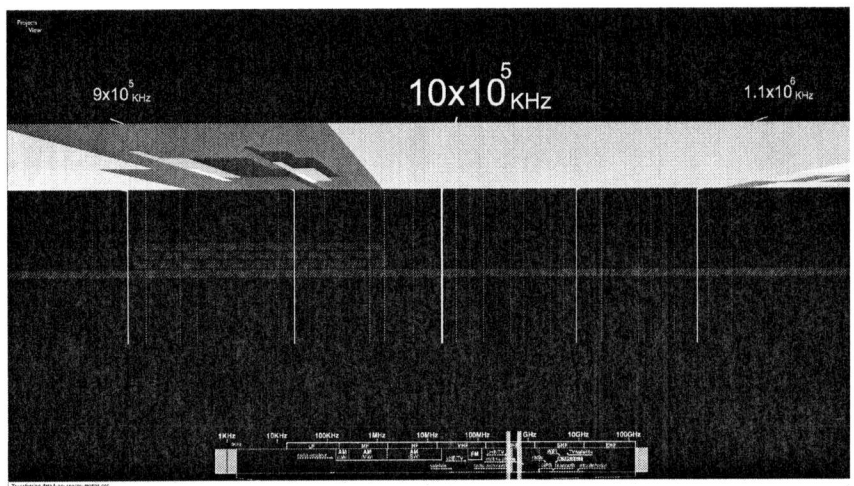

José Luis de Vicente, Irma Vilà and Bestiario – *Atlas of Electromagnetic Space*
http://www.spectrumatlas.org/spectrum/
© 2008 Santiago Ortiz. Used with permission.

can be quickly scanned. The interface underneath the primary viewport can be used to adjust the amount of bandwidth visible at a given time, allowing the user to zoom in and out and examine specific swaths of frequency. This interface is annotated with additional information about the EM spectrum, and provides a range of "landmarks" (e.g. the bandwidth employed by mobile phones)

Figure 4.

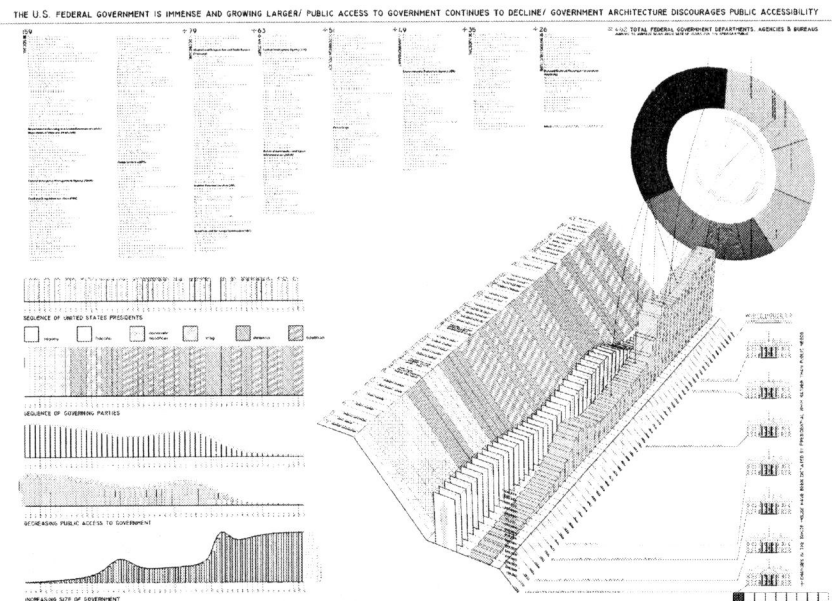

labRAD – White House 2.0 Research Infographic
http://www.lab-rad.com/
© 2008 labRAD. Used with permission.

to aid in contextualizing the information within the radio and broader EM spectrum. Providing the viewer with information about the frequency associated with mobile phones and other recognizable technologies helps make this information slightly less abstract.

While the project of representing the EM spectrum is a worthy endeavor in visualization, the Atlas also indexes a number of media art projects that rely on these frequencies. This interface provides a means of comparing the frequencies employed by creative projects in relation to the greater range of EM bandwidth as well as one another. Without having any understanding of the EM spectrum, a user can quickly identify a range of bandwidth, connect it to a common use, and then see what creative work is being done with these same frequencies. The Atlas functions not just as a visualization, but an archive.

The introduction to an award winning competition entry for the 2008 White House Redux Architectural Competition, this graphic (Figure 4) condenses a significant amount of American history into a single image. The mandate of labRAD's (designers Arielle Assouline-Lichten and Wayne W. Congar Jr.) vision of how a 21st century White House might work revolves around an exhaustive study of the American government, the legislative process, and speculation as to how public opinion might more directly influence governance.

This visualization consolidates several swaths of data, which is all represented in the main graphic on the right. First, the size of the federal workforce since 1792 is offset against public access to governance (right and left in the central bar graph). This information is referenced against the political history of America, tracking the political designation of each ruling party since 1792 (the hatched fields on the left). This data is drawn in perspective and the end of it butts up against a circular graph which communicates the relative size of federal departments, agencies and bureaus by category (i.e. infrastructure, judicial, environment, etc.). Branching off of this primary graphic are "architectural annotations" and elevation drawings recording various alterations to the White House over the last two centuries. While all of this information is combined into a composite graphic, it is also viewable as individual graphs on the left, and in the case of the information on government departments, expanded upon as a detailed list at the top.

When combined, this research yields a sophisticated analytic tool which serves as a timeline,

Figure 5.

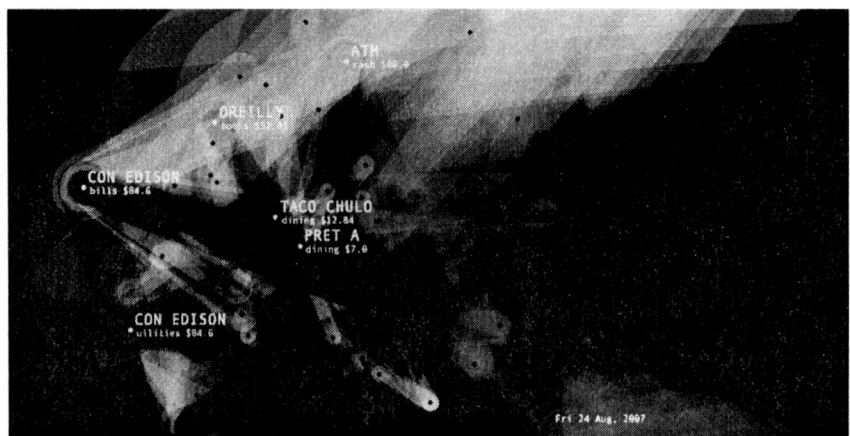

Burak Arikan – MYPOCKET
http://transition.turbulence.org/Works/mypocket/
© 2008 Burak Arikan. Used with permission.

political index and a people's history of America. This research was deployed as part of a larger narrative which positioned a 21st century White House as a nexus of visualized data which could provide the administration with a real-time window into the thoughts and opinions of Americans. The project addresses the rich history of polling and demographics and proposes that visualization can be deployed as an aid in the everyday practice of democracy.

MYPOCKET (Figure 5) is an interactive visualization documents the spending habits of designer Burak Arikan over the course of two calendar years. Arikan collects receipts on all purchases in order to build up a database of financial transactions in an effort to analyze spending habits and develop a system for predicting future transactions. On opening the Transaction Graph, the viewer is greeted by a complex weave of interconnected points that map out purchases for the most recent day archived in the system. A timeline running along the bottom of the image allows the user to scroll back in time, and doing so makes the graph dance across the page while reconstituting itself in reference to new base data. The project is a playful exercise in self-surveillance that not only aims to represent everyday patterns of expenditure and consumption as an interactive application, but also attempts to algorithmically predict future transactions based off spending habits. The statement for the project identifies these "found" future transactions, aligning the intent of this predictive venture with Marcel Duchamp as "if readymades are found in the past, predicted objects are found in the future." (Arikan, 2008)

In engaging with this visualization we are indirectly considering our own spending habits. Beyond this, the visualization elevates everyday activity (e.g. a trip to the laundromat), taking the "artifacts" from these events and turns them into raw informational material for an aesthetic experience. This brings to mind the words of David M.

Levy (2001), who in extolling the virtue of the receipt as a document worth studying reminds us of "a bigger challenge to look closely and respectfully at the lowest and homeliest of documents" and that the search for "beauty, depth and power" in receipts is most certainly a great accomplishment (pg. 8). Exploring this visualization forces the user to reconsider the aesthetic potential of banal transactions and indirectly questions the nature of commonplace financial documents such as spreadsheets and credit card statements into question.

REFERENCES

Abrams, J., & Hall, P. (Eds.) (2006). *Else/Where: Mapping*. Minneapolis: University of Minnesota Press.

Arikan, B. (2008). *MYPOCKET: Predicted Objects*. Retrieved April, 2008 from http://transition. turbulence.org/Works/mypocket/predicted_objects/

Danziger, M. (2008). *Information Visualization for the People* @ MIT Program in Comparative Media Studies Advised by Nick Montfort. Archived at: http://cms.mit.edu/research/theses/MichaelDanziger2008.pdf

De Landa, M. (2003). *War in the Age of Intelligent Machines*. New York: Zone Books.

Fry, B. (2004). *Computational Information Design*. Dissertation @ MIT Program in Media Arts and Sciences Advised by John Maeda.

Fry, B. (2000). *Organic Information Design*. Masters Thesis @ MIT Program in Media Arts and Sciences Advised by John Maeda.

Johnson, S. (1997). *Interface Culture: How the Digital Medium - From Windows to the Web - Changes the way We Write, Speak*. New York: Harper Collins.

Lau, A., & Vande Moere, A. (*2007). Towards a Model of Information Aesthetics in Information Visualization.*

Levy, D. (2001). *Scrolling Forward: Making Sense of Documents in the Digital Age.* New York: Arcade Publications.

Liu, A. (2004). *The Laws of Cool: Knowledge Work and the Culture of Information.* Chicago: University of Chicago Press.

Maeda, J. (2000). *Maeda @ Media.* New York: Rizzoli.

Maeda, J. (Ed.) (n.d.). *ACG Concepts Volume 01:Elements of Reactive Form.* Retrieved March, 2008, from http://acg.media.mit.edu/concepts/volume01.html

Manovich, L. (n.d.). After Effects, or Invisible Revolution. Presented at Danube Telelectures #4 *Remixing Cinema.* November 8th, 2007. http://www.donau-uni.ac.at/en/department/bildwissenschaft/veranstaltungen/telelectures/archiv/index.php

Manovich, L. (n.d.). Data Visualization as new Abstraction and Anti-Sublime (2002) http://manovich.com/DOCS/data_art_2.doc.

Manovich, L. (2005). The Shape of Information. http://manovich.com/DOCS/IA_Domus_3.doc

Manovich, L. (2001). *The Language of New Media.* Cambridge: MIT Press.

Panofsky, E. (1972). *Renaissance an Renascences in Wester Art.* New York: Harper & Row.

Tufte, E. (2001). *The Visual Display of Quantitative Information.* Cheshire: Graphics Press.

Whitelaw, M. (2008). Art Against Information: Case Studies in Data Practice. In *Fibreculture.* No. 11, 2008. Retrieved March, 2008, from http://journal.fibreculture.org/issue11/issue11_whitelaw.html

Wikipedia: The free encyclopedia. (2006, February 15th). FL: Wikimedia Foundation, Inc. Retrieved March 13, 2008, from http://en.wikipedia.org/wiki/Chain_Home

KEY TERMS

Data: Quantitative facts, figures or statistics provided without context.

GUI: A visual means to facilitate human-computer interaction, often employing familiar "graphic standards" such as windows, icons, menus and widgets to represent available information and actions.

Information: Organized data that increases the knowledge of the individual consuming it.

Information Visualization: The distillation of a body of data into a meaningful graphic representation.

Interface: An operable membrane that defines the boundary two entities (i.e. an operating system mediates the relationship between computer hardware and the user).

Chapter XIII
Memory Association Machine

Benjamin David Robert Bogart
Simon Fraser University, Canada

ABSTRACT

"Memory Association Machine" (also known as "Self-Other Organizing Structure #1") is the first prototype in a series of site-specific responsive installations inspired by cognitive processes. The artist provides a mechanism that allows the structure of the artwork to change in response to continuous stimulus from its context. Context is defined as those parameters of the environment that are perceivable by the system and make its place in space and time unique. "Memory Association Machine" relates itself to its context using three primary processes: perception, the integration of sensor data into a field of experience, and the free-association through that field. "Memory Association Machine" perceives through a video camera, integrates using a Kohonen Self-Organizing Map, and free-associates through an implementation of Liane M. Gabora's model of memory and creativity.

INTRODUCTION

This text describes and frames the first prototype in a body of work that aims to create artworks that find their own relationship to their context. These artifacts are embodied, meaning that they are manifested in a physical[1] form and are effected by, and effect, the world around them. The artifacts could be considered creative machines in that they transform material from their context into an original representation. The machine creates this representation through a mechanism provided by the artist. My research aims to use artistic enquiry to develop a meta-practise that binds the practises of responsive electronic media art, site-specific art, and artificial intelligence. This meta-practise includes theory from the philosophy of embodiment and is developed through the creation of embodied artifacts-as-processes—artifacts composed of computational processes. The material of the artwork is a set of computational processes that are causally connected to the physical world. The mechanism of the artwork is intended to exhibit emergent properties through the negotiation between software and physical context.

The system's processes are causally connected to the outside world through sensors (inputs) and external properties (outputs). How can an artifact—even a process—find a relationship to its context? Artifacts such as "Memory Association Machine" form an embodied relationship with their context in two ways; firstly, by being embodied so that they can both change and be changed by their context, and secondly, by having their structure altered through the process of embodiment. A more detailed account of the creation of "Memory Association Machine" is available in "Memory Association Machine: An Account of the Realization and Interpretation of an Autonomous Responsive Site-Specific Artwork" (Bogart, 2008). The process of embodiment is the negotiation between the subject and the object—the relationship between seeing and acting.

This text begins by weaving a theory of practise which has been, and is still being, developed. The practise is focused on the fundamental relationship between the artwork (artifact-as-process), the artist (author), and the world in which they are both embodied. A description of "Memory Association Machine" is interjected throughout the framing and followed by a detailed descrip-

tion of the system's architecture. Future work is presented at the conclusion of the text.

The Artifact

The formalization of the creative process is made up of two iterative sub-processes: realization and interpretation. The relationship between these processes is pictured in Figure 1. Realization is the path of intention from the artist to the world, whereas interpretation is the path of causation from the world back to the artist. This formalization is influenced by a theory of embodiment, the "intertwining", that binds the mind and body according to the phenomenology of Merleau-Ponty (1968). For each pair of actions (realization and interpretation) a frame[2] is created which colours the artifact (at that moment in time) with the tension between the expected results of realization and the following interpretation. Realization occurs when the artist chooses to effect the world in some way that manifests physically—for example, choosing the colour yellow for a particular region of a painting.

Interpretation is when the artist observes, experiences, and attributes value to the results

Figure 1. The artist realizes the artifact, which is then interpreted

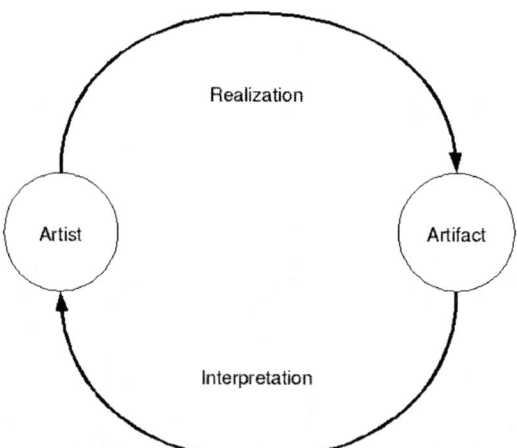

of the realization—for example, seeing yellow in a particular spot on a canvas. The perception of yellow results in the inevitable associations resulting from that stimulus, which have value in terms of the artist's experience. The painter may decide that the tone of yellow is not quite what he or she intended and adds more white to the paint. The artifact is transformed through the iterative dialectic between realization and interpretation. Future choices are made, in light of not just the current frame, but in the context of all past frames.[3]

Both the artifact and the artist's intentions are transformed through this process. The artist may interpret the artifact differently than the viewer. Artifacts-as-processes, on the other hand, are both snapshots of their own creation and enacted processes in themselves. Embodied artifacts-as-processes are machines that are constantly changing their own properties to form a relationship to, through the reflection and interpretation of, their embodied context.

Figure 1 is a simplified representation. It is not uncommon for a single artifact to contain

Figure 2. The nested realization–interpretation loop

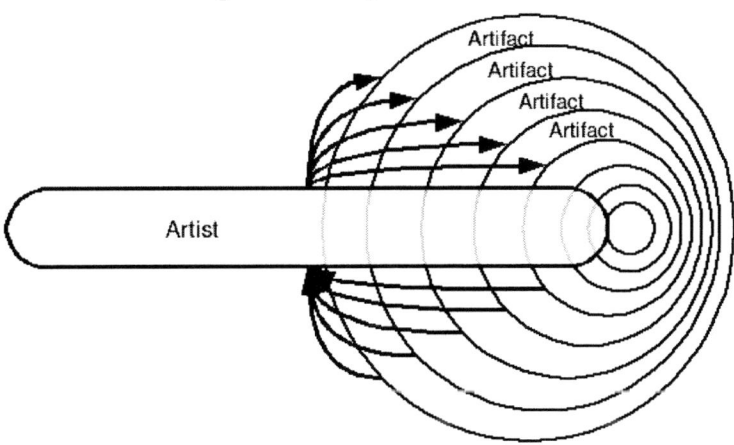

Figure 3. The realization–interpretation loop of the embodied artifact-as-process

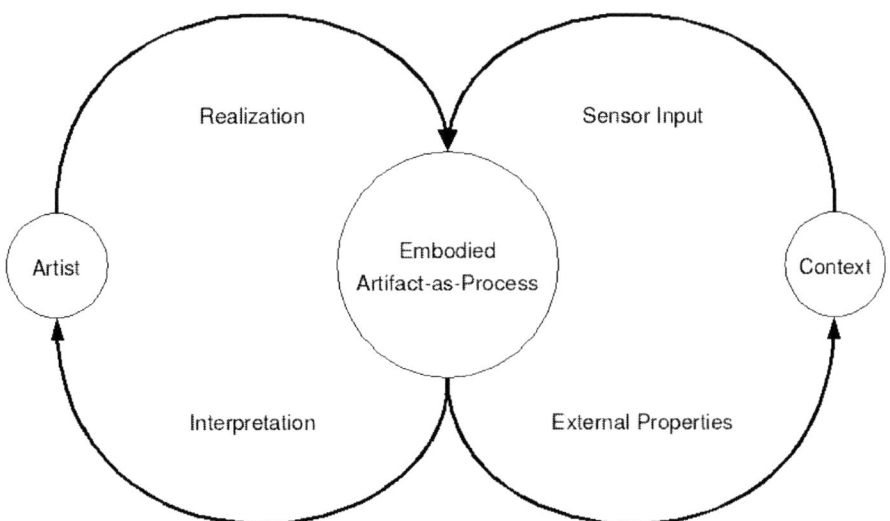

other artifacts, which in turn contain additional components, as pictured in Figure 2, where each artifact contains another artifact.[4] At each level in the nested structure there is a different realization-interpretation frame. The method of realization and criteria of interpretation may be different at each level in the structure. The time it takes to complete a single loop is entirely dependent on the nature of the creation. Some loops proceed very quickly—for example, the choice of a colour and the placement of a single blob of pigment on the canvas. Others take much more time—for example, the creation of a skyscraper.

Embodiment

"Memory Association Machine" is an embodied artifact-as-process. The embodied artifact-as-process is made up of two primary components: the software and the context. In the creation of embodied artifacts-as-processes the artist's intentions are often encoded in software (instructions for a computational process). As the software artifact is the primary medium in which I work, I consider software as my artistic material. The software material is a set of computational relations between symbols (variables). It is not a process but a static text that encodes a set of instructions. Software can become an embodied artifact-as-process when a subset of its constituent symbols become inputs and outputs that are causally connected to the physical world and perceivable by the viewer. It is this causal relationship between the world and the software that activates its instructions. The material of the work shifts from the representation of a process to action in the physical world.[5] The behaviour (the change of external properties) of the embodied artifact-as-process is a collaboration between the artist's intention, as encoded in software, and the context in which the artifact-as-process is embodied. Figure 3 depicts the realization-interpretation loop of an embodied artifact-as-process, where arrows represent causal relationships.

In contemporary artistic practise it is often the text accompanying the work that is expected to communicate the work's concept and purpose. This text is an artifact that exists independently from the artwork. Through realization and interpretation, the text, the software, and the artist's intention reach an equilibrium. What the machine means and what the machine does are unified. The software is the mechanism—but the process, executed in context, is the artwork. The result is a collaboration between the artist and the context. The role of the context must be embraced. Once the embodied system's behaviour has been interpreted, the software can be altered in order to more finely unify the artist's intention and the system's behaviour. The artist is then working in two different materials—the material of the software, which is impacted upon by the artist, and the material of the embodied process, which impacts itself on the artist. The interaction between the software and the physical world is not totally deterministic. The artist must relinquish control and allow the process to be driven equally by the software and the context. In "Memory Association Machine," the driving intention is that the artifact-as-process acts beyond of the intentions of the artist.

Artistic Enquiry

Historically, the artist is "imagined as an isolated figure of exceptional creative powers who suffers for his art." (Barker et al., 1999). There is a mythology surrounding the "creative genius" and the artifacts that the, predominantly male, artist creates. The artifact is then a record of genius which is often collected and fetishized. The artifact is manifested in the physical world. It becomes meaningful as its references are processed through the experience of the viewer. Why should the artist have more authority to attribute meaning to an artifact than the viewer?

Some approaches to art practise reject this notion of artwork as an expression of genius, and break the mythos of creativity, by shifting the emphasis away from the artifact toward the process itself (Possiant, 2007).[6] What is produced when the purpose of the work is not the creation of an art object, but an exploration of the very creative process itself? There are two products of this enquiry. The first is the artifact-as-process itself. The second is the knowledge that results from the artistic enquiry. This knowledge is manifested in the artifact-as-process and around it through documentation, rhetoric, sharing, and discussion.

This artistic enquiry is centred on the artistic practises of responsive electronic media and site-specificity. For a survey of electronic media art see "Information Arts" (Wilson, 2002). The discipline of site-specific artwork aims to create work that gives "…itself up to its environmental context, being formally determined or directed by it" (Kwon, 2004). Kwon (2004) and Kaye (2000) provide a background on site-specific practises. This body of work explores the qualities of embodied creativity through the artistic practise of developing artifacts-as-processes that find their own relationship to their context.

GROWING FORM FROM CONTEXT

In order to have "Memory Association Machine" find its own relationship to its context in a way that is not externally predetermined nor random, it is natural to look to cognitive science. The most relevant application of cognitive science happens in the discipline of artificial intelligence, which seeks to create software and hardware that exhibits some of the properties of human beings. In order for "Memory Association Machine" to relate to its context, the use of unsupervised connectionist artificial intelligence approaches is appropriate, as the behaviour of the system should not be dependent on an external knowledge-base provided to it. Since "Memory Association Machine" is an embodied system, the physical environment

becomes the "training" data for its artificial intelligence.

Methodology of Artistic Enquiry

As this research project is contextualized as an artistic enquiry, it is important to describe how the creation of these artworks is approached. The software is built piece-by-piece while the system is connected to its context. As the artwork is considered the result of the machine's negotiation with its context, the artwork can only be appropriately interpreted in its embodied context. The system's components are initially developed in isolation and attached to the rest of the system as early as possible. This method of building the system incrementally, and testing it in an embodied context at each step of development, resembles the methodology discussed in "[i]ntelligence without representation" (Brooks, 1992). The software development occurs in two modes. The first mode is an intuitive approach[7] that serves to get the basics of the system up and running—with arbitrary choices, random variables, and placeholders—so that the system can be evaluated in context, rather than in isolation. The second mode of grounded refinement involves returning to the results produced in the first mode and removing arbitrary choices by situating them in theory. For "Memory Association Machine," the source of theory comes from three main sources: connectionist artificial intelligence, specifically Kohonen "Self-Organizing Maps"; theories of creativity, specifically those present in the "Cognitive Mechanisms Underlying the Creative Process" (Gabora, 2002); and the theory that is created through the meta-practise itself. Arbitrary and random variables are either replaced with variables that refer to the embodied context, or with values consistent with available theory. The software development happens in the Pure Data visual programming system (Puckette et al., 1996), where each step of development is managed by the subversion version control system (CollabNet,

2000). Through each development iteration, the system is evaluated against various criteria. The most important criteria is my subjective interpretation of the system's embodied behaviour. The system is observed in two modes. The first mode entails direct observation of the system's external properties in context. This mode of observation tends towards a qualitative approach. The system keeps a detailed log of key variables, as it negotiates with its context, which is examined in the second mode using R (R Development Core Team, 2007). At regular intervals the system archives an image of its external properties. The archive and logs are used to expose software errors and track the system's process over time. Additionally, the logs show a record of the stability of the system as it is intended for long-term public exhibition. The artist also keeps a journal of the development process.

Why Use Artificial Intelligence?

In order to answer this question one must first define Artificial Intelligence (AI). A general definition states that AI is "part of computer science concerned with designing intelligent computer systems, that is, systems that exhibit the characteristics we associate with intelligence in human behavior—understanding language, learning, reasoning, solving problems and so on." (Barr and Feigenbaum, 1981). Although this definition does not directly mention creativity it is certainly an aspect of human intelligence. Boden (2004) argues that machines can be "considered" creative in the same way that machines can be thought of as intelligent according to the "Turing Test" (Turing, 2004). Stephen Wilson considers the relationship between AI and art:

Artificial Intelligence is one of these fields of inquiry that reaches beyond its technical boundaries. At its root it is an investigation into the nature of being human, the nature of intelligence, the limits of machines, and our limits as artifact[8] makers. I

felt that, in spite of falling in and out of public favor, it was one of the grand intellectual undertakings of our times and that the arts ought to address the questions, challenges, and opportunities it generated. (Wilson, 1995)

In this project I expect the system to be creative by it defining its own relationship to its context. Furthermore, I expect that the artwork makes creative choices that manifest themselves through the artwork's external properties, which refer to the context and represent the machine's unique integration of its experience. AI is the only discipline—with its roots in cognitive science—that explores those questions of creativity through the implementation of systems that embody aspects of the human mind.[9] For this reason, AI is the first logical discipline to consider for technique. That being said the central basis of intelligence in AI is problem solving. This research project, with its basis in artistic enquiry, does not consider creativity as problem solving. For a critique of AI's focus on problem solving, see Agre (1997).

What techniques and processes from AI could allow an artwork to form its own relationship to context? To use non-AI software techniques, I, as the artist, would determine how the work relates to its context, rather than it finding its own connection. Those AI techniques, such as Self-Organizing Maps (Kohonen, 2001), that allow the system to reorganize itself based on sensor input, are a likely requirement to build the mechanisms that allow a machine to relate itself to its context without using problem solving approaches.

THE "MEMORY ASSOCIATION MACHINE" ARCHITECTURE

"Memory Association Machine" has three states: waking, dreaming, and suspended; and is composed of two primary systems: the "Memory System" and the "Free-Association System". Both of these systems are networks of numerous identical components, or units. The selection of state is currently based on a predetermined schedule. The system is initially in a waking state. In this state the system observes its context by pointing the camera in a random direction, collecting images, and selecting a new random direction each twelve seconds. For each direction in which the camera looks, a unit of the free-association system is activated. This results in a cascade of activations in the system's previous experience.

In the dreaming state the camera ceases to explore and images are not collected. A time-seeded random value activates a unit in the free-association system each twenty-five seconds. During the dreaming state, while the system is isolated from external stimulus, the random activations propagate through the free-association network. This propagation calls up elements from previous experience which are not initiated by the embodied context.

In the suspended state the displays are put into DPMS[10] sleep and the dreaming state ends. This state exists to reduce wear on the camera and displays when there is both little light and activity in the machine's physical context.

The memory and free-association networks interact with the embodied context to generate the system's external properties. The memory system is the mechanism through which the system stores and integrates its previous experience. Central to the memory system is a Kohonen Self-Organizing Map (SOM) (Kohonen, 2001). The SOM is an unsupervised artificial neural network which acts as an arbitrary pattern classifier. The choice of which category a particular input is associated with is based on its similarity to other inputs. The patterns presented to the SOM in "Memory Association Machine" are the images captured by the camera. The result is a SOM which stores the machine's visual experience of its context and compares it with its remembered experience.

The free-association system is a network of simple units which is independent of the SOM. This network allows the propagation of signals between units—the basis of the free-association system—and is similar to a cellular automata. The activation of units in the network selects images from the system's memory. This free-association is considered a creative act, as the choices of what images to select are not predetermined, but a result of the behaviour of the entire system. This model of creativity is based on "[t]he cognitive mechanisms underlying the creative process" (Gabora, 2002).

The direction in which the camera looks is currently controlled by three time-seeded random

Figure 4. The system architecture of "Memory Association Machine"

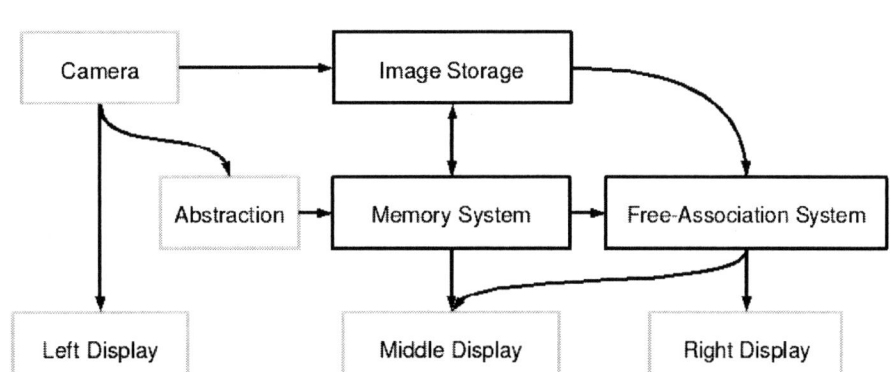

variables corresponding to the camera's pan, tilt, and zoom parameters. Figure 4 shows an overview of the "Memory Association Machine" architecture.

Memory System

Also known as a Kohonen network, a Self-Organizing Map, is an unsupervised artificial neural network (ANN) designed expressly for the purpose of classification. An ANN is an AI approach inspired by neurophysiology. An unsupervised ANN does not depend on the "correct" answer, provided by the researcher, to "learn". Rather, these networks restructure themselves based on input patterns through the process of training. These systems are characterized by being composed of numerous simple components which are massively interconnected. For a survey of ANNs see Medler (1998). The SOM is a non-linear projection of a high-dimensional data-space onto a low dimensional "feature map" that preserves topology.[11] A SOM is able to categorize an arbitrary input pattern, with a finite number of dimensions, into a finite, and fixed, number of categories.[12] The SOM is considered a "projection" as it maps values from the input space onto values in the output space. The points in the input space are the input patterns. The points in the output space are the categories.

The SOM consists of a network of units,[13] each corresponding to a potential category of input. These units are usually arranged in a 2D Euclidean lattice which serves as the output space. Through training, each of the input patterns is associated with a particular category. The input space is defined as a set of real numbers, $\Re^n$, with n dimensions. Points from the input space (the space of all possible inputs) are mapped to points in the output space (the position of all the units in the map). Points in input space are defined as $x = \{\xi^1, \xi^2, \ldots, \xi^n\} \in \Re^n$, where x is a single point defined by the magnitudes of each dimension (ξ) in the set of all possible inputs. The set

of all input patterns is denoted as x_i^j, where i is the number of input patterns, and j is the number of dimensions, where $1 \leq j \leq n$. Similar inputs are associated with similar categories. Inputs are considered similar based on the sum of the distances between their vector components, $dist(x_1, x_2) = \sum_{j=1}^{n} (x_1^j - x_2^j)^2$. The smaller the distance the more similar the inputs are.[14] Categories are considered similar based on their Euclidean distance in output space. Dissimilar inputs can end up being associated with similar categories when two or more highly dissimilar clusters of points are presented to the network. The clusters then compete for space in the finite SOM—causing folds where the Euclidean distance between units, in output space, is not proportional to the distance between the patterns they are associated with.

Each unit of the network has a number of "sensors". Each sensor corresponds to one dimension of the input space. The units have as many sensors as the input space has dimensions. Each unit has a "code-book vector", also known as the weight vector, which contains the same number of elements as the unit has sensors. It is defined as $m_i = \{\mu_i^1, \mu_i^2, \ldots, \mu_i^n\}$, where i is the index of the unit. Each unit has a "neighbourhood", denoted N_i, which is defined by the set of units within a fixed radius,[15] in output space, of the unit i. During training the values of the code-book vectors approach the values of the input patterns. The SOM training process is an attempt to change the network's internal structure (its code-books) to more closely match the structure of the input patterns. The network attempts to mimic, using fewer dimensions, the structure of the patterns it is presented with. This is the "topology preserving" aspect of the SOM. The SOM training procedure, as used in "Memory Association Machine", is as follows:

1.   Set the code-books of all units to 0.[16]
     $m_i = 0$

2. Present the network with an input pattern. (Give the units something to compare their code-books to.)

3. Select the unit that contains the code-book with the smallest sum of the square of the distances, in all dimensions, compared to the current input pattern. This is called the "Best Matching Unit" or BMU. (The BMU most closely resembles the most recent input pattern.[17]) $c = argmin_i(\sum_{j=1}^{n} (x_i^j - m_i^j)^2)$, where c is the index of the BMU.

4. Check if each unit is in the neighbourhood of the BMU.

5. For each unit in the neighbourhood of the BMU, add to the code-book vector the difference between the input pattern and the code-book multiplied by the learning rate, denoted $\alpha()$. $m_i(t + 1) = m_i(t) + \alpha(t)(x(t) - m_i(t))$, where t is the current time step.

6. Repeat from 2.

In a canonical SOM, the amount that the code-books are changed (learning rate), and the size of the neighbourhoods (neighbourhood function), both decrease monotonically over time. This is not the case in the "Memory Association Machine" implementation (to be discussed later). Training is complete when the mean of the difference between all input patterns and all units' code-books, $mean_i(\sum_{j=1}^{n} (x_i^j - m_i^j)^2)$, ceases to decrease. The procedure typically takes from hundreds to thousands of iterations.[18] As the code-book vectors are often randomized on initialization, in the canonical SOM, it is not known which inputs will be associated with which units at the start of training.

As the "Memory Association Machine" camera explores its context, the system creates a field of experience which is organized using the SOM.[19] Seen in output space, the SOM attempts to organize its experience—in the form of visual images collected from its context—into a series of regions that contain similar experiences that

are apart from areas of dissimilar experiences. The camera image is fed into the computer as a full-frame 30fps video stream. At twelve-second intervals, the 12x12 unit SOM is fed with a 100x75 pixel RGBA sub-sampled frame of the video stream. The raw values for each of those 7500 pixels, represented as four RGBA[20] floating point values, correspond to the 30,000 sensors of each SOM unit. The SOM is used to index, rather than store, the images. The BMU, corresponding to a particular image, is used to specify the index of the memory location where the image is stored. The storage area contains the same number of images as the SOM has units. If more than one image is associated with the same category, the most recent image replaces the previous. The mean time an image is held in the system is approximately two hours and a range of approximately thirty-three hours.[21] As a single image is stored for each unit in the network, the memory for unusual images is longer than for common images. Common images are seen more often and are therefore replaced more often. Images are stored at their full resolution—not the sub-sampled resolution fed to the SOM.

"Memory Association Machine" represents the memory system as an Euclidean lattice, see Figure 5. The visual representation of the system is created using the Pure Data "Graphics Environment for Multimedia" (GEM) (Mark Danks, 1995-). The display area is divided by the number of units. Each unit occupies the same amount of display space. The units are represented as circles with a Gaussian alpha channel. The "feathered" edge allows the memories, associated with neighbour units, to blend together. As a result, the structure within the images take precedence over the structure of the lattice. Only in regions associated with images that have very little spacial variation—for example, those that are out of focus—does the lattice of circles become visible.[22]

"Memory Association Machine" is intended to be in constant negotiation with its context. The SOM must be trained continuously to incorporate

Figure 5. The representation of the memory system's field of experience

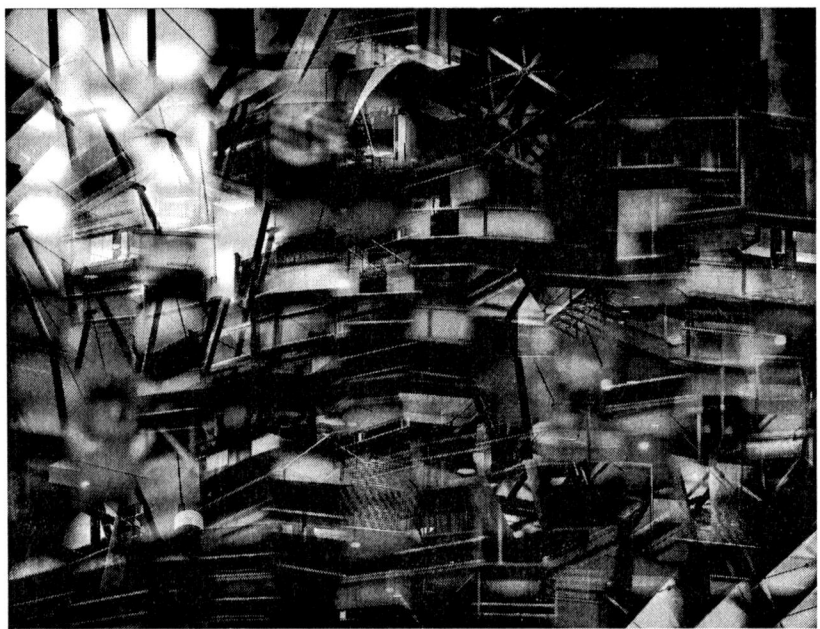

new experiences in its structure. A significant design difference between the canonical SOM and that implemented in "Memory Association Machine" is that the network is constantly converging, but is not meant to, and cannot, reach convergence. Allowing the SOM to converge implies that its process of relating to its context could be complete. As the context changes, so should the SOM structure. Continuous training is enabled by using cyclic learning and neighbourhood functions. These functions control the rate at which the self-organization evolves and is refined. They are driven by the cosine equation defined as $\frac{cos(t/100)+1}{2}$, where t is the iteration counter that wraps from 628 (» 200fl) to 0.[23] The function is iterated each twelve seconds and results in a period of approximately two hours. The cyclical functions allow the SOM to respond to a continuous flow of new input patterns and integrate them into a constantly reorganizing field of memory. Code-books are being constantly refined through the training process. This makes the initial code-books increasingly insignificant over time.[24] As the training process is continuous, the SOM is

replacing existing structure, created as a result of previous experience, with structure that reflects current experience. The network oscillates between durations of large and small change.[25]

The relatively small number of units in the SOM, combined with the high diversity of data in the system's embodied context, makes the structure of the memory representation highly complex. It is often interpreted by viewers of the installation as unorganized. The resolution of the image fed into the SOM was determined through trial and error based on the system's performance. The discrepancy between the resolution of the images used in the memory representation (320x240 pixels) and the resolution the SOM is presented with (100x75 pixels) likely compounds the problem. The memory representation takes on a quality very different than that of evenly distributed randomness.[26]

Free-Association System

Each time a new experience is perceived, it sets forth a stimulation within the content of memory.

This stimulation calls up similar (nearby in the feature map) experiences from the past. These activations stimulate other experiences—traversing the memory from the similar to the dissimilar.[27] As the traversal progresses, the energy in each stimulation decreases. Each subsequent experience is stimulated less than the previous one. As the free-association traces a branching path through the system's experience, the memories intersected by that path are visualized as a cinematic montage on the right display.

The model of stimulation and propagation is a custom network made up of Pure Data abstractions.[28] When the camera looks in a direction, before the image is stored, the free-association unit associated with the BMU (the SOM unit that most resembles the image from the camera) is sent an activation signal. This initial activation is set at the maximum value of 1. Once activated, each unit selects two random numbers between 0 and 7—corresponding to the 8 directions in which a signal can be propagated. The signal is then transferred to its neighbours[29] within that random range. Before it is propagated the signal is decreased by 20%[30] so that it falls off proportional to the distance between the initial activation and each receiving unit. Signals are propagated between free-association units using a custom message containing:

- The (x,y) position of the sending unit
- The (x,y) position of the destination unit
- The value of the signal.

The use of (x,y) pairs, rather than unique indices, was chosen to facilitate the calculation of which units are the immediate neighbours of the unit propagating the signal. Messages are sent to all units in the network, but only the units whose (x,y) indices match the destination processes the message.

There is a temporal delay in the propagation of signals. Each direction delays the signal for a time specified by a time-seeded random number ranging between 500 and 1000 milliseconds. This is done to reduce the number of activations occurring nearly simultaneously. In addition, when a unit is activated it becomes inhibited. For a duration of two seconds, that unit will not propagate any new signals. The inhibition and directional control of propagation is needed to keep the system from over-stimulating itself. Early implementations simply used up all the resources of the hardware only moments after the initial stimulation.

The cinematic montage is made up of four layers allowing four images to be visible at a time. As the free-association signal propagates through the network, the units' IDs[31] and degrees of activation are fed into the montage mechanism. The mechanism is made up of four FIFOs—each corresponding to one of the layers. Activations are stored in the FIFOs in the order of activation. Each FIFO is emptied one item at a time. For each item popped, the system retrieves the corresponding image, fades it in, delays for a duration, and then fades it out. The duration the image is visible is proportional to the strength of the signal. The stronger the unit activation, the higher the value and the longer the duration the image is visible. The duration is calculated according to $duration(s) = 1000s + 500$ where s is the signal strength and the resulting duration is in milliseconds. The opacity of the image is also proportional to its signal strength where the stronger the activation, the more opaque the associated image is ($opacity(s) = .8s$). The result is a montage of cascading sensor impressions—starting with similar and brighter images that are visible for a longer duration. As the activation decays, dissimilar impressions become darker and are visible for shorter periods.

The mechanism behind this process is inspired by the work of Liane Gabora described in "[t]he Cognitive Mechanisms Underlying the Creative Process" (Gabora, 2002). My interpretation of Gabora's theory considers creativity as a controlled form of free-association. The cascade of activations resemble how free-association could work in the human mind. In Gabora's theory, the network of memories is different in three ways when compared to the "Memory Association Machine" free-association system. Firstly, Gabora's theory considers memory as sparse, whereas the SOM organizes content into an organized spacial grid where all units are associated with some input during training—therefore containing no spaces. The folds, boundaries where nearby units are associated with very different input patterns, that some feature maps contain could be considered spaces. These folds can be seen by comparing the code-book vectors of adjacent units using the U-Matrix method (Ultsch and Siemon, 1989; Ultsch, 1993; Kraaijveld, 1992). At the time of writing the ann_som external only provides rudimentary feedback on the state of the code-books. Secondly, "Memory Association Machine" stores entire images, whereas Gabora's theory considers each memory unit as micro-features of stimulus. Thirdly, Gabora's theory refers to a "controlled" association, where the free-association in "Memory Association Machine" is an emergent result of the interactions between the system's components and context.

MACHINE CREATIVITY

Boden (2004) defines creativity as "…the ability to come up with ideas or artifacts that are new, surprising and valuable". In the research domain surrounding "Memory Association Machine," the aspect of newness is the focus above surprise and value. As "Memory Association Machine" is meant to structure itself based on its embodied negotiation with its environment, newness comes from its ability to be different for each new context. The results of its embodied negotiation create a unique reflection on its context. Additionally, the system shifts over time as its context shifts. The diversity and complexity of the real-world environment should guarantee that the system never receives an identical stimulus twice. The

value of the project is not in the machine's creative act (the systems external properties), but in the process that makes it possible. Boden specifies three classes of creativity:

- Combinational creativity is linking together known ideas that are not already associated.
- Exploratory creativity is accomplished by moving through the space of possibilities.
- Transformational creativity is the alteration of the space of possibilities.

Combinational creativity is inevitable in a connectionist network that supports learning. This is because the shift of the units' code-books change the topology of the network—combining the inputs in various ways. Exploratory creativity is also present in these systems as, through the learning process, the network is exploring the space of input possibilities. In order for a connectionist network to exhibit transformational creativity, it would have to be able to change the space of possibilities. The current combination of a SOM and model of free-association used in "Memory Association Machine" allow exploratory creativity since the free-association traverses its memory. At the very least the memory, at a snapshot in time, serves as the space of possibilities from which it can choose to be creative. Since the space of possibilities in the memory system is a constantly shifting field of experience, "Memory Association Machine" also exhibits transformational creativity through its ability to add to, and remove from, its space of possibilities over time. Even an identical memory traversal (which is already unlikely to repeat itself) would select an entirely different set of images from experience. As the SOM is a 12x12 grid of experiences, it has a fixed space of possibilities at a particular moment in time. The use of "Adaptive Resonance Theory" (Carpenter and Grossberg, 1994), "Incremental Grid Growing" (Blackmore and Miikkulainen, 1993), or "Growing Cell Structures" (Fritzke, 1991) networks could allow the memory system to create a new category for a new stimulus—without effecting the categories of previous experience. The space of possibilities would increase over time as the system gains more experience. The field of memory itself would then grow in response to the embodied context, rather than continually refining its finite number of units.

Lets consider creativity as a two-step process. Some originator, the kernel of creativity, creates a "new" item. This item then goes through a process of evaluation that filters all but the most new, surprising, and valuable items. In the case of "Memory Association Machine," the originator for creativity is the embodied context of the machine. Boden largely concentrates on the evaluative aspects of creativity and spends little time on the originator. In "Memory Association Machine," there is no mechanism that serves the role of the evaluator for the system's external properties. That is not to say that "Memory Association Machine" should not be able to evaluate the results of its own creativity, but that the evaluation should not be specified in advance, but come about as a result of its embodied process.

Boden's argument can be summed up in one statement: a creation can only be considered "creative" if it has been successfully evaluated as such.[32] Of course, these two steps are both required for a creative result. Emphasizing one over the other is to only create a partial model of creativity. A more significant error would be to reduce creativity to evaluation since, without the originator, the mechanism has nothing to evaluate. The result of the first step in isolation may not originate something *highly* surprising or valuable, but certainly could originate something *new*. If we were to execute only the second step, evaluation, then nothing would originate at all. The hierarchy is clear—creativity is most dependent on the originator, and less on the evaluator as long as "newness" is the most important aspect of creativity.

As a counter to most of the literature in the area, I aim to put more focus on the seed of creativity, as apposed to its evaluation. From its initial conception, the purpose of "Memory Association Machine" is to originate—not to evaluate. The creation of a machine that originates, without being dependent on the artist, nor being random, is the foundation of this research project. The long-term challenge is the creation of potential systems where the seed of creativity is not dependent on randomness but only on the embodied process. This aspect of the research is connected with artificial life research, which is tied to abiogenisis.[33] The originator → evaluation problem is analogous to a central concern of abiogenisis. Was it random fluctuations of early organic molecules, some form of self-organization, natural selection, or a process as of yet undiscovered that made life initially possible? The theories in abiogenisis are a potential source of technical and philosophical ideas important to the creation of artworks that relate to their context and are not predetermined in their external properties.

Machines that are Intended to be Creative

This section is a survey of selected artistic projects that, in my consideration, exhibit creative behaviour. The choice of projects is not meant to be exhaustive but to highlight a diverse set of applications of machine creativity. These projects involve both connectionist and non-connectionist approaches.

One of the most notable examples of "creative" machinery are the AARON programs written by Harold Cohen (Cohen, 1979, 1995) starting in 1973 and continuing to the present. As a collection of programs, AARON can "create" in a number of different painting styles. Each style uses a different variant of AARON that implements a different set of compositional rules. Some examples of these variants are: "abstract AARON," which creates abstract landscapes;

"acrobat AARON," which creates acrobatic figures; and "jungle AARON," which creates scenes of figures in a complex jungle ground that evoke Gauguin. AARON programs contain sets of rules that encode specific compositional and stylistic laws that are specified by Cohen. Each element of the paintings—figures, grounds, and objects—are each a representation of the model those rules encode. The choices AARON makes are generated by a weighted random number generator and constrained by rules (Cohen, 1979). The results of these choices are applied to create paintings that, in a recent version, are physically produced by a painting machine. AARON has no sensory system. It receives no feedback from the results of its actions on the canvas. The system contains an internal representation that effects the placement, pose, and arrangement of items in the picture plane. AARON is the ultimate example of modernist creation. The internal model[34] is realized in that perfect theoretical vision—regardless of the properties of the physical artifact. The only feedback between the physical artifact and AARON is through Cohen himself.

AARON could only be considered creative in a symbiotic relationship with its creator.[35] Early "paintings" were drawn by AARON but painted by Cohen.[36] Cohen does not consider AARON an artist. AARON's artwork is the result of a collaboration between Cohen and AARON. Cohen believes that this software system is a natural approach to art-making because artistic composition is rule-based. While I can agree that graphic composition, in a particular style of painting, can be considered rule-based, it does not follow that all aspects of artistic creation are. The AARON software exhibits combinational and exploratory creativity, but not transformational creativity. It is unable to compose any choice that has not already been defined in rules specified by Cohen.

In 1981 David Cope started writing "Experiments in Musical Intelligence" (EMI) (Cope, 1996) in order to deal with a creative block in his own composition. The project started as an effort to automate the compositional process using the style of Cope's own compositions to date. The software uses a variation of Augmented Transition Networks (ATNs) (Woods, 1970), which were created to model the syntax of natural languages. This is the basis of a system that models the structure of musical compositions and creates "signatures" from the common aspects of multiple compositions. The elements of these signatures are then combined, in a second process, to form a new work that exhibits the style of the source composer. Clearly using combinational creativity, the software recombines the structures it sees in source-work. Since the space of possibility is limited to the "signature", created from input data, the system is unable to perform exploratory or transformational creativity. The system is fed abstractions of compositions as source material and is unable to perceive, let alone evaluate, the results of its processes.

David Rokeby has created two works that can be considered creative.[37] The first exhibition of "The Giver of Names" (Rokeby, 1990) was in 1998 in Guelph, Canada.[38] The system perceives the outside world through a video camera pointed at a pedestal. The floor around the pedestal is scattered with children's toys that the audience is free to place in the camera's view. "The Giver of Names" attempts to give names to the objects it sees. Associating their colour and shape with concepts in its knowledge-base, the system creates a free writing passage, written in proper grammatical structure, inspired by those objects. WordNet was used as the basis of the knowledge-base and expanded using various methods—including information returned from a "reading" system which extracts the relations between words from texts. This disembodied knowledge is then linked to the machine's sensory experience of the physical world. The result is grammatical sentences, whose grammatical rules are specified by the code, inspired by the machine's sensory experience. The choice of where to begin within the knowledge-base is not a result of agency in the system, but a response to the agency of the viewers' action.

"n-cha(n)t" (Rokeby, 2001) was first installed in 2001 in Banff, Canada and builds on some of the ideas of language and interpretation that are embodied in "The Giver of Names". "n-cha(n)t" is a cluster of independent systems that are connected in a network. Each of the system units is able to both hear and speak by accessing a knowledge-base analogous to the one used in "The Giver of Names". The hearing process attempts to interpret sound from a microphone and translate it into text. That translated text then stimulates the knowledge-base, which is shared between units, resulting in a free writing passage. The passage is then spoken by voice synthesis. The hearing apparatus is a highly directional microphone that picks up sounds only in close proximity and ignores the sounds from the unit's neighbours. The units communicate their "object of interest"[39] over an Ethernet network. In the absence of external stimuli, the units tend to chant a mix of phrases and sentences in synchrony. As all units in the network share the same knowledge-base, they tend to chant when they are not disturbed. The synchrony is an emergent result of the constancy of the perceived environment and the relational database across units. The chant can be disturbed by external stimulus. When the microphone of one unit picks up a sound, the result is a shift of its trajectory in the relational database. The chant can also be disturbed when the timing between units slips.[40] Once all units are in synchrony, they chant the same sequence of words in unison. After the system has been disturbed it will eventually converge—where all units chant in unison.

Both "Giver of Names" and "n-cha(n)t" are embodied systems, as they are attached to the physical world through sensors that allow them to respond to their context. Their knowledge in the form of the relational database, on the other hand, is implanted into the system by the artist. It is not constructed through the system's experience. These systems show combinational creativity by pulling words from their knowledge-base to create texts. The associations in the knowledge-base change in response to sensory experience. The change remains in the system for a short period and dissipates over time. It is unclear if Boden would consider these systems capable of exploratory creativity. Although the vocabulary of these systems is fixed, the associations between those words are in flux. The space of words that follow is transformed in terms of both the words that preceded it, and the sensory experience of the machine. The space of possibilities is fixed at one moment, but the associations can change. The result is a space that changes from one moment to the next. It does not appear possible that the system could use a grammatical structure that is not encoded in the system.

George Legrady's "Pockets Full of Memories" (Legrady and Honkela, 2002) was made possible by a commission from the Centre Pompidou Museum of Modern Art in 2001. The project was revisited in 2003 and exhibited in the Dutch Electronic Arts Festival in Rotterdam, Netherlands. "Pockets Full of Memories" is one of the few artistic projects that makes use of a connectionist network. The system uses an implementation of the Kohonen SOM to organize content provided by the audience. The installation consists of a large projection and a number of kiosks with flat-bed scanners. The audience is encouraged to scan an image of some artifact in their possession. The kiosk then prompts the participant to answer questions about the meaning of the artifact. The answers to those questions are stored in a database and bound, as meta-data, to the images from the scanner. This meta-data is fed into the SOM and the categories of the images are visualized in the projection. Artifacts attached to similar attributes are plotted closer together than artifacts with dissimilar attributes. The SOM used in this project is a collaboration between George Legrady and Dr. Timo Honkela, who conducts research into artificial systems to study cognitive processes.

CONCLUSION & FUTURE WORK

At the time of writing, "Memory Association Machine" is running in a long-term installation at Simon Fraser University, where it is the platform of development for the author's M.Sc. thesis work. This stage of development concludes the first phase of intuitive development. The next stage will be to reflect on the behaviour of the system, and use that knowledge go back through the software to reconsider the arbitrary and intuitive choices.

One of the arbitrary choices that has a significant effect on the system is the choice of learning and neighbourhood functions. These functions were selected in the initial intuitive phase. The ann_som implementation will be altered to include a function to set all the weights in the network to random values. Currently, the only way to randomize the weights, using ann_som, is to feed the object with a random value for each sensor. This is highly impractical in the case of "Memory Association Machine," where approximately four million random values need to be set.

The major remaining random variables include the direction in which the camera looks and the direction, and timing, of the propagation of free-association signals. A simple stimulus-response model of attention could be used to drive the direction in which the camera looks. This could be initiated by ambient sounds in the environment, which could inspire the attention of the camera. The long-term goal is that the direction in which the camera looks is a result of the free-association itself. The mechanism that feeds the system new stimulus would be driven by previous stimulus. It is not clear how this could be accomplished at this time. In a future version, an ideal mechanism to control the propagation of the free-association signal would be to degrade the activation signal by an amount proportional to the input space distance between the code-books of source and destination units. The result would be that free-associations within clusters of memories would last longer and tend to terminate once they near the folds in the SOM. A reinforcement model could allow folds to be traversed, as some units would increase, rather than degrade, the activation signal. With this constraint in place, the activation could be set to propagate in all directions.

This project is an attempt to weave a meta-practise binding technique from AI with site-specific and responsive electronic media art practises. The meta-practise is grounded in the phenomenology of Merleau-Ponty and is developed through the production of artworks (embodied artifacts-as-processes) that are meant to find their own relation to their context. The material of these creative machines is a fusion of the software (written by the artist) and the behaviour of the system (in collaboration with the artist).

REFERENCES

Agre, P. (1997). *Computation and Human Experience*. Cambridge University Press, ISBN 0521386039.

Barker, E., Webb, N., & Woods, K. (1999). *The Changing Status of the Artist*. Yale University Press. ISBN 0300077424.

Barr, A., & Feigenbaum, E., (Eds.) (1981). *The Handbook of Artificial Intelligence, 1*. Morgan Kaufmann.

Blackmore, J., & Miikkulainen, R. (1993). Incremental grid growing: encoding high-dimensional structure into a two-dimensional feature map. In *Neural Networks, IEEE International Conference on, 1*, 450-455).

Boden, M. A. (2004). *The Creative Mind: Myths and Mechanisms*. Routledge; London, 2nd edition.

Bogart, B. (2008). *Memory association machine: An account of the realization and interpretation of an autonomous responsive site-specific artwork*. Master's thesis, Simon Fraser University.

Brooks, R. A. (1992). Intelligence without representation. *Foundations of Artificial Intelligence, 47*, 139-159.

Carpenter, G.A., & Grossberg, S. .(1994). *Adaptive Resonance Theory.* Boston University, Center for Adaptive Systems and Dept. of Cognitive and Neural Systems.

Cohen, H. (1995). The further exploits of aaron, painter. *Stanford Humanities Review, 4*, 141-158.

Cohen, H. (1979). What is an image? In *Proceedings of IJCAI.*

CollabNet. (2000). *Subversion.* http://subversion.tigris.org.

Cope, D. (1996). *Experiments in musical intelligence.* AR Editions.

Danks, M., Geiger, G., Zmölnig, J.M., Clepper, C., & Tittle, J. II. (1995). *Graphics environment for multimedia,* http://gem.iem.at/.

Fritzke, B. (1991). Unsupervised clustering with growing cell structures. In *Neural Networks. IJCNN-91-Seattle International Joint Conference on, 2,* 531-536.

Gabora, L. M. (2002). Cognitive mechanisms underlying the creative process. In T. Hewett & T. Kavanagh, (Eds.), *Proceedings of the Fourth International Conference on Creativity and Cognition,* (pp. 126-133).

Kaye, N. (2000). *Site-specific Art: Performance, Place and Documentation.* Routledge,. ISBN 0415185599.

Kohonen, T. (2001). *Self-Organizing Maps.* Springer. ISBN 3540679219.

Kraaijveld, M.A. (1992). A non-linear projection method based on kohonen's topology preserving maps. Pattern Recognition, II. *Conference B: Pattern Recognition Methodology and Systems, Proceedings, 11th IAPR International Conference on,* (pp. 41-45), Aug-3 Sep. doi: rm10.1109/ICPR.1992.201718.

Kwon, M. (2004). *One Place After Another: Site-Specific Art and Locational Identity.* MIT Press. ISBN 026261202X.

Legrady, G., & Honkela, T. (2002). *Pockets full of memories: an interactive museum installation. Visual Communication, 1*(2), 163-169. http://vcj.sagepub.com/cgi/content/abstract/1/2/163.

Medler, D.A. (1998). A brief history of connectionism. *Neural Computing Surveys, 1,* 61-101.

Merleau-Ponty, M. (1969). The Visible and the Invisible, trans. Evanston: Northwestern University Press.

Possiant, L. (2007). *Media Art Histories,* (pp. 229-250). MIT Press.

Puckette, M. (1996). *Pure data.* http://puredata.info.

R Development Core Team. (2007). R: A Language and Environment for Statistical Computing. *R Foundation for Statistical Computing,* Vienna, Austria. http://www.R-project.org. ISBN 3-900051-070.

Rokeby, D. (1990). *The Giver of Names.* http://homepage.mac.com/davidrokeby/gon.html.

Rokeby, D.. (2001). *n-cha(n)t.* http://homepage.mac.com/davidrokeby/nchant.html.

Turing, A. (2004). *The Essential Turing: Seminal Writings in Computing, Logic, Philosophy, Artificial Intelligence, and Artificial Life, Plus the Secrets of Enigma, chapter Computing Machinery and Intelligence* (1950). Oxford University Press, USA.

Ultsch, A. (1993). Self-organizing neural networks for visualization and classification. *Information and Classification,* (pp. 307-313).

Ultsch, A., & Siemon, H. (1989). *Exploratory data analysis: Using kohonen's topology preserv-*

ing maps. Technical Report 329, University of Dortmund, Germany.

Wilson, S. (2002). *Information Arts: Intersections of Art, Science and Technology.* MIT Press.

Wilson, S. (1995). Artificial intelligence research as art. *Stanford Electronic Humanities Review, 4*(2).

Woods, W.A. (1970) Transition network grammars for natural language analysis. *Commun. ACM, 13*, 591-606, 1970.

Zmölnig, J.M. (2001). *ann_som: Component of the Artificial Neural Network library for Pure Data.* http://puredata.info/Members/dmorelli/ann/?searchterm=neural.

KEY TERMS

Artificial Intelligence (AI): A discipline of computer science that seeks to create systems that can exhibit abilities similar to those of human beings (e.g. problem solving).

Connectionism: A thread of AI that is characterized by the development of systems, which are composed of numerous simple units that are massively interconnected and inspired by neurons.

FIFO (First in First Out): A type of buffer or stack where items can be removed in the same order in which they were added.

Self-Organizing Map (SOM): Also known as a Kohonen Network: an unsupervised artificial neural network designed to be an arbitrary pattern classifier.

Site-Specific Art: An art-form where the artwork is installed in a particular context and is meant to impact the viewer's reading of the work.

ENDNOTES

[1] The phenomenological assumption is that the physical world is shared between subjects.

[2] The "frame" is not referring to frame analysis, but a boundary that changes the meaning of what it contains simply through the act of isolating it from context.

[3] Past frames are not just referring to those in this particular project, but to frames in all previous creative projects conducted by this artist.

[4] A single artifact may also contain a number of components at the same level not depicted in this figure.

[5] Action is considered the system's behaviour that results in the change of its external properties.

[6] Possiant discusses the shift from artifact to interface. An interface, by definition, implies a process. Her argument then additionally supports a shift from artifact to process.

[7] The intuitive approach involves the use of tacit knowledge in order to get a component functioning in the embodied system as quickly as possible.

[8] Note that Wilson's characterization of creators as "artifact makers" indicates that even in the technological arts the object is often the central focus, rather than the process the artifact implements.

[9] Robotics approaches to AI are also sources for potential technique.

[10] VESA Display Power Management Signalling

[11] "Topology" can be considered spacial structure or geometry.

[12] The number of categories and dimensions are specified before training proceeds.

[13] Units are also known as "nodes" and "neurons."

[14] This measure is based on a direct pairwise comparison between the corresponding

components of the input patterns. The measure of similarity is then unable to see two inputs, with identical blocks of values occupying different dimensions, as similar.

15   This is the neighbourhood function as used in MAM but is more commonly implemented as a linear fall off or Gaussian function.

16   Often the code-books are set to random values. According to Kohonen, this was initially done only to demonstrate how robust the SOM is—even when used with arbitrary initial conditions. Training can be accelerated by doing some pre-processing on the input patterns to determine initial code-books. See Kohonen (2001) Section 3.7. At the time of writing ann_som (Zmölnig, 2001) does not implement an internal function to set code-books to random values.

17   During the first iteration, due to the ann_som implementation, if the code-books of all units are equal, the unit with the largest index is always chosen as the initial BMU.

18   Training time is dependent on the number of sensors and the size of the network.

19   The Pure Data ann_som external (Zmölnig, 2001) is the SOM implementation used.

20   The alpha value is passed to the SOM even though it is fixed at 1 (opaque in GEM colour units) for all pixels in all frames. This is due to the high CPU usage of extracting the alpha values from the 30,000 element message in Pure Data. A possible solution would be a custom external to remove every 4th element in a message.

21   These values include the time images are kept during the dreaming and suspended states.

22   This is somewhat rare and always limited to a regional cluster—due to the SOM.

23   The result of a neighbourhood size that ranges from 1 to 0 (inclusive) is that the code-books of few units are updated for each iteration. The initial choice of this range was based on the assumption that ann_som was

normalizing the neighbourhood to the size of the network, which has proven not to be the case.

24   In the case of MAM, which trains over hundreds of thousands of iterations, the initial code-books are of little consequence.

25   Large changes result from a large neighbourhood and learning rate—small changes from a small neighbourhood and learning rate.

26   For more information on the quality of the feature maps produced by MAM see Bogart (2008).

27   The traversal depends on the state of organization of the SOM.

28   Abstractions in Pure Data are roughly analogous to functions in procedural languages.

29   These are the neighbours of the free-association units, which are totally independent of the SOM neighbourhoods.

30   The signal degradation decreases to 10% during the sleeping state. This is to encourage longer free-associations in that state.

31   The ID in this case is the unique index, not a pair of (x,y) indices.

32   For Boden, the creativity is the result of the two-step process, so before the evaluation the "newness" should not be considered creative at all, but simply as an unclassified response.

33   The study of the origin of life.

34   AARON's rules could be considered a model of creative intention.

35   Every "creative" machine depends on its creator to be creative.

36   Cohen has been known to paint a drawing differently than specified by AARON.

37   The characterization of these systems as creative is the interpretation of the author. The artist's intention was not centrally the construction of creative machines.

38   A prototype of "The Giver of Names" (Rokeby, 1990) was also exhibited in Toronto, Canada, in 1997.

39 The "object of interest" is the concept in the knowledge-base that is currently being stimulated.

40 "Slips" occur because the timing mechanism of each unit is intentionally not synced to its neighbours, which results in some temporal drift.

Chapter XIV
EVO–PARK:
Designing Better Architectural Projects Using Participated and Interactive Genetic Algorithms

Stefano De Luca
Evodevo, Italy

Eugenia Benelli
Evodevo, Italy

Francesco Altarocca
University of Rome, Italy

Dario Dussoni
University of Rome, Italy

ABSTRACT

Designing good and sound architectural projects is a hard job. Generally these kinds of projects involve many stakeholders, everyone with his/her own aims, and designing activities could be very difficult to face. We can make this process easier using an "autonomous genetic design facilitator" and "collective subjective designer" to realize projects that meet needs of every part involved in. The chapter describes a methodology we suggest as a process of urban parks design. This methodology can be adapted to other situations considering many variables (and consequently a huge amount of possible solutions) and many specific needs to satisfy.

We believe that people need and have a right to determine and shape their own environment

(Christopher Alexander)

1. INTRODUCTION

Modern democracies are built on the representative model, where citizens delegate decisions to elected people that will act on their behalf, governing on the base of greater knowledge, specific skills and so on. Decision making is submitted to all the members of society.

On the other hand, there is a definite trend towards direct participation of citizens, that wish to influence directly the agenda, the planning and the choices. Yet as far as in the 1963 Lipset asserted that "the stability of an industrialized democracy [...] legitimates open participation by all groups in the economy and the polity" (Lipset, 1963, p. 521).. Citizens participation is considered an adaptive mechanism that can "contribute to the stability and legitimacy of the larger system" (p. 529). Putnam reports that "strong traditions of civic engagement [...] are the hallmarks of a successful region" (Putnam, 1993).

Governments are interested in engaging citizens to enrich the democratic participation, taking their ideas and points of view; in this way, governments, especially local ones as cities, towns and counties, can shorten the distance between citizens' needs and the implemented planning, can improve the quality of the actions receiving suggestions from the people that live in those territories, can reduce the potential conflicts that may arise from decisions, can augment the networks and shared vision that are an important part of social capital. So there is greater and greater interest in e-participation and e-government, at International and National level (OECD, Organization for Economic Co-Operation and Development, 2003), (OECD, Organization for Economic Co-Operation and Development, 2006), (De Pietro & al., 2003), (Partecipando, 2006), (Urbact Project, Partecipando, 2006). "Social capital is not a substitute for effective public policy but rather a prerequisite for it and, in part, a consequence of it." (Putnam, 1993).

"Sustainable cities need to reimbed themselves in ecological and cultural systems. And such reimbedding cannot take place without people's participation in designing, planning and managing the resources that support urban life." (Urbact Project, Partecipando, 2006) In order to promote a sustainable and democratic urban development, it's essential to monitor urban regeneration processes by participation. Many participatory processes have been implemented in different ways and at different levels to let the inhabitants value and decide together with technicians, administrators and elected representatives the development of the places they live in (Urbact Project, Partecipando, 2006).

But how can a citizen control the decision making? And how can a citizen be informed of the problems? And how is it possible influence urban planning without a specific technical knowledge?

Indeed, a decision involves different dimensions, each one dealing with lots of parameters, so that the main goal (for instance, a new museum building) will be described in term of finance, architectural technologies, aesthetic and activity programmes of the museum, its impact on the urban context, the time needed to deliver a good and so on.

One of the important challenges of e-democracy is to involve people in participating in the design of buildings features, as new schools and museums and in the landscape design as well. These processes that are usually very expensive, impact on the citizens' life and in the city skyline, are subjected to an aesthetic judgment by all the users (who will live in those places). Hence the planning and realization of an architecture can more easily yields critics and then it is higher the need of a participation of the citizens, given the high impact that a new project has on their lives, and it is very important to build places where people are pleased to live in.

However, given the technicality of the design, it is difficult for people to propose a "good" idea that can be translated into a real project that can satisfy all the mentioned needs, in particular, financial constraints and technical feasibility.

Our work is about a decisional support system that can guide the citizens in choosing personalized and interesting solutions respecting technical constraints set by the experts.

The system is based on four interrelated components:

- Citizen participation;
- Public management programs as stated by the local government;
- Planners design solutions;
- Urban education, in terms of design and real constraints involved in, and process awareness.

Each element is influenced by the others, and the assumption is that each element can benefit and

be enhanced by the interrelation. The schema is derived from the one described in the paper about integrated urban environmental management in Porto Alegre (Menegat, 2002), but we included explicitly the planners, that become mediators between the government goals and the citizens' needs and give the tools that will be used to build the participated design, as we will describe.

This form of participated design requires a new approach by the government, that have to accept risks of a design mediation, by the architects, that have to design with an open approach that leave many decisions in the hands of the citizens, and a new approach of the citizens too, that can participate but have to understand the projects elements (and in the process, they receive keys for their urban education).

The interaction model can be implemented by iterative public meetings or aided by software.

We can find more details of the first choice in the Porto Alegre experience (Menegat, 2002). The Prefeitura Municipal launched a sequence of "city conferences" there, in 1993. The conferences involved thousand people and debated the central question of the new urban and environmental master plan. The process of discussion and de-

Figure 1. Interaction model

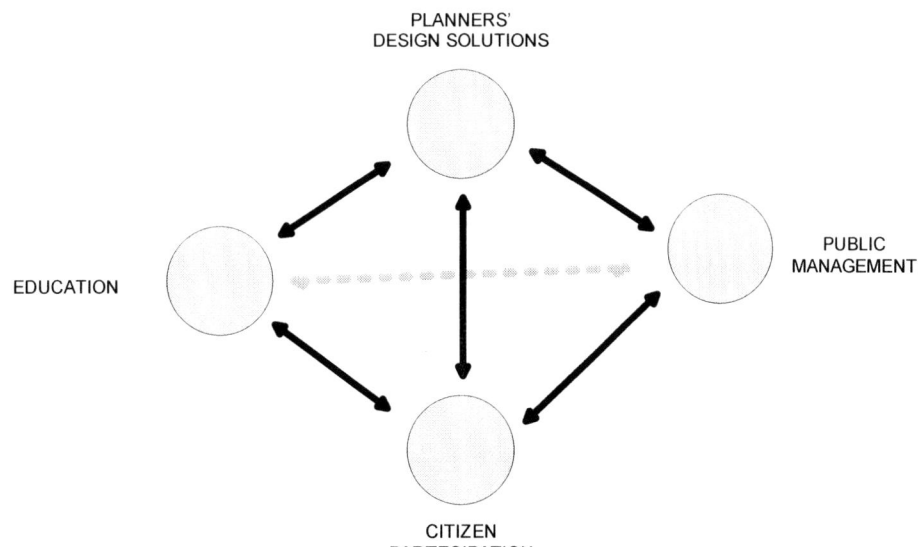

cision making followed an annual cycle of two main stages, first defining priorities and proposals for public spending in plenary assemblies, in which all the people could participate; and second drawing up the budget proposals and expenditure plan in which the priorities approved by the citizens should be developed for a submission to the state legislature as the municipal budget It was a time consuming process that required high organizational efforts (e.g., coordination of focus groups at different urban scales) and expensive communication issues (as advertising and promo tools and communication media plans).

Dealing with the second choice interaction, one of the first experiences in software support systems are Interactive Genetic Algoritms - IGA (Takagi, 2001), (Llorà, Sastry, & Alias, July 8–12, 2006). IGA derive from the usual genetic algorithm but the users has the possibility of give feedback to the generated solutions, i.e., to vote them. (More on IGA to follow.)

However, the IGAs are usually employed when the system can "create" a totally new solution, without external constraints, and the solutions are generated by software only, and people can only give evaluations (about the aesthetic, the interested and so on) but not design of the new solutions.

The main drawback of existing interactive genetic algorithms is that the system can produce not feasible solutions (e.g., too much expensive, not so interesting in aesthetics terms, not feasible by law and so on), so that decision makers cannot employ them. Actually, IGA can converge into an interesting solution, but not always feasible, because does not match all the set constraints. This may produce a negative boomerang effect, because at the beginning you ask people to give feedback in participative process, and then you cannot follow their indications. This means that at the end of the process the realized project (e.g., the *built* park) will be different from the people proposals, and will not meet people's expectations. This mismatch will disappoint them, that is exactly the contrary of the initial purpose!

For that reason, we propose a new method which guarantees good and sound solutions for urban landscape design. Using our design methodology, implemented in our product *evo-park*, each yielded solution is a good one, and the boomerang effect doesn't apply. The main idea is to deploy the system on Web, where interested people or stakeholders (i.e. people that live or work in a definite area) can give their own contribute submitting new projects and evaluating the existing ones, created by people and by the system according to users' votes.

Summarizing, the main goals of the system are supporting stakeholders in their activity of expressing preferences and at the same time improving collective abilities in term of creativity and design.

The chapter is organized as follows: section 2 describes how collaborative design has been developed in architectural design processes; section 3 is an overview of genetic algorithms (GA) and in particular about interactive GA; section 4 is about our proposal of Participative Interactive GA and its implementation; finally, a few conclusions are presented.

2. COLLABORATIVE DESIGN IN ARCHITECTURE

Collaborative work has increasingly become a widespread form of work of great importance and frequency in urban strategies, business, academic contexts and communities in the last years (Axerold, 1984). The knowledge behind what makes for a successful collaboration has also improved, structuring healthy processes to overcome unforeseen challenges and difficulties that may arise (Bobbio, 2004), (D'Albergo & Moini, 2007).

The development and the evolution of collaborative processes have been applied to many urban objectives: to offer space and tools to analyse public and private proposals for housing people, to elaborate and diffuse data and information about the neighbourhood, to involve locals and train some of them in renewal and restoration activities, to design green and public areas (Giangrande & Mortola, 2000) (Alexander, Ishikawa, & Silverstein, 1977).

2.1 What is Collaboration

Collaborative design consists in a complex framework of management methods, group dynamics, information sharing and communication strategies that are responsive to the needs of the community it is serving (Johnston & Hicks, 2004). It involves individual and collective energies enabling people to operate in the natural and built environment in creative and innovative ways. A healthy collaboration can also make a significant impact beyond the mere purpose of the project, frequently building social capital between the stakeholders and strengthening the community.

A *collaborative process* is generally facilitated by groups of technical experts (architects, researchers, sociologists, local administrators…) who help the participants to build a shared language, to define goals and objectives, to identify emerging problems (Alexander, 1987), to use evaluation methodologies and to allow the process to naturally evolve (Salingaros, 2005).

These practices improve the community members' sensitivity regarding the environmental balance and quality of their neighbourhood improving the quality and transparency of communication between administrators and citizens, from a political and technical point of view. They increase neighbourhood identity, contrasts environment decay and empower people to participate actively in the transformation processes of the territory.

2.2 Collaborative Work History

Collaborative design approach starts from some interesting architecture and urban planning experiences during the 60s and the 70s such as the American *advocacy planning*, Christopher Alexander's *Pattern Language*, experiences of *self-construction* applied in the developing countries and the *community planning* in Europe. Giancarlo De Carlo has been the most important figure who worked in Italy on self-sustainable planning during this period (De Carlo, 1973).

2.3 The Collaborative Process

Collaborative strategies are a way of understanding and possibly controlling a complex system; they build something that is functionally and structurally coherent with emergent properties. The results of the collaborative processes can be unforeseen and exciting.

The classical key phases in the process are:

1. The construction of the *diagnosis map* - that diagnoses where problems and potentials exist and predict the expected outcome of the collaborative activities;
2. The *Visioning methodology* – that can configure a probable development of the area in 20/30 years clarifying the possible planning strategies;
3. The construction of a local *Pattern Language,* which is the basis on which the project is built – "A pattern is a careful description of a perennial solution to a recurring problem within a building context, describing one of the configurations which brings life to a building." (Alexander, Ishikawa, & Silverstein, 1977). The definition of a network of patterns on different scales is a powerful tool for controlling complex processes and find solutions to the diagnosis map problems.
4. The evaluation and validation of the final design solution.

Architects and planners facilitate the collaboration processes in phase 1 and 2 by themselves. Our methodology supports them during phases 3 and 4: in phase 3 our system aids the local patterns visualization and their combinations, in phase 4 we empower citizens in the evaluation process.

2.4 Collaboration Developing Areas

The traditional tools of collaborative work are: surveys, physical models, collages, design games, public meetings activities, neighbourhood parties, street performances, press releases, but this kind of methods reach only a part of the relative community because people's presence is required for them to be effective.

There are some crucial factors that have to be taken in consideration:

- **The decisional process accessibility:** People who constantly take part in the Community laboratories activities are only a very low percentage if compared to the total;
- **The definition of the collaborative work rules and interaction is still difficult:** There's a lack of effective communication and shared language among the stakeholders;
- **The definition of procedures and actions to validate and build the projects:** There are still some uncertainties in the action planning.
- **The definition of the design scale:** There is no established connection between the building scale and the urban scale in the collaborative process.

The new information and low cost communication technologies are currently changing the collaborative practices, adding new available tools to aid community participation and to solve part of the emerging problems. Business, professional and social relationships are moving to virtual meeting places enabling more people to be connected and interact: this is an important opportunity of evolution for collaborative work.

Some multimedia experiments, for example, have been carried out in the city of Rome in the 90's to enable people to mature an opinion about the city transformations in progress (Mortola, Fortuzzi, & Mirabelli, 1995) (Sivo, 1995). Local councils have developed websites to create a synergy between some community laboratories, to communicate activities and to facilitate interchanges of local experiences. In alternative to traditional paper documents the experimentation have focused on multimedia technologies for a richer and more comprehensible communication and on hyper-textual technologies.

The goal of this experience has been to inform a wide amount of people of the current urban planning, but has left the collaborative design work to the traditional public meetings in the laboratories, involving a small group of citizens.

The technological challenge now is to communicate to all the people directly implicated in the neighbourhood and **let them interact directly on the design phases of the project** (the construction of local patterns and the evaluation/validation phases) **to mark, determine and shape their own environment in real time.**

3. GENETIC ALGORITHMS AND INTERACTIVE GENETIC ALGORITHMS

3.1 Introduction to Genetic Algorithms

Genetic Algorithm (GA) is a technique useful to find solutions to search, optimize, design, model, support entertainment and solve artistic problems. The idea is to replicate the Darwinism biological mechanism for population progress to improve initial random solutions; in GA we speak about inheritance, offspring, mutation, crossover, selection etc. etc. (Goldberg, 1989), (Goldberg, 1999), (Eiben & Smith, 2003).

If we want to find a solution to a problem, we can use GA applying generalized steps that can be useful to solve other problems too. First step, we encode the problem as a *chromosome*. This is usually a vector containing numbers or strings, although it can be formed by other more complex structures. The chromosome represents a solution, where its elements (*genes*), are parameters of the solution; a chromosome is the representation of an individual, i.e., its genetic code that can be translated therefore in the *real* individual, the *phenotypic* representation. Chromosomes are the genotypic definition of the individual ("he has green eyes, black hairs…"), that need to be translated in the phenotypic representation, e.g., a specific person (that can be bald!).

For example, if we want to find the maximum of a function $y = f(x_1, x_2, ..., x_n)$, where the x and the y are real numbers, the chromosome will be composed by *n*-genes, each one representing a number. If we are optimizing a path from a city *a* to a city *b*, the chromosome can be the path formed by the traversed cities. For instance, to find the path from Rome to Berlin, we can have a chromosome <*Rome, London, Barcelona, Berlin*> and another <*Rome, Verona, Monaco, Berlin*>.

As you can see, the fact that we have represented the solution in this way does not mean that each solution is a good one; indeed, the first is bad and the second is good. We can represent this difference measuring the *fitness*, i.e., how valid is the solution or, using the biological metaphor, how the individual is adequate to the surrounding environment.

Now we are ready to use the genetic approach to evolve the solutions of our problem by creating an initial population composed by several chromosomes. Usually, the initial population is created with randomized chromosomes. It is important to stress that GA works with a population, and not a single solution: each individual competes with all the other in the chance of reproduction, giving better solution as in nature we get more adequate species.

Once we generated the population, we can apply selection and the genetic operators to create new individuals (and hence a sequence of new populations) with hopefully better solutions to our problem, because they are produced by the stochastic operators. The literature of GA is full of operators, but usually all the GAs use selection, recombination and mutation:

- **Selection** is used to maintain the best individuals, i.e., to give better reproduction possibility to the most fitted solutions; there are lots of way to accomplish this task, as the roulette wheel or tournament (described as follows), but the general idea is to give better reproduction rate to individuals with higher fitness. Selection choices who will reproduce and who will terminate.

- **Recombination** is a genetic operator (or better: a class of operators) that combines different solutions producing new ones, as in the biological sexual reproduction; the most used operator is cross-over, that mix two chromosomes generating two new ones, mixing the first part of a chromosome and the second part of the other chromosome and vice versa: for example, from <*a,b,c,d,e*> and <*1,2,3,4,5*> we obtain <*a,b,c,4,5*> and <*1,2,3,d,e*>; in this way, the new solutions arise from the parents but are different and, again, hopefully better.

- **Mutation** introduces new solutions modifying some genes into a chromosome, i.e., modifying a parameter of the solution.

Figure 2. Genetic algorithm cycle

```
t =0;
InitializePopulation P(t);
Evaluate P(t);
while not done do
      t =t +1;
      P'=SelectParents P(t);
         Recombine P'(t);
      Mutate P'(t);
      Evaluate P'(t);
      P=Survive P, P'(t);
end while
```

In Figure 2 there is the pseudo-code of a classic GA's algorithm.

A first simple formal definition of a GA system can be a triple

$$P = \langle L, E, I \rangle$$

Where L is the set of system layers, E is the set of layer elements and I is the set of individuals in the system.

Let b be a mapping of all elements of E in L.

$$b : E \rightarrow L$$

If $|\operatorname{Im} g(b)| = |L|$ we get a $|L|$-partition of E set. Let $l_j = \{e \in E \mid b(e) = j\}$ be a set of layer elements, then all $l \in L$ have one or more elements, $E = \bigcup_j l_j$ and $l_i \bigcap l_j = \varnothing$ for every $l_i, l_j \in L$.

In other words, we can consider every element e belonging to one and only one layer $l \in L$. The result of this operation is a partition of E set in exactly $|L|$ subsets of E.

An example can be useful to understand abstract concepts.

Image a park (the system) divided in four layers for projecting issues.

Example 1

L = (water, green, sand, children's games)   (1)

Every layer has its project variants. Therefore, the E set is composed by the following elements sets (divided by layers):

Example 2

$$l_w = (water - 1, water - 2, water - 3) \qquad (2)$$

$$l_g = (green - 1, green - 2, green - 3)$$

$$l_s = (sand - 1, sand - 2, sand - 3)$$

$$l_s = (sand - 1, sand - 2, sand - 3)$$

$$l_g = (games - 1, games - 2, games - 3, games - 4)$$

In that manner we can define the E set, its partitions and the b function. In fact, $E = l_w \bigcup l_g \bigcup l_s \bigcup l_c$, the intersection of every pair of elements set is disjointed and the function is implicitly derived from partitions. As we saw in the example, layer cardinality can be different: green layer has three elements while children's games layer has four elements.

It's simple to calculate all the parks that can be built:

$$\prod_{j=1}^{|L|} |l_j| \qquad (3)$$

I is the set of individuals that populates the system, as we have defined it before. Every individual is univocally identified by theirs elements:

$\langle x_1, x_2, ..., x_{|L|} \rangle$, where $x_i \in l_i$ for $1 \leq i \leq |L|$ and $l_i \subset E$.

Let c be a *cost* function:

$$c : E \rightarrow \mathfrak{R}^+$$

and w_i be a weight (positive real number) for every layer in L. Now the fitness of an individual $p = \langle e_1, e_2, ..., e_{|L|} \rangle$, $p \in I$, can be defined as follow:

$$f_o(p) = \sum_{j=1}^{|L|} \frac{1}{c(e_j)} \cdot w_j \qquad (4)$$

Greater is the cost of an individual's element lower is the fitness. If all w_i for $1 \leq i \leq |L|$ have the same value, every layer gets the same "importance" in the system. On the other hand, if one layer is more important than another one the first has a higher weight.

A simple example can fix concepts introduced before.

Example 3

Let $P = \langle L, E, I \rangle$ be a genetic algorithm system, let L and E respectively be as in Example 1 and 2. Moreover, let cost function c be as Table 1, let I be as Table 2 and consider all the weights equal to 1.

In the example found in Table 2, fitness values are calculated using (4), we consider ind#6 and ind#10 receiving respectively the maximum and the minimum fitness values.

$$f_o(ind\#6) = \frac{1}{c(water-2)} \cdot w_w + \frac{1}{c(green-1)}$$

$$\cdot w_g + \frac{1}{c(sand-1)} \cdot w_s + \frac{1}{c(games-2)} \cdot w_c$$

$$f_o(ind\#6) = \frac{1}{1.2} + \frac{1}{3.0} + \frac{1}{0.2} + \frac{1}{2.0} =$$

$$0.833 + 0.333 + 5 + 0.5 \approx 6.67$$

$$f_o(ind\#10) = \frac{1}{2.0} + \frac{1}{5.0} + \frac{1}{0.9} + \frac{1}{5.5} =$$

$$0.5 + 0.2 + 1.111 + 0.182 \approx 1.99$$

The number of all possible parks is calculated using (3):

$$n = l_w \cdot l_g \cdot l_s \cdot l_c = 3 \cdot 3 \cdot 3 \cdot 4 = 108$$

This value can consistently increase if the number of elements for every layer is greater. Consider a system of 5 layers for each of them there are 8 different choices. The number of possible outcomes is $5^8 = 390625$. If there are more layers but less choices for layer the number possible solutions is less than previous example: $8^5 = 32768$.

Table 1.

element	cost
water-1	1.0
water-2	1.2
water-3	2.0
green-1	3.0
green-2	5.0
green-3	0.5
sand-1	0.2
sand-2	0.8
sand-3	0.9
games-1	1.0
games-2	2.0
games-3	8.0
games-4	5.5

Table 2.

individual	water layer	green layer	sand layer	games layer	fitness
ind#1	water-1	green-1	sand-1	games-3	6,46
ind#2	water-2	green-1	sand-3	games-2	2,78
ind#3	water-1	green-2	sand-1	games-3	6,33
ind#4	water-2	green-3	sand-3	games-4	4,13
ind#5	water-3	green-3	sand-2	games-1	4,75
ind#6	water-2	green-1	sand-1	games-2	6,67
ind#7	water-1	green-1	sand-1	games-1	7,33
ind#8	water-3	green-2	sand-1	games-2	6,20
ind#9	water-2	green-1	sand-2	games-3	2,54
ind#10	water-3	green-2	sand-3	games-4	1,99

On this structure it is possible to use genetic algorithms operators such as selection, crossover, mutation.

The system lifecycle can be described by the Algorithm 1.

3.2 Interactive GA

The most critical element of GA is the fitness function, that evaluates how good is a specific individual in the environment. This function outcome is a number. Its semantic tells us how much a particular element of the system is "good" in term of capability to survive and reproduce in an environment, i.e., the design has to be appreciated by people. Defining such function is sometimes very hard. Fitness in IGA is different from fitness in GA because it is necessary to consider, totally or in part, the user evaluation, that introduce new problem, as noise related to the variability of users' preference (Guo, Gong, Hao, & Zhang, 2006).

In our task, the main drawback is that we can't define a totally objective *Evaluate* function to compute the fitness value. In fact GAs are suitable to systems that can calculate population's fitness in a objective manner. If the system is biased on people's judgments this kind of system doesn't

work. We must introduce a new fitness function so that the system can be influenced by subjective variables (stakeholder's votes). Such systems can be defined as Interactive Genetic Algorithm (IGA) as reported in (Cho, 2002), (Hao, Gong, & Huang, 2006), (Quiroz, Louis, Shankar, & Dascalu, 2007), (Brintrup, Ramsden, & Tiwari, 2007). This technology is usually employed to design artistic artifacts or industrial design.

It's necessary to pay attention to many issues to make this methodology work properly. We have to deal with a space of solutions that can be rapidly growing and with huge opinion data. Managing this number of "variables" is often hard or too expensive.

It's also necessary to introduce new subjective variables that can catch and summarize people needs and that is able to lead the evolution of the system through the right way.

Referring to the formal definition we give in previous paragraph, I is the set of all individuals in the system. Let $i \in I$, i is defined as a tuple of $|L|$ elements $(x_1, x_2,...,x_{|L|})$ where every x_j, $1 \leq j \leq |L|$, $x_j \in L_j$. All possible individuals in the system are therefore $\prod_{j=1}^{|L|} | l_j |$ where $l_j = \{e \in E \mid b_j(e) = 1\}$.

IGAs have been widely and sucsessfully used in many field i.e. graphic art and computer graphic

Figure 3.

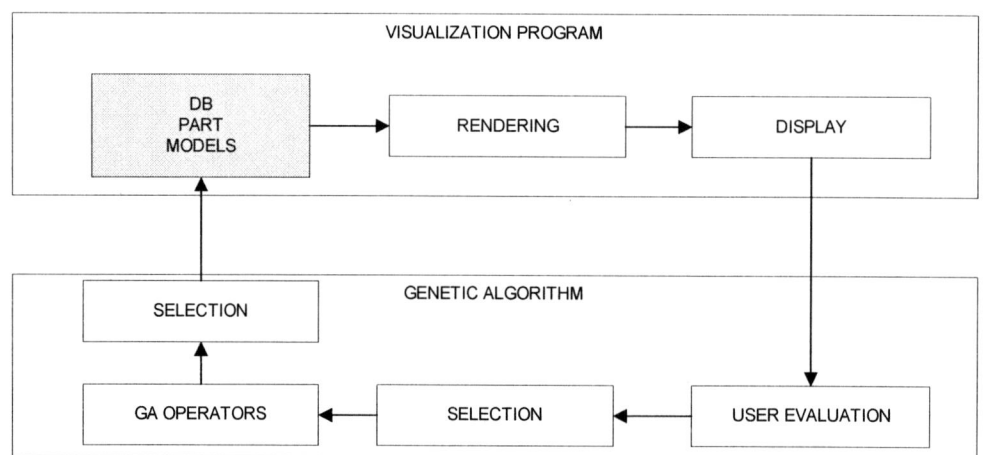

animation, 3D computer graphic lighting design, music, industial design, face image generation, database retrieval, art education, control and robotics, etc. as reported in (Takagi, 2001). Another interesting work (Cho, 2002) reports two IGA systems. The main idea of these systems is the susbtitution of a classic fitnes function with one came from users. The following figure summarize the lifycicle of IGA systems. Genes and chomosomes are stored and encoding in a database and then a start population is initializated.

The system diplays indivuduals and user expresses their own personal judgments. At this stage, GAs operators work using user's fitness instead of classical fitness funcion.

4. ASSISTED DESIGN PARK MODEL

4.1 Participated and Interactive Genetic Algorithm

To use Interactive GA in the contest of participated design of an urban park, and more generally in architecture, we need to permit to the users to vote the design and to design themselves too. We modify the IGA schema in order to accomplish user participation, and we refer to this technology as *Participated and Interactive Genetic Algorithm (PIGA)*

There are many differences between IGA and PIGA systems. First of all they have different general goals (Cho, 2002):

* IGA assists domain experts in theirs designing tasks;
* PIGA helps expert's poll in finding people needs during design process, moreover it allows people to actively and democratically participate in designing activities.

IGA systems characteristics can widely vary. Many experiences that use interactive evolutionary computation systems (a superset of IGA)

are report in a comprehensive survey (Takagi, 2001).

From the technical point of view, PIGA differs from IGA for some aspects. Follows a brief list of main differences:

* IGA only relies on subjective user opinions, PIGA on opinions and objective fitness;
* PIGA allows users to design their own project while IGA doesn't do it;
* PIGA system uses inheritance for calculating the offspring subjective fitness part.

Practically, PIGA system puts together some important designing process factors: experts pool experience and knowledge (process starters), people experience, collective wishes and needs, heterogeneous skills, continual improvement and innovation (Goldberg, 1999).

In this section there are collected all relevant implementation details of our system prototype. During design phase a lot of importance is focused on reusability (in similar contexts) and modularity (interchangeable strategy, pluggable algorithms, parameters, …).

We assume, for practical reasons, that every layer is autonomous, therefore there aren't constraints on two any elements belonging to different layer. Yet, as in participated park design, some real situation needs to relax this assumption. This problem can be addressed in different ways. A simple one is to use an incompatible information table and adding extra costs to incompatible solutions or rejecting incompatible solutions.

We model our PIGA system using four main entities:

* **PIGA Layer:** Every instance of this "class" represent a logical or physical layer in which the system is divided;
* **PIGA Element:** Represents an element that can be used to fill a layer;
* **PIGA Project:** Is an individual in the system;

- **PIGA Subjective variable**: Corresponds to a characteristic on which stakeholder can express personal judgment.

4.2 Voting Parks

Consider a system that contains a number of arbitrary park projects designed by the planners team (process starter). The first operation that a user has to do is to vote individuals (landscape projects) in the system. The user has to express his preferences regarding subjective variables defined by the system administrator. Follow some instances of subjective variables:

- **Aesthetics:** How much a park is proper by an aesthetical prospection;
- **Interest:** How much a park is close to user's interests;
- **Activity programmes:** Distribution of activities provided by the park plan according to audience needs;
- **Global opinion:** A whole opinion of a park.

Obviously it's possible to include more subjective variables but it's important to keep down the number of variables in order to simplify the voting process and avoid user's bother. All such variables, with their weights, concur to create a global judgment on all individuals in the system (subjective fitness).

There are at least two issues to deal with in this first phase: the number of individuals to display and how to select them. The number of individuals to vote depends on many concerns: how the user is motivated to give feedback, how many solutions can be displayed in a screen, how many subjective variables user must judge, and so on. The issue concerning selection involves other phases of the process, we will talk about this aspect next.

Basically we can identify three main approaches, every one with pros and cons:

- **Vote and design once:** The user is required to expresses his preferences for another existing project before designing a new park. This approach avoids a "lazy" or "smart unfair" user proposals made without voting other projects for his own interests and vice versa. However, this method needs an authentication system that avoids bad behaviours and limits audience and therefore user's subjective contribution.
- **Vote and design independently:** User can vote and design parks without restrictions. A greedy or shy[1] user can artificially alter system evolution outcomes. Simplicity and freedom are main advantages of this way.
- **Vote many times, design once:** The user is required to expresses his preferences on many existing projects before designing a new park. A particular instance of this approach can be realized voting solution more than once during a time period. This approach force user to vote many times and therefore the system can be more accurate in giving subjective evaluations. However, main drawbacks of this method are: realization of an authentication system (as in first approach) and users motivation and enthusiasm (ask to users to vote many time can be annoying and frustrating).

We can imagine advantages and disadvantages in permit the user to vote again after some time, in case the user has more knowledge about the programme. We are investigating now in many new approaches and variants to the those described before to support ever better the participatory process.

How to choose the best suitable approach depending on many variables as instance: how many people are involved in the process, where the system is located (i.e. can be striking to put kiosks for voting and projecting in the place where the project will be realized), how complex are the projecting and the voting phases.

Vote many times, design once approach best fits to our goals because we need both users' votes (many users judgements) and we also want to involve actively people in the participated design process. Therefore, only after user has voted the projects he can access to design the park. Follow a user's design session.

It's necessary to take care of design phase because there are many issues related to it. During design process the system assists user to avoid critical situations. There are many critical scenarios. Follow some instances:

- **Vote and design the same park:** User designs a park that is already voted in the previous phase. If this operation is allowed, user can "vote" a park twice, on the contrary it is possible that the system forces the user to design a park that doesn't meet his needs. We believe that the first solution is better because the user can design a park very close to his ideal park anyhow. We also avoid behaviors that attempt to pump a particular solution.

- **Design an already existing park:** The park designed by user is already in the system – it is already designed by another user or the system has created it during evolving cycle. If it occurs, the system doesn't create a duplicate but considers this action as assigning maximum vote to all subjective variables.

In both cases our prototype uses Web Service Ajax Design Pattern (Mahemoff, 2006) in order to detect critical scenario and respectively deny park creation and assigns positive feedback to an existing solution.

If previous scenarios don't apply, the system saves the user's park and assigns to it the maximum value on all subjective variables.

Scenarios described here don't cover all possibilities. Imagine a park was deleted by pruning algorithm during GA's lifecycle (we discuss details of the pruning algorithm later). The first and simple solution is to deny the design of a deleted park. Even if this can be consider sound, it has many drawbacks:

Figure 3. How the layers are used to produce a new design

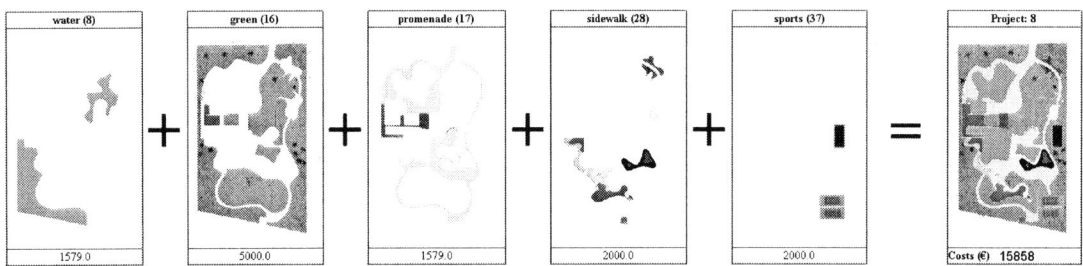

Table 3.

sequence	#1	#2	#3	#4	#5	#6	#7	#8	#9	#10	#11	#12	#13	#14	#15
sequence 1	1	2	2	1	3	1	3	5	5	4	4	4	2	5	5
incremental mean s. 1	1.00	1.50	1.67	1.50	1.80	1.67	1.86	2.25	2.56	2.70	2.82	2.92	2.85	2.93	3.13
sequence 2	5	5	2	5	5	2	4	2	1	1	3	3	4	1	4
incremental mean s. 2	5.00	5.00	4.00	4.25	4.40	4.00	4.00	3.75	3.44	3.20	3.18	3.17	3.23	3.07	3.13

- It limits the crossing power (crossing a good park with a bad one could produce a better solution than crossing two good solutions);
- No new chances are given to not optimum parks.

These considerations gain more importance if we consider the possibility of an unlucky sequence of vote. Consider the two following sequence of vote for a particular park.

Both sequences have same votes but the first one always has a mean less than 3, the second one is always greater than 3. Therefore if the system threshold to eliminate a solution is set to 3 the first sequence is always (except for the last one) a candidate for the elimination process. In the last sequence the same park is never a "rubbish park". Another important issue is related to how many times the system is up and running. Suppose that a participated system requires a long term design phase and users can vote and design many times. Inhibiting the design of deleted parks is not a good idea because people can change opinion and can design or vote a discarded solution.

Another good question involving deleted parks and votes is if it is useful or not keeping deleted parks votes. We think that if a system has a long lifecycle, storing this information could strongly damage rehabilitated parks. In other cases old information are useful for a complete, democratic, fair and stable judgement.

It's always useful to choose a variance depending on business instance needs.

Selection Techniques and Reproduction

Many system steps of the overall system require an individual selection algorithm. Moreover, for any singular task we can change selection algorithms and use a mechanism for plug the proper algorithm in a simple manner (Strategy Design Pattern). Design Patterns Patterns (Gamma, Helm,

Johnson, & Vlissides, 1995) are an interesting field of software engineering and provide simple and versatile solutions to well-known software design problems.

There are many approaches and algorithm variants for selection tasks, we provide a brief list (cf. (Eiben & Smith, 2003) for a more general description):

- **Roulette wheel:** All individuals have a chance to be randomly selected (the most is the fitness value the most is the probability);
- **Tournament selection:** n individuals are randomly selected from system population, it is gotten the best from this set (individual with a greater fitness);
- **Rank selection:** Population is ordered by fitness. If n is the number of individuals, n is assigned to the individual with the greater fitness, $n-1$ to the second individual and so on. It's applied the roulette wheel selection method on new values (instead of fitness). This technique is useful when population fitness can change widely: roulette wheel penalizes too much low fitness individuals.

SELECTING PARKS FOR VOTING

Selecting parks to display is sometimes tricky. It's important to introduce solutions with better fitness because they probably are candidate for success, but also we must pay attention to give all individuals some opportunities to be voted. Moreover an approach that only uses better individuals can lead to a system that converges to a sub-optimal solution. So, in this phase, it's important to select those that haven't received enough votes. Similar approaches are used for selecting parents for reproduction, therefore we could also use tournament selection or roulette wheel for the voting system. Implementing such policy can be very tricky. Testing activities show that system generated solutions[2] are too penalized and when the system starts there is no vote information about the park, so subjective fitness for all the parks in the system is 0. We prefer to use a completely random strategy in order to "ensure" a fair selection method.

SELECTING PARKS FOR REPRODUCTION

Classic genetic algorithms employ selection methods that rely on fitness values. In PIGA systems we must be careful to apply such algorithms because fitness function is strongly affected by subjective judgments and we need enough votes before using them in a reliable way. We have introduced subjective fitness normalization using a parametric sigmoid function. Sigmoid function is used as a factor: greater is the number of votes an individual get, greater is the subjective fitness coefficient. This weight avoid that a solution that received not much optimum votes can be better than a good solution (in the long term).

Our system, PIGA, therefore can conjugate project constraints (hard-cabled in the set of possible solutions made by experts' pool), stakeholders' opinions (subjective project's votes), keeping project's costs lower (the higher is the cost of a project the lower is the fitness), avoiding boomerang effect (all projects are feasible, so that the system converges to solution projected by experts and voted by people who will "live" the project), artistically contribute to find a better solution using GA operator (applying crossover to better solutions and altering some characteristics).

We start with a set of possible layer configuration and a set of project designed by experts pool to reach all these goals. After the starting phase of process, the system administrator asks all stakeholders to participate. It's important to involve people in this process (i.e. people by the area where the park will be done), otherwise the project can be useful only for a small amount of population or can be not useful at all.

Figure 4. How the user designs a new park

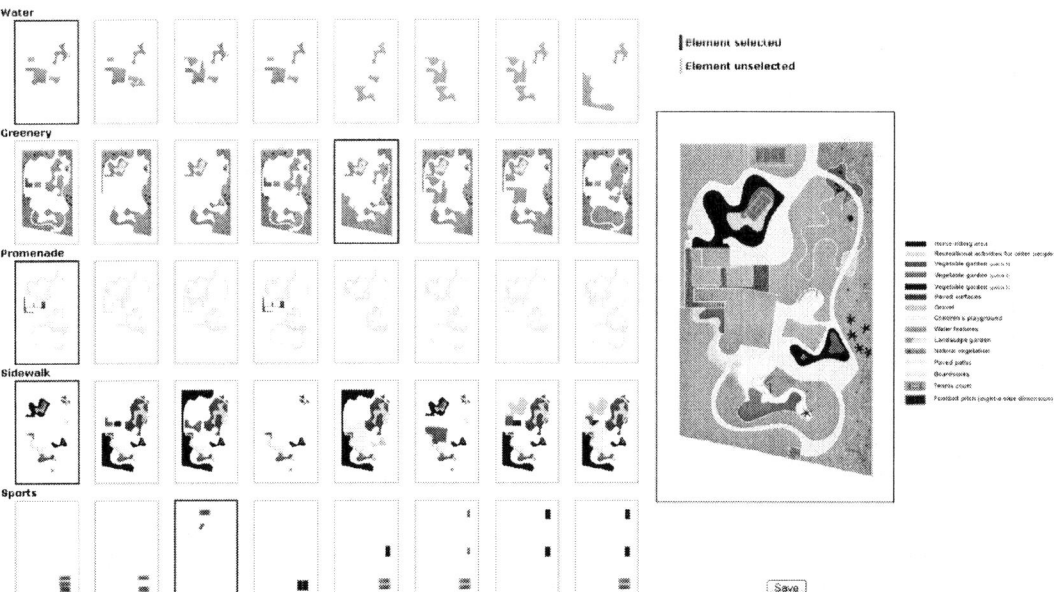

Basically, people receive a token that allows to:

- Vote other predetermined number of project in the system;
- Create their own project.

It's also important to select solutions that have received few votes because, otherwise, they have less chance to be selected by a pure roulette-wheel algorithm . On the other hand we must award solutions that get better votes (and then select them to allow user to vote them). Moreover we also need worse solutions that can help GA to produce optimum solutions instead of local maximum solutions. In the implementation section it's described a simple software engineer method that gives flexibility need to our purpose.

The main issue to deal with is to map population elements and their characters (phenotype) in their representation (genotype) so that applying GA's operator is easy and the population elements are modular and interchangeable.

We define a project as a *set of layers for local patterns*. Every layer has a number of different elements to be employed. For example a park can be composed by four overlapping layers: vegetation layer, water features layer, paved paths layer and children's playground layer. Each one of these layers can be built using many different components and each solution is compatible, or not, with the other layers components. Figure 5 depicts a user's project session. In this case we have worked on the evolution and mutation of a Burle Marx design for the Burton-Tremaine Residence garden (1948, Santa Barbara, California, unrealised):

Once the user has projected his own park, he/she has to vote some other project in the system. He/She expresses judgments in relation to some defined subjective variables (i.e. aesthetics, activity

Figure 5. The user gives votes to existing designs

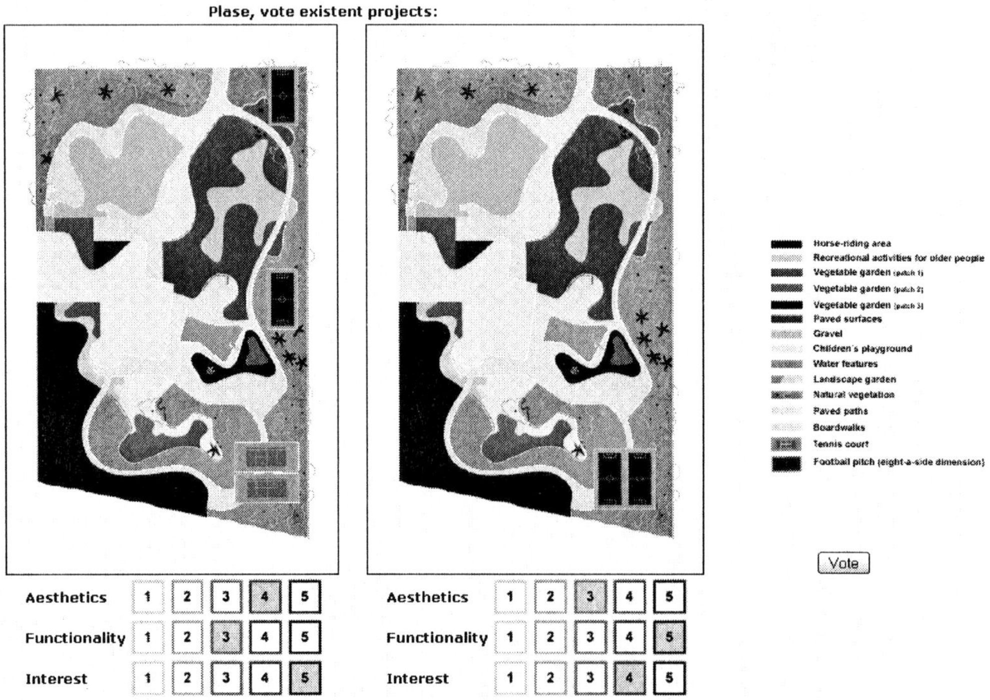

248

Figure 6. Administrator view of voted designs

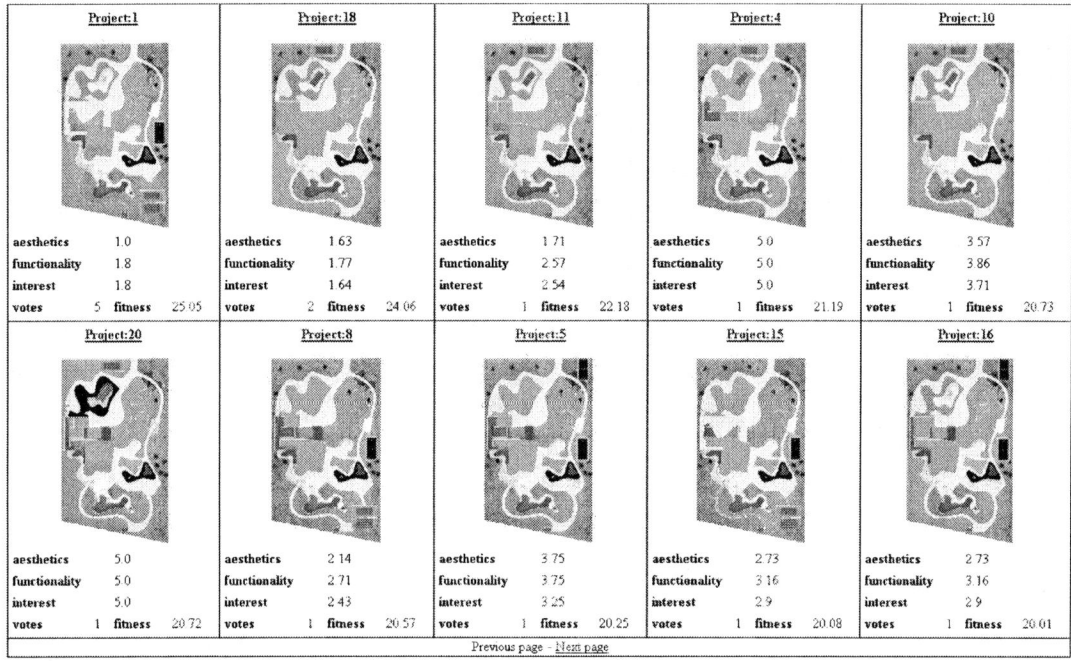

Figure 7. User's design session

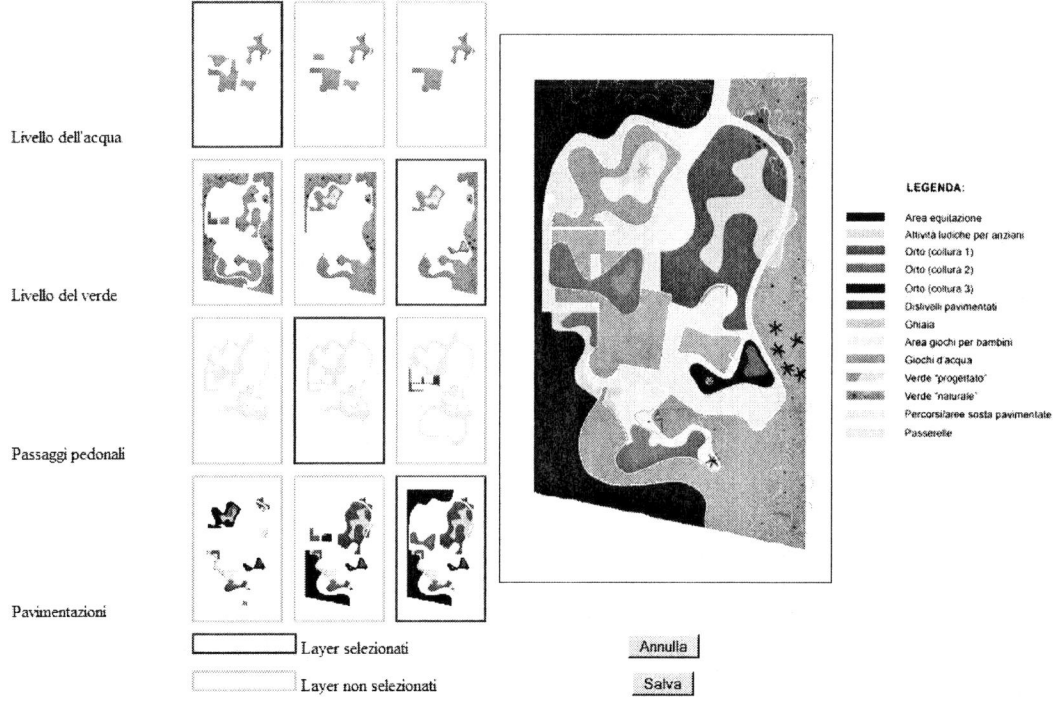

programmes) rating each variable with values within 1 and 5. These votes contribute to define the project fitness.

4.3 Calculating Fitness

In this paragraph we describe different component of fitness used in our PIGA system. We define the fitness of a population individual p as the sum of objective and subjective variables.

$$f(p) = fs(p) + fo(p) \qquad (5)$$

Objective variables are related to cost (an extra cost for incompatibility with other element in other layer), so the objective fitness is defined as the sum of inverse proportional of the cost of each layer multiply for the layer weight (1). In order to calculate the subjective fitness is necessary to retrieve the mean value for each subjective variable. Each variable has a weight, therefore the subjective fitness is the sum of all weighted votes (2). In this formula we have insert a factor to steady the subjective fitness. It depends on the number of votes a solution receives. The function used is a parametric sigmoid. It avoids that an individual who has been "well" voted only once or few times is better than another individual who received many "optimum" votes but gets a lower mean.

Referring to the formal definition we give in previous paragraph, let n be the number of votes a individual gets and let v be the number of subjective variables defined in the system, we can calculate both parts of fitness function as follow:

$$f_o(p) = \sum_{j=1}^{|L|} \frac{1}{c(e_j)} \cdot w_j \qquad (6)$$

$$f_s(p) = \left(\sum_{i=1}^{v} w_i \cdot \left(\frac{\left(\sum_{j=1}^{n} v(p)_{ij} \right)}{n} \right) \right) \cdot sigmoid(n) \qquad (7)$$

4.4 How PIGA works

After the vote, the system selects two projects in the database and applies the crossover operator to them. The system applies a mutation to offspring with a predetermined probability. Two new projects are inserted in the system. The system limits the projects size in order to avoid the presence of too many projects in the database. Every time the maximum size is exceeded the system selects the worst solutions and eliminates them. The last operation takes care of keeping low candidate solutions (in comparison with the search space) but at the same time avoids to eliminate projects that haven't got many votes.

The aforementioned techniques that use the sigmoid function contribute giving stability and convergence to the system.

We have tested and validated this technique using a Focus Group. Subjective Test and Convergence Test results encourage us to go through in this research area.

CONCLUSIONS AND FUTURE WORK

Our work started with the goal of supporting citizens in designing their cities, their skylines and the urban tissue where they live in. We feel that to let people to give feedback on projects designed by other people is not enough for improving the participation.

Evo-park approach this important theme using a form a creative computer science, genetic algorithms modified to permit votes by people (hence *Interactive* GA) and individual creation (hence *Participated Interactive* GA), while assuring the decision makers that each design is good and sound and that will be possible to realize.

We experiment with Evo-park and the results are good: even when the combinatory explosion of local patterns is very high, the system converges towards a good, shared design.

Now we are working with several cities to use the system in real park building and reengineering projects, and so we aspect to improve the model and the software.

Further development of Evo-park will be focused in improving visualization of the designs, in particular on 3D rendering of each designed park. This will be useful for citizens to visualize their own projects and to evaluate other people ones.

Another interesting research field is the simulation of "habitability" of the park, i.e., how different users can found interesting to stay in: how children feel that the park is interesting for their entertainment, how sportmen and sportwomen enjoy running and doing sport, how families find interesting to spend some hour in the park. To evaluate this emotional feeling, and the interrelation between people (e.g., if the children play structure are too near the cycleways, children can disturb the cyclists) we are planning to use an Agent Based Simulation (Gimblett, 2002) that we successfully used to model natural hazards.

REFERENCES

Alexander, C. (1987). *The New Theory of Urban Design.* New York: Oxford University Press.

Alexander, C., Ishikawa, S., & Silverstein, M. (1977). *A Pattern Language.* New York: Oxford University Press.

Axerold, R. (1984). *The Evolution of Cooperation.* New York: Basic.

Bobbio, L. (2004). *A più voci. Amministrazioni pubbliche, imprese, associazioni e cittadini nei processi decisionali inclusivi.* Napoli: Presidenza del Consiglio dei ministri, Edizioni Scientifiche Italiane.

Brintrup, A., Ramsden, J., & Tiwari, A. (2007). An interactive genetic algorithm-based framework for handling qualitative criteria in design optimization. *Computers in Industry* , (pp. 279-291).

Cho, S.-B. (2002). Towards Creative Evolutionary Systems with Interactive Genetic Algorithm. *Applied Intelligence* , *16*, 129–138.

D'Albergo, E., & Moini, G. (2007). Il potenziale trasformativo delle pratiche partecipative : tre casi a confronto. In E. D'Albergo, & G. Moini (Eds.), *Partecipazione, movimenti e politiche pubbliche a Roma.* Rome: Aracne.

De Carlo, G. (1973). *L'architettura della partecipazione.* Milano: Il Saggiatore.

De Pietro, L., & al., e. (2003). *Linee guida per la Promozione della Cittadinanza digitale: E-democracy (in Italian: Guidelines for promotion of digital citizenship: e-democracy).* FORMEZ.

Eiben, A., & Smith, J. (2003). *Introduction to Evolutionary Computing.* Springer Verlag.

Gamma, E., Helm, R., Johnson, R., & Vlissides, J. (1995). *Design Patterns: Elements of Reusable Object-Oriented Software.* Boston, MA, USA: Addison-Wesley.

Giangrande, A., & Mortola, E. (2000). *Progettare con la comunità.* Università Roma Tre (DiPSA) and Comune di Roma (USPEL).

Goldberg, D.E. (1989). *Genetic algorithms in search, optimization and machine learning.* Addison-Wesley.

Goldberg, D.E. (1999). *Genetic and Evolutionary Algorithms in the Real World.* IlliGAL Report No 99013, University of Illinois at Urbana, Department of General Engineering.

Guo, G., Gong, D., Hao, G., & Zhang, Y. (2006). Interactive Genetic Algorithms with Fitness Adjustment. *Journal of China University of Mining and Technology* , *16*(4), 480-484.

Hao, G., Gong, D., & Huang, Y. (2006). Interactive Genetic Algorithms Based on Estimation of

User's Most Satisfactory Individuals. *Intelligent Systems Design and Applications, 2006. ISDA'06. Sixth International Conference on, 3.*

Johnston, E., & Hicks, D. (2004). Speaking in teams: Motivating a pattern language for collaboration. *Interdisciplinary Description of Complex Systems , 2*(2), 136 – 143.

Lipset, S.M. (1963). The Value Patterns of Democracy: A Case Study in Comparative Analysis. *American Sociological Review , 28*(4), 515-531.

Llorà, X., Sastry, K., & Alìas, F. (July 8–12, 2006). Analyzing Active Interactive Genetic Algorithms using Visual Analytics. *ACM GECCO '06.* Seattle, Washington, USA: ACM.

Mahemoff, M. (2006). *Ajax Design Patterns.* O'Reilly Media.

Menegat, R. (2002). Participatory democracy and sustainable development: integrated urban environmental management in Porto Alegre, Brazil. *Environment & Urbanization , 14*(2).

Mortola, E., Fortuzzi, A., & Mirabelli, P. (1995). Communications Project of Designing with Multimedia Interactive Tools. *ECAADE Conference Proceedings, Multimedia and Architectural Disciplines.* Palermo, Italy.

OECD, Organization for Economic Co-Operation and Development. (2006). *Citizens as Partners: Information, Consultation and Public Participation in Policy-making.* OECD.

OECD, Organization for Economic Co-Operation and Development. (2003). *Promise and Problems of E-Democracy.* OECD.

Partecipando. (2006). Retrieved 2 25, 2008, from European Project "Partecipando": http://urbact.eu/projects/partecipando/home.html

Putnam, R. (1993). The Prosperous Community: Social Capital and Public Life. *The American Prospect , 13*(1), 35-42.

Quiroz, J., Louis, S., Shankar, A., & Dascalu, S. (2007). Interactive Genetic Algorithms for User Interface Design. *IEEE Congress on Evolutionary Computation,* (pp. 1366-1373).

Salingaros, N. (2005). *Principles of Urban Structure.* Amsterdam, Holland: Techne Press.

Sivo, G. (1995). Intervention at meeting. *I Laboratori di quartiere nella città di Roma.* Rome, Italy: Orme.

Takagi, H. (2001). Interactive Evolutionary Computation: Fusion of the Capabilities of EC Optimization and Human Evaluation. *Proceedings of the IEEE , 89,* 1275–1296.

Urbact Project, Partecipando. (2006). *European Handbook for Partecipation.*

ENDNOTES

[1] Greedy user designs many projects but doesn't express other projects' judgments. Shy user votes other projects without proposing their own park.

[2] System generated solutions don't have received any votes. Their subjective fitness is lower than parents, because is calculated as the mean of subjective parent's fitness, and has sigmoid coefficient low .

Chapter XV
Mathematics, Computer Mathematical Systems, Creativity, Art

Sergiy Rakov
G.S. Skovoroda National Pedagogical University, Ukraine

Viktor Gorokh
G.S. Skovoroda National Pedagogical University, Ukraine

Kirill Osenkov
G.S. Skovoroda National Pedagogical University, Ukraine

ABSTRACT

The chapter discusses the possibilities modern IT opens for Mathematics and its applications to real life, in particular to Art – by an example of automated construction of caricatures. The research approach in math education which is based on computer modeling as a methodology of a competency approach is discussed as well as the future educational environment with free access to Internet resources and possibilities to operate software in distributive mode. Many examples are given – from simple illustrations to new math facts invented with original dynamic geometry package DG and proved with the help of computer algebra system Derive.

INTRODUCTION

Creativity as a Common Root of Producing (Discovering) New Knowledge in Math and Science and Producing New Artifacts in Art (Philosophical aspects)

The tradition to oppose Math and Sciences to Art based on dichotomy "non-creative (algorithmic, reproductive)–creative (non–algorithmic, productive)" is wrong and may be concerned only in the framework of usage of Math, and Science and Art artifacts but not in the process of their creation; the processes of constructing new artifacts (both new Math and Science knowledge and new Art products) have the same base – CREATIVITY, and both these processes are similar in both cases (concentration, meditation, modeling, associative and critical thinking etc.). In wider context CREATIVITY is the main goal of a human being, CREATIVITY is the only thing that God can justify him/her (e.g. it is his/her anthropodecium (Berdyaev N., 1931)) – each personally or a mankind in a whole; only CREATIVITY transforms a human life into a human being and approach a human to God (Berdyaev N., 1931). It does not matter in which area you are working (are you a scientist, artist, engineer, a worker, a teacher etc.) – the only thing that matters is: are you CRE-ATIVE or not. CREATIVITY is a type of God for mankind, the only God and the entire God, the Trinity of God: the origin, the existence and the soul of being. For mankind, human activities, make sense only if they have a CREATIVE matter, are conditionally divided into distinct branches: Technologies, Science, Philosophy, Art, Ethics (in particular religion), each of which plays more or less a definite role in definite historical periods of any individuum or community (tribe, nation, country etc.). During the historical periods one of these branches in their leading role replace one other. Nowadays it seems that technologies, in

particular IT, play the main role in civilization's progress, stimulating other spheres: Science, Art, Ethics. This is why modern society is called an Information Society (I-Society). But technology itself can't be the goal of civilization's progress and it is obvious that this type of society will be changed by the knowledge society (K-Society), in which the main systems leading role would be played by Science, inflating and stimulating other branches of human activities. Poverty of mankind, growing the number and the scale of global ecological problems should demonstrate bounding of such models of society as I-Society and K–society and maybe Art and Ethics will be the base of a new type of society – Soul society (S–society) and the concept of a Sustainable society (economic development, social development and ecological stability) is a step in this direction. IT helps to create crucially new possibilities in all other spheres of a human being: human activities, human work, human leisure etc. In particular IT creates a base for new possibilities in Art, Science, Math. etc. (for example computer graphics is beneficial in a lot of IT applications). In Figure 1 the "magic curves" are shown and all of them are the graph of the one (!) parametric equation

$$\begin{cases} x = \sin(p\sin(t)) \\ y = \sin(p\cos(t)) \end{cases}, t \in [0, 2\pi[,$$

depending from the only parameter – real number p. What are these graphical compositions? Mathematics? IT? Art? The reader is asked to answer these questions for himself/herself and to try to estimate the approximate values of p corresponding to each of the Pictures 1– 6[1]. Maybe these pictures are the synthesis of all these – somewhat an entity of Math, IT, Art etc. And Math, IT, Art etc., are all only the sides of this entity which may exist separately only in our heads – if our heads are not too creative, but in creative heads they all compose this entity.

Figure 1.

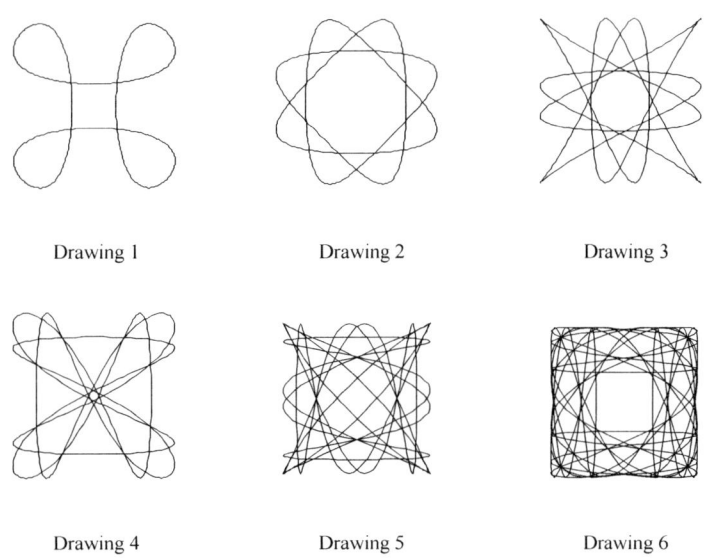

Drawing 1   Drawing 2   Drawing 3

Drawing 4   Drawing 5   Drawing 6

Figure 2. Dynamic drawing "Magic Curves" in the DG environment

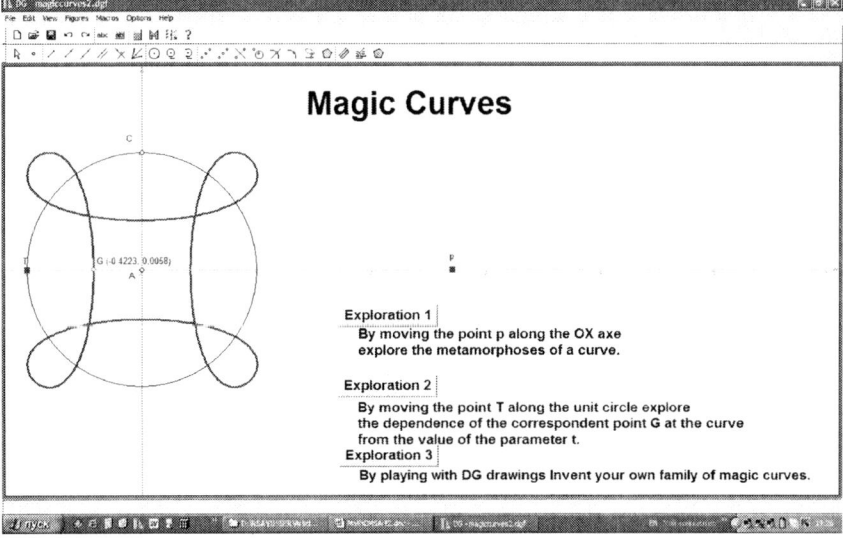

IT IN MATH RESEARCH WORK AND MATH EDUCATION

The Competency Paradigm of Math Education

The modern paradigm of education as a whole and of Math education in particular is the competency paradigm: it means not only to teach the developed completed math theories and demonstrate known facts, but also to show the process of constructing new math knowledge as a response to the requirements of human practice. It is not a new approach in education, it is based on the background of problem solving, inquiry learning, social constructivism, heuristic method of teaching and learning, developing teaching and learning etc. Each highly educated and responsible teacher intuitively used ideas of the competency approach in education in all the times. We can recall for example the experience of famous Leonard Euler with his 90 volumes of Math proceedings. He not only discovered a lot of principal results practically in all areas of Math, but he reflected his process of discovering of new facts as a collision of ideas, the chain of trials and mistakes, insights and routine demonstrations, induction and deduction, generation of hypothesis and constructing the counter examples to them, posing and reposing new problems, systematization of new facts and exploring the particular cases etc. But only IT gives hope to bring this experience of competency paradigm into real classrooms, to transform the classroom into an educational laboratory – the model of the real Math Research Laboratory in which new math knowledge is producing (subjectively new or objectively new does not matter in this context). It becomes possible thanks to the development of powerful and user-friendly mathematical packages which can workout all the routine work of math exploration in automatic mode and present the results in expressive analytic, table or graphic form suitable for analysis and making decisions.

Competency educational paradigm in the framework of school education means that the main objective and mission of it is developing students' competency which should provide them with the ability of self-realization in society and through this self-realization to promote the development of the society in the spirit of science, democracy, humanism, and tolerance. The school's graduate competency should correspond to the social competency standards and ought to be developed only by the student him/herself under the guidance of a competent teacher. Thus the teacher competency is a key point of the implementation of the competency educational paradigm in real school practice.

The hierarchal structure of the competencies includes the following levels: *key competencies* – meta-subject and subject-subject competencies and which are developed in the education as a whole; *branch competencies* – competencies that are to be developed by the students in studying the subjects belonging to some educational branch (math competencies, science competencies, etc.); *subject competencies* – components of the branch competencies correspond to the concrete subject.

An individual (in particular a student) can develop competencies only by himself/herself through active and productive work (and not only through the educational work), through the personal creativity, through the personal experience and through the critical analysis and assimilation of the social experience — in other words through their own individual active being.

There are different approaches in defining the key, branch and subject competencies in different countries and it is natural because of the level, the cultural traditions and objective of Math education (as example) in different countries are different. Nevertheless from the other side, all the national subject competencies and in particular the national math competencies should have the common framework, which reflects the common ideas and principles which define the place of math in human culture, the domain, methods and applications of math knowledge and each country only chooses its own priorities and essences in them and of course can define their own additional components of the subject competencies.

The structure of math competencies in the Ukraine was defined in the following way (Rakov S., 2005):

- **Math procedural competency (or math algorithmic competency):** Ability to solve the standard (typical) math problems;
- **Math logical competency:** Ability to use deducing as a method for proving or disproving a math hypothesis;

- **Math technological competency:** Ability to use Computer Math Systems (CMS) for solving math problems, construct computer models for math problems and explore them;
- **Math research competency:** Ability to use the Math methods for exploring the individually and socially valuable problems;
- **Math methodological competency:** Ability to estimate benefits and restrictions of the Math method for exploring and solving individually and socially valuable problems.

The proposed structure of the math competencies corresponds to the principles of the Math literacy developed in the framework of an international study of the quality of math education PISA (Program for International Student Assessment) which was launched in 1993 under the support of OECD (Organization for Economic Collaboration and Development).

The matter of Math literacy is (by the OECD/PISA): ability of a person to identify and recognize the role of Math in the modern world, ability to make correct and adequate Math assertions, ability to perform Math activities corresponding to his (her) current and future individual needs as creative, constructive, conscious person and citizen.

Math literacy includes (by the OECD/PISA): ability to think mathematically; ability to argue mathematically; ability to construct mathematical models; ability to pose and solve math problems; ability to present data; ability to manipulate with math constructions; ability to math communications; ability to use the math tools.

The Math literacy (by the OECD/PISA) includes the following three classes of Math competencies: *1 Class: definitions, reproduction, calculations* (ability to reproduce the math constructions, give definitions to the math objects, make calculations); *2 Class: structuring and integration for problem solving* (ability to use different approaches and different branches of Math,

ability to structure (decompose) the problem onto some more simple problems, to solve them and integrate the results into the solution of the initial problem); *3 Class: mathematical thinking, generalization and insight* (ability to "mathematize" the situation and use Math for problem solving; to analyze the solution, to interpret results; to construct models and strategies; ability not only to solve the given problem but as well to pose them, present the math argumentation, including deductive proof and generalization).

In what way Math competencies can be developed by students? The only possible way lays in a framework of a CREATIVITY atmosphere in the lessons which each student actively participates. Not only new knowledge and skills are the results of teaching and learning but as well the process of discovering new knowledge and algorithms for solving different classes of problems is the matter and the desired result of such lessons. The role of a teacher is dramatically changed from the sage at the stage to the guide by the side – as the well known educator folklore says. The matter of correspondent research approach in Math education and its IT support in Computer Algebra System and Dynamic Geometry System packages are discussed in the next paragraphs. Now some examples of productive usage of computer models in Math education for supporting different aspects of learning modeling and learning explorations as a part of learning research work are presented.

Problem 1. The Falling Ladder Problem

This problem is used in school geometry courses in such context: what curve will the middle point of a segment line describe when it is moving in such a way that its ends remain at the two mutually perpendicular lines? As often seen in practice – the main problem for students is "to see" (or insight) that in this case the locus is an arc of a circle, it is rather surprising, and when this insight "is caught" the logical (deductive) proof of that fact

Figure 3. Dynamic drawing "The falling ladder problem"

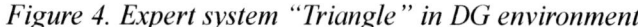

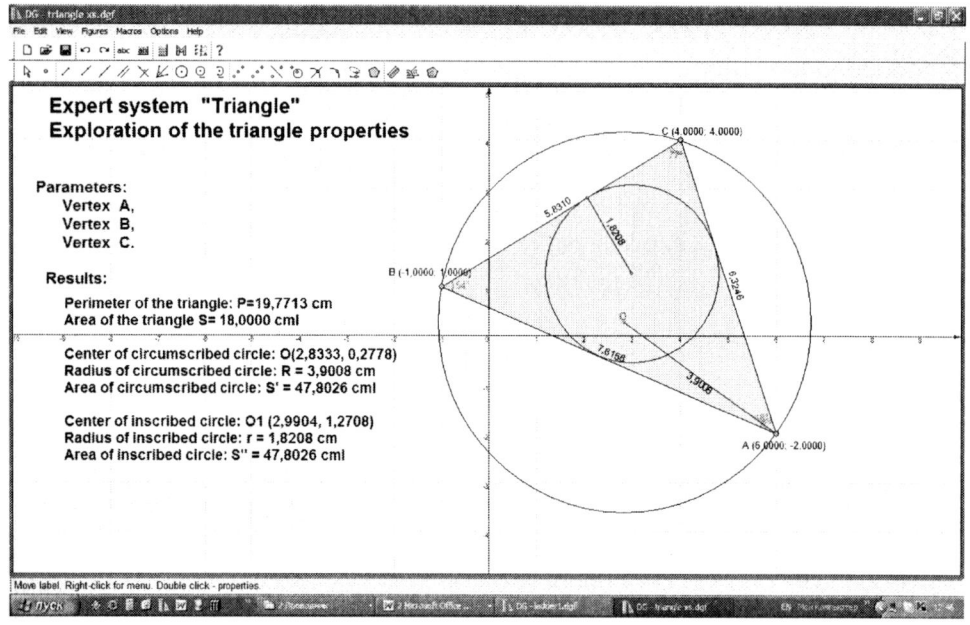

Figure 4. Expert system "Triangle" in DG environment

is much easier. On the other hand the fact that the locus is an arc of a circle can be checked with great accuracy by using the tools of the package and in such experimental way, find (conjecture with a great level of probability) the parameters of a desired circle: its origin and radius. Of course after that, the logical deductive proof is crucially needed otherwise it is not mathematics, but some kind of Science[2]. After that the locus of a bucket in the cases of different placement of a bucket at the ladder can be explored or the case when the "wall" is not perpendicular to the "ground".

Problem 2. Construct the Expert System "Triangle"

With the Expert System "Triangle" a student can construct a triangle of any shape and evaluate any parameters of this triangle, and thus can really use this dynamic drawing as an expert system with the practical or research aims. Such expert systems should replace drawings with paper and pencil technology, because of many benefits of electronic drawings (accuracy – all parameters can be measured or calculated with the precision of the computer's arithmetic processor, dynamism – changing the parameters of a drawing you can explore not only a single instance of a figure but the whole class of figures).

In a similar way students can construct corresponding Expert Systems for any geometric object and use it in a similar way[3].

THE RESEARCH APPROACH IN MATH EDUCATION

The research approach in math education assumes more or less the special activities which can be formalized in some special steps in the framework of each math course, each math topic, each math problem (of course these steps are only desired and in real practice the teacher should always only keep them in their mind and choose the appropriate form, time and place for them).

The theory of social constructivism plays the special role in a research approach in education which is based on the works of the famous psychologist L. Vigotsky concerning the educational community, the main characteristics of which are: *diversity of talents* (each student & each teacher is talented and this is recognized and used as a treasure in educational community); *collective resonance* (you told me the message and I heard

it, enriched it with my vision and returned it to you and you heard and so on, in other words: look and see, listen and hear, …); *collective reflection* (the collective resonance reflection, which is important as in the process of problem posing itself as well as at the end of this process) (Vigotsky L., 2004).

The components of research approach in math education may be formalized as follows:

Mathematization of a problem: the problem or problem field should be the result of the analysis of a suitable context which is interesting and understandable for the students (real life context, internal math context, applying to literature or describing the historical circumstances under which the problem was posed or may be posed). Mathematization of a problem is closely connected with the next components of RA and plays an important role in the open problem approach in math education.

Problem Posing

The role of a teacher is to arrange work in a form of a dialog, sometimes or always as a kind of a mindstorm in posing different problems in the context of a discussed problem field. As it was mentioned before, the crucial point are the ideas of L. Vigotsky concerning the matter of the educational community. Improving of the educational system in a framework of the competency paradigm on the base of research approach in education may be realized as the integration of main outcomes of the psychological and pedagogical researches and practices.

Modeling and CMS Modeling

Modeling and models in mathematics and in math education can be different in nature and can be used in different ways, in the context of the current paper it is important to keep in mind some aspects:

- *Construction of the model (in particular CMS-model) should be done by student himself/herself* (at least the student should be clearly acquainted with the matter of model and check the model by himself/herself at the particular special cases of a model for ensuring its correctness);
- *Modeling process is creative, inventing process and corresponds both to the ideas of constructivism* in math as well as to the ideas of social constructivism in education;
- *The model of an object differs from the object* (simpler as a rule);
- *The modeling process helps to clarify, refine, formalize etc. the object* of modeling and through these to promote the conceptualization process;
- *The models in CMS correspond to the type of the human cognitive math models* (analytical (functional), predicative, dynamic geometric etc.);
- *Models play important role in research approach in Math education* and could be productively used in each step of the mathematical research work: conceptualizing of a problem, conjecturing, experimental checking of hypothesis, constructing counter examples, constructing problem solvers etc.

Conjecturing (Hypothesing)

Productive conjecturing is the element at the core of creativity and plays an important role in problem solving, open problem solving (Pehkonen E., 2004) and consequently in research approachs in Math education.

Proving (Disproving) a Hypothesis

Proving or disproving conjectures (through constructing counter examples) are important, moreover the characteristic property of mathematics and thus in mathematical education (mathematics

is the branch of science using deduction as the only way for verification of statements). Computer models and especially dynamic computer models can help construct the counter examples to the hypothesis as well, and can help to find out regularities which lead to the logical (deductive) proofs (cf. Hemmi K., 2006).

Applications

The practice (real and imaginary) is the only origin of problems, productive ideas for their solving and domain for applications of constructed (or discovered or invented) theories. It is interesting and natural to ask him/herself about possible applications of new knowledge and the CMS gives the unique possibilities to construct the automatas for automatic solving of problems of a given class.

Systematization

Systematization means implementing new knowledge into the individual knowledge systems: finding connections with known facts, searching for generalizations, analogies, investigation the limits of the discovered facts, posing of new problems, reflection at the process of problem solving and analyzing productive ideas and traps in this way etc.

All the steps of mathematical researches mentioned before are in close correlation with the objectives, thinking skills and methods of math education in different countries, but in real practice they are not actively used yet. For example real Math educational practice was studied in Finland and the Ukraine in a framework of their comparison. Four lessons in Finland and four lessons in Ukraine were visited, recorded on video and carefully analyzed in context of usage of research approach and its IT support. Unfortunately it was reported that nevertheless IT environments in visited classes (9th Grade, city, regional, suburb schools chosen in a random way) was at a rather advanced level (PC, MM-projector, Interactive Boards, Internet, and availability of the variety of Computer Mathematical Systems (CMS), the CMS was not used in practice in these lessons. The examples of possibilities for improving research approach components with IT-support of the visited lessons is given as follows, (in the context of visited lessons and discussed problems) as an argument for using the research approach with IT support in CMS environment for improving Math educational practice at the competency base.

Problem 3

Let the quadrilateral ABCD be a parallelogram. Let the point F divide its side AB in ratio 2:3 and the point G divides the diagonal AC of the parallelogram in ratio 1:2. Explore, do the three points D, G and F lay at the same line or not.

The process of solution for Problem 3 at the lesson was the technical checking (rather professional) that the vectors *DG* and *DF* are not collinear and the students were successful in realization of this solution in their majority.

Comments to Problem 3 from the Point of View of Research Approach with ICT Support

Let's look at all steps of *Research approach in Math education* (Rakov S., 2005) in the context of this problem.

Mathematization of a Problem

The problem could be posed as a problem field about the geometric configuration in a more common way, for example: maybe it would be natural to tell about the origins of this problem (real life context, reference to literature or describing the circumstances under which the teacher decided to propose this problem to class):

There are three points in a parallelogram ABCD: vertex D, point F at the side AB, point G

at the diagonal AC. What questions can be asked about this geometric configuration?

Problem Posing

Maybe the problem proposed by the teacher at the lesson would be the most natural for the class (it is quite natural to define a point at the given segment by the ratio in which it divides the segment and ask when these three points belong to the line), but no doubt that mentioned problem would be the only one problem among the other interesting problems for example: evaluate the value of the angle $\angle DGF$ or find the interval of values of the angle $\angle DGF$ when the point G moves along the diagonal *AC*.

Let's concentrate at the original problem 1.

An interesting discussion could be started around the initial parameters of a problem. A set of initial data is so small: only ratios of two points in which they divide the side and the diagonal of a parallelogram. What does it mean? – It means that this problem is concerned not only with the concrete parallelogram or about the family of the parallelograms of the same shape (similar parallelograms); it's concerned with all the parallelograms and the only matter is the ratios in which the given two points divide the side and the diagonal of a parallelogram.

Modeling

We can't define the parallelogram with only two parameters: the point at the side and the point at the diagonal which divide them in given ratios. So we'll take the additional suitable parameters for defining the parallelogram itself and solve this problem. If the additional parameters will disappear in the result – it will be the proof that it is correct for all class of parallelograms, if not – maybe we'll obtain the conditions on the parallelogram under which the conjecture remains true.

It is interesting that the vector language puts this step of the problem solving (necessity to choose the additional parameters) at the background of our reasoning, makes this fact more explicit and it is both the advantage and the disadvantage of the vector method in this case.

Modeling in the case of this problem may be understood as finding the conditions for collinearity of two vectors DG and DF in analytic form and exactly this was done at the blackboard at the lesson in a classroom.

The conjecture that the answer really doesn't depend from the type of parallelogram means: the given ratios define the collinearity or non-collinearity of vectors *DG* and *DF,* and in other words it means in particular that the ratio *r1* in which the point F divides the side *AB* defines the ratio in which the diagonal *AC* is divided by the intersection point of *AC* and the segment *DF.*

Computer Modeling

The computer modeling in the DGS environment is concrete and it assumes defining all the free parameters of the model. On the other side the possibilities to change the parameters of the geometric model in a dynamic way gives us an opportunity to explore a great amount of realizations of a model and so we can talk about exploring the class of the models, in our case about exploring the class of parallelograms. So we construct a Dynamic Geometric Figure (the shorthand DGF) the concept which is used in the context of DG for computer models developed in its environment.

Picture 1 shows the screen of the DG environment with the correspondent DGF, but a little bit common setting is given (the point E of a segment EF was put at the side AD of the parallelogram, so with the DGF we'll have possibilities to explore the initial problem as the particular case of more common problem which is given at the DGF screen).

Conjecturing

Dynamic measurements and dynamic texts allow users (in particular teachers or students) to manipulate the model and to insight hypothesis and check them at the great amount of particular cases

Figure 5. Dynamic drawing for the problem 3

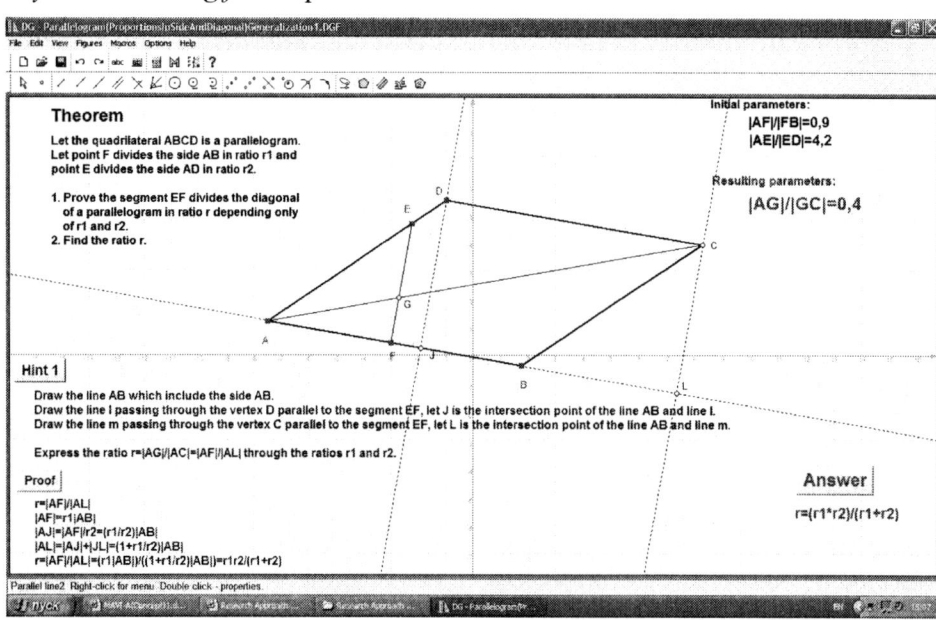

(it is one of the main ideas of dynamic geometry[4]). At the picture it is seen that all the parameters of a configuration can be changed: the length of the sides of a parallelogram and their place as well as the position of the points E and F on its sides[5]. Manipulations of independent vertices A, B, D which define parallelogram of any shape[6] show that initial conjecture and its generalization (the case of point E moving along the side AD) is confirmed by the experiments.

Proof (Disproving)

Of course modeling and computer experiments and experimental checking of the conjecture are not the logical deductive proof (and can't replace the deductive proof[7]), but manipulations with the computer model due to the high accuracy as well as a great amount of the particular cases can help to find the counter example to the conjecture or help to find regularities which leads to the deductive proof. In our case the additional drawings shows in the Picture 1 may help to find out by experiments that the length of the segment AL depends only from the length of the side AB, ratios rl and r2 and then it is not difficult to express this fact analytically. The Hint, Proof, Parameters and the Answer of the discussed DGF are organized as the textboxes controlled by the corresponding buttons.

Interpretation

This work is needed first of all in the case when the initial problem was not formalized, moreover if it was a real life problem etc. (in these cases at least the interpretation in terms of initial domain is needed). But in any case it is useful to try to find new interpretations of obtained results in other branches of mathematics as well as in new real life situations.

Unfortunately the authors did not find interesting interpretations of the result mentioned before but no doubt the collective mind could do it.

Thus the authors propose readers to think about a real–life or any other associative interpretation of this result[8].

Application

One of the objectives of math is to work out the algorithms for solution of typical problems. The traditional math education has been organized as teaching and learning of known algorithms for solving the known typical math problems. The new approaches in math curriculum based on the competency paradigm of math education assume finding the solution of a new problem and it is the matter of the research approach in math education. But after discovering the solution of a new problem, it is natural to explore its properties and then implement this algorithm on a computer for automatic solving of the problems of a given class. This step is very important in math education due to a lot of important methodological, methodical, educational aspects. We do not discuss these questions in this article and only mention that developing such a computer solver is very creative and at the same time understandable and challenging problem for students. This student constructive work in developing (inventing) the computer solver reflects as well the productive work in Math (in both Pure Math as well as in Applied Math) as well as in Computer Science.

At the end of this discussion we highlight that dynamic models presented at the pictures 1 and 2 really can be used as solvers of the corresponding generalizations of the initial problem: the user can interactively define the parallelogram of any shape, choose the positions of the points E and F and then read in the textbox *Resulting Parameters* the resulting ratio with any precision up to 11 digits.

Systematization

This step is the step of reflection: in what way the new knowledge can be implemented in the existing individual knowledge system. What connections are there with known facts? Are there any generalizations of new facts, any analogies in other math domains, what new knowledge gives in particular cases etc?

One of the interesting closely connected facts is: does the assertion of the theorem remain true if the point F belongs not to the side AD of a parallelogram but the side CD. The DGF below shows that the computer modeling confirms this fact. We leave the deductive proof of a correspondent theorem to the reader.

It is interesting to return to the first ideas concerning the matter of the fact: the ratios in which points E and F divide the sides of the parallelogram define the ratio in which the intersection point of the diagonal of a parallelogram and the segment EF divides the diagonal. One of the productive ideas may be embedding the parallelogram into the space (3D space) and then project it orthogonally on the plane. In this case the properties of points which are defined by the ratios in which they divide the segments of initial parallelogram will remain valid for the projection of this parallelogram at the projection plane (this figure will be the parallelogram as well). The following productive idea may be to change the initial parallelogram onto the simplest parallelogram, for example, the square.

Moving along these ideas it is easy to come to the new turn of our research work: it is sufficient to prove that the parallelogram of any shape may be obtained as an orthogonal projection of the square at the plane. Then the ratio in which the segment EF (with given ratios in which points E and F divide the sides AB and D of the square) divides the diagonal of the square will be common ratio for all shapes of the parallelogram.

Comments to the School Visit 1 from the Point of View of RA with ICT Support

1. It seems to be that the proposed discussions and approaches would be interesting and corresponds to the level of Math culture of the majority of students, but it would be really productive only if the ideas of RA in math education would be acquainted and accepted by students and teachers. The ICT support in DG environment technically was possible in the visited school (they have the computer class and courseware DG), but the computer class is in the teacher of informatics response and is in use for informatics almost all the time.

Figure 6. Generalization 2 of the initial problem

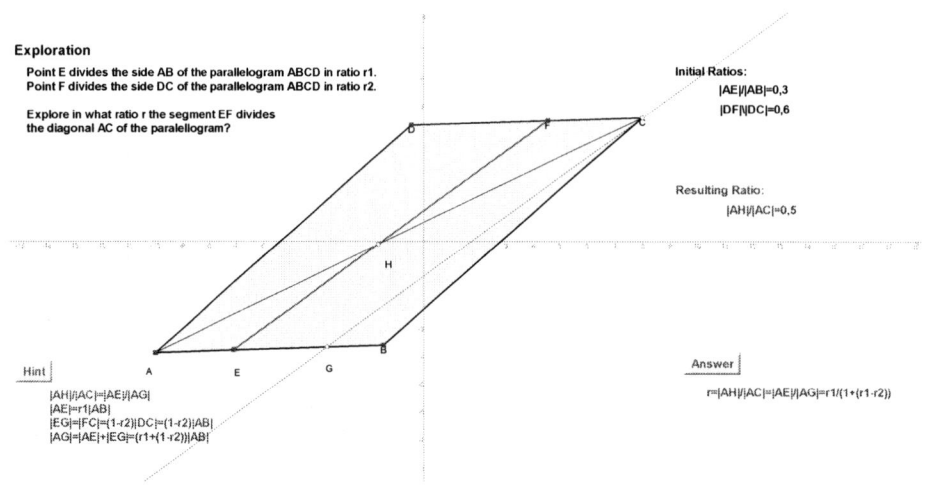

2. We did not discuss the question how the DGFs presented at the Pictures 1 and 2 were developed. Of course, the teacher could prepare them beforehand but it is much more interesting, useful and creative for students to prepare those on their own (remind the principle *learning by doing*). It is interesting in any branch of mathematics (arithmetic, algebra, probability, statistics etc.) but it is of special interest in geometry because the DG packages may be considered as the interpretation of the geometry. It is important at each stage of learning geometry. At the beginning of the course DGs help to explain and check the correctness of algorithms for constructing the geometric configurations (for example, an algorithm for constructing a square, equilateral triangle etc. can be easily checked by moving the basic points – is the shape of the figure changed or not?). At more high grades DGs help to make any kind of math activities more effective, productive, attractive etc., in particular the aforementioned components of the RA in Math education.

Information Technologies, Mathematics and Art

IT has changed all kinds of human activities. Mathematics is not the exception - it becomes technologically dependent. The most important changes take place in the process of doing mathematics - discovering new facts and their proof. Special mathematical packages CMS (Computer Mathematical Systems) equip the user with suitable environments for arranging computer experiments in the problem field with the aim of finding mathematical regularities on the first step of exploration and then support the process of proof with the powerful opportunities of computer algebra. No doubt, the future of mathematics is symbiosis of a human being with computers. CMS usage becomes the inalienable component of mathematical culture.

Innovative trends in mathematical education lay in the framework of a constructive approach - involving students in the process of constructing their own mathematical system, which consists of mathematical knowledge and mathematical beliefs (Pehkonen, E., & Rakov, S. , 2005). One of the most effective ways of the realization of a constructive approach, is the method of learning explorations when students explore open-ended[9] problems or problem fields on their own. This can be considered a model of the professional mathematical work. Therefore it is natural to use IT in mathematical education exactly in the same manner: computer experiments as the source of powerful ideas, computer algebra as a tool for the deductive method. Using IT in arranging learning explorations and carrying out the proofs can not only do this work more effectively but acquaint students with the modern technologies of mathematics and it is the only way to develop math competencies.

As an example of such work a chain of problems (or problem field) is proposed, which initially started with one well-known problem (Boltyansky V.G., 1984). This paragraph reflects the personal experience of authors in the productive usage of CMS for mathematical work: at first the pleasure of playing in DG environment[10] was resulted with conjecture of its first generalization, which then they have successfully proved with the help of Derive (Rakov S., Gorokh V., 1998). But it was only beginning – from that moment new and new generalizations and associated problems were arising, the last of which concerning caricatures had appeared in the fall of 2005.

Thus the current article is in some sense a report of long time exploration of one problem field concerning geometric figures and their transformation. The authors successfully used this material in the framework of computer practice with students of the Math teacher department in Kharkiv National Pedagogical University named after G.S. Skovoroda and this gives us hope that it can be interesting for a more wide audience.

Initial Problem and Problem field

Two squares $A_1QC_1D_1$ and $A_2B_2C_2Q$ with the centers O_1 and O_2, and common vertex Q of the identical orientation are given. Let L and N are the midpoints of the segments A_1A_2 and C_1C_2. Then the quadrilateral O_1LO_2N is a square as well.

Prove this statement and propose its generalizations.

Computer Experiments in Problem Solving

Experiment 1

At first we construct a computer model of a given problem in the DG environment and then play with it. Our **Model 1** is dynamic («alive») in the sense that we can dynamically change its parameters. All the initial points *(A_1, Q and B_2)* are movable (they are marked with red squares) – they can be moved with the mouse. Playing with this model we really see that the third (depending) quadrangle is a square (we can become convinced in it by view or by measuring the parameters of a figure with the *DG tools* - ruler and compass). Of course our experimental assurance needs the deductive proof, it is only the hypothesis, but this hypothesis is of a great confidence. We put aside now the attempts to prove the hypothesis and continue our computer experiments.

Experiment 2 (Disjointing the Vertices)

Is it really important that the initial squares have the joint vertices?

Unfortunately, the previous model cannot be modified for these purposes. Therefore we must repeat the algorithm described before with a little change in the second step - the initial points will be four independent points B_1, D_1, B_2, D_2, the diagonal opposite points of the future squares[11]. We omit this construction and leave it as an exercise for the reader. Playing with **Model 2** give us

experimental check that in this case the resulting quadrangle is a square as well what gives us the base for formulating the next hypothesis:

Generalization 1

Two squares $A_1B_1C_1D_1$ and $A_2B_2C_2D_2$ with the centers O_1 and O_2, of the same orientation are given. Let A_3, B_3, C_3, D_3 are the midpoints of the segments A_1A_2, B_1B_2, C_1C_2, D_1D_2. Then the quadrilateral $A_3B_3C_3D_3$ is a square as well.

Experiment 3 (the Middle Points Release)

Why must the vertices of the resulting figure be the midpoints of corresponding segments?

Maybe the result remains valid in the case of arbitrary points, which divide the segments joining the correspondent points in a given proportion?

Modify our previous model. For this, a new macro would be needed. This construction will be a macro *Divide* which divides the segment in a ratio given by the segment and a point on it. Such a macro can be defined at the base of the *Thales theorem* for example. Then the construction of the **Model 3** we modify with this macro defining an arbitrary point A_3 on the segment A_1A_2 and then defining all the other points B_3, C_3, D_3 with the macro *Divide* as the points which divide the correspondent segments in the same ratio.

Now we can change the sizes of the squares with the mouse and their mutual positions and the place of the division points as well. As in previous case we can easily convince ourselves by the experimental check that the resulting figure is a square. Computer experiments with **Model 3** give us the base for the next generalization 2.

Generalization 2

Two quadrates $A_1B_1C_1D_1$ and $A_2B_2C_2D_2$ with the centers O_1 and O_2, of the same orientation are given. Let A_2, B_2, C_2 and D_2 be the points of the

Figure 7. Model 1 (Initial Problem)

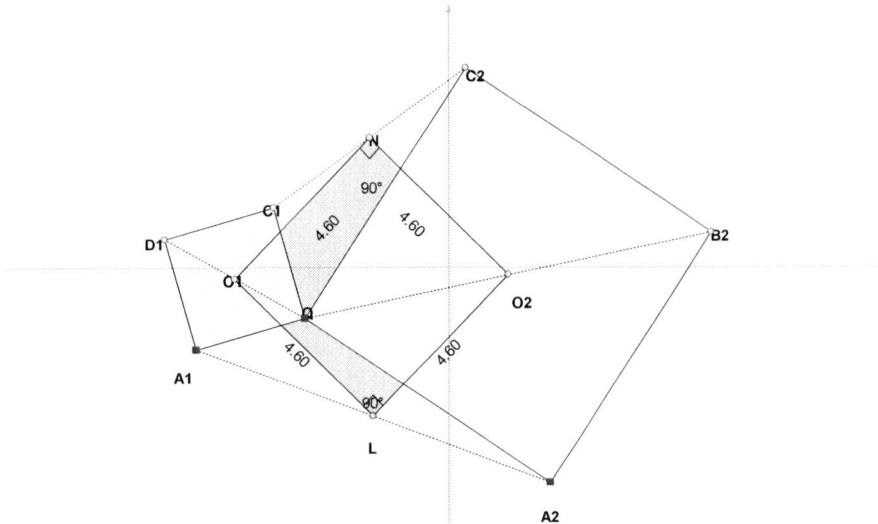

segments A_1A_2, B_1B_2, C_1C_2 and D_1D_2 respectively which divide them in the same rate. Then the quadrangle $A_3B_3C_3D_3$ is a quadrate as well.

Proof in the Derive Environment

Let $O_1(a, b)$ and $O_2(c, d)$ are the centers of the squares $A_1B_1C_1D_1$ and $A_2B_2C_2D_2$. Denote vectors $O_1A_1 = n_1 (p, q)$, $O_2A_2 = n_2 (r, s)$.

Declare the vector constants in the *Derive* package (using commands *Declare, Constant*):

#1: O_1 := [a, b]
#2: O_2 := [c, d]
#3: n_1 := [p, q]
#4: m_1 := [-q, p]
#5: n_2 := [r, s]
#6: m_2 := [-s, r].

Express the coordinates of the vertices of the original squares (using the same commands *Declare, Constant*):

#7: A_1 := O_1 + n_1
#8: B_1 := O_1 + m_1
#9: C_1 := O_1 - n_1

#10: D_1 := O_1 - m_1
#11: A_2 := O_2 + n_2
#12: B_2 := O_2 + m_2
#13: D_2 := O_2 - m_2
#14: C_2 := O_2 - n_2 ;

Declare the function *Divide*, which returns the coordinates of a point dividing the given segment in given ratio:

#15: *Divide (x , y , t) : = (x + ty) / (1 + t).*

Declare coordinates of the points A_3, B_3, C_3, D_3 which divide the segments A_1A_2, B_1B_2, C_1C_2, D_1D_2 in the ratio *t : 1 :*

#16: A_3 := Divide (A_1, A_2, t)
#17: B_3 := Divide (B_1, B_2, t)
#18: C_3 := Divide (C_1, C_2, t)
#19: D_3 := Divide (D_1, D_2, t).

Find the difference of the vectors A_3B_3 - D_3C_3:

#20: *(B_3 - A_3) - (C_3 - D_3).*
Simplify this expression:

#21: *Simplify #20*

#22: *[0, 0]*

Consequently, the quadrilateral $A_3B_3C_3D_3$ is a parallelogram.

Evaluate now the scalar product of the vectors A_3B_3 and B_3C_3:

#23: *Simplify $((B_3 - A_3).(C_3 - B_3))$*

#24: *0*

Thus vectors B_3A_3 and C_3B_3 are *mutually perpendicular, so* the quadrilateral $A_3B_3C_3D_3$ is a rectangle.

Compare the length of the sides A_3B_3 and B_3C_3 of this rectangle. For this purpose declare a new function *SModV* (**S**quare of the **Mod**ule of the **Vector**):

#25: SModV(x): = x.x

#26: Simplify (SmodV(B$_3$ - A$_3$) - SModV(C$_3$ – A$_3$))

#27: 0

We have finished the proof: the rectangle $A_3B_3C_3D_3$ is a square. Remark that the square $A_3B_3C_3D_3$ can degenerate in a point.

We have proven the hypothesis with Derive! Really we made all our transformations in symbolic form not for any particular case but for the common case.

Further Generalization

The reader can easily restate the Generalization2 in terms of arbitrary similar quadrilaterals, construct corresponding Model 4, play with it and as a consequence formulate the following generalization:

Generalization 3

Let F_1 and F_2 are two similar figures in the plane of the same orientation (it means that there ex-

ists a similarity f of the first class, which maps the figure F_1 onto the figure F_2. For each point X_1 of the figure F_1 and a point $X_2=f(X_1)$ of the figure F_2 define a point X, which divides the segment X_1X_2 in a ratio t $(X_1X : XX_2 = t)$ Then the figure F, which consists of all such points X, is similar to two original figures or consists of the single point.

Proof of the Generalization 3 in DG Environment

The radius-vector of a point M will be denoted as in further.

Let X_1 and Y_1 are two arbitrary points of the figure F_1. Let $X_2 =f(X_1)$ and $Y_2=f(Y_1)$.

Denote through X and Y the points, dividing the segments X_1X_2 and Y_1Y_2 in a ratio t.

Then:

$$\overline{X} = (\overline{X}_1 + t\overline{X}_2)/(1+t),$$

$$\overline{Y} = (\overline{Y}_1 + t\overline{Y}_2)/(1+t).$$

As a consequence we obtain:

$$\overline{XY} = (\overline{X_1Y_1} + t\,\overline{X_2Y_2})/(1+t).$$

Therefore we have:

$$\left|\overline{XY}\right|^2 = (\overline{X_1Y_1}^2 + 2t\,\overline{X_1Y_1}.\,\overline{X_2Y_2} + t^2\,\overline{X_2Y_2}^2)/(1+t)^2.$$

Let be the angle between the ray X_1Y_1 and its image under the similarity f. This angle is a constant because the map f is a similarity of the first class. With respect to the relation , where k is the coefficient of the similarity we finally obtain:

$$\overline{|XY|}^2 = \frac{1 + 2tk\cos\phi + t^2k^2}{(1+t)^2}\left|X_1Y_1\right|^2.$$

Consequently , where C is a constant.

Thus, the figure F is similar to the figure F_1.

Remark

In the most common case (**Generalization 3**) the proof was done without computer help and was more simple and natural then any previous variants. It is a rather general situation because from the common point of view the unimportant details are disappeared and the matter of the fact became more clear and obvious. Nevertheless the computer experiments have played the substantial role in the process of this chain of generalizations. By the way the proof discussed before could be done on a computer as well, but all the analytic calculations were so simple that computer help was unnecessary.

Further Generalizations

As it is clearly seen from the proof given before the fact remains valid in the case of arbitrary dimensions (not only in the case of *2D*, but *3D*, *4D* etc. as well).

The successfully proved fact can be used in computer animation for modeling continuous transformations of figures in plane or space.

Unexpected Turn: Geometry and Caricature

It seems that there are no more productive ideas for further generalization of the Generalization 3. But what about the case when figures *F1* and *F2* are not similar?

Let $\phi : F_1 \to F_2$ is an arbitrary bijection (1-1 correspondence) and $\varphi[F_1] = F_2$. Define a family of maps φ_t, depending from parameter $t \in R$ by the rule:

$$\varphi_t(x) = (1 - t)x + t\varphi(x).$$

We have: $\varphi_0[F_1] = F_2$; $\varphi_1[F_1] = F_2$. If map φ_t is a similarity then φ_t is a a similarity by the Generalization 3 and thus $\varphi_t[F_1]$ is similar to F_1 for any $t \in R$. In the common case figure F_1 is

continuously transforming into F_2 when is changing from 0 to 1.

Interesting problems concerning map φ_t (they would be better called problem fields) may be the following (they can be proposed for readers as interesting topics for own projects):

1.	Develop the math theory of caricatures.
2.	Design a software application for making interactive caricatures.

The next figure demonstrates the simple model for making caricatures. One of the possible ways for constructing caricatures may be the following: choose some set of characteristic points for a face (points of angles of eyes and mouth, ends of brows etc. etc.) preparing the statistical average of 100 (or 1000) images of characteristic points of different people, and proportionally changing features of the characteristic points of a person (for whom we construct the caricature) from the corresponding points of an average image.

Other Problems

Now we'll discuss some new results in plane geometry (by our mind), which were discovered through the computer experiments in DG and then proved in Derive by the authors just at the same manner mentioned before. The details of the corresponding experiments and proofs will not be discussed because of the lack of space (the methodology and technique could be shared from the previously discussed problem).

Theorem 1

Let three equilateral triangles $A_1B_1C_1$, $A_2B_2C_2$ and $A_3B_3C_3$ of the same orientation are given. Let the points P, Q, R are the middle points of the segments C_1B_2, C_2B_3 and C_3B_1 respectively. Then the triangle PQR is quadrilateral if and only if the triangle $A_1A_2A_3$ is quadrilateral.

Remark 1

The necessary part of the Theorem 1 is well-known (see for example (Skopets Z.A., 1990, p. 100)), the sufficient part seems to be new.

Theorem 2

Let two equilateral triangles $A_1B_1C_1$, $A_2B_2C_2$ of the same orientation are given. The equilateral triangles $A_1A_2A_3$, $B_1B_2B_3$ and $C_1C_2C_3$ of the same orientation are built on the segments A_1A_2, B_1B_2 and C_1C_2 respectively. Then the triangle $A_3B_3C_3$ is equilateral.

Theorem 3

Let two squares $A_1B_1C_1D_1$ and $A_2B_2C_2D_2$ of the same orientation are given. The equilateral triangles $A_1A_2A_3$, $B_1B_2B_3$ and $C_1C_2C_3$ $D_1D_2D_3$ of the same orientation are built on the segments A_1A_2, B_1B_2, C_1C_2 and D_1D_2 respectively. Then the quadrilateral $A_3B_3C_3D_3$ is a square.

Remark 2

As it could be seen the two theorems mentioned before are the particular cases of the following theorem that could be proved in Derive for the common case as well.

Theorem 4

Let n regular m-polygons at the plane $A_1^1 A_1^2 \ldots A_1^m$, $A_2^1 A_2^2 \ldots A_2^m$ and $A_n^1 A_n^2 \ldots A_n^m$ of the same orientation are given. Then if two n-polygons $A_1^1 A_2^1 \ldots A_n^1$ and $A_1^2 A_2^2 \ldots A_n^2$ are regular then the all remain n-polygons $A_1^i A_2^i \ldots A_n^i (i = 3, 4,\ldots,m)$ are regular as well.

Geometric Transformations, DG and Linkages

It is a very interesting and productive work in context geometry, its relation to the real world and applications to constructing linkages, playing with them and exploring properties of geometric transformations. With the help of DGS, in particular in DG environment it is possible to construct an electronic model of linkages. The following screen copies illustrate such constructions and activities with them.

The DG implementation of geometric primitives and transformations offers the user to construct simpler but more powerful "electronic linkages". Such possibilities are mainly the result

Figure 8. Model 4 (Interactive construction of caricatures with DG)

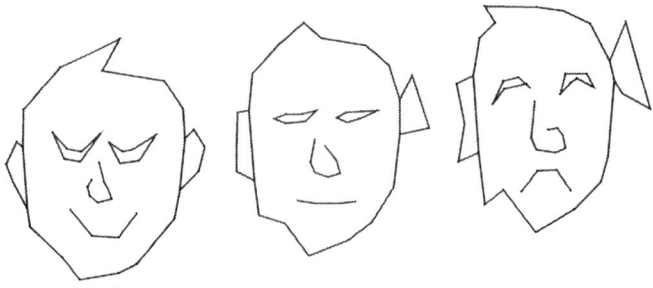

t=0.49

of one internal property of a point at the DG segment: it divides the segment at the invariant ratio (it does not change under the transformations of a segment). This property inspires a special type of electronic linkages, which can be called the "rubber linkages" (because of the aforementioned property of an "electronic segment" can be modeled with homogeneous rubber strips or rubber sticks with a mark (as a model of a point) on them). For example the simplest "Rubber Dilator" is a segment with a point at them. It is interesting that "Rubber Tools" can have the entire plane as the domain[12] (obviously, real tools have bounded domains).

Problems

1. Construct the possibly simplest "Rubber Symmetrisator", "Rubber Reflector", "Rubber Translator", "Rubber Rotator", "Rubber Dilator" with the entire plane as a domain.
2. Find as much as possible real life problems, which can be solved approximately with the help of linkages.
3. Construct your own linkages and explore its properties.

IT OLYMPIADS FOR STUDENTS

More than 20 years the Olympiads in Informatics for students of schools and universities in the Ukraine are held. Moreover now there are National Competitions and Olympiads in different branches of Informatics: Programming, System Programming, Computer Science, Theoretical Informatics, Applied Informatics, Web Design etc. By the tradition the majority of the problems for Applied Informatics Olympiads are concerned the Creative Informatics for Mathematics: the participants are proposed to develop the tool for computer modeling in the Math field. We give below some examples of such problems which were successfully solved by the participants of the Olympiad and their solutions can be used as a prototype for real packages with appropriate comments. We assume that these problems could illustrate a variety of IT usages in real life and on the other side could inspire readers to their own inventions in creatively using IT.

Problem 1. (Archeologist AWP (Automated Work Place))

An archeologist excavated some pieces of an ancient artifact. Invent a program for restoring the

Figure 9. Model 5 (Theorem 3)

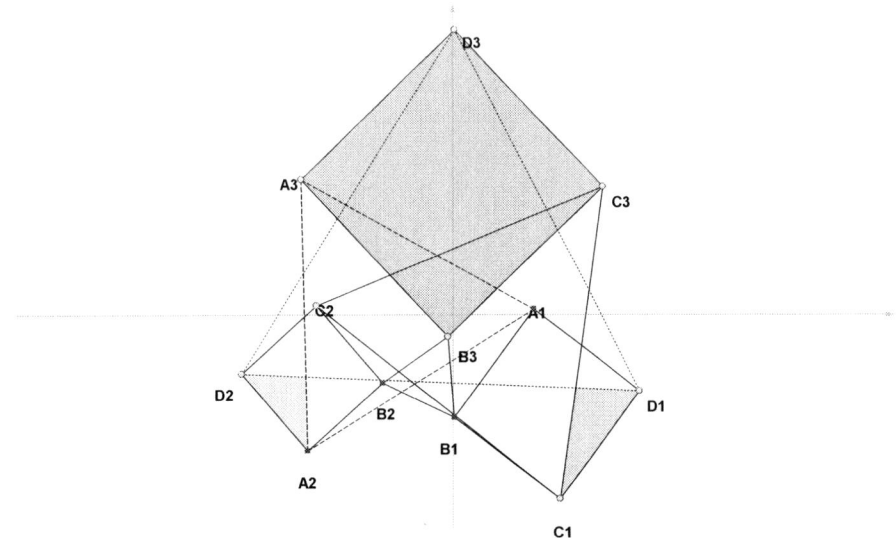

Figure 10. Symmetrizator – linkage for drawing symmetric figures

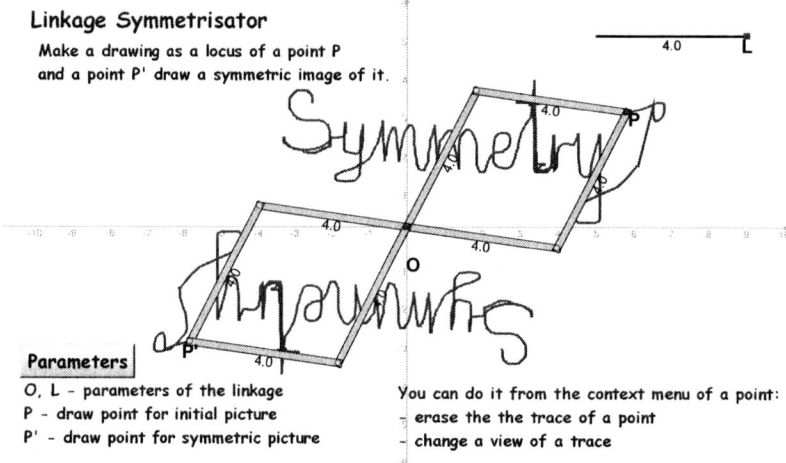

Linkage Symmetrisator

Make a drawing as a locus of a point P
and a point P' draw a symmetric image of it.

Parameters

O, L - parameters of the linkage
P - draw point for initial picture
P' - draw point for symmetric picture

You can do it from the context menu of a point:
- erase the the trace of a point
- change a view of a trace

Figure 11. Reflector – linkage for drawing a reflected figure

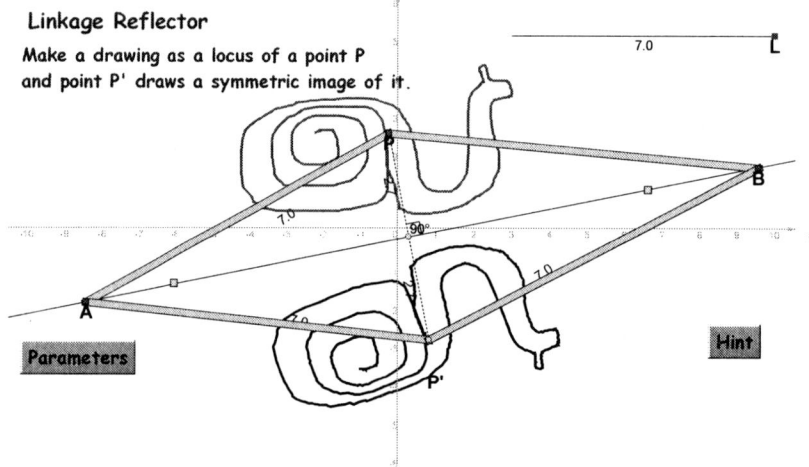

Linkage Reflector

Make a drawing as a locus of a point P
and point P' draws a symmetric image of it.

Parameters

Hint

Figure 12. Translator – linkage for drawing the translation image of a figure

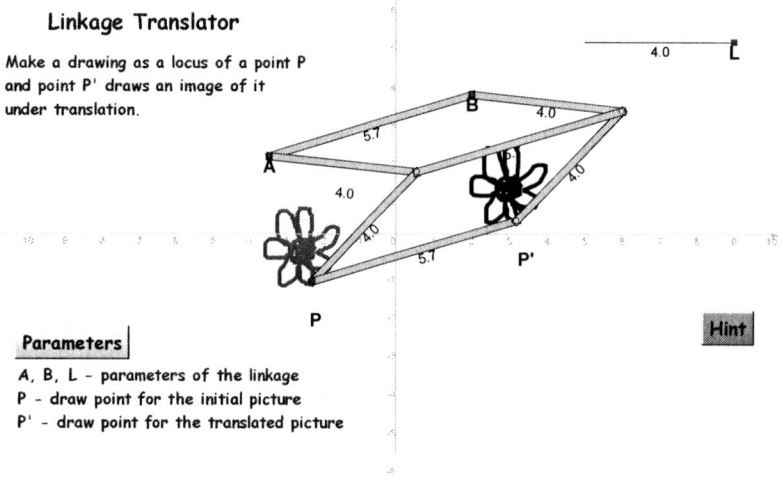

Linkage Translator

Make a drawing as a locus of a point P
and point P' draws an image of it
under translation.

Parameters

Hint

A, B, L - parameters of the linkage
P - draw point for the initial picture
P' - draw point for the translated picture

Figure 13. Linkage rotator for drawing the rotation image of a figure

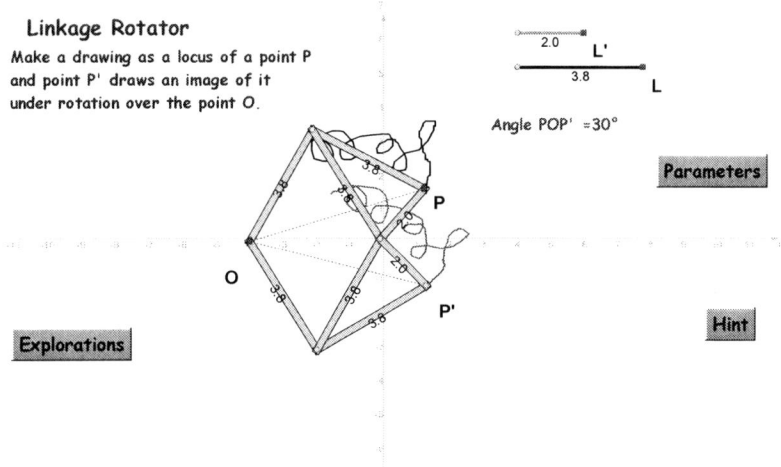

Figure 14. Linkage dilator for drawing the dilation image of a figure

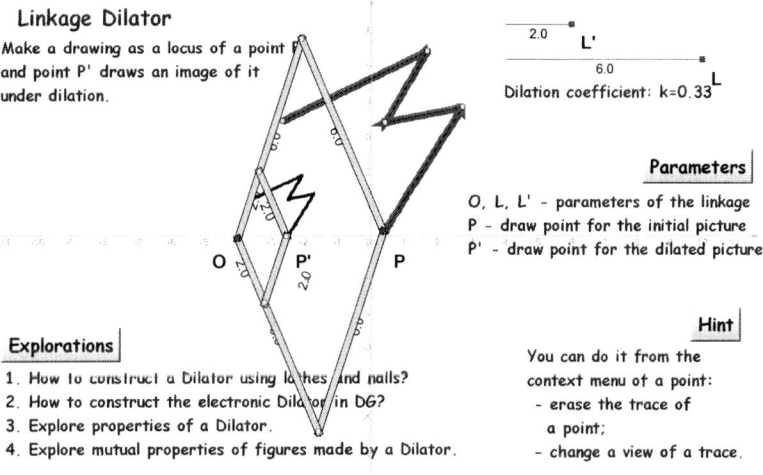

artifact (by scanning the pieces and automatically finding the position of putting the pieces together). For simplification explore the case of two pieces of artifacts in a plane which are bounded by the polygons with vertices in points with rational coordinates.

Problem 2.
(Khinchin – Kadets Problem)

The law of large numbers reflects the stochastic nature of the real world and more and more is used in various areas of math research penetrating even in such areas which are at the first glance far from stochastic. One of such branches of mathematics is functional analysis and Khinchin inequality:

$$\frac{1}{\sqrt{2}} \le HK(\{x_i\}_{i=1}^{n}) \le 1$$

where:

- n – an arbitrary integer,
- $\{x_i\}_{i=1}^{n}$ – arbitrary real number sequence,

Figure 15. Rubber dilator

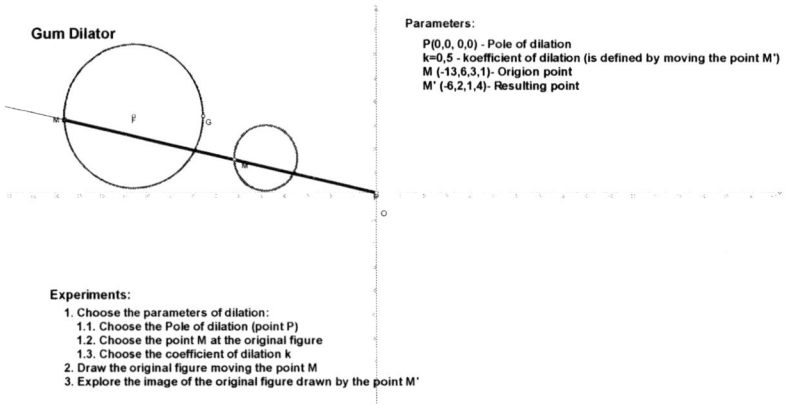

Gum Dilator

Parameters:
P(0,0, 0,0) - Pole of dilation
k=0,5 - koefficient of dilation (is defined by moving the point M')
M (-13,6,3,1)- Origion point
M' (-6,2,1,4)- Resulting point

Experiments:
1. Choose the parameters of dilation:
  1.1. Choose the Pole of dilation (point P)
  1.2. Choose the point M at the original figure
  1.3. Choose the coefficient of dilation k
2. Draw the original figure moving the point M
3. Explore the image of the original figure drawn by the point M'

$$HK\left(\{x_i\}_{i=1}^n\right)=\frac{\dfrac{1}{2^n}\displaystyle\sum_{\{\varepsilon_i\}_{i=1}^n,\ \varepsilon_i=\pm1}\left|\sum_{i=1}^n\varepsilon_i x_i\right|}{\sqrt{\displaystyle\sum_{i=1}^n x_i^2}}.$$

Problem:

1. Develop the program HiKadets for calculating value of function $HK\left(\{x_i\}_{i=1}^n\right)$ for arbitrary non-zero sequence $\{x_i\}_{i=1}^n$ for arbitrary $n \le 20$.

2. Evaluate the value of a function $HK\left(\{x_i\}_{i=1}^n\right)$ for stationary sequence $\{x_i\}_{i=1}^n = \{1\}_{i=1}^n$ as long as possible.

3. Find the limit $\lim\limits_{n \to \infty} HK\left(\{1\}_{i=1}^n\right)$.

Remark

The right part of the inequality is well known and is a consequence of the Caushy–Bunyakovsky inequality. The left part of this inequality with some positive constant K was discovered by the famous Ukrainian mathematician Y.Khinchin. Another famous Ukrainian mathematician M.Kadets for many years explored the value of this constant and assumed that its value is $1/\sqrt{2}$. This hypothesis was proven in 1980th by the Polish mathematician S. Sharek when he was a student of a Warsaw

University. The experience of the Olympiads in Informatics showed that there are no problems, which were posed by the Scientific Commission of the Olympiad, and which was not solved by some participants. Moreover, if the problem is correctly posed, but was not solved by the commission – the participants can solve it (the commission try this risk experiment on the Problem 2 (part 3) below. For a month after the Olympiad two participants answered the question in two independent ways – the answer is 2/π.

Problem 3. (Disjointing the Pieces of the Cut Figure)

A circle at the plane with the center in the origin of the coordinate system and radius 1 was cut onto two parts by the break line with vertices in the points with rational coordinate. Develop a program for defining whether it is possible or not to disjoint the pieces of a circle one from other by moving it at the plane; if it is really possible find such a move.

Remark

Problem 3 has a dramatic history which is not finished yet. This problem was proposed to the participants of the III Soviet School Olympiad in

Informatics (Kharkiv, 1991). The Scientific Commission of the Olympiad worked for 5 days in discussions, selecting and polishing home propositions of its members. Problem 3 was recognized as one of the best proposals and it was accepted for Olympiad. It was assumed in discussions of Scientific Commission that disjoining should be done only by the parallel move but this condition was not included in the final problem formulation. And the day of Olympiad has come... Several participants proposed some optimal solutions for disjoining by parallel move. But they did a lot of more: they discovered exciting examples of cuttings for which there are no disjointing by parallel move but there exists disjoining by rotation; they discovered another example where disjoining can be made only by the composition of rotation and parallel translation and can not be done by a single transform (rotation or parallel transform). In common the problem is not solved up to date, while the most qualified geometers of Kharkiv mathematical school tried to solve it.

Maybe Problem 3 should be proposed to the participants of the International Olympiad in Informatics for its solution?

CONCLUSION: PHILOSOPHY OF NEW TOOLS AND A NEW FORM OF EDUCATION: CREATIVITY, COMPETENCY PARADIGM, RESEARCH APPROACH, COMPUTER SUPPORT, DISTRIBUTED EDUCATIONAL COMMUNITIES

A lot of philosophers, politicians, educators actively discuss a question of the ways of the mankind development. Through the misty glass of science the frame of a future became more and more clear, because their principles are natural moreover necessary steps not only for development but as well the surviving of mankind at this crucial step of transmission to the global world:

- Information society
- Knowledge society
- Sustainable society
- Soul Society

Each of these types of societies is closely connected with inventing the new tools and new forms for Research Work, Education and Art. Moreover the success of developing these types of societies is closely connected with them especially with new tools and forms of education.

This classification is rather formal: Information Society, Knowledge Society, Sustainable Society, Soul Society – life is always more complicated than any formal schemas. Nevertheless this classification reflects the tendencies of human society progress, reflects changes in human priorities which in turn reflect the changes in growing power and character of human being and growing power of environmental problems challenging human being. Education plays a more and more important role in society, in its progress, in its stability. Education for all, long life education, e-education, competency paradigms of education as a reply on the challenges of developing the K-Society, research approach in education as a methodology of a competency education, IT as a technological support of research approach in education, Computer Mathematical Systems (CMS) as a tool for research approach in Math education (computer modeling, computer experiments, data gathering and processing, etc.), Dynamic Geometry Systems (DGS) and Computer Algebra Systems (CAS) as a most universal and powerful CMS, perspective universal Mathematics packages, which should integrate and mutually empower DGS and CAS.

But in what ways the forms of education change? What kind of forms will correspond to the new dynamic situation in which each individual should learn new and new areas of knowledge, relearn and renew old areas, moreover participate in creation of new knowledge. The school based on the ideas J. Comenius of class-lesson system can't solve all the problems rising by the new dynamic reality. It is oriented on the cloning of specialists (a teacher (a professor) is assumed, knows the absolute knowledge and his/her problem is to transmit this knowledge to the heads of his/her students in the best way) and such paradigm really worked productively for centuries because life was rather stable and knowledge received (developed or transmitted (?)) was sufficient for a successful life. But now the situation was dramatically changed and now the competency paradigm (with its domination of knowledge and skills *in new situations*) is changing the knowledge paradigm for previous type of societies (with its domination of knowledge, skills to make adequate decisions and act *in standard situations*). That is why metaphorically saying "the knowledge of multiplication table doesn't help[13]"

Ideas of new educational forms should be based on the new possibilities given by IT and should include the ideas of initiative and responsible self learning, collective forms of learning based at the ideas of social constructivism and free educational communities (in particular free virtual educational communities). Wide usage of computer modeling, research approach in education, using Internet as a global knowledge repository and a powerful tool for arranging the virtual educational communities with distributed managing the appropriate packages and all type communications. The role of a teacher in this situation should be dramatically changed from *the sage at the stage* to the *guide by the side*, as the education folklore says. The educational community (virtual or real) should be the research educational community. The real step in this direction to new forms of education is the Laptop Class Project, the matter of which should be possibilities for members of the educational community to communicate with others in virtual educational environments at any moment, arranging educational research work at the base of using appropriate educational and research software and involving in this community all the previous experience of mankind through the ac-

cess to the Internet repository of knowledge. The one crucial point of such education environments should be an individual laptop (or palm-computer or some computer of new type, which should be readily available to the user at any moment, which should have a wireless connection to the Internet and should be the only gate to the user's personal IT resources). Another crucial point of this educational community should be the possibilities to use the needed software in a distributed way. The great challenge for the educational community of the world is to invent new forms of educational work which allow using all the possibilities of IT in full range. We hope that the research approach with IT support, is moving in the right direction.

POST SCRIPTUM: SOME CRAZY IDEAS ABOUT YOUTH, CREATIVITY, SCIENCE, MATHEMATICS AND IT

Maybe time has come to think about new forms for engaging youth for solving actual problems in Math, Computer Science, Science, etc.?

Creativity is the organic attribute of a individual's thinking. The main goal of education is to help students (create such conditions) to realize their creativity in the best way (at least do not kill or depress it).

It seems creativity can't be taught, it is the immanent property of intelligence and the most productive creativity time is in youth. Maybe the main problem of mankind is to find out ways to formulate the problems for the youth, and to understand and listen to their answers and productive ideas?

Maybe the role of adults in this world should be shifted in the direction of a interface between the creative "youth world" and conservative adult world (as a translator from the primitive language of adults (A–language) onto the creative language of youth (Y–language))?

Maybe IT, computer graphics and dynamic computer modeling give the opportunity for creating such Y–languages?

Maybe the most perspective way lies in developing the possibilities to arrange free creative research communities with the help of IT and Internet, which should include the use of appropriate IT tools in distributive modes?

Maybe time has come to engage the youth in solving these creative and future oriented problems?

As mentioned before, it is not the opposition of the world of Youth and the world of Adults but attempts to structure them in a more natural and productive way. These two worlds should accept and respect each other, support and complement one other, and stimulate and challenge one other. These two worlds correspond to the two main branches of human beings as a whole, and in science in particular: creation and systematization – each are needed. It is not correct to arrange competition, or to consider which is more important, because each can not exist without the other.

REFERENCES

Berdyaev, N.A. (1931). *Self investigation. The attempt of philosophical autobiography.* Paris, 1949 (Chapter VII)(in Russian).

Boltyansky, V.G. (1984). About one parquet. *Mathematics in School, 1,* 65–66.

Ernest, P. (1985). *Social Constructivism as a Philosophy of mathematics.* Albany: State University of New York Press.

Hemmi, K. (2006). *Approaching proof in a Community of Mathematical Practice.* Doctoral Dissertation, mathematical Department, Stockholm University.

Pehkonen, E. (2004). State-of-Art in Problem Solving: Focus onOpen Problems. In *ProMath Jena 2003. problem Solving in math Education* (eds. H.rehlich & B.Zimmerman), (pp. 93–111). Yidesheim:Verlag Franzbecker.)

Pehkonen, E., & Rakov, S. (2005). Comparative Survey on Pupils Beliefs of Mathematics Teaching in Finland and Ukraine. *Teaching Mathematics and Computer Science, 3*(1), 13-33.

Rakov, S.A. (2005). *Math education: competency approach with ICT support.* Kharkov, Ukraine: Fakt.

Rakov, S., & Gorokh, V. (1998). Information Technologies in Geometry (an example of generalization of one well known problem about squares). *Bulletin User Group of Derive, 31,* 25–30.

Skopets, Z.A. (1990). *Geometric miniatures.* Moscow, Russia: Prosveschenie.

Vigosky, L.S. (2004). *Psychology of human development.* Moscow, Russia: Smisl (in Russian).

KEY TERMS

Computer Algebra System: A computer algebra system (CAS) is a software program that facilitates symbolic mathematics. The core functionality of a CAS is manipulation of mathematical expressions in symbolic form. The symbolic manipulations supported typically include: simplification to the smallest possible expression, substitution of symbolic, partial and total differentiation, symbolic constrained and unconstrained global optimization,solution of linear and some non-linear equations over various domains, solution of some differential and difference equations, taking some limits, some indefinite and definite integration, including multidimensional integrals, arbitrary-precision numeric operations etc. *http://en.wikipedia.org/wiki/Computer_algebra_system*

Dynamic Geometry Systems: Dynamic Geometry Systems (DGS) or Interactive geometry software (IGS, also called "dynamic geometry environments", DGEs) are computer programs which allow one to create and then manipulate geometric constructions, primarily in plane geometry. One starts construction by putting a few points and using them to define new objects such as lines, circles or other points. After some construction is done, one can move the points one started with and see how the construction changes.*http://en.wikipedia.org/wiki/Interactive_geometry_software*

Research Approach in Education: Research approach in education is a methodology of competence paradigm of education based at the principles of discovering new knowledge through learning explorations which reflect the ways of research work. Research approach in education is the reflection of educational systems on the challenges of new type of societies – knowledge society (K-society) and sustainable society (S-society) and widely use the possibilities of information and communication technologies (ICT) in computer modeling and computer based laboratories (CBL) with appropriate software for automatic and automate experiments.

Social Constructivism: Constructivism is an epistemology, a learning or meaning-making theory, which offers an explanation of the nature of knowledge and how human beings learn. It maintains that individuals create or construct their own new understandings or knowledge through the interaction of what they already know and believe and the ideas, events, and activities with which they come in contact (Cannella & Reiff, 1994; Richardson, 1997). Knowledge is acquired through involvement with content instead of imitation or repetition (Kroll & LaBoskey, 1996). Learning activities in constructivist settings are characterized by active engagement, inquiry, problem solving, and collaboration with others. Rather than a dispenser of knowledge, the teacher is a guide, facilitator, and co-explorer who encourages learners to question, challenge, and formulate their own ideas, opinions, and conclusions. "Correct" answers and single interpretations are de-emphasized. *http://www.ericdigests.org/1999-3/theory.htm*

ENDNOTES

[1] In any Dynamic Geometry Package (Sketchpad, Cabri, DG, Gran, Cinderella, Next etc.) you can play with these fun permanent metamorphoses of the curves of this family by moving the slider corresponding to the parameter p.

[2] Really the only characteristic which differ Math from science is the method of checking the truth: in Math it is deducing from axioms by given logical rules, in Science – the real experiment.

[3] It is important to mention that the process of constructing the Expert Systems is creative, competency activity which helps to catch the correspondent Math object and its properties.

[4] The second (or maybe the first initial idea was to not store the drawing as a picture but as a sequence of the commands which constitute the construction algorithm)

[5] The specifics of the known dynamic geometry packages offer the *rubber geometry effect*: the point of a segment preserves the ratio in which it divides the segment (like a rubber thread with a knot on it).

[6] With corrections on the digital nature of computer data processing.

[7] The only specifics that differs Math from other sciences is its method of verification of true statements – logic, more precisely – the usage of deductive method.

[8] There are two well-known metaphors of math professional work: *tailor*, who produce suits by order or *atelier*, which produce

[9] suits beforehand the real order in a hope that customer will find a suit which suits him. As it is always in real practice both of these functions (metaphors) are presented in professional work of each mathematician but proportions vary. In our case we propose readers to work as an atelier salesperson and to try to "sell" the production of the atelier to some potential customers.

[9] All the problems in Math can be presented in a form $(\forall x \in X) A(x) \Rightarrow B(x)$, where the predicate $A(x)$ is a condition and predicate $B(x)$ is a conclusion of a problem. If there are some undefined elements in predicate $A(x)$ – the problem is called an open begun problem, if in $B(x)$ – open ended problem, if in both – the open problem or problem field. Solving an open problem in Math education are activities that are the closest to the real Math research work.

[10] DG is the original dynamic geometry package which was developed by the authors of the article (Rakov S. – scientific advisor, Gorokh V. – main mathematician, Osenkov K. – main programmer). http://dg.osenkov.com

DG is a certified product, is recommended for schools by Ministry of Science and Education of Ukraine, is a component of standard educational Math tools in school Math Computer Laboratory (teacher work place + interactive board + 3 student work places).

[11] It is useful to construct a macro SquareByOppositeVertices (the square can be constructed by two opposite vertices in a unique way) beforehand.

[12] The domain of the linkage is a set of points, the correspondent images of which can be constructed with this linkage.

[13] When the hero Kay of a wonderful tale by H.C.Andersen "The Snow Queen" tried to find any tool against the Snow Queen he didn't remind any prayer and only could remind the multiplication table but it couldn't help him and it was beginning of all his further problems.

Chapter XVI
Teaching Artful Expressions of Mathematical Beauty:
Virtually Creating Native American Beadwork and Rug Weaving

Jim Barta
Utah State University, USA

Ron Eglash
Rensselaer Polytechnic Institute (RPI), USA

ABSTRACT

Students who may typically view mathematics as a sterile and disjointed subject are learning new skills and concepts using a suite of virtual design tools to create artful expressions. Students being instructed in the use of these tools can artistically explore artifacts illustrating several Native American cultures as they learn mathematical concepts and apply them to replicate the art or create their own interpretations. Readers of this chapter will gain insights into cultural definitions of "art", "mathematics", and "technology" in various Native American communities, culturally effective instruction for Native American students, the role of technology in enhancing mathematical understanding of students using Native American Art (bead and loom work, wampum, and rug weaving), and illustrative applications of technology connecting art and mathematics instruction using virtual design tools.

Primitive art, embracing all of the diverse traditions of native North America, remains mysterious to the contemporary Western mind, and in its mystery, its "otherness," has lain much of its superficial allure. But the observer who looks beyond the exotic surface of so-called "primitive art" to the aesthetic complexity and the alternative philosophies underlying it sees a new potential, a vast store of untapped creative vision. Such art can have significance for our beleaguered civilization and frighteningly depersonalized world, offering hope and an expressive alternative. (Wade, 1986, p. 25)

INTRODUCTION: ART AS DEFINED BY CULTURE

There may not exist a suitable universal definition of "art" because of the difficulties of one person, community, or culture imposing their set of aesthetic standards on objects, actions, or activities created by people who share a different cultural practice and paradigm (Sturtevant, 1986). Art is defined by the culture creating it; ranging from the ornamentation of utilitarian objects, to decorative items for trade or sale, to artifacts created for sacred rituals. Art can be an expression of identity representing multiple social dimensions of a cultural community: a collective unity, an exclusive elite, or an individual rebel. An examination of any culture shows that the creation of art can often convey a sense of the people, their traditions and their values, even just by images alone.

Many scholars report that "Art" like "mathematics" is viewed as a verb rather than as a noun in many Native American cultures. According to this view, Art and math are not viewed as stand-alone subjects but rather seen as a process by which an item, artifact, skill, or technique is manifested. Haberland (Haberland, 1986, 121). explains, "The character of [indigenous] art can only be understood through its function within the life of [indigenous] peoples." In many Native American cultures even items of daily use were/ are artistically decorated or designed to describe personal and collective ownership, membership, and representation. Artful items were/are seldom duplicated as the artifact was often seen as very personal and in some cases sacred or at least spiritually representative of the relationship between seen and unseen forces and entities.

On the other hand, there are disadvantages to framing indigenous art as radically different than that of western art; particularly if that difference is characterized as highly localized or provincial. Colonialist discourse often claimed a broad cosmopolitan outlook for the colonizers and a narrow parochial view for indigenous people; this justified the colonists as parental guardians over the "children of the forest." The contrast continues in our neo-colonial context. Appiah (1992 pp. 137-8), for example, describes a 1987 New York exhibition of African art in which a panel of nine experts were asked to comment on photos of all the various pieces. The curator notes one exception: she limited the only African artist, sculptor Lela Kouakou, to commenting on art from his own ethnic group (Baoule), because "field studies have shown that African informants will criticize sculptures from other ethnic groups according to their own traditional criteria...." Appiah then juxtaposes this limiting prohibition with commentary from one of the putative unbiased experts on the panel, billionaire art collector David Rockefeller: "I own somewhat similar things and I have always liked them.... It would look good in a modern apartment or home." "I have to say I picked this because I own it. It was given to me by president Houphouet Boigny of Ivory Coast." "The best pieces are going for very high prices.... And that's a fine reason for picking the good ones rather than the bad." Clearly Rockefeller is no less biased by the "traditional criteria" of his tribe--worship of money and home décor--than Kouakou would have been by his own criteria. The idea that indigenous art "can only be understood through its function within the life of indigenous peoples," while western art speaks to a universal audience, is not as compelling as it may at first seem. Indigenous minds are not trapped in local frameworks: they too deal with universal themes of heart and mind. The German Bauhaus artists were famed for their use of Euclidean geometry; should we think any less of the fractals in African cornrows or logarithmic curves on Maori rafters?

These differing interpretations of art are shared to provoke the reader into questioning their own perspectives and relationships with art, math, and technology so they may better understand the purpose and necessity of teaching math from a different perspective; namely one that incorporates more fully the culture(s) and the communities of those they teach.

NATIVE AMERICAN STUDENTS AND CULTURALLY EFFECTIVE INSTRUCTION

School, as a vital aspect of everyday life rather than a separate reality, begins to incorporate those very activities, which are often overlooked and undervalued by emphasizing them within the instruction. Such efforts help to scaffold the instruction, not merely in the activities described, but also by including less obvious yet equally important values, traditions, and cultural nuances of those who are familiar with the activities portrayed in each community.

The role of context and culture is vital in helping define how a teacher makes their instruction purposeful and meaningful for their (Native American) learners. Yet, for many students from Native and non-Native communities alike, mathematics reflects more of a spectator-sport, one to be watched from the sidelines, rather than an experience in which one's engagement both illustrates the ways their "community" shapes and is shaped by their involvement. Barta and Brenner (2008) wrote:

Any mathematics is part of a larger knowledge system (Barnhardt & Kawagley, 2005) that determines the nature of the problems to be solved as well as the methods to be used. When such instruction is omitted as is typical in contemporary classrooms, tensions which may impede the learning of some students may arise. Students whose cultural identity is shaped through their enculturation in communities where worldviews are different and distinct from the Eurocentric view comprising the western formalized system of school must often struggle to interact to a system of education where the philosophical and peda-gogical foundations of their cultural traditions are disrespected or ignored completely.

The use of cultural references and examples in mathematics classrooms increase as teachers begin to see the value of framing the mathematics they teach in a context in which the students are familiar (Lipka, 1998). Such integrations work to validate local knowledge and mathematical applications implemented within the student's community that are typically omitted from curricular inclusion. Gerdes (1988) warns of the harmful effects that may result when educators selectively create divisions between daily life events and what is taught in school. D'Ambrosio contends that mathematical applications and 'ways of knowing' which students naturally obtain when entering schools, or what are known as "spontaneous matheracies," can be devalued, overwhelmed, and supplanted by the decontextualized and depersonalized "learned matheracies" of typical school curriculum (D'Ambrosio, 1985). When students are not allowed, or encouraged, to co-construct mathematical knowledge they run the risk of discounting who they are and what they know. Furthermore, they may grow to lose interest in learning mathematics or question their ability of being successful in using mathematical skills (Kamii & Dominick, A., 1998).

Culturally situated curriculum is much more than a superficial attempt to enliven instruction; rather it indicates the complex nature of teaching and learning itself. This instruction integrating ethnomathematical dimensions reflects a shift in the educational status-quo which typically occurs in the mathematics classroom and calls for a discussion of challenging political issues of power such as, as "whose" way of math is it, and how are "they" being represented in the mathematical instruction. While artifacts, traditions, and behaviors of a learning community are obvious components of instruction, subtle pedagogic techniques are also required which attune to and further support the natural learning styles and tendencies it students.

The history of Native American education is indicative of early attempts to civilize Native children and enculturate them into western society. Systems of education involving native youth existed long before the forced intervention by outsiders wishing to civilize and improve on native culture. Education from a traditional perspective meant knowing oneself and one's place in the world. With this knowledge came the obligation of acting responsibly for the sake of the tribe, family, and individual. What was learned was done within a context provided by interaction in the environment respectful of the multiple dimensions of being – spiritual, mental, social, physical (Cajete, 1999).

Early contemporary western educational practices attempted to remove the student from their traditional practices. Rather than learn about one's people and tribal ways of life from an elder, Indian children were forced to learn about the ways of others in an alien context and environment from teachers knowing little about the culture or its influence on the teaching and learning of children. Reyhner writes, "Had the goal of coercive assimilation been reached, there would be no recognizable Indian people today" (pg. 33)

Sadly, today much of the same ideology and racism and that were used to justify the educational enculturation process still exists in our schools attended by Native American students. The trend to separate the child from his or her culture for what passes as "public education" continues; yet this trend must be reversed if students are to realize their full potential. Culturally situation instruction designed to integrate art, mathematics, and technology can be one approach.

CULTURALLY SITUATED USES OF TECHNOLOGY

Mathematics and technology have been present in Native American cultures since long before Euro-

pean contact. We will describe the technological aspects of these traditional arts—beadwork, for example, as the pixel-based display of its era—and show how some of the same mathematical challenges that are encountered by contemporary engineers were engaged by native artisans (e.g. the "staircasing" problem). Finally we show how the use of computer technology (the web-based suite of Culturally Situated Design Tools) can simulate these traditional technologies, allowing students to translate from the math embedded in these artistic practices to the mathematics they need to learn in the classroom.

Students are learning complex mathematical concepts incorporating culturally inclusive art and activities via computer technology. Teachers and their students are learning important math skills while exploring Native American beadwork, wampum, and Navajo rug weaving using the Culturally Situated Design Tools. The Culturally Situated Design Tools project (see http://www.rpi.edu/~eglash/csdt.html), funded by the NSF, Dept of Ed, and HUD, has developed computer simulations of indigenous arts. These simulations enable students to learn mathematics through experimentation with these knowledge systems. In our initial evaluation the use of these design tools showed statistically significant improvement of minority student math achievement (Eglash et al 2006). At the same time, they offer new ways to explore cultural connections to mathematics.

One of our first design tools was the Virtual Bead Loom (VBL). The opening webpage for the VBL allows users to select from four categories: Cultural Background, Tutorial, Software, and Teaching Materials. The cultural background section opens with several examples of four-fold symmetry in Native American design. Before reading the text, teachers can ask students to look at the designs and describe them; such discussions offer opportunities to introduce symmetry as a term and concept. The text describes how four-fold symmetry is a deep design theme in many Native American cultures, and is evident not only in a wide variety of native arts, but also indigenous knowledge systems such as base four counting, four-quadrant architecture, the "four directions" healing practice, etc. A second web page shows how such structures are analogous to the Cartesian coordinate system. Finally, the webpage introduces the Native American bead loom as another example in which we find an analogue to the Cartesian grid.

The tutorial begins by showing how the VBL simulates a traditional Native American bead loom, and reviews the various tools. We typically skip the tutorial, since the applet interface is fairly intuitive. Interacting with the interface may provide the best learning opportunity for aspects

that are not gained merely intuitively. The VBL applet features a Cartesian workspace in which designs are created, and tools, which appear to the right of the workspace (figure 1). The simplest tool, for creating or deleting single beads, is at top. Students enter the X, Y coordinate pairs, select a bead color, and press the "create" button.

Other tools include lines, rectangles, triangles and iterative patterns. The tools can be used in combinations to simulate traditional beadwork patterns, to allow students to create their own designs, and to engage in a variety of specific standards-based math learning exercises. The VBL was created in collaboration with teachers and students at schools serving the Shoshone-

Figure 1. The virtual beadloom

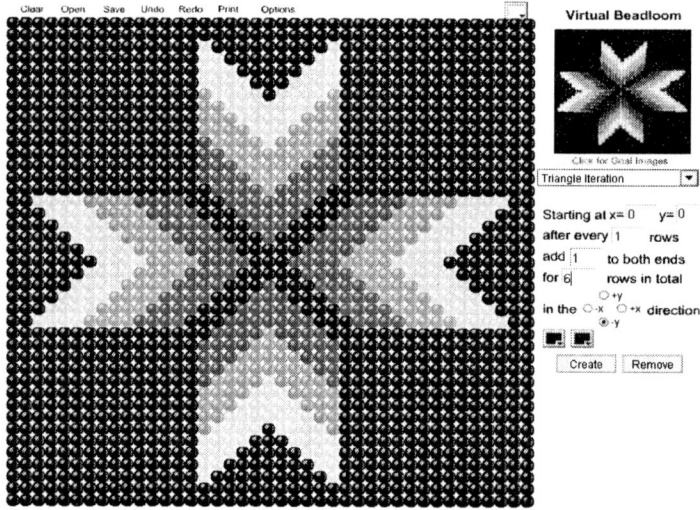

Figure 2. Virtual and physical beadwork

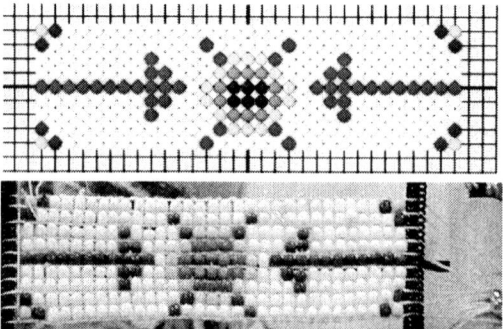

Bannock reservation in Idaho, the Northern Ute reservation in Utah, and the Onondaga Nation reservation in New York, and has been an enthusiastic success. It has also been used with a wide variety of other ethnic groups, including white students and teachers, with great enthusiasm. Teachers have found a wide variety of concepts and skills, primarily from analytic geometry, that can be taught using the VBL. Another exciting aspect is the ability of students to develop physical beadwork based on their virtual designs (figure 2), often by linking an art class with the math class. Additional samples of student work, evaluation tools, teachers' lesson plans and other materials appear in the "Teaching Materials" section of the webpage.

A wide variety of mathematics learning strategies can be approached through the Virtual Bead Loom. However some of the most interesting results have been cultural. In one exercise involving a mixed age group from the Shoshone Bannock reservation school, we found that older children tended to produce designs that more

closely resembled traditional beadwork patterns, and younger children created more "playful" designs. This was consistent across two design tools, the VBL and a basket simulation.

The age difference may have been due to the context of a science summer camp, in which the older students were constantly asked to set an example for the younger, but it also matches research that suggests that minority ethic/racial self-identity makes a dramatic (i.e. non-linear) shift in this age range (cf Forbes and Ashton 1998). The names that students used to title their work also revealed playful creativity, including irony or parody, even for the older students who were more serious about the concepts of Indigenous knowledge. The design in figure 3, for example, was titled "hands old" because it was created in the early morning at a camp at 9,000 ft in Idaho's Sawtooth range.

Thus the meanings of the designs are not predetermined by the software; it is up to the user to "appropriate" the technology (Eglash et al 2004). In use with non-native students, for example, we also see the expression of local cultural meanings. Figure 4, for example, is a design created by an

Figure 3. Design by student from Shoshone-Bannock reservation school, age 17. "Hands Cold."

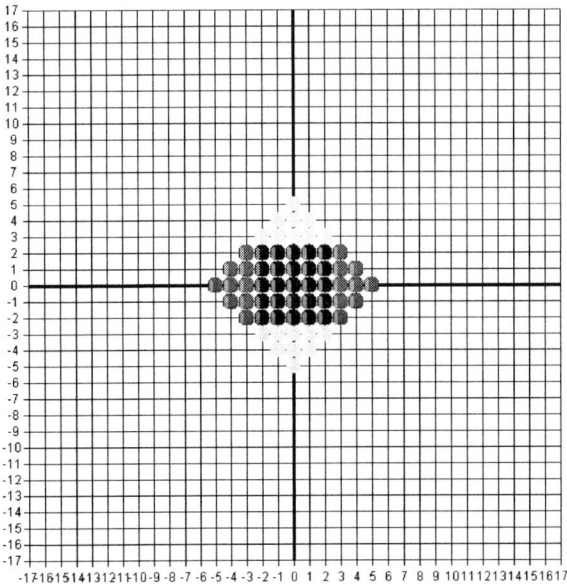

Figure 4. Design by an African American student in upstate New York, showing the 9/11 tragedy

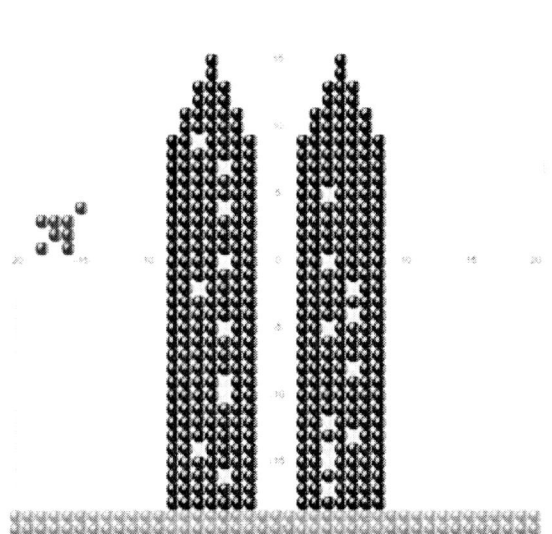

Figure 5. Design of a Puerto Rican flag (minus its star) on an older version of the VBL

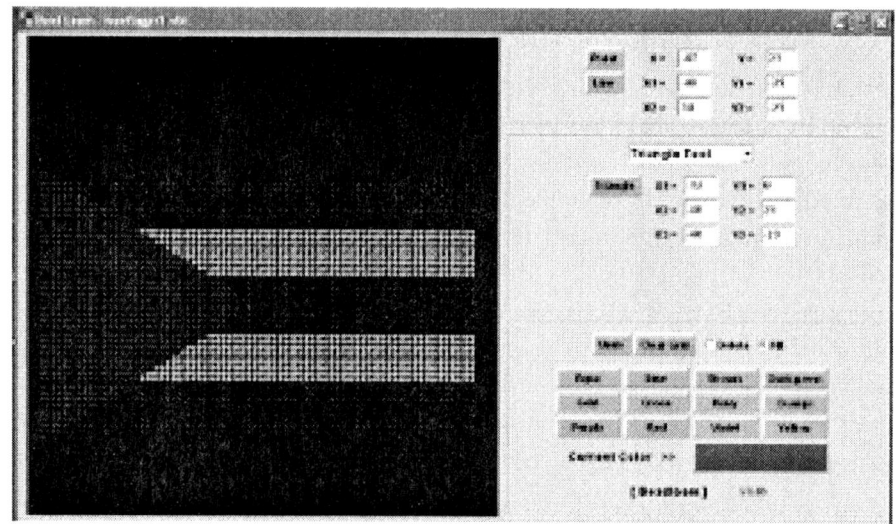

African American student in upstate New York, showing the 9/11 tragedy.

Such appropriations can have a significant mathematical learning component. For example, one student of Puerto Rican heritage (also in New York) decided to create a beadwork image of the Puerto Rican flag, which includes an equilateral triangle. At first he tried to create an equilateral triangle by having the same number of beads on each side, but that did not work because the beads along the diagonal are spaced farther apart than the beads along the vertical or horizontal. He finally arrived at a solution by using the ratios of a 30-60-90 triangle to arrive at a discrete approximation (figure 5); a challenge that he might have balked at had it simply been assigned to him.

THE NAVAJO RUG WEAVER

Recently our team has used the term "ethno-computing" to describe the attributes of these knowledge systems that better fit analogues to computing than to mathematics. The use of iterative patterns in the VBL, for example, bears a direct relationship to iteration in computer programming (Eglash et al 2006). Another such opportunity to explore ethnocomputing arose when we received an invitation from the math coordinator at Rough Rock School in the Navajo Nation to develop a similar design tool allowing their students to simulate Navajo rugs. At first we thought the design would be very simple: just replace beads with a small image representing a single "weave" (that is, a single thread of weft over a strand of warp).

Upon examining the rugs we realized that would not be possible: the weft alternates up and down (see blue highlights in figure 6), because every other strand goes in back of the warp, and in the row above it the alternation is the opposite,

Figure 6. Design of a close up of a Navajo rug, showing vertical alternation of the weft

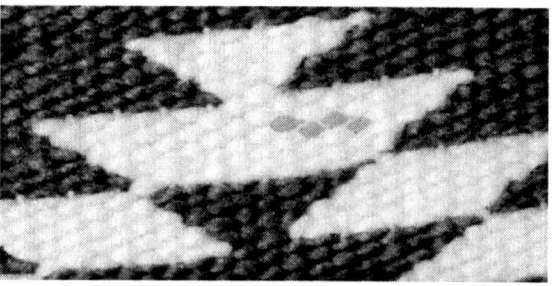

with every other strand going in front of the warp. Since an important part of the weaving process compresses the rows together (the weaver pushes down with "comb"), the gaps (where the weft goes behind) from each row above and below are filled, thus producing an up and down alternation. Thus we could not simply map each individual weave to integer intersections on a Cartesian grid, as we did with the bead loom. One alternative was to simply map the weave into non-integer spaces between the grid intersections, but that would destroy much of its utility for the math teacher. Our solution was to simulate using two steps: first the user maps out a weave pattern using only the grid intersections (Figure 7a), and then the user presses a comb icon, and the applet fills in the gaps (Figure 7b), thus creating a completed weave simulation. Not only did this satisfy both the math teacher and our own interest in a good visual simulation, but also it also better represented the actual process of weaving.

Like the bead loom, the computational aspects of weaving pre-date Europeans. In devising a particular slope, for example, both weavers and bead workers alike will use a particular algorithm (over one up two for a steep slope, over one up one for a more shallow slope, etc.). However this seemed to be more intense in the case of rug work, since weavers needed to keep track of many different colors of thread simultaneously and thus developed "counting patterns" for particularly complex weaves. One weaver interviewed said that her mother had developed a weave for hexagonal patterns but refused to tell her without payment; she then foiled her mother's attempted sale of this algorithm by reverse-engineering (deriving the counting pattern by examining the original weave). This was recounted in a humorous way, but it hints at an indigenous information economy not unlike that associated with contemporary computing. Presumably the more simple patterns are "open source."

CONCLUSION

The natural interplay between art, mathematics, and technology (seemingly novel in schools) was in fact a common and important cultural component in many traditional Native American communities. Perhaps, school today may be one of the few places where such sharp distinctions between subjects and bodies of knowledge are as emphatically maintained. The traditional curriculum and its related pedagogy found in schools divide learning and subjects into non-related compartments and make relational or integrated thinking or activities harder to integrate. The ideas and activities shared within this chapter can provide teachers other options for incorporating art, mathematics, and technology in a blend of purposeful use. When such teaching and learning occur, what one studies and the reason for knowing it becomes apparent. Students can acquire deeper knowledge and understanding as they create artful expressions of mathematical beauty through their use of the virtual beadwork and rug weaving tools.

Figures 7a. Design of a close up of a Navajo rug, showing vertical alternation of the weft

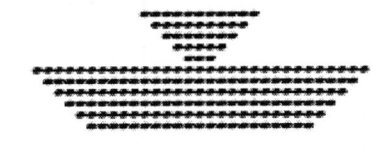

Figures 7b. Next the user presses a comb icon, and the applet fills in the gaps, creating a complete weave simulation

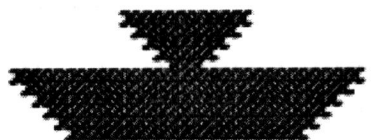

ACKNOWLEDGMENT

This material is based on work supported by the NSF under Grant No. 0634329.

REFERENCES

Appiah, A. (1992). *In my Father's House.* NY: Oxford.

Barnhardt, R., & Kawagley, O. (2005). Indigenous knowledge systems and Alaska native ways of knowing. *Anthropology and Education Quarterly, 36*(1), 8-23.

Barta, J., & Eglash, R. (2008). Seeing With Many Eyes: Connections Between Anthropology and Mathematics. In B. Greer & S. Mukhopadhyay (Eds.), *Ethnomathematics.* Oxford, UK: Routledge Publishing.

Cajete, G. (1999). *Igniting the sparkle: An indigenous science education model.* Skyland, NC: Kivaki Press.

D'Ambrosio, U. (1985). Ethnomathematics and its place in the history and pedagogy of Mathematics. *For the Learning of Mathematics, 5,* 44-48.

Eglash, R., Croissant, J., Di Chiro, G., & Fouché, R. (Ed.) (2004). *Appropriating technology: Vernacular science and social power.* Minneapolis, MN: University of Minnesota Press.

Eglash, R., Bennett, A., O'Donnell, C., Jennings, S., & Cintorino, M. (2006). Culturally situated design tools: Ethnocomputing from field site to classroom. *American Anthropologist, 108*(2), 347-362.

Forbes, S., & Ashton, P. (1998). The identity status of African Americans in middle adolescence: a reexamination of Watson and Protinsky - 1991 - response to M.F. Watson and H. Protinsky. *Adolescence, 26,* 963.

Gerdes, P. (1988). On culture, geometrical thinking and mathematics education. *Educational Studies in Mathematics, 19,* 137-162.

Haberland, W. (1986). Aesthetics in native American art. In E. Wade (Ed.), *The arts of the North American Indian: Native traditions in evolution.* New York, NY: Hudson Hills Press Inc.

Kamii, C. & Dominick, A. (1998). The harmful effects of algorithms in grades 1-4. In L. J. Morrow & M. J. Kenney (Eds.), *The Teaching and Learning of Algorithms in School Mathematics. 1998 Yearbook.* Reston, VA: National Council of Teachers of Mathematics.

Lipka, J. (1998). Expanding curricular and pedagogical possibilities: Yup'ik-Based mathematics, science, and literacy. In J. Lipka, G. V. Mohatt and the Ciulistet Group (Eds.), *Transforming the culture of schools: Yup'ik Eskimo examples* (pp. 139-181). Mahwah, NJ: Erlbaum.

Rehyner, J. (1992) *Teaching American Indian students.* Norman, OK: University of Oklahoma Press.

Sturtevant, W. (1986). The meaning of native American art. In E. Wade (Ed.), *The arts of the North American Indian: Native traditions in evolution.* New York, NY: Hudson Hills Press Inc.

Wade, E. (1986). The arts of the North American Indian: Native traditions in evolution. In E. Wade (Ed.), *The arts of the North American Indian: Native traditions in evolution.* New York, NY: Hudson Hills Press Inc.

KEY TERMS

Culture: Learned and shared behavior of a society; its ways of life and ways of looking at the world.

Culturally Responsive/Integrated/Situated Curriculum: Curriculum created to include (cultural) connections and context(s) reflective of the community of learners it is designed to serve.

Culturally Effective Instruction: Instruction incorporating ways of knowing and learning of the community of the learners it is designed to serve.

Ethnocomputing: The study of the cultural aspects of computing, and the computational aspects of culture.

Indigenous: Synonymous to terms such as American Indian, Native American, First Nations, Native Hawaiian, or Alaskan Native; the first people who originated in a particular geographic region and are considered for their historical connection to it.

Indigenous Information Economy: Indigenous information economy - An system of trade or sale in traditional knowledge itself: for instance the value of algorithms (functional applications) for weaving or beading that were traded or sold in these communities. Typically, information economy is considered as something that only arose with the advent of computers.

Reverse-Engineering: Creating an algorithm or understanding a process by examining the outcome or artifact that it produced.

Chapter XVII
Visual Analytics and Conceptual Blending Theory

Mia Kalish
Diné College, USA

ABSTRACT

One visualization in Diné philosophy is four small dots arranged in a circular sequence at 90°, 0°, 270°, and 180°. Each position is associated with a time of day, a season, a color, a type of stone, a time in the lifecycle, and a process of living and learning. I use Conceptual Blending Theory to explore this complex information space of small spatial stories that combine to form an "information system of information systems." This approach to visual analytics uses reduction to human scale, which easily adapts itself to automated analysis and data configuration. This process reveals a previously unseen world and contributes new ideas to understanding both the creation of new visualizations and the decomposition of existing visualizations. This verifiable methodology can validate the steps in the decomposition process itself and also be used to predict the content of missing data.

INTRODUCTION

Contemporary approaches to information visualization focus on finding ways to present new information using contemporary information technology. These representations range from text-based visualizations that present thematic information in an easy-to-access format, such as management dashboards, to visualizations of more complex systems of data, such as weather patterns. Comparatively, orthographies are visual representations of language. Hebrew is considered to have a deep orthography because it is written without vowels while Spanish is a shallow orthography because everything one needs to know is visible on the surface. Using this metric, text based visualizations may be considered "shallow orthographies" while "deep orthographic visualizations" are graphical visualizations where a great deal of knowledge is necessary to grasp the full import of the information. Examples of these include scientific mappings for wind, weather, hurricane isobars, and the implications of environmental measurements on marine and animal life. In general, the contemporary approach to visualization has been focused on the development of these visualizations, both shallow and deep, to represent information in a forward direction, that is, to develop representations of complex and developing knowledge that is important today, and upon which new knowledge and new understandings will be developed tomorrow. Concomitant to the process of visualization development has been the emergence of what is known as "visual analytics" that:

focuses on human interaction with visualization systems as part of a larger process of data analysis. Visual analytics has been defined as "the science of analytical reasoning supported by the interactive visual interface" (IEEE Visualization, 2006). Its focus is on human information discourse (interaction) within massive, dynamically changing information spaces. Visual analytics research concentrates on support for perceptual and cognitive operations that enable users to detect the expected and discover the unexpected in complex information spaces. (Wikipedia, 2007)

In this chapter, I explore the complex information space of an ancient visualization, one that is easily represented mathematically as both a fractal and as a set of sets (Kalish, 2007). However, these representations, despite their mathematical elegance, do not seek to examine how the Diné (Navajo) philosophical and cultural knowledge emerged as a complex information space. Answering this compelling question, which clearly involves creating a visualization of the complex of "perceptual and cognitive operations" employed not to "detect the expected and discover the unexpected" but to create the information space itself, is a reach back into the unknown, using modern analysis techniques and technology to reveal structure and process. Unraveling creation of an existing and heretofore little unexplored information space can lend new understanding to the processes through which data becomes

information, and information becomes condensed into small, easily referenced and equally easily comprehended visualizations. In its simplest form, the visualization of the complex Diné philosophy is four small dots, arranged in a circular sequence at 90°, 0°, 270°, and 180°. Each of these positions is associated with a time of day, a season, a color, a type of stone, a time in the lifecycle, and a process of living and learning. The question is not so much *How did they get this way?* but *What tools can we use to explain the development of this complex, mathematical information space?*

I use Fauconnier & Turner's (2002) Conceptual Blending Theory (CBT) to explore with the reader the complex information space of Diné philosophy. CBT was developed to create a structured, semantic analytic for understanding the process of metaphor creation. But where in classic CBT, representational blend structures are generally individual linear objects that enter linear, single and multi scope integrations, in Diné philosophy, small blend structures that are actually small spatial stories combine to form larger blend structures that are both fractal in nature, and are at once reduced to small, comprehensible visualizations that are simultaneously rich information systems in themselves, thus forming an "information system of information systems." This composite form implies that a non-linear blend process must emerge, and that there must be an underlying cognitive mandate that governs the process itself. This mandate is *transformation* in Diné terms, and *reduction to human scale* in CBT terminology. While all scientific visualizations seek to reduce data to human scale, to create for humans apprehensible models of the complex data spaces that are becoming more common as technology and knowledge grow exponentially, in many of these approaches human scale is a parameter rather than a process.

In this chapter, I share with the reader an approach to visual analytics that uses reduction to human scale as part of the visualization process, and which also can easily adapt itself to automated analysis and data configuration. This process, itself a visualization, reveals a world previously seen by only a few, and one which, in its very simplicity and elegance, will contribute new ideas to understanding both the creative process for new visualizations and the decomposing process for existing visualizations. I will also show how using a verifiable methodology can validate the steps in the decomposition process itself and also be used to predict the content of missing data.

USING STORY TO FRAME THE ANALYSIS

Knowledge, as it is used and reused, becomes progressively more abstracted and more automated. Details become more assumed and consequently more invisible. Examples of this are ubiquitous and range from the ease with which we speak and drive cars to the understandings and interpretations of surds and integral signs. Ancient visualizations function in the same way as mathematical symbols, standing for a complex of meaning, understanding, process and evaluation. Disparities in language and expectation (Powell, 1880) contributed to the enormous loss of ancient knowledge in many cultures across the globe, and have left us in the position we find ourselves today, which is that of trying to understand ancient knowledge through the reacculturalization of its symbols and forms, of its visualizations that are the representations of its own human cognitive and cultural knowledge.

Understanding visualizations requires an understanding of what can be called "the organizing story." Story itself is an organizing form, one that in the integration of structure and the audience's expectations guides both teller and listener not only towards the climax, but also through the twists and turns, the small paths leading to the end, or in this case, the next beginning. The beauty of story is not that it is technologically complex and sophisticated in its construction but that it is accessible across populations. People love stories. Each of us, as we write or speak, tells a story of our knowledge and our own understandings. Each mathematical symbol also tells its own story: the integral sign not only tells the story of continuous summation of small slices of area under the curve defined by the derivative, but also of the fits and starts in the evolution of calculus.

Every visualization must have at least one organizing story, for while the form provides structure and organization, it does not supply meaning. The organizing story is not "the story of . . ." nor is it one of the many possible versions that can or have been constructed using the many possible cultural and dramatic possibilities available in the story-telling context. Instead, the organizing story is akin to the integral sign, where the form communicates expectations of procedure and content, but the details of limits and function are supplied in each specific application. Similarly, meaning derives from the story that uses the form, and from the integrated cultural understandings and expectations. The Diné visualization has one organizing story, following the rays of the sun as the Sun-God carries it across the sky, but many, small, spatial stories simultaneously enrich the organizing story and establish the cultural foundation. The organizing story is 3-dimensional, and both elaboration and the creation of higher-order relations are topologically analogical to the 3-D form. Being able to render the visualization accurately is dependent on hardware sophistication. The human mind easily visualizes the sky or the globe of the earth, but computer renderings are dependent on graphics and software capabilities, and the visualization designer's ability to conceptualize interactions in 3-D.

The construction of Story is an active and aware process in which the author selects forms and makes choices that create the meaning: "In constructing stories . . . authors attempt to convey their intentions by selecting incidents and details, arranging time and sequence, and employing a variety of codes and conventions that exist in the culture" (Carter, 1993, p. 6.).

The process is recursive because Telling the story and Evaluating its results are part of the Construction process. Since Storytellers refine their telling based on audience response and an assessment of how well they met their goals (Intentions), we can add a fourth Process. and then when Story can be more formally defined, it might look like this, where *A_* identifies an Action.

A_Construction:
    A_Design→ (A_Select→ Purpose; Form; Events; Details)
    A_Arrange → A_OrderByTime (Events; Details)
    A_Tell → A_Express (Construction → A_Employ(Codes; Conventions; Modes(Oral; Text; Gestures)))
    A_Evaluate → Assess(Response) → Plan(Refinements)

Here Story is represented as four major actions, guided by two constraints. The constraints are less obvious than the actions, so stating them explicitly, they are C_Storyteller's_Intentions and C_Prior_Knowledge_of_Storyteller's_Audience. Both of these constraints guide all four Story processes. The Storyteller's_Intentions guide the Design of the story in the choice of the Events and Details, but audience constrains the Sets from which those choices can be made.

Story as a Global Medium

[The] complexity of commonplace reasoning [was] discovered when researchers in artificial intelligence unexpectedly encountered extreme difficulty in their attempts to model it explicitly. This extraordinary complexity previously associated only with highly expert thought, turns out to cut across thinking at all levels and all ages. (Fauconnier & Turner 2002, pp. 17-18)

The value of story for scholars is its "extraordinary complexity," which has spurred a new theoretical interest and caused the concept of story to move out of the domains of myth, folklore, and entertainment, and to emerge as its own knowledge domain. Story is dynamic; it is in many ways dependent on cultural understandings, but it structures this variability into forms that comfort the audience into knowing that they can understand. The story may leave them gasping for breath, but it will be because they understand the story, not because they are lost in the translation. Story is also the mechanism by which conceptually large aggregations of knowledge across time, space and complexity are made apprehensible, that is, reduced to human scale.

Story—the sophisticated human mental ability to conceive of orchestrated suites of events, objects, agents, and actions—is much older than any sophisticated mathematical accomplishment. Our advanced abilities for mathematics are based in part on our prior cognitive ability for story. There are basic human cognitive operations that make it possible for us to invent mathematical concepts and systems. One of those operations is the fundamental human operation of story—of understanding the world and our agency in it through certain kinds of human-scale conceptual organizations involving agents and actions in space." (Turner, 2005, p. 4)

Stories of how a people understand the world incorporate the events and details of the cultural and disciplinary ontologies, that is, what exists in the world as we see it. These stories also incorporate the processes of cultural and disciplinary epistemologies, or the relationships between Turner's "agents and actions in space" and time. Cultural epistemologies are ways of knowing about the cultural ontologies, and stories, whether written, oral, pictorial, kinesthetic, or computer-based representations in song and movement, are the ways in which the knowledge is represented, shared and preserved. In many cultures, songs, stories, and oral histories have been passed unchanging for millennia, the collected knowledge of the people, not only of their world and their agency in it, but also of their identity and the ways they give meaning to their lives.

These stories, songs and oral histories are often small spatial stories used as inputs to create larger and more comprehensive stories, or blends, because people are especially well-equipped to understand the world by creating mental blends based on small spatial stories. These small spatial stories are then constructed into conceptual integration networks in which an increasing number of the inputs are small spatial stories. Western conceptualization often separates objects from events, and the small spatial stories created are those in which these objects are actors that perform actions. This separation is not typical of all languages. In software, for example, it is impossible to separate agent from action and still have an executable instruction.

The Origin of Horses: a Diné integration of several small spatial stories, was recorded in Crystal, NM in 1929, and translated by Edward Sapir:

From there, [Turquoise Boy] again started off, they say. As he was still walking along, (something) white fell blanket-like to the ground to him, they say. When he looked upward, here (many) horses stood in a line, they say. The Sun stood first in line with a turquoise horse, they say. After him, the Moon stood with a white shell horse, they say. After him, (one of) the Mirage folk stood with a mirage horse, they say. After him, (one of) the Mist folk stood with a mist horse, they say. (1942/1975, p. 117)

In this story, Turquoise Boy is human and is at the center of the story, as humans are in the Navajo conceptualization of the Universe. The four small, spatial stories are identified by the both each Supernatural and each Supernatural's horse. For traditional Navajo people, the celestial bodies, here the Sun and Moon standing with their horses and the Mirage and Mist folk, "are really a class of living beings sharing genealogical, emotional and physical proximity with human beings" (Griffin-Pierce, 1992, p. 112).

The organizing story is analogous to the theoretical framework that establishes the concepts, constraints and corollaries to the research question that are implicit because of the framework chosen, but absent from process is the goal of reducing size and complexity to human scale. In contrast, reduction to human scale is essential in the concepts of story and CBT, for it is how vast knowledge becomes not only apprehensible by humans, but also how it is easily and reliably perpetuated. Understanding how this works in the current context requires an understanding of the Diné organizing story, and also of the process by which knowledge complexes are reduced to human scale, for explicating the abstracted Diné model will require a reversal of the reduction process. An explanation of the Diné organizing story will be presented first, followed by discussions of reduction to human scale, conceptual integration, also called blending, and the generalized CBT process.

The Diné Organizing Story

The Diné organizing story, the basic cultural metaphor, is that all life comes from the sun, and so all living things will follow the sun's path. Before they knew horses, the Diné story told how each day, the Sun-God would walk across the heavens, carrying the sun on his back. In the evening, he would hang the sun on a peg, so that it could cool off, and spend time with his family, resting and preparing for his return to the east. When he was ready, he took the sun from its peg, hid it so its light would not shine, and returned to the east in the darkness (Welker, 1996).

This organizing story is an integration network that has a several small, spatial stories, already reduced to human scale, (1) of the Sun-God traveling across the sky, lugging the sun; (2) spending the evening with his family; and, (3) returning to the east without being seen. The sky is a topology, divided into dawn, evening, and darkness, each of which relates to a physical position. Cultural stories that use this structure are common to many cultures. Story forms based on topologies and directions use what is known formally as the method of loci, the setting "up a simple trajectory of attention across a set of features. . . of the environment" (Edwin Hutchins, quoted in Fauconnier & Turner, 2002, p. 72). The method of loci creates a visual mnemonic in the connection between stories and physical locations. Developing good metaphoric constructions is not easy, but once the culture has derived and polished them, they are easily learned.

In Navajo, ééhózin, loosely translated as "knowledge," is the awareness of both the thing and its symbolic representation. The Navajo aste'hastqin (First Man) "[u]nlike Adam, . . . did not go about naming things (creating symbols); he went about learning things (interpreting reality through already established symbols)" (Witherspoon, 1977, p. 43). In Navajo, form has existed from the earliest time:

The first underworld had substance but it possessed no inherent form. Because the capacity to originate and impose form is inherent in the intellect, First Man and First Woman imposed form onto the substance of the first world. Things do not think; symbols do not think – man thinks.
(Witherspoon, 1977, p. 44)

Traditional Diné thinking is holistically dynamic; it is based on Diné cosmology, and sees the world as a constant flux of holistic relationships with humans integral to the dynamics of the Universe in its entirety. Each individual occupies the center of his or her world, and is connected through feet to the Earth and through the top of the head to the sky. These are three small spatial stories: the relationships of Earth and Sky; human to Earth; and, human to Sky. These stories

are then blended create a fourth that shows the connectedness of Earth, Sky and People. The integration of small spatial stories is easily seen in terms of Diné relationships in this description of the relational process of the ceremonial song, *Nohosdzaan Biyin*:

The song then goes to the darkness which is part of the nightly cycle and to the dawn which provides purity and revitalization of all natural organic processes. From the dawn the song goes to the Yellow Evening Twilight and on to the Sun with an acknowledgement of the life that it brings. From there the song goes on to name four pairs of constellations of the seasonal skies. . . . The song then comes back to acknowledge the growth of all vegetation, fertilization and cross-pollination, expressed through white corn, yellow corn, pollen and the peak of the ripening process, anilt'ánii. Traditional people say that we are the manifestation of the totality of all things mentioned in the song and our relationship to all things is expressed through our centeredness to all things. It is through this manifestation and relationship that beauty and harmony exist all around us.
(Begay & Maryboy, 1998, p. 280)

The Navajo view of knowledge, particularly of traditional or ritual knowledge, is that ritual knowledge cannot be discovered or created, although it can be purchased. It can be shared but not destroyed. Knowledge is manifested externally through the process of creation, and this process is endless: "When asked what they were planning in the sweathouse, the [Supernaturals] said, 'We are planning to extend knowledge endlessly'"(Witherspoon, 1977, p. 33).

The fractal concept is a conceptual integration of the two small spatial stories, one in which "each part enfolds or encapsulates the whole" and the other in which knowledge extends endlessly. This integration network is precisely the visual definition of fractals when viewed as a recursive blending of elements. The Diné patterns are similar in that each created instance is itself a complete representation of the whole. Each instance is conceptually robust and maintains all connections. Each instance is itself an information system, and as integrated blends, form an information system of information systems.

The Diné Visualization

The Diné visualization is cosmological. Unlike water, which is in constant motion, and sand, which can be blown away, the stars are immutable, visible forever, and were used by First Woman to embody the laws that would govern mankind. Associated with the Sun, the Moon and the constellations are supernatural beings, the Diyin Diné, who have left the earth and are invisible. The sunwise circuit is considered proper in most ceremonies (Griffin-Pierce, 1992) and gives rise to a network of small spatial stories in which Sun-God is the agent. Sun-God leaves his house in the East at dawn, hovers overhead at midday, returns to his house in the west at twilight, and ultimately disappears into the darkness. In another network of small spatial stories, the spiral motion of the "eight main stellar constellations [is] geometrically related to the geophysical environment of the Diné Four Sacred Mountains" (Begay & Maryboy, 1998, p. 20) and conceptually integrated, these stories form a small integration network that defines the boundaries of Navajo traditional lands.

The CBT forms show the cultural blending of knowledge in reverse; geometric semantics define elaborated relationships; and, completion and elaboration processes work to include the cultural stories represented by the visualization. A version of the Diné visualization, with blended Direction, Times of Day, and Season, is shown in the Quadripartite Seasonal Model in Figure 1. This figure also illustrates the visual fractal: a small dot expands to a circle; four more dots appear, each at a directional point; these dots expand to circles, and the process continues. Cultural completion supplies the categories and

Figure 1. Quadripartite seasonal model

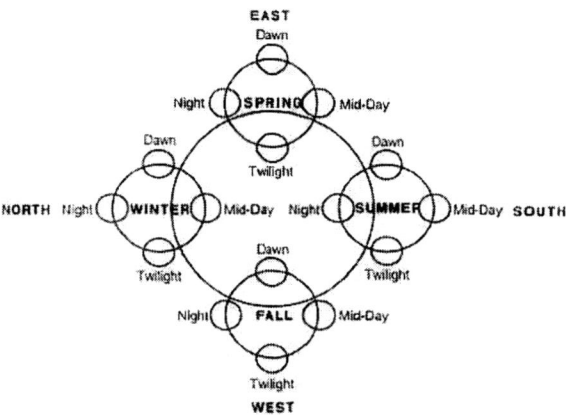

the labels, while elaboration in the blend creates the relationships.

Attributes and values are associated with what will become "quadrants;" those things that are inherent are frequently assumed, and consequently are invisible or unstated. These attributes and values will have particular relationships or cultural meanings, different from the physical agency of the mountains and humans. Besides color, the attributes include Seasons; Times of Day, and, Processes in the Human Lifecycle and in the Diné Cycle of Knowing.

The cognitive and kinesthetic processes in the Diné cycle of knowing are sophisticated and integrate with experienced or learned past and expected future. What is not clear from this description, but is characteristic of the process model, is that past, present, and future are present simultaneously, and one can move among them at will. Together, these processes, their relationships, and their dynamic interactions are recursive, fractal, and in constant flux.

Nitsáhákees: A mental image provides an awareness of need and the cognitive process of thinking begins. . . . Relationships: east, the color white, the process of dawn, the season of spring, and the process of birth and age of childhood.

Nahatää: A planning process through which experience and data furnish information that is integrated with stabilized knowledge from the previous process of nitsáhákees in order to act for a life-protecting and life-enhancing purpose. Implicit in the planning process is constant reciprocal communication within an integrated whole. …Relationships:: south, the color turquoise-blue, the process of midday, the season of summer, the age of adolescence.

Iiná: A process that is characterized by vitality, dynamism, living, in a regenerative and cyclical manner. This is a process of application to life and implementation, and is thus operational. … Relationships:: west, the color yellow, the process of evening twilight, the season of autumn, the age of maturity.

Siih Hasin: This is a transformational process of developing a strong sense of confidence and assurance, derived from having firmly established proper thinking, planning and living. It is a level of personal growth accompanied by strong feelings of self-identity, self-esteem and satisfaction of accomplishment through the stabilization of one's existence. . . . Relationships: north, the season of winter, the process of old age. (Begay & Maryboy, 1998, p. 338-339)

These processes are analogous to Story processes that use the storyteller's intentions, are constrained by cultural knowledge and references, and are recursive.

Nitsáhákees (A_Design) → **Nahatää** (A_Arrange) → **Iiná** (A_Tell) →

Siih Hasin (A_Evaluate)

The Diné process is recursive, so we can express this as:

Siih Hasin (0) → Nitsáhákees (1) → Nahat'á (1) → Iiná (1) →
Siih Hasin (1) → Nitsáhákees (2)

This representation is misleading, giving the impression that action proceeds linearly from lowest to highest cycle. and in sequence from process to processes. In reality, this is not so. In a single phrase in Diné Bizaad, one can be in "the past, the present and the future, all at the same time" (Sherwin Bitsui, reading at Diné College, January 24, 2007). The structures of what form the emergent structure in the blend are what make it possible both to be in all times simultaneously, and to move freely along, in and out of different domains. The process can be visualized as being hosted in a massively parallel architecture where the last line could more accurately be written as:
Iiná (x) → Siih Hasin (x) → [Cycle (here $\pm n$)].

Creating visualizations from American Indigenous cultural understandings raises critical issues about how cultural representations are created in the first place. Some artistic conventions, like those of both the Diné and the Inuit, are disanalogous to Western conventions:

Many of our own pictorial conventions began in Europe in the fifteenth century. They were influenced by, and reinforced, particular mathematical ideas. The study of optics, going back to Euclid in 300 B.C.E., depended on rays traveling in straight lines between the eye and objects being viewed. The fifteenth-century concern was where those lines would intersect a perpendicular plane placed between what was being viewed and the viewer. Time became frozen, and hence the issue of time and the issue of motion were eliminated. Not only was time frozen in what was being viewed, but the position of the viewer was also frozen. The entire picture could only show what a viewer could see from a single fixed place in a single fixed instant; one could not show simultaneously what could be seen from above, below, behind, and inside as well as outside. Related to the fixed vantage point in front of the picture plane are the horizon line and "vanishing points" of a picture. These are where sets of parallel lines would appear to us to meet. Based on them are the sizes of objects that are to be seen as being at different distances from us. To show recession, the objects diminish proportionally. Westerners, whether they be creators or viewers of pictures, are taught these conventions. Others are used by other peoples. Clearly the Inuit do not share our conventions. For them, time and space remain unified and the contents of the picture is not confined to what can be seen from a single fixed position in space-time. An event through time can be depicted by showing its interrelated parts, each in its most significant aspect, despite the fact that they could not necessarily be seen at the same time nor from a single place. (Ascher, 1991, p. 136)

Visualization research must recognize that the Western Time/Motion metaphors are not universal: the Inuit, Diné and Aymara (Núñez & Sweetser, 2006) have different Time, Space and Motion metaphors that represent themselves as different visualizations. This not only requires awareness in the development of culturally situated visualizations, but opens a realm of opportunities for the development of new visualizations as theoretical and scientific knowledge expands. Viewed philosophically as ontological and epistemological realities, these concepts are somewhat mind-boggling since they seem to imply realities that are not only simultaneous, but that can be occupied simultaneously. However, this same seeming paradox is not so paradoxical when reconceptualized as a massively parallel machine where individual Execution-Agents, each part of a single machine, occupy individual pieces of shared realities simultaneously and communicate with each other along relational lines that may be dynamic or predefined. Given paint brushes, these Execution-Agents could paint all the parts of a picture at once, giving the appearance of a single "individual' in many places simultaneously.

CONCEPTUAL BLENDING THEORY

Conceptual Blending Theory is an extensive set of principles and processes that has emerged from the early work on metaphors by George Lakoff and combines concepts of Charles Fillmore's frame theory and Fauconnier's earlier work on structural analysis and knowledge representation, especially as it related to structures and schema. In *The Way We Think* (2002), Giles Fauconnier and Mark Turner discuss CBT in detail, analyzing metaphoric and analogical forms from a multiplicity of domains and disciplines. The concepts of CBT have inspired many people and spurred an abundance of research in conceptual, textual and computational analysis.

The primary goal in CBT is to achieve human scale, accomplished by the construction of stories, and more specifically, small spatial stories, that tell of human agency and action in a space or locale (topology), organized by a series of events, processes or times. From the integration of these small spatial stories, relationships emerge that may be implicit or explicit, and result from the human ability to complete patterns based on similarities and differences, proximities, or forms. The Quadripartite Seasonal Model (Figure 1) is a visual representation of a small part of the complex cultural construction of understandings based on the organizing story of following the sun's rays across the sky, a form rich with kinesthetic and cultural meaning.

The Sticky Wicket in the Visual Blend

[Although] the miracles of form harness the unconscious and usually invisible powers of human beings to construct meaning . . . form does not present meaning, but instead picks out regularities that run throughout meanings. Form prompts meaning and must be suited to its task. (Fauconnier & Turner, 2002, p. 5)

Form is interesting, and can provide structure for visualization, but in most cases, form is not *the meaning*. Meaning is bound in the integration network that results from cultures' creations of suitable blends. Networks that result from multiple, successive inputs to the blend are often the result of pattern completion. The initial Diné pattern is one in which four loci are related to four directions. Viewed as a blend, this pattern functions as a template and does not have to be reinvented. It specifies the general form of the projection, and the culture recruits and adds additional inputs through the creative, cognitive process known as "running the blend." The Diné culture has already run the blend recursively many times for many, many inputs, such that the entire integration network and its derived relations are immediately available from the simple Diné model.

Achieving Human Scale

Achieving human scale is the way that as humans we make big ideas understandable and memorable. In the reduction process, information is compressed, becoming invisible, in the creation of simplicity, but since the process of running the blend is – and must be – reversible, the compression details become visible again when the blend runs in reverse. In graphics, this is called a zoom, and takes the visual parameters + (plus) and – (minus). Thus while the integration process of CBT may not be a common concept in visual analytics, the display function is, and like all good metaphors, provides a cognitive model of the process with its associated meaning.

The process of achieving human scale has five procedural sub-goals that have direct analogies in visual analytics. These are:

1. Compress what is diffuse
2. Obtain global insight
3. Strengthen vital relations
4. Come up with a story
5. Go from Many to One (Fauconnier & Turner, 2002, p. 312)

Compressing what is diffuse is an issue of scale: molecules of a gas diffused in an ambient atmosphere can be compressed by increasing the pressure, forcing them into a smaller and more apprehensible space. Visually, zooming in compresses and zooming out decompresses.

Global insight is often accomplished by the relationships that emerge from completing patterns, as society has begun to understand the effects of acid rain, the greenhouse effects in the atmosphere, and global warming. The enormity of global warming is reduced to human scale in human-interest stories of how increasing insurance rates are affecting people who live along the coasts. This story serves to strengthen vital relations like Change, Cause-Effect, and Identity/ Uniqueness.

There are 15 vital relations that are transformed to achieve human scale: Change; Identity; Time; Space; Cause-Effect; Part-Whole; Representation; Role; Analogy; Disanalogy; Property; Similarity; Category; Intentionality; and, Uniqueness. Time, Space, Cause-Effect and Intentionality are can be scaled, or can be syncopated, where only a few key points are retrained; Syncopation is common in the Diné model in its compressed form. Vital relations can be compressed into other vital relations, as for example Analogy into Identity or Uniqueness, Cause-Effect into Part-Whole, and Identity into Uniqueness. Each resultant blend alone must provide enough information to support the unpacking process, the reconstruction of the generic and input spaces, the cross-space mappings that connect like elements from different domains, and, the network of connections. Here, pattern completion, specifically based on visual form, will be the primary unpacking process and various cultural data will be used to identify the elaborations and small spatial stories.

Stories supply meaning for the compressions, relations and individual blend elements. Stories may be input spaces themselves, or they may emerge in the blend as a result of the transformations. Meaning for the Diné visualization comes from many cultural stories and not only perform the simple task of "telling a story" but also illustrate and give meaning to the relationships among all things as they define the Diné Universe. Small spatial stories often establish the Identity and Uniqueness vital relations, as in the *Origin of Horses*, where single representatives from the Mist and Mirage folk stand with their individual horses.

The Blend Process Sequence

The blend process is recursive and steps can be executed in any order. Although it appears from the list that follows that one must have an established mental space before matching can commence, in fact, initiation of a matching process can be used to define the characteristics of the mental space that will enable completion of the matching process.

1. Setting up the mental spaces
2. Matching across spaces
3. Projecting selectively into a blend
4. Locating shared structures
5. Projecting backwards to inputs
6. Recruiting new structures to the inputs or the blend
7. Running various operations in the blend itself (Fauconnier & Turner, 2002, p. 44)

We have few systems today where visualizations are organizing elements in the blend, but in this analysis, Model Abstract (Figure 2) is blended

to create the Diné Generic Input Space (Figure 4b) that functions as the organizing space. This method represents a shift in visualization approaches, because the visual form itself is used to guide emergence, rather than the more common approach where textual or numeric data are recast into what are thought might be appropriate visualizations.

DECONSTRUCTING THE VISUALIZATION

It is a great virtue of conceptual integration that we can create new blends at human scale, by using something that is already at human scale as one input to the blend and by performing various kinds of conceptual compression on structure within inputs and on relations between inputs. (Turner, 2005)

Here I tell two visual stories of four dots (Figure 2a). One version tells of the emergence of the visual metaphor for the traditional Diné way of learning (Figure 2b), and the other of the CBT process model that creates the visual framework for the emergent meaning (Figure 2c).

Both of these models have been created as blends of

1. The four dots topology;
2. A geometric form – circle or diamond - that connects them; and,
3. A set of arrows that indicate directional movement of the process.

Figure 2. Model abstract

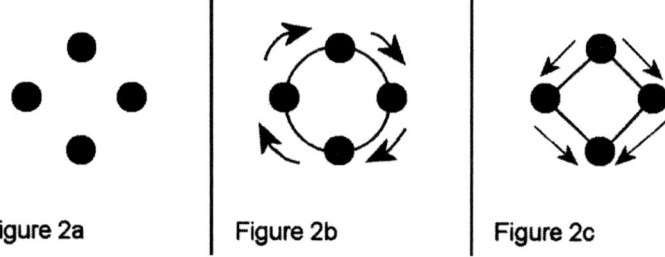

Figure 2a   Figure 2b   Figure 2c

In the blended visualizations, identity emerges through the psychological principle of "good completion" and the movement in these models could be represented as computer animations.

In deconstructing the blend process, the original Model Abstract is the Generic (Organizing) Space and defines the end point, or "goal" or running the blend backwards. Visually subtracting the four dots from the final blend leaves the circle and diamond Connection Paths, and the Directional Process Arrows that form Inputs 1 and 2. Since running the blend backwards should result in the normal blend form as the final stage in the process, this form is used to present the results of the deconstruction (Figure 3, *Comparative Visual Blends*). However, visually, these forms could be inverted, or, the directions of the process flow arrows could be reversed, to show the active process. As always, these visualizations can be automated to be both dynamic and interactive, giving people a great deal of flexibility in exploring the models and their results.

Organizing Stories for the Models

These two models seem similar: the only differences appear to be whether the lines are curved (Diné model), or straight (CB model). Actually they quite different, and they invoke one of the primary questions that must be asked about visualizations: *What do you see?* The geometric semantics of the Diné model do not indicate where the process begins, but do imply that the process is cyclical. CB model geometric semantics indicate that the process begins at 90° and proceeds simultaneously through 180° and 0/360° with the Blend emerging at 270°. Recruiting two small spatial stories into the blend can make it more human. In one, a person has a thought, brings her hands together to create that thought physically in the world and then, places the result on display at her feet, proud of her work and sharing it with others. In another story, people at county fairs and dog shows stand behind their entries, using position to create an Identity vital relation between the two. The blend can recruit the elaborations of sharing success and accomplishment, and responding to judges' requests and questions. These stories work because the visualization resembles the human form, and we know that people have heads, hands, and feet. These are related by Analogy to the Generic, Input and Blend space topology.

Blend Recursion and Elaboration

Figure 4, *Blending the Diné Generic Space*, shows the Organizing Fractal (refer to Figure 1) and the beginning of the blend process. The choice of where to initialize blend recursion is somewhat arbitrary, given that the Basic Generic Space has already been identified as a representation of the Diné blend process. In Figure 4b, the

Figure 3. Comparative visual blends

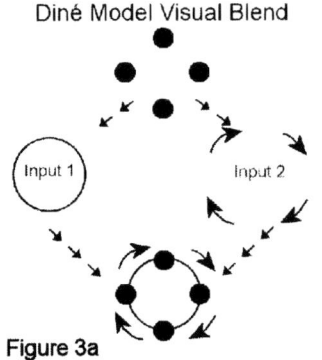
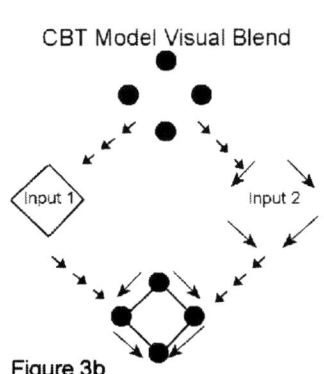

Basic Generic Space is Blend[0] and functions as the generic, organizing space. Blend[1], which establishes the relations between the directions and the sequential index, is a robust blend that becomes the next Generic Space, establishing the fundamental blend organizing story. In Figure 5, the Generic Space will be blended with two basic concepts, Directions, and the Sequence of rotation and reference, to create the Diné Generic Space, Blend[1]. The choice of subscripts was made to focus attention on the most powerful blend in this process, Blend[1], which has been elaborated with both the cultural organizing story and the indexing for computer-based dynamic visualization.

Maintaining structural integrity requires that each input set have one association for each direction. The Sequence information in Figure 4b shows that the starting point of the process is East, and that the process proceeds sunwise to the South (2), West (3) and North (4). The index eliminates any pre-ordering requirement: if items in the input are indexed, they will be associated with the specified locus, but if not, will be associated in the default order. In the Diné Generic Space (Blend [1]), Direction is the cultural index, but visual analytics is a computer-based discipline and in the blend process, the numeric index has derived a Bridge Function, connecting ancient and contemporary technologies. The Bridge Function "emerged" when the group index was projected into the blend from the Directions Input Space. The Bridge Function was not present in any of the inputs, and so is called Emergent Structure in the formal model. Emergent Structure is generated through composition of projections from the input spaces; completion, in which frames and scenarios are independently recruited; and, elaboration, also known as "running the blend." Recruiting an elaboration where the Bridge Function emerges in the association of a computer-friendly index and cultural directions requires that we have an organizing story that has recruitable elaborations. Such an organizing story would be the visualization meta-story, or Visualization Organizing Story (VOS), this name both more informative and an example of compressing what is diffuse into an Identity vital relation.

The Bridge Function is extrinsic to the Diné visualization itself, but serves as a mechanism for automating running the blend. As previously noted, the blend must be able to be run in both directions. Input spaces are "packed" to create a blend, but the blend must also support unpacking, or running the blend backwards. During unpacking, input structures that have been merged become visible again. In the Diné Generic Space

Figure 4. Blending the diné generic space

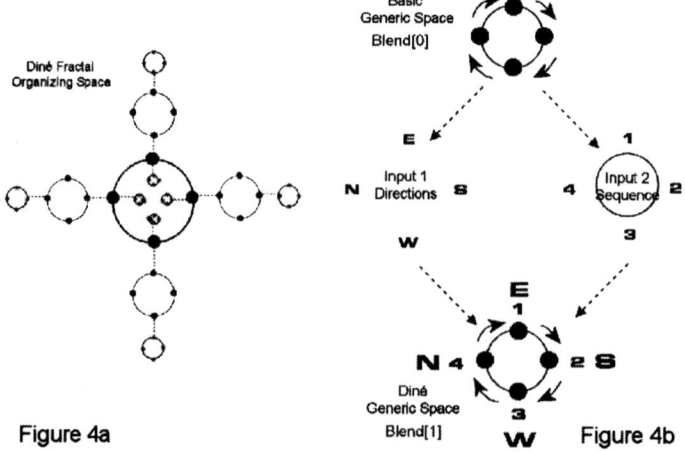

Figure 4a

Figure 4b

(Blend[1]), the indices and directions were blended with the visual abstract model and new relations were established between *1* and *east*; *2* and *south*; *3* and *west*; and, *4* and *north*. Relations were also established between the *1- east* relation and the point at 90°; between *2-south* and 0/360°; *3-west* and 270°; and *4-north* and 80° and these relations are Identities.

Zoom Out: Compression

In the blend, the association of Directions and topology have maximized the vital relations of Change, Time, Space, and Part-Whole. Cultural completion and elaboration have maximized Identity, Uniqueness, Representation, Role, Analogy, Property, Similarity, and Category. These are not apparent from simple observation of the form itself. Each node is a positional Identity for the properties associated with it, while the overall Blend is culturally holistic, and elaborates the Cultural Organizing Story (COS), as for example where the blended directions delimit the Diné World. Color is very detailed, and changes with both Time of Day and Season. Although individual colors are the compressions of the many different light blends seen on any given day, because of the fractal form of the visualization, during unpacking they can be expanded infinitely to precisely represent a specific color at a specific time. The Times of Day listed are compressions of the spans of the Time and Space in the Sun-God's movement across the sky, but also because of the fractal process inherent in the form, can be expanded recursively to rerepresent the diffuse, individual and precise pre-compression times. Human lives span many decades, with multiple repetitions of Years, Days, Times of Day and Colors of light, but in the compression, the Life Cycle becomes four specific segments, each with its individual characteristics. Lifecycle processes, viewed from the center, establish personal Identity; but also compress the vastness and details of Change, Time and Space to achieve human scale. The com-

pressed, elaborated visualization is a mnemonic that functions to call meaning to mind and so requires a meta-story, the organizing story of the visualization in the world. Thus it can be seen that there are two organizing stories speaking here. The Cultural Organizing Story (COS) tells us about the information in the visualization itself while the Visualization Organizing Story (VOS) tells us how we use the visualization in the world. The visual organizing space is the fractal model where each small dot expands to a circle on which four more dots appear each at a directional point and expand into four more circles (refer to Figure 1). Thus what the COS does for meaning, the visual organizing space does for the VOS.

Since Blend [1] already includes Index and Directions, the next tasks are to blend the Color, Time of Day, Season and Lifecycle categories. The relation between the four elements and their category is itself a Many to One compression that forms a organized Category from an otherwise diffuse information domain. Each of the columns in Table 1 contain the elements for the four Input categories. Since these are already in order, we can use default association. The relationship codes listed after the blend categories are mnemonics that identify the compressed blend categories. Because the compression is dense, the 2-dimensional format requires alternative representations for the data. Relationship Code 1 in Table 1 identifies the blends present in the organizing space in Figure 5a. Relationship Code 2 identifies all the categories that have been compressed after the blend in Figure 5b completes. These relationship codes will be used to complete the elaboration in the next section, Zoom In – Elaboration

Figure 5 shows how blending Seasons and Times of Day with the Diné Generic Space results in the recreation of the Quadripartite Seasonal Model (Figure 5a; see also Figure 1). Figure 5b shows the blend of Color and Lifecycle and also the increasing complexity of creating visualizations for dense compressions. In Figure 5b, the relationship acronyms from Relationship Code

Table 1. Blend input spaces

Index	Direction	Time of Day	Season	Relation-ship Code 1	Color	Lifecycle	Relation-ship Code 2
1	east	dawn	spring	1EDS	white	process of birth and age of childhood	1EDSWB
2	south	midday	summer	2SMS	turquoise-blue	age of adolescence	2SMSTBA
3	west	evening twilight	autumn	3WEA	yellow	age of maturity	3WEAYM
4	north	night	winter	4NNW	black	process of old age	4NNWBOA

1 in Table 1 identify the positions in the model abstraction.

Zoom In: Elaboration

This section is where we illuminate the story elaborations, associating the coded relationships from the compressions in the previous blends to the cultural stories. In actual practice, cultural stories are simultaneously incorporated into the blends as part of the elaborative process. The four sacred mountains are topologically simultaneous with the directions, and were not shown in the blends although they could have been. Instead, they will be associated by elaboration. Cultural stories for the seasons, the colors, and the creation of First Man and First Woman will also be linked through elaboration. Many of these stories address more

than one relationship, and so need to be linked to them all; this will be accomplished by relating the story to Relationship Code 2, defined for the emergent relationships in Blend [3].

The stories in Table 2 add the names of the mountains to the relationships, as well as the colors and purposes of the shells and stones associated with good living in the culture. The Supernaturals (Holy People) are agents in the small spatial stories that include not only the descriptions of their travels to the mountains around the perimeter of Diné lands, but also their activities upon arriving at their four destinations, and the reasons for those activities. Because the relationships relate all the elements to each other, elements from any group can connect to the corresponding element in any other group.

Figure 5. Four-category blend

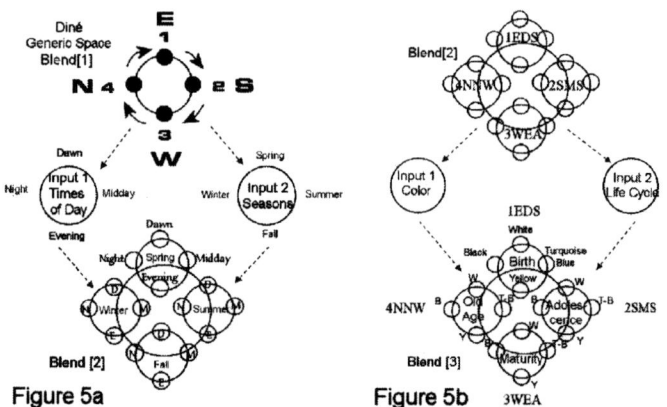

Figure 5a

Figure 5b

Table 2. Direction/Mountain stories

Relationship Code 2	Story
1EDSWB	The Holy People traveled to the mountain of Sisnaajiní (Mount Blanca) following the rays of the rising sun. When they arrived, they dressed the mountain with white shell so that we would have good thoughts throughout our day (Torres-Nez, 2004, p. 2).
2SMSTBA	The Holy People then traveled to the south mountain of Tsoodzil (Mount Taylor) continuing to follow the rays of the sun. When they arrived, they dressed the mountain with turquoise to keep us healthy and help us develop clear goals in our lives (Torres-Nez, 2004, p. 8).
3WEAYM	The Holy People then traveled west to the mountain of Dook'o'oosliid (San Francisco Peak) continuing to follow the rays of the sun. When they arrived, they dressed the mountain with abalone shell to help create an understanding and appreciation of life and our place in it (Torres-Nez, 2004, p. 22).
4NNWBOA	The Holy People then traveled [north] to the mountain of Dibé Nitsaa (Mount Hesperus), continuing to follow the fading rays of the sun. When they arrived, they dressed the mountain in black obsidian for bravery and positive self-image to help protect us from danger (Torres-Nez, 2004, p. 26).

Table 3, the Season Stories, introduces the concepts of the balance between male and female that are intrinsic to the Diné model. These concepts are much broader than semantic references to sex and gender. Conceptually, elements that are categorized as "male" are associated with the static dimension of reality, while those that are categorized as "female" are associated with the dynamic dimension of reality. The categorizations should be interpreted as the transformational movement from the static to the dynamic, and then from the dynamic back to the static in anticipation of beginning a new cycle.

The stories of the Seasons exactly correlate with our human understandings of our lifecycles. Because of the relationships and increasing elaborations, birth and the beginning of life are associated with dawn, with white, and with spring, summer and autumn with female and the growth and maturing of life, and, winter with male and completion of the lifecycle.

Compression and conceptual mapping are well-known in the Diné culture, although they are not discussed in this way. The elaborative stories in Table 4, written by Diné medicine man and Elder Wilson Aronilth, Jr., illustrate the Diné awareness of both these processes, as well as of the analogical relationships between temporal events and human accomplishment.

A different mapping relates the cycle of Day to the worlds in the Diné Emergence (Creation) story, which achieves human scale when mapped in this way:

The cycle of day is also a ritual drama "which begins with a return to the underworlds and a reenactment of the undesirable events and resulting conditions found in the episodes of cultural heroes and concludes with a triumphant restoration of Hózhó [the dynamic, active life-beauty] and a glorious reemergence into the beautiful, colorfully-lighted, cosmos of the fifth world. (Witherspoon, 1977, 144)

Other cultural stories of the emergence relate the chronological sequence of the Creation to colors, to the topological structure that is now familiar in the visual metaphor, and to the cre-

Table 3. Season stories

Relationship Code 2	Season	Story (Witherspoon, 1977, p. 143)
1EDSWB	Spring (male)	birth or beginning of life
2SMSTBA	Summer (female)	growth of life
3WEAYM	Autumn (female)	maturing of life
4NNWBOA	Winter (male)	death or decay of life

Table 4. Time – life cycle stories

Relationship Code 2	Color	Story
1EDSWB	Day	A page in life "where a good deed is done each day" (Aronilth, 1991, p. 85).
2SMSTBA	Week	A chapter "where progress and advancement are accomplished" (Aronilth, 1991, p. 86).
3WEAYM	Month	"a unit in our life when goals and objectives are planned out" (Aronilth, 1991, p. 86).
4NNWBOA	Year	"a volume of our life in which a great deal of knowledge, wisdom, skills and a good positive foundation can be accomplished (Aronilth, 1991, p. 86).

ation of First Man and First Woman, who were the inventors of form:

The First World, Ni'hodilqil, was black as black wool. It had four corners, and over these appeared four clouds. These four clouds contained within themselves the elements of the First World. They were in color: black, white, blue, and yellow. . . In the East, at the place where the Black Cloud and the White Cloud met, First Man, Aste'hastqin, was formed . . . Now on the Western side of the First World . . . there appeared the Blue Cloud, and opposite it there appeared the Yellow Cloud. Where they came together, First Woman was formed. (O'Brien, cited in Witherspoon, 1977, p. 140)

From the relationships established between the balance of male and female, the seasons, and the stories before, the understanding emerges that the First Man was formed in the area delimited by the male clouds, and First Woman was formed in the area covered by the female clouds. The cultural story reveals the deep interrelationships and philosophical sophistication typical of Diné

culture. The story also reinforces the base 4 normalization of the recursive cycles.

Table 5 directly relates colors to the different worlds in the Emergence story. However, while in many cultures, colors seem to have a fixed meaning, in Navajo this paradigm does not hold. Color meanings are derived from their context. In the analysis of this visualization, I have used the color relationships that tend to be constant such as that between color and direction, color and time of day, and color and season. The described associations should not be assumed to hold true across all possible association in Navajo philosophy.

THE FUTURE

The existence of a good blend can make possible the development of a better blend. (Fauconnier & Turner, 2002, p. 24)

While automating CBT functionality is nontrivial, due partially to the required transformation from the purely semantic domain to the procedural, and in large part to the more sophisticated require-

Table 5.Color stories

Relationship Code 2	Color	Story
1EDSWB	White	Color of the present or Fourth World, a combination of all the colors
2SMSTBA	Blue	Color of the Second World
3WEAYM	Yellow	Color of the Third World
4NNWBOA	Black	Color of the First World; symbolic of the return to darkness and the chaotic but formative conditions of the underworld (Aronilth, 1991, p. 54-56)

ments of the databases necessary to support facile integrations, the technology necessary to support this evolution is available today. CBT automation offers vast possibilities for the development of curricular and pedagogical materials that incorporate cultural and disciplinary knowledge as part of the material construction process. Such automation also offers sophisticated analytical support for the understanding of complex knowledge, information, and processes through the interactive deconstruction-reconstruction process. The verifiability of the approach works to eliminate incorrect or insignificant conjectures, and reveals those sites where additional information is required to complete a scenario. However, for these benefits to be realized, the processes and more specifically the language by which such automation will be achieved must be designed.

Next Steps

In typical approaches to visualization, text-based procedural or object-oriented programming languages are used to create the visualization. This requires management of the many details necessary to create even the simplest visual form. Analysis of the Diné visualization has demonstrated the benefits of using visual components as part of the implementation semantics. From an external perspective, this is a fairly simple task, one that has been repeated billions of times by millions of people. In contrast, the typical software implementation approach is to proceduralize presentation of the visual idioms. Functions exist in many languages to draw the circles and lines that are necessary in the Diné visualization, but this approach loses the information embedded in the idiom, information that is dependent on locus, domain, and process. Clearly, an approach that includes the semantic information combined with the related semantic, cultural and cognitive information would be a more productive approach in terms of creating visualizations quickly, easily and robustly. However, getting to there requires

a change in paradigmatic perspective. This idea is not entirely new, and has been discussed in terms of expanding the data types that data bases are able to handle. Practical applications allow the incorporation of video and other non-text items. The idea of using graphical forms to paint Escher-like designs is also not new. What is new here is the proposition that visualization components function as robust objects that include not only the graphical information, but also the various embedded knowledges in the component information systems, and this requires a different approach to the design of the language. Additionally, underlying functional paradigms should retain the familiarity that users have developed in window environments, and includes such necessary options as click-and-drag to resize or rotate, double-click to open new layers, back icons or click events to close.

CONCLUSION

Conceptual Blending Theory has been used to explore the sophisticated and often complex and cyclical cultural understandings that are embedded in the Diné visual mnemonic. The elements of form and the compression process in the Zoom Out section make a much neater and more organized presentation of information than the storied elaborations in the Zoom In section. The forms are characterized by the reduced amount of information that makes categorization possible, while the elaborative stories are rich in detail, information and relationships. These two states are orthogonal to each other, but characterize primary issues in visualization, where decisions must always be made between a sufficient reduction of detail necessary to achieve global insight, and rich enough elaborations to support the users in their efforts "to detect the expected and discover the unexpected in complex information spaces." The Diné cultural stories presented here are typical of the complex information spaces in a variety of domains in their sophistication and information density.

The relationships between the Identities in the categorical forms and the elaborative stories suggest an algorithmic approach to creating a simple 3-dimensional visualization that would allow users to explore the Diné information spaces. Such an approach would be investigatory in nature, since current Computer Science research into CBT applications is still exploring constraints and paradigmatic approaches. Development of an investigatory Diné visualization would be facilitated by the strong structural support inherent in the form, and also by the normalization to four elements per category. Challenges are the recursion, which has the attendant issues of specification, that is, providing mechanisms for users to choose depth and information type, and, the visualization itself. As became evident in the development of the blending figures, when more information becomes embedded, alternative representations must emerge to allow access to both the direct data itself, and subsequently, the emergent relations and recruited elaborations. The cultural stories reveal multiple interconnections and relationships, precluding use of the direct relationship indicators that have been employed in visual interactions with relational databases. The Zoom Out function that reveals categorical associations, and the Zoom In that provides access to the elaborations seem promising, and align well with both the Cultural and Visualization Organizing Stories.

REFERENCES

Aronilth, Jr., W. (1991). *Foundation of Navajo culture*. Navajoland, USA: Wilson Aronilth, Jr.

Ascher, M. (1991). *Ethnomathematics: A multicultural view of mathematical ideas*. Boca Raton, FL: Chapman & Hall/CRC.

Begay D., & Maryboy, N. (1998). *Nanitáá S Nanitáá Sąąqh Naagháí Nanitáá Bikáeh Hózhóón, Living the Order: Dynamic Cosmic Process of Diné Cosmology*. Unpublished doctoral dissertation, California Institute for Integral Studies, San Francisco, CA.

Carter, K. (1993). The Place of Story in the Study of Teaching and Teacher Education. *Educational Researcher, 22*(1), 5-12.

Fauconnier, G., & Turner, M. (2002). *The way we think: Conceptual blending and the mind's hidden complexities*. New York: Basic Books.

Griffin-Pierce, T. (1992). The Hooghan and the Stars. In R. A. Williamson, & C. R. Farrer (Eds.), *Earth and sky: Visions of the cosmos in Native American folklore* (pp. 110-130). Albuquerque: University of New Mexico Press.

IEEE Visualization. (2006). Call for participation. In *IEEE Symposium on Visual Analytics Science and Technology 2006*. Retrieved 11/14/2007 from http://conferences.computer.org/vast/vast2006/

Kalish, M.K. (2007). *Navajo immersion mathematics: Culturally grounded 5th grade mathematics curricular and pedagogical materials study*. Unpublished doctoral dissertation, New Mexico State University.

Núñez, R.E., & Sweetser E. (2006). Aymara, where the future is behind you: Convergent evidence from language and gesture in the crosslinguistic comparison of spatial construals of time. *Cognitive Science, 30*(2006), 401-450.

Powell, J.W. (1880). *Introduction to the Study of Indian Languages with Words Phrases and Sentences To Be Collected*. Washington: Government Printing Office.

Sapir, E. (1975). *Navajo texts*. New York: AMS Press Inc. (Original work published 1942).

Torres-Nez, J. (2004). *Beesh Łigaii in Balance*. Santa Fe, NM: Museum of Indian Arts and Culture.

Turner, M. (2005, July 12-15). *Mathematics and Narrative*. Paper presented at Mathematics and Narrative, Mykonos, Greece. Retrieved April 5, 2006, http://www.thalesandfriends.org/en/papers/pdf/turner_paper.pdf

Welker, G. (1996, February 8). Song of the horses. In *Indigenous people's literature*. Retrieved 2006, April 29 from http://www.indigenouspeople.net/songhors.htm

Wikipedia contributors. (2007). Terminology. In *Scientific visualization*. Retrieved November 14, 2007 from http://en.wikipedia.org/wiki/Scientific_visualization

Witherspoon, G. (1977). *Language and art in the Navajo universe*. Ann Arbor: The University of Michigan Press. (Original work published 1977).

KEY TERMS

Blend: (n), A new representation of knowledge in which form and meaning are inseparable created from the inputs and reduction processes; (v.t.), the creation of a small spatial story from the blending of conceptual inputs and a transformation using one of the 15 vital relations (Change; Identity; Time; Space; Cause-Effect; Part-Whole; Representation; Role; Analogy; Disanalogy; Property; Similarity; Category; Intentionality; and, Uniqueness).

Compression: Incorporating details such that they become invisible in the final blend or visualization. Compression is frequently accomplished using one or more of the 15 vital relations.

Conceptual Blending Theory: The name given to the processes described by Giles Fauconnier and Mark Turner that offer a scientific explanation for the ways in which humans cognitively create metaphors. In the CBT process, pieces of different concepts are transformed to produce the final metaphor. A key component of the Theory is the importance of both the recursive nature of the blend process and the ability to run blends backwards and forwards, forwards to produce the metaphor, backwards to allow the inputs to remerge as integral objects. The omni-directionalality emerges from the understanding that blend form and blend semantics are integral and may not be separated from each other. Blends also do not need to start from a specific starting point, but can run from any point. Further, blends may move from one blend process to another, simply by choosing a different blend path; this feature makes chasing information back to its ultimate source a corollary of blend analytics.

Diné & Diné Philosophy: "Diné" literally means "The People" and is typical of the names early American-continent inhabitants gave themselves in their own language. The term "Navajo" is also used, but the choice to self-identify as Navajo or Diné is a personal one. Little is written of Diné Philosophy, both because the knowledge is restricted, and because the transmission follows the oral tradition. The information provided in this chapter is public in that the visualization itself and different pieces of the embedded information have been published in other documents. My choice to use this visualization as an illustration was guided both by its inherent sophistication and by my own sense of justice that calls for the recognition and understanding of all peoples' knowledges.

Emergent Meaning: Semantic information not present in any of the blend inputs; often recognized implicitly rather than explicitly.

Information Space: A collection of information that is not limited by source, form, process, semantics, or application.

Information System: An organized collection of information that combines as an integral object to form "information systems of information systems" In the context of this chapter, these systems contain much more information than is visible on the surface in that they are observably comprised of the components that were recruited into the blend, but are also comprised of the invisible blends and transformations used in the construction processes themselves, and also of the information necessary to run the blend in reverse.

Organizing Story: The information system that contains the elements from which the storyteller may choose. The organizing story is amorphous in that while it includes the events and important concepts, it does not restrict the sequence of events, the choice of possible events, or the particular way in which each storyteller may choose to present his or her version.

Reduction to Human Scale: Creation of understandable representations of ideas that are too large or too complex are difficult for humans to grasp, by integrating actors and events. Such blends are already evident in the Diné organizing story, where the Sun-god (human) carries the sun (object) across the sky.

Small Spatial Stories: Stories where actors with human characteristics move themselves through time and space, thinking, acting, and manipulating with hand and tool.

Section III
Implications of Technology, Social Dynamics and Culture

Chapter XVIII
The Challenge of Enculturation on Art

Lindsay Grace
University of Illinois and Illinois Institute of Art, USA

ABSTRACT

Enculturation is the act of passing cultural ideologies from one person to the other. It is what breeds innovation instead of new creation. It is the disease of derivation, instead of the birth of creativity. This chapter assumes the practical perspective of critical anthropological distance to understand the culture of art. Such critical evaluation should illuminate the distinct characteristics that encourage patterns. In the tradition of anthropological and sociological study of existing culture, this chapter seeks to illuminate the distinguishing characteristics of contemporary art production and offer perspective on the critical creative process. It takes new media art as its case study because it serves as a cross-cultural intersection of scientific invention and artistic innovation.

BACKGROUND

There is little novelty in the concept of enculturation. It exists in a variety of disciplines and social situations. Research into specific enculturation practices, patterns, and effects ranges from the typically sociologic, Best Practices for Enculturation (Boyle, P., & Boice), to the extraordinary, Alan Bishops Mathematical Enculturation: A Cultural Perspective on Mathematics (1991). Perhaps as part of a growth in the accessibility of information, the cross pollination of formerly specific cultural aspects has witnessed a consistent growth in contemporary history. In kind, the interest in culture and intercultural study has grown. In Robertson's often cited *Globalization: Social Theory and Global Culture*, he writes "by now it must surely be clear to most sociologists that in contemporary sociology and social theory that there is an awakening . . . of interest in the social relevance as well as the intrinsic significance of culture and cultural change" (Robertson, 1992, p. 32).

The new media arts, although not the only artistic endeavor effected by the growth of enculturation is an excellent subject for the evaluation of this process. New media art is a distinctly acculturated art practice. It sits at the nexus between a wide, and often changing variety of artistic cultures. Because of the interdisciplinary nature of this art, the new media artists often float between science and art disciplines. It is this edge at which intriguing acculturation of science and art occurs. This is what Lev Manovich (2002) described as "the computerization of culture" which "not only leads to the emergence of new cultural forms such as computer games and virtual worlds; it redefines existing ones such as photography and cinema" (p. 9).

Responding to substantial changes within two cultures is an everyday fact of the new media arts. They must be acculturated in order to exist. The culture of technology, and that of art, is the subjects and tools integrated into new media art. Again, in Manovich's (2002) words, "the gradual computerization of culture will eventually transform all of it" (p. 6). Here, that which existed in extra-technical space of artistic endeavor, finds itself blending with the technical. Two cultures are acculturated to make a third. The resultant cultural accumulation is the focus of this chapter.

The cultural aspects, here described, are limited in scope to those characteristics best understood as part of western tradition of art making and evaluation. The reasons for limiting the scope include the popular dominance of these approaches, the author's proximity to these cultures, and the abundance of writing about them. While it would be interesting to develop sharp contrasts, and use other cultures as a kind of touchstone, the resulting analysis would be the subject of a book, not merely an expository chapter. Instead, I invite readers to develop and investigate their own hypothesis about the relationship of art, culture and education after reading this chapter. This writing serves only as a starting point from which to begin such investigations.

BACKGROUND: RESEARCH IN MULTIPLE DISCIPLINES

Anthropological and sociological analysis of the culture of art is not a novel practice. Bourdieu's Distinction: A Social Critique of the Judgment of Taste serves as an argumentative base within the social sciences for the relationship of cultural standards and art value. The book includes an analysis of the economy of cultural worth that proves relevant more than twenty years after its publication. In the introduction of his work he states "scientific observation shows that cultural needs are the product of upbringing and education:" (Bourdieu, 2007, p9). It is this observation that functions as one of the foundations for this chapter. If culture is provided by education, than an analysis of education may illuminate how the creative process is affected.

In concert with the academic, there is much written by contemporary essayists about cultures of art production and their relationship to society. One such book, Lipstick Traces: A Secret History of the Twentieth Century, by Greil Marcus, serves not as academic support of the focus of this chapter, but as evidence confirming the permeative character of distinct cultural ideologies. Marcus serves a history that, perhaps unintentionally, supports the claims of this chapter. As told, punk music, even in its purported efforts to develop an anarchist, anti-establishment movement works within deeply engrained cultural ideologies. Here, the master example of the punk group, the Sex Pistols, provides evidence of deep enculturation among artists working against cultural standards. Marcus's history demonstrates the affinity between the worlds of punk musicians, Situatonist manifestos, and Dadaist poetry (1990). Marcus demonstrates, through a disparate collection of artists' mediums and ideologies, that there are common cultural threads that exist in an intersection of time and social situation (1990).

The range and variety of writing about the culture of art stands as evidence of not only its existence, but of its complicated nature. It is not enough to describe it once. It is instead investigated, and reinvestigated as the subject of academic and non-academic critique. It rests between that which can be assessed through scientific evaluation, and that which requires entirely qualitative research.

BACKGROUND: INNOVATION OF INVENTION

To understand enculturation as it relates to creative efforts, it is first important to distinguish innovation from invention. Innovation is the subset of creativity which emulates, and derives. Invention is the creation of entirely new products, ideas or the like. This dichotomy between invention and innovation finds the most usage in neither science, nor in art. This distinction is more often made in two unlikely complements, business and education. Betty Edwards', Drawing on the Artist Within: A Guide to Innovation, Invention, Imagination and Creativity champions specific educational approaches that foster development of these distinct areas of creative process (Edwards 1986). To understand the distinction from an artistic analogy, consider innovation kin to drawing a familiar form from a reference, while invention is subject-less creation of form without reference.

Business, particularly as it orbits Information Technology and Intellectual Property, sharply defines the distinction between invention and innovation. In a National Institute of Technology Standards paper assessing the business of technology, the authors describe the difference between invention and innovation. In this paper, Branscomb and Auerswald, write "*invention* is distinguished from an innovation by its character as pure knowledge. The direct products of a technological invention are not goods or services *per se*, but the recipes used to create the goods and services" (2002). Supporting distinctions are provided by both the international business community, as in the paper "Sustainable Innovation as a Corporate Strategy" (Khan, Al-Ansari, 2005), and in the domestic business writing provided by the Industrial Research Institute (Roberts 1988)

This writing does not seek to be a critical evaluation of the art making process, but instead an opportunity to step away from current patterns for a kind of social-scientific evaluation of process. If an anthropological distance is encouraged, then there is an opportunity to understand the *why* in certain cultural patterns.

THE CULTURAL CHARACTERISTICS OF ART PRODUCTION

In anthropological study, it is important to identify particular cultural aspects in order to analyze their relationship to themselves and others. As such, the following few paragraphs enumerate the primary characteristics of the modern, American and European art production process. This study also seeks to consider critical aspects of what the Western world distinguishes as commercial and non-commercial art work. Much has been written about the tension between these two arts. It is true that some artists intersect at the fulcrum of these two, and that all art may be considered commercial through some perspectives. It is, however more important for this study to rely on the traditional definitions that conveniently define distinct barriers between that which is commercial and that which is not. Here commercial arts are those which were made with the distinct goal of selling a product, while non-commercial arts are all other academic, outsider, or artistic productions not intended to sell a specific product.

CULTURAL CHARACTERISTICS: CRITIQUE

Critique as a formal process predates the invention of new media arts. Traditionally it is the primary evaluative means under which art is shaped and aligned. There is a specific language to critique, that like most sub-specialty diction, purposes to increase the clarity of communication by defining specialized terminology relevant to critique. These terms include subject, medium, and others. Such terms have been absorbed into popular use as people outside the art community are exposed to the art production process through courses, behind the scenes content, interviews, and their own art critiques.

One hint at the cultural history of the language of critique suggests its cultural presupposition. The specific language of critique, the diction with which one communicates about a work is designed not to be interdisciplinary, but to be the opposite. The words with which we critique are meant to be specific, useful in abstract across domains, but in particulars only in analogy. Consider the challenge of describing the characteristics of a new media piece. Unlike the traditional arts, which have predefined terms for their descriptions, most of the language of new media critique is borrowed. For Manovich's, The Language of New Media, terminology must be adapted from computer science (2002). For others, it may be the opposite. According to Kesseler & Bergs (2003), society responds to its technology to produce language developed from the use of new media. The indication from either perspective is that there is insufficient language to describe new media. Interestingly the critique is preserved as an effective tool; although the tool's standard form of communication falls short of critique's needs.

CULTURAL CHARACTERISTICS: CONSENSUS BUILDING

The culture of contemporary art production favors consensus building. An integral part of the critique process is a shared explanation of one's art. The formal critique, for example, often begins with an artist sharing their work with an audience, explaining their goals and objectives and then listening to a panel of specialists who offer their feedback on the artist's perceived fidelity to such goals. Under this approach, the badge of a successful work is in the consensus of the critics. If the respected professionals consider the work to be strong as a group, then it is understood that the critique went well.

Critical evaluation of the last few observations of this process should concern both the philosopher and the sociologists. First, the individual response is merely a part of a group, which in democratic societies may be perceived as an ideological success. Yet, sociologists recognize that groups do not always make the same decisions that individuals make. Groups make the decisions groups make. The consensus built by the evaluating mass, may not be consensus delivered by individuals evaluating individually. In particular, the observation of one evaluator, may lead successive observation and comments made by those who follow. Even the most seasoned critics fall off topic because of intellectual curiosity or social dynamics.

Now consider that there abound on the Internet informal and formal copies of this process in varied scale. At nearly every point at which an artist can share their work, an opportunity for others to evaluate their work exists. These include rating scales, comment boxes, and popularity ranks. From Deviantart.com images to Youtube.com videos, the artists on the Internet are encouraged to use consensus as an evaluator of their work. Yet, through cultural value systems, it is understood that these places lack the academic rigor required for serious consideration in the art world.

CULTURAL CHARACTERISTICS: ACADEMIC APPROVAL AND PROCESS

Academics are institutionalized artifacts of the development of art works, art production and development. There has been much written in and out of the academic domain about the value of academic art training and development, the survey of which is too large to present in this document. Instead, it is most important to identify that the structure of academic institutions has been a significant aspect of the development of most artistic critique and production.

Art schools, both commercial and non-commercial, are culturally recognized institutions for the development of artistic skill. The promise of art schools vary from the practical objective of placing students in art making jobs to the abstract of improving artistic process and rigor. The latter is more often the promise of traditional art institutions, while the former is promised by career oriented commercial schools. While the subjects of both institutions may vary, the process of education is typically very similar. They each rely heavily on the enculturated aspects of art production: critique and consensus building.

Art schools, however, add the third dimensions of art instruction to art production. Identifying, evaluating, and promoting art instructors varies by institutional level. At the primary level, art education is primarily executed by individuals trained in arts education. These are individuals who have devoted some post secondary level of their education to understanding how to educate others in the process of creating artists' works. At the secondary level there may be a hybrid of art education trained artists and practicing artists, whose primary focus is arts education. Finally, at the postsecondary level, the educators typically shift from people trained in educating artists, to people trained or practicing in particular arts. Most postsecondary education is provided by artists, not by art educators. This is of particular dilemma at institutions that are seeking to provide an education that is about process, not production. To be a good practicing artist, as evaluated by these institutions, you need not be well-versed in the variety of artistic process; you need only know that artistic process that has proven successful for you as an artist. Success in the postsecondary

education academic universe is a practice in one's own practice, not in the exploration of others. Excluding the dominant practices of art historians, who do routinely investigate art process, art production instructors at the post-secondary level do not need to survey, they need to produce.

Beyond the particular goals of their institutions, what many good art professors have proven is that they know how to navigate the academic terrain. They are then good teachers for future teachers, but they are not necessarily good teachers for future practioners.

This is not an attack on the institutions, or the institutional machine. Instead it is an observation about the culture. To teach computer programming, one must understand computer programming. To provide effective critique, one must understand art, not necessarily how art is produced.

This may be a result of the cultural reliance on production process and a separation of art from the everyday. Where some academic enterprises are seeking to refine, evaluate and critique process, many artists are asked to explain, and repeatedly invent the process through which they make their art. This process orientation is not historically the same in Asian cultures where "appreciating art was often integrated into everyday life rather than being considered something special"(Grau p. 279). This perhaps, is because "a distinction between art, entertainment, and commercial products did not exist"(Grau p. 279). Consider the work of Takashi Murakami, for example, whose factory oriented production process does not devalue the value of his individual works within Asian cultures. In such an environment the line between commercial and non-commercial art is blurred, and as result process becomes ancillary.

The use, again, of websites, actually moves the western cultures toward the eastern evaluation criterion. In the online gallery spaces, where artists share their work of various digitized mediums, there exists an *almost marketplace* of art. The art is shared, it is evaluated, and it is given accolades, but it is rarely given money. The interesting thing here is that work published on the Internet, by amateur and professional artists is rarely provided compensation for the quality of the work, although the gallery through which it is shown is wholly reliant on artist contribution. The marketplace of art is then neither Eastern, where commercial art and non commercial find similar compensating models, nor in the Western traditional, where the compensation models are sharply contrasted between commercial and non-commercial. Of course, the Internet provides a space where geographic distances mean little, and culture mixes more readily.

CULTURAL CHARACTERISTICS: THE GALLERY

The traditional gallery represents the artistic marketplace of western art. It serves as the place for display, recognition and compensation for artistic works. A gallery's reputation is built on its curatorial practice. A gallery that is selective demonstrates a rigor that the culture of art identifies as important. Thus, a web based gallery, with an all artists welcome approach provides little recognition for the artist's work. Acceptance into a non-selective gallery means little in art culture's value system. Wide, positive review, on the other hand, does provide a specific brand of recognition. The dilemma is then that web-based galleries provide a wide audience, and the possibility of wide recognition, where the traditional gallery space offers its opposite. Traditional galleries, because of the physical truth of offering only limited hours of access, limited capacity, and limited location, provide recognition through acceptance and promotion. To be shown at the Art Institute of Chicago proves greater accomplishment than Google video, simply because the artist has matriculated through successive evaluations. In short, the process of achieving the highest honors in art production is often not only a test of one's ability to produce art, but to navigate the process of art promotion. An artist who is shown in a well-respected gallery may not be the greatest artist, but they are good at getting into well respected galleries.

These claims are old claims, but what is important is to understand the cultural implications. Consider the well published fact that many gallery spaces are not conducive to the display of new media art (Grau 2007 p. 251). Where then does new media art exist? It generally must find its gallery space in mass media. This often means web- based galleries or specialty arenas like nightclubs and raves. The result is a tightly coupled affinity between the patterns of new media art and the practices of the commercial world. The venues for display of new media art tend to coincide with commercial spaces. These include public spaces. Coupled with designs consistent relationship with commercial arts, the relationship mirrors itself across multiple artistic disciplines. The nightclub DJ works with the VJ artist, for example.

The web-based gallery, on the other hand creates a scenario of self-replication. Since the curatorial measure of success on the web is largely popularity, artists receive positive feedback when their work is highly popular. Popularity is not necessarily derived from quality, but simply from novelty, whether positive or negative. The first naked body on the Internet, for example, would be very popular, but in a sea of naked bodies, there is a competitive space. The resulting scenario is that images of largely attractive or unattractive bodies become the novelties that draw attention. The positive is popular for its general appeal, while the negative for the appeal of its repulsion, or the rarity of its experience. This same scenario plays itself out on the repositories of video prevalent on the web. In a short period, producers of film, for example, note what has succeeded or failed, and seek to drive even further in the direction of their experience. If a tribute montage of popular film draws attention, the amateur chooses a film they believe to need better tribute, and soon enough they are providing their own tribute montage. The result is a collage of references, a series of simulations, and self replications. These creative efforts are poured into trying to make something like something else, or decant the best of their subject. They are, innovating on an existent work. While the artistic quality of their creations is varied, the core is still the same. They start from common points of departure, not new locations. Since raw popularity rules such spaces, innovation succeeds because it is based on the familiar.

It is also worth noting that this pattern is a marketplace pattern. It is the analogous to the cycle of video games clones, of replicas, and of re-release.

CULTURAL CHARACTERISTICS: DISTINCTIONS BETWEEN ART, DESIGN AND SCIENCE

The culture of art understanding defines divisions between the arts and sciences. It later offers a third, born from these two. We call it design. Design is the progeny of art and science, not fully science and not fully art. Yet, critical examination of this enculturated aspect leaves the science of art and the art of science production bastardized. Such a division, or the very desire for need for such divisions, leaves the new media artist in a no man's land between the two great disciplines. An electronic device designed by an artist to produce new musical tones, is, by contemporary definition neither wholly art nor wholly science. It is a product of design, with an artistic aspiration, but a functional purpose, that is implemented through scientific process. The question is then, where does this device rest. Is it at home in a gallery? Gallery spaces have a very particular "look but don't touch" policy, which orphans this device among paintings, video, and sculpture. Is it at home, or does it belong in a scientific journal? It lacks the rigor of scientific

research, and the practical application required of many journals. Lastly, design offers a home for it; save for the fact that design offers only the devices used by its ancestors, gallery or journal. This is the fate of many new media arts. They are destitute in the marketplace of ideas, because they are mixed breed arts. They exist, but their existence frustrates the cultural definition of pure science and pure art.

Coupled with the lack of appropriate gallery space, the new media artists find themselves working closely with the commercial arts. The graphic designer works with the web designer to create a consistent *look and feel*, for example.

THE PATTERN OF INNOVATION AND THE GENIUS OF CREATIVITY

The cultural history of art production in the Western world is old. It saturates the everyday, expressing itself in crayon drawn works posted on refrigerators, or the layout of furniture in a room. The world is understood by our society as a gallery, a forum for critique, a science or an art, a response to consensus, or matriculation through an academic right of passage. The transparency of these perspectives makes them unquestionable truths. Their adoption is how everyday life becomes enculturated.

Unfortunately, enculturation can act like an infection. It gets under the skin of creative efforts and spreads, suppurating overused ideology and reeking of what has been. The most creative art is that which is not infected by what has been explored. It is not within the affectations of art instruction, of rewards for simulating accepted methodologies and means of evaluation. Quality creativity is outside culture. It exists in the outsider, who visits but for a moment, to exclaim their brilliance, and then secede from society to perhaps exclaim again. The history of art innovation often breeds and champions the opposite. It loves the orphaned progeny of a dominant school who returns from hiatus with the spoils of another culture's approach. These approaches are then adopted and become part of the general enculturation infection.

Consider the common pattern of a new technology. When the technology is first used, it is novel, exciting and even inspiring. The result is a barrage of imitations. The pattern falls on the adage, *the sincerest form of praise is in imitation*. The visual novelty of the motion picture Matrix included the concept of bullet time. This technique originally employed adaptation of stop-motion animation concepts and multiple cameras, to display a live-action subject in mid-action during a slow motion 360 degree dolly (Orek 1999). Within a year of its release, multiple films and video games used the same visual effect, in tribute, jest and homage. The effect traveled from professional to amateur, down a hierarchy of technology. It is now a standard which has remained, especially among video game designers. It is adopted, and now rests under all creative efforts as a possible solution. It floats in the sea of creative devices, to be innovated upon by later artistic efforts. Like zoom or pull focus, it is likely to be reinvented, through derivation, by countless artists.

The same pattern is displayed ostensibly in the history of Apple computer-inspired design. At the introduction of the translucent iMac in 1998, its aesthetic became a hallmark of not only the company's design, but of industrial design in general. The clunky lines of an electric grill become the sleek, white lines of a George Foreman grill. Websites seek white and plastic glows. The world innovates on a central design.

HOW CULTURE IS GIVEN

There is little that can be produced in a society that is not polluted by its own rewards systems. What is produced is solicited, not by the patron, but by the rewards of socially accepted production. A child that draws is rewarded for better

simulating, and as such seeks to better simulate. If they understood their subject, say a person, a plane, their pet dog, and the need to envision it in a way other than what is prescribed, they are given transparent praise. Ultimately, their work is immediately compared to the previous renditions of similar subjects. This is a form of enculturation, for culture is not innate, it is given.

The world for these artists exists then only in comparison. It exists because we know how its elements interact. We do not know one thing without its other. We do not know the world without the signs of the world. We do not know existence, without the way in which our senses react to that world. We are then in constant opposition. We can imagine, only in that we imagine what can be described by a relationship to what exists. We can describe only in our words, we can imagine only in what we understand through those words. Our perception is a menagerie of comparisons, a constant description of what is and what could be compared to what is. We do not understand the nether regions of possibility, because we have been blinded to it. We see what our infected senses allow us to see. Perspective exists because we have been taught perspective. Balance exists because we have been taught balance.

If the world was not prescribed to us, we might prescribe solutions that are not reinventions of what we have already experienced, they may actually be new materializations. As one new to philosophy might ask, what is the color blue if you and I don't agree to call that particular combination of light properties blue? I cannot see through your eyes, I do not understand what blue means in the synapses of your brain, but you and I have agreed to call it blue. So too, you and I have agreed to call it quality, or to call it rational because we have both been infected with the same doctrines. We both limp about our worlds with the same wounds. We both have the same dumbness because we have both experienced the same enculturation.

The infection is only spreading. We are indoctrinating remote societies with ideology, inoculating them from potentials outside our constructions. However, it is in those remote societies that the greatest revelations may be derived. The purest notions, those that are least derivative, are those which have not been exposed to the subject, and thus can not mirror it. Insight and creativity are often available through the uninitiated.

The central dilemma of enculturation in the arts demonstrates itself as a subject of critique for audio arts of the last century. As artistic composers in the 20th century evaluated the possibility of sound beyond definition, and experimented with extra-musical sounds, they have come across this dilemma. How do you define sound, when sound has already been defined? All that is left is redefinition. To redefine is to innovate. It is taking an element that already exists, the definition, and refining it. Redefinition is a derivation from a skeleton, and a reassertion of that definition. To redefine space, is to reorganize it or re-appropriate its original intention for a new intent. Redefinition does not exist without the former state, and as such is permanently bound to it. Without reference, such effort ceases to be. These are all bound in innovation.

Yet, creativity lacks that other world changing event – the invention. Invention is typically considered the domain of science. To invent is to create what has not been. When the light bulb was invented, light did exist, but indoor electrical light did not. To invent is not merely to derive from what has already existed, an addition, or reapplication, but to make what did not exist, exist. If the wheel is an invention, the unicycle or bicycle, are innovations on the wheel. Without the wheel neither vehicle exists, but without bicycles, the wheel will remain. Innovative art exists as a stem on the trunk of historically invented art. Innovative art only self-replicates, in a kind of recursive motion, inventing ever smaller, ever more self referential innovations on the same in-vention. Such art is a fractal, capable of creating new through revisions of the former.

Invention is the seed from which that fractal is born. Invention was the ability to preserve the intersection of time and light into photographic image. Innovation created moving film and color photography. Invention changes the world; innovation is what moves it in specific directions.

Imagine then that innovation has as its motivator, reason. This is, after all, a fundamental aspect in the current standards of art and design. If reason does motivate innovation, than the innovation works toward its purpose. An innovation in automobile design creates a more fuel efficient car. An invention in the world of automobiles may obliterate the need for automobiles. If an innovation destroys its subject, it not considered an innovation. It becomes something new, or invention.

Invention may serve as an anarchist's tool, shaking up a status quo without specific intent. An innovation, on the other hand, can serve to strengthen the status quo, or a particular position. Consider the innovation to make a car more fuel efficient. A hybrid automobile is an innovation in automobile design. It encourages the continued use of autos, but does not negate them. No new concept for the use of automobiles can, by definition, make automobiles obsolete. They can only re-appropriate its uses. Overtime the electric light bulb did make its predecessor, the candle, obsolete for the bulk of its uses. The electric light bulb is not an innovation on the candle, nor on fire, it is an invention which subverts the use of candles and of at least one use of fire.

The politics of preserving enculturation then seem apparent. If an automobile company inspired the invention of something beyond the automobile, it could destroy its own industry. An electric light bulb would not be invented by a candle maker. Yet, both would encourage the innovation on each. Improved candles sell more candles, and improved cars sell more cars. To preserve itself culture must encourage innovation over invention.

Innovation has historically persevered. It was apparent in the use of crosses in architecture. When the cross became too prevalent, the symbol was substituted, but still people championed a language of symbols. Ever wonder why history seems to read like a steady line, with short, stubby, offshoots? This is because the vast majority of people working on that art have been innovating on the same theme. They are trained, and when they are not trained, they are still trained. To see an artist make money, or to see an artwork become famous, means that the artist has received training. Even the outsider artists had a kindergarten art teacher, or an uncle who provided them positive or negative feedback on their creation. To be a part of society in any way means that one is enculturated.

The intention of innovation is what helps the process of enculturation. That which is invented is not purposed, and its lack of purpose therefore fails to support any dominant regimes. An art teacher for example, cannot endorse an invention simply because it fails the rules of art design. An invention does not contribute in an obvious way. It does not start from a familiar point and move forward. An invention creates a radical shift, to which the world must often catch up. The discovery of new land does not result in the immediate exploration of it, simply the acknowledgement that it does exist. So too, when faced with inventive technologies, the world does not respond by embracing it, it waits to understand what it may subvert.

This may be the reason the world is bereft of invention, and full of innovation.

This is the danger with experience. Experience remains, even if it is forgotten. A positive experience is a positive experience and it reinforces continued behavior. Consider the experience of the senses. If a person is born blind, they may not know shadow, and they may not want for it. Yet, the person who is sighted notices a shadow's absence in even the simplest drawings.

To invent, the inventor must be blind to specific experiences.

To define the experience to which one must be blind in order to invent would be to miss the concept entirely. It is not something that necessitates a closing of one's eyes in order to prevent enculturation from happening. Instead, it requires a willing exploration into what is not being explored. To be an inventor, one must look for the spaces that are lacking definition, which like the anthropological world have not been explored.

DEFINING THE MAP

Anytime that a people feel they have explored everything or read the limits of their world, an adventurer discovers something new. When the world was flat, it determined to be round. When the world was the center of the universe, it was discovered to be but a speck in a sea of galaxies. This is the pattern of humanity discovery. A void is absent, discovered, and filled. The dilemma for many artists, is then defining what has not been explored. Again, the problem is that enculturation dictates their map of the world. They understand the world as being somewhere between art and science. They understand the world as defined by Euclid, Plato or Paul Rand. The world is what they have experienced. This is a kind of natural topography that people form. They want to learn, so they travel the outlines. The world begins to take definition. As they make that definition, the shape forms. Those who look to find color beyond the lines of the shape are dismissed for the threat they create. The reward for exploration is small. The adventure is marginalized.

To expand the metaphor, we do not send inventors out to sea to bring back the spoils of unexplored land. We send innovators, to confirm our maps, and support our understanding that we have seen the limits. Consider the experience of electronic musician and composer of Musique

Concrete, Pierre Schaeffer. After a lifetime of pursuing musical sound beyond the conventional definition of music, he "returned to the notion that no music was outside of conventional musical sound" (Kahn 1999 p.110). In short, he felt he discovered nothing beyond the map which had already been drawn.

The assumed corollary is too simply derived. If we need a Christopher Columbus to find colonies and define the missing parts of the map, what will the result be? Will it follow the result of colonization, polluting and terrorizing native concepts? Will it infect and perhaps even destroy the power of enculturation? Again, if enculturation is an infection, does a kind of cure exist for it? These questions rise from an enculturated mind. If we understand something we seek to find analogy in it. Analogy is what makes understanding easier, but it may not make it better. The entire world of sciences is polluted by analogies, that when properly examined admit their own breakdown. The world of art, instead finds analogy in relationships. Each art work, the current science of art critique suggests, is really a part of a vast network of ideas and previous art works. Yet, what was the last new invention of new art critique? When the last time science was designed an evaluative approach outside of the scientific method?

The problem is that to invent such things would mean deriving them. To know that you are reinventing scientific method means that you have experienced the scientific method. Instead the inventor of something deemed a new scientific method, will be analogized to the scientific method. Human understanding is bound to what has been understood. This means that enculturation is not avoidable. It is not something that can be born; it is something that must exist.

The next most likely place to seek the inventor is to seek those people who are so out of touch with what is happening that they may be considered absent from the culture entirely. In common terms, these are the *crazy* people. They are the clinically, psychologically absent. The ones to which we assign the label of extra-ordinary in the extreme. Perhaps if we seek to understand their behaviors, creations, exercises outside of analogy, and outside of base, they may actually have something we deem valuable. In such a world the genius may need to be crazy. They may need to be detached from standard experience in a way that makes them capable of invention. If they are not capable of invention, they may still offer perspective which is not obscured by the lineage of cultural standards.

Yet, current definitions of quality necessitate a more practical solution. Besides, logical, thoroughly enculturated individuals are not capable of looking beyond the enculturation to see value in the observations of the extra-cultural. People cannot see crazy as a probable source of invention, as once we could perceive of only a geocentric universe. The result for some must then exist in more practical solutions that are practicable by participants who are enculturated. One such practicable solution that finds much proof of success in art history is called the orphaned protégé.

THE ORPHANED PROTÉGÉ

Art history has proven the story of Oliver Twist, or Little Orphan Annie, to be allegory. Those separated early from their roots, and then magically reunited offer tremendous insight. Art adores the Picasso, who trains classically, discovers what existed, and declares it new. An artist is revered if they seek what exists, and innovates upon it. They adore that which finds, steals and brings it home (Tator, et al. 1998). They love that which turns into analogy, the former invention, to be packaged as innovation.

Another art-historical approach is the temporary escape by the trained artist. Such pilgrimages include cross-cultural immersions, retreats, or investigations. The most extreme involve a complete divorce from enculturation by use of drugs, in the pursuit of synesthetic episodes.

Their work, although stemming from a practice rejected by many contemporary artists, does provide an opportunity for artists to explore beyond that which culture has defined. Their short-lived explorations often resulted in production of some extra-cultural ideas, but at the cost of their own health. Unfortunately, this kind of self sacrifice is commonly the fate of the orphaned protégé.

While an orphaned protégé situation may be practiced, it is not one that people readily engage. It offers tremendous benefit to the process of art production but with obvious cost to the artist. There exist, still more practicable solutions that offer diminished results, but move toward the reversal of enculturation.

CONCLUSION: HOW TO EXPLORE INVENTION

There are several practices which will help the creative individuals escape the conventional effects of enculturation described in this chapter. The following suggestions seek to present starting points for at least the derivation of new process that move toward invention instead of innovation. The irony, that these processes in themselves are innovations, only indicates the pervasiveness of true invention.

The simplest of all techniques to encourage invention are reversals or omissions of the specific practices already employed. If for example, critique is extensively employed in a specific environment, removal of a critique-based evaluation may alleviate that strain on the creative process. Of course, the absence of specific cultural aspects does not remove the culture, but it does remove the reinforcement of that culture. If a person's daily experience does not include specific cultural aspects, there is more opportunity to deviate from its definitions. Such absence is a mini-departure. It is both a safe and inexpensive journey, with little risk, and minor reward.

An alternative solution, if we accept that enculturation is a process of exposure, is to limit exposure entirely. The hope is that some invention may be derived from inexperience. Following the teaching philosophies often visited in the 1960s, it takes more work to un-teach what has been learned than to simply avoid teaching them entirely. The omission of traditional winner-loser competition in primary school is such an example. Although the subject of much formal and informal critique, supporters and non-supporters are most concerned with its effect. There is no doubt, for those who care about such omissions, that there will be an effect. One side argues that it is utopist fantasy void of real world analogy, the other champions its' confidence building and support. No end of the spectrum argues its lack of effect. It would then follow logic, that if an essential aspect of the modern art culture is widely omitted, it may find similar effect.

Begin a critical forecasting of such results by removing a gallery orientation, from primary through secondary education. When art is shown, but not produced, students may not consider social response in the production of their art. With the risk of not being placed on the refrigerator, or on the walls of the classroom removed, students might take greater risk. Speculatively, art production for such students might become a very personal process. Art may also realign toward a different kind of purposeful production. Instead of accomplishing the goal of recognition; it might serve an invented purpose. Invention of the light bulb for example, was not a project derived for the purpose of recognition; it was designed for the purpose of light. By removing the gallery as the central aim of achievement, the achiever may work toward another intended goal such as understanding a process or creating a previously non-invented item. Instead of demonstrating understanding that will gain the approval of the reviewer, the work may allow the student to gain understanding.

Consider another, more moderate situation. Imagine a critique under which a college art instructor increased the consensus or critique base from students in the same discipline to students in multiple disciplinary bases. Envision a situation in which all critics are encouraged to critique the art from the perspective through which they were trained. By increasing the consensus base, the kind of incestuous enculturation that occurs when art students critique similarly trained students is mediated by fresh perspective. This may be particularly effective if the student critiquing has no immediate experience with the work, and works from their non-art perspective. Instead of asking artists if the art is good, ask the non artist what they experience. Such activities also work against the encultured divisions of art and science, and deteriorate the construction of distinctions.

While it is true that grander results may be borne from further points of departure, they are also not well suited for the inexperienced. It is unpractical to assume that students, for example, will take to an entire omission of art-cultural aspects positively. Such students must continue to act within the rest of the world, which as described, is rich with cultural expectation. Such students would become a subculture which might be capable of innovating, but only on what was provided in substitute. Thus if the students are heavily encultured with an alternative mode of teaching which turns away from the standard practices, they may merely be enculturated with the alternative mode of teaching instead. Every culture is a culture, and its effects remain. The force of daily operation within cultural constraints forms the process and attention of the artists within them.

It would then seem clear that there is a kind of stalemate. If there is not an opportunity to build a separate culture, and the current culture leaves the artist in an innovative rut, what is left? The answer is what it has often been – what we seek not to acknowledge. If an absence from society

yields inventive solutions, than one must be absent to facilitate that process. Although contrary to the institutionalized spirit of academic practice, it is a viable solution. Yet, it is not an anti-academic endeavor. Consider the idea that new media artists must be both scientist and artist. If the new media artist wants to invent, instead of innovate, they could avoid the environment which seeks to develop their creative efforts in traditional ways. Simply, they should invest in scientific academic pursuit, and remain absent from academic art. They will likely become enculturated with a process that will acculturate with their existing artistic practice. Instead of pursuing the same goals that each peer artist has admired, they will be exposed, and hopefully enculturated, with the admirations of scientists.

The forward-looking critique of such a process yields a new dilemma. If every artist is engaged in non-art academic training, then there would become a new culture of non-artist trained artists. That is, the culture would then become a culture of mixed cultures. The results are a kind of fractal recursion where subdivision yields but a finer image of what already was. However, for a brief moment, such departures yield a new form. They serve to invent themselves.

Some of these techniques have been practiced in whole or in part by alternative colleges or by art schools seeking to encourage invention. However, much of this effort has been in pursuit of specific critique of the art evaluation process, or of education itself. Instead, it may help to evaluate the pattern of social and artistic enculturation to find patterns that alleviate these problems. A school, for example, that removes grades but maintains a strong focus on gallery showings and publication has simply changed the focus of their grades. To step outside of enculturated concepts requires a complete removal of the enculturated concept. One is not healthy unless the infection is completely removed.

REFERENCES

Bishop, A. (1991). *Mathematical Enculturation: A Cultural Perspective on Mathematics.* Melbourne, Australia: Kluwer Academic Publishers.

Bourdieu, P. (2007). *Distinction: A Social Critique of the Judgement of Taste (*Nice, R trans). Cambridge, Ma: Harvard University Press.

Boyle, P., & Boice, B (2002). Best Practices For Enculturation: Collegiality, Mentoring, and Structure. *New Directions for Higher Education, 1998,* 87-94.

Branscomb, L.M., & Auerswald, P.E. (2002, November). *NIST GCR 02–841: Between Invention and Innovation.* Retrieved February 27, 2008, from NIST Advanced Technology Program Web site: http://www.atp.nist.gov/eao/gcr02-841/contents.htm

Edwards, B. (1986). *Drawing on the Artist Within: A Guide to Innovation, Invention, Imagination and Creativity.* New York, Ny: Simon & Schuster.

Grau, O. (Ed.). (2007). *Media Art Histories.* Cambridge, Ma.: M.I.T. Press.

Kesseler, A., & Bergs, A. (Ed.). (2003). *New Media Language.* New York, NY:

Khan, D. (Ed.). (1999). *Noise Water Meat.* Camrdige, Ma: M.I.T. Press.

Khan, M.R., & Al-Ansari, M. (2005). Sustainable Innovation as a Corporate Strategy. *Triz-Journal, 2-1,* Retrieved February 27, 2008, from http://www.triz-journal.com/archives/2005/01/02.pdf

Manovich, L. (2002). *The Language of New Media.* Cambridge, Ma: M.I.T. Press.

Marcus, G. (1990). *Lipstick Traces: A Secret History of the Twentieth Century.* Cambridge, Ma: Harvard University Press.

Orek, J. (Director) Matthies, E (Producer) (1995). *Making the Matrix* [Motion Picture]. New York, NY: Home Box Office

Roberts, E.B. (1988). Managing Invention and Innovation. *Research–Technology Management, 31,* 11-29.

Roland, R. (1992). *Globalization: Social Theory and Global Culture.* London: Sage Publications.

Tator C. (1998). *Challenging Racisms in the Arts.* Toronto, CA: University of Toronto Press.

KEY TERMS

Acculturation: Change that occurs through the modification, adoption, or adapting a neighboring cultures characteristics.

Art Culture: The value systems, beliefs, and habits of the society of artists.

Enculturation: The transparent integration of a culture's beliefs into one's personal beliefs

Extra-Cultural: The values, beliefs and habits which rest outside the perception, or acknowledgement of the given culture.

Innovation: The derivation of creative efforts that builds, modify, or is directly reliant on an existing artifacts.

Invention: The creation of objects, ideas, or other items not previously conceived.

Inventor: An individual who defines a scientific or artistic approach.

Innovator: An individual who continues a tradition within the sciences or art.

Chapter XIX
The Philosophies of Software

Lindsay Grace
University of Illinois and Illinois Institute of Art, USA

ABSTRACT

Software is philosophical. Software is designed by people who have been influenced by a specific understanding of the way objects, people and systems work. These concepts are then transferred to the user, who manipulates that software within the parameters set by the software designer. The use of these rules by the designer reinforces an understanding of the world that is supported by the software they use. The designer then produces works that mimic these same philosophies instead of departing from them. The three axes of these philosophies are analogy, reductivism, and transferred agency. The effects on computer-based artistic expression, the training in digital art production, and the critique of art are evaluated in this chapter. Tensions between the dominant scientific approaches and the dominant artistic approaches are also defined as destructive and constructive practice respectively. The conclusion is a new critical perspective through which one may evaluate computer integrated creative practice and inspire fresh creative composition.

INTRODUCTION

There is a simple logical proof that describes software's relationship to philosophy. Software is designed. Design prescribes philosophies. Since software is designed, it must also dictate philosophy. The existence of these philosophies, their sociological effects, and the need to critique these philosophies is the focus of this writing. This writing does not seek to define ontologies of philosophies, nor does it seek be an exhaustive examination of the many philosophies that have been institutionalized into the practices of developing and using software. Instead, this article seeks to highlight the existence of a few important philosophies in an effort to encourage practitioners to critically examine their relationship to software and its effects on their practice. In particular, critical assessment of software philosophies engenders fresh approaches to universal, original and effective design.

There are several existing areas where philosophy exerts an influence on software. Each of these areas is not only affected by inherent philosophies, but each area inspires the growth of their individual philosophies by the design and use of their systems. In some cases, the philosophy intersects to create a fulcrum on which multiple assumptions about the construction of the world express themselves. The following sections attempt to outline a few of the major philosophical undertones of common software applications as they relate to the Design of User Interfaces, Avatars, and the use of object orientation.

Careful examination of software decants the following key philosophical elements:

- The heavy use of analogy
- The application of reductivism
- An emphasis on transferred agency

Each of these elements directs users toward specific modes of operation, problem solving and creative efforts. This chapter concerns itself with the identification and evaluation of the philosophies resulting from the use, either successful or unsuccessful, of software built with these elements. The final section of this writing highlights how these philosophies instruct software users.

BACKGROUND

For some, Philosophy is a term that should not be paired with software. Within this subset, philosophy is abstract, whereas computer software design is science. Granted that there are scientific underpinnings to software, it is important to recognize that software is used in increasingly abstract ways. It is used to create art, it is used to communicate, and it is used as an integral part of daily work that involves abstract thinking.

The philosophy of software is a topic of research and rhetoric in many disciplines. Although not always considered a philosophical examination, practitioners of law, education, commerce and nearly every software-effected discipline have discussed a kind of philosophy of software. These concerns include intellectual property rights, electronic learning, and the design of systems. The philosophical and commercial work of the Free Software Foundation, for example, is directed toward the specific effect software production philosophies have on the quality of software produced. Theirs' is largely an examination of how software production is practiced, not an examination of how software effects production. This writing seeks to expose the effects of philosophies so ingrained in the production of software that they are seemingly transparent. In the oft-used paraphrase of Marshal Mcluhan, we shape our tools and then our tools shape us (1994).

It is important to note that this discussion excludes an examination of hardware's role. This is because hardware finds design from the realities of physical sciences, where software finds design from logic. The critical evaluation of this logic decants priorities, ideologies, and value systems. Simply stated, it influences the foundations of existing philosophies. Those philosophies are encoded in the language and structure of software, and interpreted by the user.

Investigations into the effect of language on people's ability to understand specific ideas are perhaps more akin to the focus of this chapter. In Noam Chomsky's (2006) essays, (complied later in the book Language and Mind), he encouraged critical assessment of the relationship between linguistics and philosophy that opened for examination whole processes of communication and approach. The role of linguistics as a tool through which we produce and communicate meaning is similar to the role of software. Software serves as the tool through which much daily production and communication occurs, making it the lingua-franca of operation in most of the western world. It is the basis for communication and the primary vessel that facilitates communication. Programming languages are also the tools we use to direct the development of a solution.

The wide canon of writings describing human-computer interaction also serve as a solid foundation for understanding the tight relationship between psychology and sociology in the development of software. Ben Shneiderman's Software Psychology: Human Factors in Computer and Information Systems is a good starting point. Work in Human Computer Interaction is an easy avenue through which to discover the philosophies of software. It is at this edge of computer science that its many philosophies are first experienced. Human Computer Interaction is also an approachable subject for anyone who has ever used software, as the user has experienced at least one end of this relationship. As such, this writing extends much of its critique to the decisions made about how people must interact with computers.

Lastly, and perhaps most importantly, readers may find valuable related critique in the writing of Jaron Lanier. In particular his essays, One Half of a Manifesto, and Digital Manifesto, evaluate the relationship of computer technology to the society in which it is developed (Lanier, 2006). Where Lanier's concern is with the overarching social effects of computer development, this text investigates the narrower effect of software design on the process of problem solving.

ANALOGY

Design of User Interfaces

Software user interfaces are largely constructed through metaphor. The desktop of an operating system, the buttons, files, sliders, and others tools are digital implementations of real-world objects and interactions. As theorists have outlined, many of the metaphors are reinterpretations that fail to be wholly faithful to the representation of their original. The desktop, for example, is a shallow metaphor because it exceeds and misaligns the attributes of a real-world desktop. Among its weaknesses as a metaphor is the fact that actual desktops have three dimensions, edges, physics, and other elements not present in any of the dominant operating systems. Even more recent attempts, such as Bumptop, choose specific attributes to perpetuate the use of such analogy (Agarawala 2006). The result is a filtering of subject. This filter exposes the author's values and understanding.

This selected representation of specific elements through metaphor identifies the first hint at the philosophy in user interface. If there is a reinterpretation of the physical object into digital space, then the designer has selected from a list of properties those items that best meet the designer's understanding of the physical object. The result is a simplification, or a kind of wire frame, that exhibits only the items that are most valuable in identifying an object. This value is defined by the one who implements the interface.

Analogy itself is neutral, but the application of analogy is not. In writing, analogy is a rhetorical device employed by the author to make a point. Analogy is a device of argument. In the writing of software, analogy continues to make claims. It highlights what is important, and shadows what is not. Yet, unlike writing, software claims are not the subject of critique. Software is understood because it is a tool, and is not designed to withstand critical assessment as its primary function. Yet, an analogy speaks volumes about both the item critiqued, and the author. Every analogy resounds with the author's value system, simply because the process of analogy requires the author to identify wheat and chaff. That which is discarded is of no value to the author, yet in other contexts, that which is discarded is most valuable.

Consider the window as analogy. The dominant property of the operating system window is its ability to display content within it. The analogy is simple to understand. Windows in buildings show the content of the world; software windows display their subject content. Yet, that is not all windows do. Windows insulate, move, and provide multidimensional information about time of day, weather conditions and more. Windows also do more than open and close. Of the many properties of windows, only one dominates the analogy. Other properties are discarded, to simplify the analogy.

In its early fabrications, the graphical button is simplified to an item with two states, on and off. It is differentiated by size, color, and more recently shape. In these two simplifications, the philosophy of design manifests itself. From the dominant theories of computer science, the button is simplified into binary states. The button is either on or it is off. Yet, a literal representation of a button would allow for range. Do buttons in real world machines merely have an on and an off, or do they also exhibit other properties based on length of time depressed, speed of depression, and number of clicks? Interestingly, the extremely pervasive intermediary between user interface and user, the mouse, does exhibit these properties: click and hold and double-click. However, once the button becomes digital such properties are selected against. These selections permeate successive generations of software and in turn shape the way in which buttons are understood. The rarity of timer-based buttons, switching buttons or dial and push buttons, all historically useful physical interfaces, is a hint at the forgotten population of interface elements.

Instead, the designer of interface routinely works within the understanding of interface defined by their predecessors. Web Design and Business systems are particularly prone to such tendencies, as their production times are shortened in the race to bring the product to market. Yet, these are everyday interfaces, like the microwave and the television. The everyday interface may easily be among the first human-computer interactions for an individual. This initial experience will likely define an individual's expectations, and more importantly, their understanding of interface. The everyday interface is the gateway to software philosophy.

Philosophical Contradiction: The Acceptance of Non-Truth

To become aware of the philosophies of software, it is important to become aware of its values. Windows and buttons are shallow metaphors. They require substantial suspension of disbelief or belief when used. That suspension is aided by obscuring the user's conventional understanding of the item. Suddenly one window cannot be seen through another. Somehow, two buttons can exist in the same space. One such classic example is the original Apple Macintosh trash can interface element. The trash was a place to discard old files, but it also serves as the means of ejecting a diskette from the machine. To use the operating system, the user accepted the contradiction that disk retrieval occurs through the discarding process.

In these environments, the permeation of analogy inflicts perceptual and conceptual contradictions that the user must accept in order to use the system. Paradoxically, the interface becomes a world of same, same but different rules. Those rules are managed by an inherent value system, which enforces what is important to perceive and what is not. Conventionally, if a user tries to make a button stick down, they will fail. If a user tries to hold multiple buttons they will fail. Interface is thus, prescriptive. It instructs its value system by creating a reward system. If the user accepts its value system, the reward is success. If the user fails to accept its value system, the user is punished with impotence in the software environment. Beyond the interface, this instruction abounds in software from video games to business applications. Efficacy in a software environment cannot be achieved unless the individual subscribes to the applications rules. If those rules contradict each other or the individual's understanding of the world, then the individual must still accept them or surrender his ability to act within the software environment.

The dominant human-computer interaction model demands this prescription, but it is not the only approach available. Some video game environments have, in their constant treasure hunt for play, made spectacular inroads into non-prescriptive human-computer interaction. Sandbox games or environments in which the toolset may be constructed and manipulated by the user's definitions offer fascinating anti-prescriptive opportunities. Their potential for educators, and for critical assessment of process, is inspirational. Although the application of this method in non-play environments continues to be limited, sandbox environments such as Gary's mod are quality examples and offer alternatives to the most common modes in use today.

The Origins of Analogies

In order to critique the analogy, it is important to understand its genealogy. In some cases such choices are the result of iterations on the same initial design. In other cases, these choices are derived from an interpretation of other disciplines. The analogies of the early painting programs demonstrate a clear translation of paint technique and color theory. Each time, there are notable exceptions that illustrate a system of values in the application. Software that may be analogized to painting processes have focused on the brush and vehicle, but not the surface to which they are applied. Users may choose brush size, pattern, vehicle, and others yet the fundamental choice of the character of the surface to which the virtual paint is applied remains the same. The selected properties mimicked by the application do not compliment the process of painting, but they are prescriptive.

Such environments also preempt the possibility of alternative models of creative process. The process, its tools and their relationships are predefined. Consider how difficult it is to create a work in the tradition of Jackson Pollack, for example. His work relied heavily on a multidimensional approach to paint application. Most applications focus on the tool and the narrow application of intended use. These applications do not employ algorithms that calculate brush momentum, gesture or brush material. The applications selectively mimic the process. The resulting omissions exclude specific forms of art process because they are beyond the software designer's initial definition of painting.

In compliment, selected art theory is applied to even the most mundane graphical user interface elements. The fundamental notions of graphic design extend into button design. Half of what is understood as graphic design's basic elements voices itself in the differentiation of buttons. As interface technology improves, it implements more of these elements. Shape, texture and hue, for example, are now standard button attributes where once they were not.

Yet graphic design prescribes only one practical means of understanding its subject – in finite space and time. The philosophy of graphic design is tethered by a long-standing relationship to permanent state production. In its philosophy, an item is created in a specific moment and has a set of attributes that remain true for the objects existence. Size, shape, texture, lines, hue and others are permanently identifiable. This is not true of interactive design, which is fundamentally impermanent. An interactive design has varying properties depending on multiple dimensions that include time, space, and event but are not limited to them. Interaction design may also be understood as a confluence of product of actor, state, use case and more according to the Unified Modeling Language (UML). If UML, a language designed to assist in the design of large software systems, describes interaction more completely than the software used to create interface, there seems to be a schism between understanding and implementation. The likely reason is reductivism.

Reductivism: The Finite State Machine

Reductivism defines a historical art movement of painting and sculpture that emphasized simplification. It is also a dominant practice in the construction of computer solutions.

The practice of analogy-based software implementations is reductive in nature. The relationships, interactions, and processes executed by the computer are reduced to their simplest forms. The fundamentals of computer science call for its authors to reduce their subjects to their interpreted atomic forms. For example, a button becomes an element with only two states, or a customer becomes a number.

The finite state machine, or FSM, is an excellent starting point for analyzing the effect of this reductionist philosophy. For computer scientists the finite state machine serves as a model of behavior. It decomposes its subjects into a finite number of states, then or and transitions between those states and actions. The finite state machine is a popular approach to engineering computer logic, including artificial intelligence and human-computer interaction.

Philosophically, the finite state machine defines computer science. Its first step is to make the seemingly complicated simpler. Like much of the scientific approach, it begins by identifying the fundamental elements of its subjects and then it constructs a simulation of behavior from those elements. This construction occurs through a process of deconstruction, where anything to be modeled is first dissected and labeled. Driving a car for example, becomes a flowchart of red light checks and speed monitors effecting driver and car.

All of the FSM modeling process hinges on the appropriate identification of states, transitions and actions. If an element is left out, or it is not related correctly, the resulting software fails to complete its accomplished goal. The process relies heavily on proper decomposition. If the subject is cut in the wrong way, the software may fail to be an accurate model. Yet, the subject must be reduced in order to fit the limitations of computer science. If it is to be modeled in software, the dictum reads, it must be simplified. Again, as in analogy, this reduction necessitates a set of selections. The initial designer of software must decide which elements remain, and which do not.

Quality software production, as currently defined, borrows its understanding of process, states, and the atomic units of its subject from study of the specific situation. A system designed to model chemical interactions, for example, will be informed by research in chemistry. This works particularly well for scientific disciplines, which share in the reductive approach to understanding problems. However, what happens to the disciplines that are not reductive? If a discipline or philosophical understanding of a discipline is not reductive, then computer science may fail to apply its theories.

Consider the contemporary definition of art, which does not assert itself as an understanding of singular necessary components. By some, art is understood as a deliberate arrangement. Deliberate can't be defined as a state, a transition, or an action. Deliberate arrangement is the quality of an action, but the FSM has never been well suited for qualities. The qualitative, that which rests on a non-finite judgment, is not simply categorical. It cannot be quantified, and as such becomes an unmodeled element. In most cases it is simply truncated as a non-calculable. When qualities are judged in computer science, they must still be converted to quantities. Consider the basic algorithm for drawing a curved line on a standard monitor display. If the curve is to be bitmapped for display, software must decide which square pixels will be lit to establish the curved line. All pixels are arranged in columns and rows, so a curve must be estimated. Simply, a curve must be forced into the categorical and decomposed into rows and columns. The result is something that looks like a curve, until it is scrutinized carefully.

It is like a curve, but it is a simplification of a curve. The bitmapped curve serves as a proof in computer software design terms. It is analogous to the way software models subjects that are not scientific; it ignores that which does not fit its designed intention. The result is that the bitmap serves as a strong analogy, and will be used. The vector does not, and will be ignored.

Dominant computer software development techniques, such as object orientation, procedures, and others fail to sufficiently solve many problems that are wholly qualitative. In order to apply procedures, for example, the subject must still be decomposed. This is where a very wide gap distinguishes itself. If the subject of a software-ization is not reductive, it will more often be turned into reductive elements in order to fit dominant software philosophies.

For computer scientists, the many dimensions of interaction can be encoded into state machines. A common scientific view of interaction requires three processes; define a timeline on which to design, define a screen dimension, and create an event model. Yet, even when all three of these dimensions are combined into a single piece of software they prescribe a specific understanding of the world. Event models describe actions that are rigidly categorized to support the computers understanding of interaction. Events are trapped singly, as an intersection of space and time. Space is defined in absolute terms, as coordinates in a grid.

This model, included first by the software developer, is then adopted by the user of the software, a designer. The designer, eager to build with the tool they have been given, must accept this model in order to operate within the software's constraints. The designer must define their understanding of interaction to accommodate the software's abilities.

Before long, the other means of interpreting user input or of describing relationships simply eludes many designers. The other interpretations are not beyond comprehension, they simply fall in the shadow of other's successes. Much like the history of the electric car, which is as old as its combustion based peer, other software philosophies languish. They languish in the absence of research, and in the distraction produced by the show-stealing conventional approaches. After all, the logic reads, these approaches have served us well. Still, it is easy to be critical of such interpretations. In the case of interface, isn't space also understood in relative terms, as in proximities, neighbors, and distances? Aren't there degrees to interaction that indicate situation? The use of such software drives the user away from these questions. Instead, a predefined level of granularity, deemed appropriate by an initial designer, is accepted as useful truth. Eventually, a literal understanding of interaction is supplanted by a modeled understanding of interaction.

Finally, it is important to understand that the finite state machine sits near the intersection of linguistics and computing. Students who learn programming language design, for example, learn the fundamentals of regular grammars and the Chomsky Hierarchy. The hierarchy was theorized by Noam Chomsky, the linguist and philosopher. Chomsky's work provides the philosophical basis for the way many computer languages interpret instructions. Simply, every programming language that employs Chomsky's Hierarchy is employing Chomsky's philosophies. These are philosophies of communication, human behavior, and human relationship.

Reductivism: Object Orientation

Translating the designer's needs into binary terms that the computer understands is an artifact of digital design. The current philosophy states that for software to work, all things must be reduced for codification. This is due in part to finite state machines, but object oriented development, with its hierarchies and inheritance, also drives software development and use.

Software applications are typically developed under the master philosophies of object orientation and inheritance. This philosophy prescribes that there are distinct entities, which when categorized can be forced into an ontology that adequately describes all expected situations. The assumption apparent in this approach works well for producing specific types of software but is exasperating when evaluated from the creative perspective. The philosophy reads that the world is comprised of a finite number of blocks through which there are an estimable set of permutations. If so, this determination predicts a calculable end to creative potential. There are only so many ways that each object can be constructed in this finite world. Wouldn't that then leave the creative world toward an enormous game of Sudoku, where each artist is merely attempting to complete the missing permutations?

A challenge facing software developers can then be traced to the fact that they are taught to compose solutions from dissected components. Object orientation champions the process of simplifying and labeling. These simplifications are shallower than those of the analogies dominating interface because they have become the truth of the software system. Developers begin building their solutions from the already modeled objects that existed before they began their project. A conversation between two computers for example, may have been modeled as a group of listener and speaker objects. Through years of use, that definition of conversation becomes the only definition of conversation. If many applications have been built against that object model, and if there are no problems that arise from that understanding, then it is understood to be an accurate model. Yet, if everyone is working of off that model, and the understanding of that model is passed all the way through to the user, then how will problems occur? Who is left to reinterpret it? The model becomes truth.

Interestingly, science informs science, as computer science finds and defines its solutions by definition provided through other sciences. The terminology in object orientation, for example, is clearly borrowed from genetics. This is a type of incest, which begets solutions that serve themselves. To understand computer science, one must accept its cousins, namely genetics and math. But both these sciences, by their own admission, are incomplete and arguably self-affirming. Genetics has many questions in heredity to answer and assumptions to debunk. Contemporary mathematicians are in the middle of a fundamental reevaluation of math itself (Barrows). As Barrows (1992) described in his critical history of mathematics "our picture of the most elementary particles of matter as little billiard balls, or atoms as mini solar systems, breaks down if pushed far enough, so our most sophisticated scientific description in terms of particles, fields, or strings may well break down as well if pushed too far" (Barrrows, 1992, pp. 21). If "mathematics is also seen by many as an analogy" (Barrrows, 1992, pp. 22) then isn't the construction of software solutions the organization of analogy on an analogical foundation?

Also forgotten is the idea that science, as has historically occurred, can borrow from art or other non-scientific approaches. The producers of several great works in the mathematical realm include multi-practitioner philosophers ranging from Plato to Bertrand Russell. Like computer scientists who attempt to fit solutions into the analogy of object orientation, these practitioners use analogy to explain their philosophies. Yet, not surprisingly, these analogies do break down when pushed too far. Artists, after all, are encouraged to find this breaking point.

The results of this breakdown, the disparities between the pre-defined modular units of a software application and an individual's desired solution, occur routinely. When identified, the common resolution for such problems is to use the existing model to construct unexpected results. If, for example, an artist's 3D software creates only 4 primitives, they are instructed to create a

5[th,] previously undefined primitive by using the original 4. Yet, the 5th pyramid may not actually be a 5th pyramid; it may be a model of the 5th pyramid, lacking some of the 5th primitive's real properties. In casual software language, this is a workaround or hack. The solution is an un-planned retrofit of the solution provided. Too much need for hacking typically indicates insufficient design, yet for creative enterprises, the hack is often the fundamental work unit. New media artists, for example, are fully immersed in the process of hacking, simply to create their proposed solutions. In more practical terms, 3D animations are performed on stages, with rough simulations and environments, like backdrops on stages, and are often not represented in three dimensions, but in two dimensions. This fact then hints at an insufficient solution. The current solutions fail to meet individual's needs.

Returning to the 5th primitive, it is not important, in the eyes of the computer scientist, because it is not part of the original definition of its subject. The world, as defined in the initial software design, does not contain such objects. This is the case with non-Euclidian spaces, like the Mobius Strip or Klein Bottle, which, outside the original models of geometry created by software architects, are very difficult to construct in computer software. The artist is made impotent in a world of digital imagery that precipitates from a chosen philosophical approach, here Euclidian geometry, in to the representation of their image.

In a broader scope, this codification permeates our approach to solving many problems. Simply, Computer Science philosophies deteriorate our understanding of the world. It champions low fidelity, by encouraging the simplification of data, relationship, and multidimensionality. A good computer scientist, as the mantras dictate, converts complicated problems into a subset of simple ones. The mantra ignores its opposite. No computer scientist seeks simple problems and complicates them. Yet, artists often seek simple problems and complicate them. From this

perspective, the computer scientist is trained in the act of decomposition. The artist is trained in composition.

As an example, war has a simple solution, stop fighting. The artistic philosophy seeks to unearth the complication in the solution. The artistic philosophy mandates a complication – why is it so hard to stop war, who is involved in war, what does it mean to stop war. The computer science philosophy, instead, seeks to simplify the problem of war so that it may be codified into algorithms. For the philosophy of computer science war is a collection of attributes, mini problems, hierarchical structures, and structures which, like atomic structures, combine to create a whole. Computer Science suggests that it is the responsibility of the designer to interpret those atomic parts before construction. The instructed exercise is simplification - moments become minutes, individuals become groups. To approach the resolution of a problem in any other way on a computer is futile.

The first governments were built on the identification of appropriate purpose of government. Yet that understanding changes over time, and the models changed with their understanding. For some, kingdoms turned to democracies based on assessment of need. Kingdoms, as defined by their kings, were for the benefit of their subjects. They were understood to provide a necessary top-down approach which enforced perpetual, informed management. As some governments evolved, they found a less hierarchical design met their needs. Through a series of wars, hierarchies were overturned for democracies. The democracy continued to offer perpetual, informed management, but the management moved from single silos to more complete, multi-dimensional perspective.

What these civil histories offer, even described in scientific terms, is a model for the potential evolution of software design. The kings of software design hear dictatorial truths about the process of creating solutions, confronting revolutions. Alternative models, such as the growth of open source software or the democratization of information encouraged through various web-based tools like Google maps, indicate a change in way software is constructed. The change is somewhat democratic. Where there was one architect, there are ten. Where there was one algorithm, there are now three algorithms, and twice as many authors.

In order for artists to continue their history of revolution from within the digital domain they must operate outside the inherent philosophies of the software they use. The artists must operate beyond the defined class with identifiable property and objects. They must find a creative space that does not dictate a master–slave relationship between hardware and software components. To accomplish this requires far more initiative and conviction than one might assume. Even the seemingly democratizing force of web art is inherently ruled by philosophies of super-users, IP checkpoints, and a cascade of style inheritance.

Fundamentally, codification means a reinterpretation from spectrums to silos. The only thing that changes is the resolution of those silos. Silos become wider, or thinner, but they continue to be silos. The process of codification is richly philosophical, requiring judgment, selection and interpretation. Critically, the decisions to codify are defined by science itself, leaving little space for other approaches. Yet, the opportunities for extending the reach and power of software may exist beyond the walls constructed by the dominant approach.

Reductivism: Examining the Reductive Language

The language of computing is binary. Its language does not operate on ranges, but its resolution is able to exceed human perceptual range. In the display of graphics, for example, color calculating algorithms are capped to the 65 million colors that are understood to be the perceptual range of human beings. This decision presumes many assumptions – there is no need to code for anything but the average person. The science of human color perception is complete, so development beyond perception is unimportant. Out of historical context, these assumptions seem reasonable. In the context of history, they are ideological. Did we once believe the world was flat? Did we once believe the entire world had been mapped? Did we believe the human body was made of humors? Critically, the act of simplification, the philosophy of deconstructing and codifying, abounds in the software we use.

TRANSFERRED AGENCY

The Use of Avatars

In recent years, avatars have become the dominant device for movement in 3D virtual spaces. They are a logical extension of the mouse pointer. What the mouse pointer, is to the index finger in a 2D user interface, the avatar is to the body in 3D space. Both, however, offer an inherent ideology – the user needs agency through a third party.

Good design has evolved from the pointer-facilitated navigation to the touch-screen. The result is a system that is easy to use and requires little training. It is, to use an often dangerous term in human-computer-interaction, intuitive. Touch screen use in automobile navigation systems or automated teller machines is likely easy to use because it removes an intermediary. People do not expect, or necessarily want, the computer to act for them. They want to act.

In education this is a point of critique. Should the teacher assist the student in coloring within the lines, or allow the student the freedom to color outside them. In software, the bias is toward prescription, not exploration nor adaptation.

The use of avatars in games is an interesting illustration of this tension. Games are designed to be immersive. They seek to envelope the user in a manufactured experience through a series of simulations and real world analogies. To seem threatening, fire, for example, must emulate the properties of fire in the real world.

A game avatar is a copy of self, in another environment. It is a live broadcast, with self as subject combined with a fiction. Paradoxically, the avatar is the person and it is not. If the player understands the character is *not self*, then they may sacrifice some components of the immersive experience. If they believe it *as self*, then they must subscribe to an arresting philosophy.

The logic is as follows. The avatar is not self. The avatar has great efficacy in the world. I do not have efficacy in the world. I can use the avatar as a tool to gain efficacy in the world. Efficacy is gained through the use of tools. The avatar reinforces the use of tools for agency, not just augmentation of agency. The tools within software encourage their necessity. Using software reinforces, at least philosophically, the need for software.

The dynamic of avatar based software is one of master to slave. The slave, or avatar, is only useful when they are faithful to the commands of the master, or user. An avatar that fails to heed commands is buggy or useless. The prescribed use of avatars indicates that avatars must take commands and take the user's risks.

Avatars are also identified by an outward appearance. In the multimedia world this is a balance of sight and sound. Yet, this is another shallow interpretation of identification. If science has been wrong in its perceptions, how can perception be decanted to what is reproducible in vision and sound? The dominant theory is that more lifelike performance is delivered from increased data resolution. If pixel resolution extends beyond our understanding of human perception, then theory dictates that it will be perceived as a real image. Yet, theory does not offer resolution to the dilemma that perception is a multi-dimensional equation. To sell a better image as "more real" is to sell a larger steak as more cow. The whole of perception extends beyond the silos of sight and of sound. One philosophical view is to consider sight and

sound, not as the computer treats them, but as they may be in the real world – a codependent harmony. The deconstructive tendency of Computer Science encourages the use, treatment, and display of perceptual elements independently.

Although quickly discredited in scientific communities, the notion of self and of image of self has been argued to a point beyond what science perceives. Psychological and sociological basis both find themselves as under-represented social science minorities when analyzing avatar implementation. It might be argued that avatar compliments, however, do provide social-psychological informed equivalents in crowd simulation and other models of social behavior. Yet, critical review of these systems finds an initial iteration that is first informed by physics simulation and then roughly layered with pseudo-socio-psychological logic. This logic is, of course, reduced to mini-module logical expressions that reduce the social sciences to computable patterns. Simply, the aforementioned social science minority finds representation in an avatar world, but it is neither significantly represented nor wholly represented. Where Newtonian physics find pervasive application, the soft sciences are episodically integrated into the 1st order science worlds. As anecdotal proof, I offer the disposition of a course introducing the relationship between psychology and computer programming:

"It was hoped that this course would have encouraged participants to view software engineering as a human activity, as well as a more formal discipline. However, the course was cancelled by the powers-that-be after its first semester with lack of student interest being cited as the reason." (Lenarcic, 2004, pp. 257)

The exclusion of social sciences in the development of software systems, even systems that seek to emulate human behavior, is seemingly absurd. How does the science that researches behavior find itself noticeably absent from the science seeking to simulate its philosophies? How did social sciences come to occupy the radical edges of computer science when the computer was designed to serve human needs?

Most likely, the reason is in the philosophy of software. Social sciences are scant to reduce problems to a few small causes. Good social science, the ideology dictates, recognizes the complexity of relationships between the myriad of factors causing specific situations. Good computer science seeks to reduce those factors.

Consider the Boids algorithm. It is an attempt at coding the movement of animals. It reduces the intelligence of movement to 3 key factors, separation, alignment and cohesion. The result is a believable simulation of flocking movement. Successive iterations expand on or use these factors. Yet the foundation for this behavior is primarily physical. A social science description of the primary factors might begin with factors such as intention, drive, desire, and social affinity. Since neither implementation is actual executed in a physical space, but in a theoretical virtual space, both has as much applicability as the other. Yet, the 3D world is driven by a previously existent definition of its world based on a definition of 3D spaces. Hence, an animator finds themselves demonstrating emotion through physical gestures, and later working in sound. The entire basis is physical, with emotion and the soft sciences deriving their reception from perception. Could it be that there are other modes of received information? Social scientists have conducted experiments with alpha waves and other tools ad infinitum, which computer software has completely left out of its definition of world. The direct result is that they have been omitted from the possible expressive means of the users of their software.

Consider the dilemma of relationship building. Computer science decants human-relationships as a computable system of categories. Matchmaking systems find matches through data in a relational database. Social networking systems do the same, and offer computed scores as feedback. Users acquire thousands of friends with little regard for their qualities. Friends become binary, they either are friends or they are not. The quantity is what matters, the quality is non-calculable. Even when the number exceeds what might be considered the literal definition of relationship, the counter keeps climbing. If social science research calculates the maximum number of human relationships at 150 (Dunbar, 1992), computer science defines it as infinite. A scientific peek under the skin of such systems hints at some astounding suggestions. Are friends to be collected, like points in a game? Are friends to be removed for poor performance? Are friends or partners to be determined by categorical matchmaking? Is the sociology and psychology of friend-making simply an equation of demographic data? Do these systems hinder exploration outside the software designed silos? Who directs a search in these environments, database tables or human need?

What We Learn from Our Software

Try evaluating the user interface as whole. Interfaces encapsulate a variety of anti-explorative philosophies. They teach users how to be author led. The environment of an interface is strewn with expected paths, wrong turns, and caution signs. Interface is defined with a push-pull between human and computer.

The need to simplify is emphasized in the abstracted icons and half-analogies of many digital interfaces. Software limits resolution, determining the adequate detail to which a designer designs. The interface of design software emphasizes its own approach. It requires its world to be understood though the same system under which the designer must design.

By using these interface elements, authors are prescribing to these philosophical decisions. Interestingly, few authors of interactive works think critically about these rules to which they are subject. They simply understand the world in which they create to provide rules under which they must work. If these rules are oppressive they do little to thwart them. Instead, they make every effort to use them, and the artificial divisions that are constructed for them. These systems teach even the most revolutionary how to operate under constraints. That the rules of this interchange between human and computer are unchallengeable is another fundamental given assumed by both author and user.

However, this is not solely the fault of the creator of these systems. The author received instruction, and more than likely that instruction included mandates about when one control is used over another. A common introduction to interface design often includes a list of controls and when to use them.

Graphical user interface instruction focuses on what is, not necessarily what should be. Interface design is routinely taught as an exercise in organizing pre-defined interface elements, not as an exercise in creating new interface elements. Designers, in particular, are driven toward a junkyard mentality, acquiring interface elements as they are offered by computer scientists that design them. Yet, to do so is much like painting with only primary colors. If it weren't for the cloud of innate philosophies in software, designers would see the potential in blending interface elements. In its simplest, there would be use for a drop-down list button or a check-box-button-image-list. Instead, many designers are using an out of the box approach, creatively employing use of the set interface elements provided with their chosen application.

Designers also have tremendously untapped potential in the custom design of interface. This extends beyond the common use of specialized controls, since controls themselves are one of many solutions to the dilemma of soliciting feedback from a user. Using the analogy of the real world, progressive designs have requested gestural input. Gestural input represents a fundamental shift simply because it breaks from of the computer as machine analogy that permeates control oriented software. Universal design has also adopted audio interface as a hybridized solution to the

dilemma of users with encumbered hands. These two approaches demonstrate a human-computer interaction that attempts to emulate human-human interaction and it could be argued, deteriorates the analogy of graphical user interface as machine interface. Yet, these solutions do not move far from analogy. Speech communication, could for example, be derived from human-human interaction. Such design is then not limited to the creative limits of the designer, but by their ability to find analogy in relationships other than human-machine interface.

As an example-limited design, Microsoft PowerPoint has often been described as a limiting force in presentation. As Tufte (1993) suggests in the Cognitive Style of PowerPoint, it guides discussions in linear paths and decants content into simple bullet points. It also changes the way users organize information, as its attempt at simplicity guides the formation and organization of information. Even its organization of information indicates a value system, where, for example, style is simplified to color, the software precludes the use of angles, and changes the order in which commands might be executed. Users of PowerPoint then become subjects of the PowerPoint philosophy.

What Software Teaches Users

Sociologists believe that specific social systems effect the way that their members perceive and act. The corollary is that members of specific technical systems, through which they work and socialize, will have a similar experience. The software we use on a daily basis effects the way we understand and act in the rest of the world. Our systems have already defined new language, like emailing to im'ing. These new verbs define asynchronous modes of conversation. They describe conversation initiation without invitation. They describe new ways to converse.

The effect extends well beyond language. Software systems define the way in which we interact.

Where once machine mediated communication was full-duplex, allowing the communication of our message at the same time we are listening to our message, many popular electronic modes of communication are not. Asynchronous systems, although in some ways technologically minor, are the dominant mode of electronic communication. Email, message boards, blogs, and critiqued posting are all single duplex message systems. These software systems encourage users to talk, and then listen. An email message is sent, and the user waits for the response. Other responses to other questions may arrive in the interim and messages may simple go unresponded. The social equivalent is talking into a crowd with no expectation of your message being heard. In the case of community posting websites the notion of communication is profoundly alienated. Communication becomes a system where one talks and others wait to find that which interests them enough to bother talking. A user posts a movie, and other users browse the catalog of movies, and may decide to critique the recently posted movie. There may be no response at all, akin to speaking in a classroom and getting no response from teacher or student. There may be a flood of responses, but those responses may drift from the subject of the movie to the appearance of the poster. This is similar to a presenter receiving critique on their posture, instead of the content of their presentation.

The notion of being conversant, listening and talking, are replaced in these environments by a new model of conversation. This model is information provider heavy, and information consumer bereft. The consumer famine, is not for lack of information, it is for lack to consume. Software systems make it easy to ignore and even easier to skip. These are the conveniences of software systems. They are also philosophical loaded.

In the standards of design, systems that do not offer the autonomy to choose what to see and what not to see are failing the user because they fail to provide what the user needs. It is considered draconian to dictate, or to take control of the information received by the user. Yet, are there not compulsory experiences that require the attention of the user. Is a film not an entirely different experience if the viewer skips through the center of it, or watches it while watching another? Are the world's greatest speeches effective as web broadcasts?

One philosophy dominates the software industry, and that philosophy is that freedom of choice is positive. This is perhaps a remnant of a developed society which champions choice, or the opposite, a society which has reveled in the choices provided by its systems championing that which it believes empowers it. Does it improve the movie viewers experience to be able to move through the film or is that choice a remnant of an analogy to an archaic device whose translation disagrees with the philosophies of software design and development. Media player software, for example, uses the same control concepts as a cassette recorder. Yet, isn't digital media unbound from the limitations of its predecessors? Aren't the choices provided by digital media players analogy based, but use deficient? The choices they provided are not necessarily appropriate; they are informed by previous systems. The choices given are no better than they choices we had. In such cases the choices given are not given by some well-designed analysis of need, they are given by precedent. Choices given by precedent may become superfluous choices.

What is the need for a choice when its results are negligible? A conventional tape player offered play, stop, pause, fast forward, and rewind and eject. Any media player that provides all six options offers superfluous choice. Eject lacks proper analogy in the digital domain because files are switched, not removed and replaced. Pause lacks application because digital files are either played or stopped; there is no need to leave the tape head pressed against the tape to preserve the current time slot. Yet design dictates that this choice, in particular must be preserved. The result is that many media players retranslate the stop. Stop becomes stop and reset to the beginning, where pause remains the literal stop playing media. The result is a misappropriation of concept. Where once there was analogy, there is an artificial preservation of choice and reapplication of concept. Stop is redefined. Stop means stop and reset, pause means stop. The relationship to the philosophy of communication returns. If one pauses a message, it lingers. If one stops a message it must be started again, for there is no communication that continues where it was left off. Here the nature of communication is being dictated through choice. Communication, whether it is an asynchronous messaging system or the message in digital media, has a few properties defined by software. These include:

- Messages may be broken into segments; the whole is equivalent to its parts.
- Messages are navigable
- Messages are sequenced by quantifiable units

All of these properties are dominated by contemporary computer science approach. The message can be reduced to its smallest parts and in doing so the message can be simulated. Speak with many artists and this can be an inflammatory idea. Can a film be understood by anyone of its 60 minutes? Can an oratorical discussion be preserved as a list of items which can be skipped through, sorted by topic, or returned by topic relevance? Is this essay reducible to a single paragraph? Is a mash up of sound bites representative of its subject?

The destructive power of computer science reduction demonstrates itself. That which was whole becomes parts. What does such daily behavior teach people about their world? Does it encourage us to long for an opportunity to fast-foreword through our monotonies? Does it encourage us to find the shortcut and get to what matters to us? Does it teach us to perceive the parts instead of the whole?

Consider other philosophical approaches to message. What does a non-navigable message communicate? What does the definition of quantifiable units do to the subject? Can message units also include objective items outside the analogy of time-based linear systems?

What Software Teaches Masters and Novices

It is important to remember that software systems have the ability to confirm our perceptions. When the application of an idea proves successfully, it proves itself. If editing a movie in a linear editing system like Adobe Premier proves effective, we are encouraged to subscribe to its philosophies. It is only when the system fails to be successful that we begin to be critical of it. We then ask the important critical questions. Why didn't it work? What is the software doing, that it should not? Where is the incongruity between my understanding of the situation and the system itself?

These are the questions that more often arise from either the masters of the systems (e.g. expert users and hackers) or from the uninitiated who have not been indoctrinated with the philosophical grounds of the application. The hacker exploits the shortcomings of the software system, its lack of proper granularity, or its inability to handle specific situations.

The beginner experiences the software without confining definitions. Yet, each software system requires the beginner to understand its definitions, whether original or derived, for use. If the beginner fails to understand, they fail to use the software, at least in its intended use.

This creates an interesting situation. Those people most capable of critical assessment of software philosophies are those at its ends. People who have never used it and people who have learned it very well have the best perspective to provide critique. The distribution of those two populations varies widely between software systems.

The result is a varying quality of criticality. The more specialized system receives the least number of highly expert, highly critical assessments. The least specialized and highly used application does receive critical assessment at the expert level, but little assessment at the beginner level. Herein **lies** the dilemma. Experts offer their critique from within the constraints of the system. They can do comparative analysis, and understand the shortcomings of an application from within the application's design. Beginners are outside the application and its inherent philosophies. They have not been indoctrinated with the rules of use. Their unfamiliarity gives them an important perspective from which to critique. As software designers, much of the critical assessment comes from the expert user. Critique of systems by beginners is instead often used to understand the critical gateway to indoctrinating the new user into the software systems philosophy. The goal of many software design assessments is not to radically alter the approach, it is instead to confirm, refine, and improve. This is another philosophy exuberantly promulgated in the philosophy of software. Systems design is not in need of revolution it is only in need of constant revision.

CONCLUSION

The world of software design is due for a reinterpretation of its values in the same way that historical societies have revolutionized themselves by deep assessment of their universal assumptions. To even describe such revolutions as *next generation* is a failure in critical evaluation of the pervasiveness of these assumptions. Users of technology have been a part of a wide sea of universal assumptions that have at their heart clear philosophical character. As the population of software users increases, the visibility of design flaws has naturally increased. They make themselves apparent in every days design challenges, the daily hacks created not by poor design, but by insufficient design philosophy. Any user of software has experienced these philosophical disconnects. Yet, because of the philosophical disparity in arts and education in particular, the difference between software philosophies and practice philosophies becomes most clear. Art practice champion's approaches are somewhat ignored by software. Education seeks a more complete approach than what computer science deems practical.

The need to look at the design philosophies inherent in software is a real. The science is maturing from a childhood stage of rule accepting, to an adolescence of rule-bending. Its historical structures are showing their wear as its users rock its pillars. In order for new work to break free from the loops of software design, artists can explore opportunity in the undefined regions of software implementation. The map is incomplete. These new approaches are not limited by art creation. They include examining alternative programming paradigms, such as declarative programming and building software apart from prepackaged design suites and application programming interfaces.

REFERENCES

Agarawala, A., & Balakrishnan, R. (2006). Keepin' it real: Pushing the desktop metaphor with physics, piles and the pen. *Proceedings of CHI 2006 - the ACM Conference on Human Factors in Computing Systems* (pp. 1283-1292).

Barrow, J. (1992). *Pi in the sky: Counting, thinking and being.* Clarendon Press.

Chomsky, N. (2006). *Language and mind.* New York, NY: Cambridge University Press.

Dunbar, R.I.M. (1992). Neocortex size as a constraint on group size in primates. *Journal of Human Evolution, 22,* 469-493.

Lanier, J. (2003). One half a manifesto. In J. Brockman (Ed.), *The new humanists: Science at the edge* (pp. 233-262). New York, NY: Barnes and Noble.

Lenarcic, J (2004). Behavioral Issues in Software Development: The Evolution of a New Course Dealing with the Psychology of Computer Programming. *Journal of Issues in Informing Science and Information Technology*, (pp. 247-252).

McLuhan, M. (1994). *Understanding Media.* Cambridge, Ma: MIT Press.

Nielsen, J. (1993, November). *Iterative Design of User Interfaces. IEEE Computer, 26*(11), 32-41.

Reynolds, C.W. (1987). *Flocks, Herds, and Schools: A Distributed Behavioral Model, in Computer Graphics, 21*(4) (SIGGRAPH '87 Conference Proceedings), 25-34.

Shneiderman , B. (1980). *Software Psychology: Human Factors in Computer and Information Systems.* Boston, Ma: Winthrop Computer Systems Series.

Tufte, E. (1993). *The Cognitive Style of Power-Point.* New London, CT. Yale University Press.

KEY TERMS

Avatars: A digital representation of a user in a virtual environment.

Finite State Machine: A model for the construction of computer programs that defines solutions using a discrete set that may be reduced to a finite set of configurations.

Human Computer Interaction: The discipline considered with optimizing the dialogue between human being and computer.

Metaphorical Design: A design approach employing analogy in the design of software solutions.

Reductivism: Defines both an historical art movement and the practice, in software design, of simplifying complicated relationships in order to encode them into computer science terms.

Software Philosophy: The collection of values, ideologies, and perspectives made apparent in the design of a specific software solution.

Object Orientation: An extremely popular software design approach employing classes to encapsulate specific entities into logical, modular units. Object orientation encourages code reuse.

Unified Modeling Language: A formal standard specification, visualization, construction, and documentation of software.

Chapter XX
Technological Social–ism

Judson Wright
Pump Orgin Computer Artist, USA

ABSTRACT

Culture is a byproduct of our brains. Moreover, we'll look at ways culture also employs ritual (from shamanistic practices to grocery shopping) to shape neural paths, and thus shape our brains. Music has a definite (well researched) role in this feedback loop. The ear learns how to discern music from noise in the very immediate context of the environment. This serves more than entertainment purposes however. At a glance, we often can discern visual noise from images, nonsense from words. The dynamics are hardly unique to audial compositions. There are many kinds of compositional rules that apply to all of the senses and well beyond. The brain develops these rule sets specific to the needs of the culture and in order to maintain it. These rules, rarely articulated, are stored in the form of icons, a somewhat abstracted, context-less abbreviation open to wide interpretation. It may seem somewhat amazing we can come up with compatible rules, by reading these icons from our unique personal perspectives. And often we don't, as we each have differing tastes and opinions. However, "drawing from the same well" defines abstract groupings, to which we choose to subscribe. We both subscribe to and influence which rule-sets we use to filter our perceptions and conclusions. But the way we (often unconsciously) choose is far more elusive and subtle.

INTRODUCTION

Language may have both a hard-wired component in our DNA, and a learned component (Chomsky, 1977). Neither is operable without the other. Or at least we don't get language without both. This is a debatable theory, yet very useful to us. If spoken languages could be thus constructed/understood, it seems sensible that non-verbal languages could also follow this organization. Fundamentally each are means of using symbols to represent ideas we want to transfer from our minds into another's (Calvin, 1996a).

Furthermore, it appears likely that where music operates neurologically on a (non-verbal) linguistic level, it too is organized in this dual fashion. Music also obeys both fundamental laws and is influenced by the immediate culture, while influencing it. Music serves cultural cohesion on a neuro-level (shown in many modern studies thanks to the fMRI (Levitin, 2006; Doidge 2007). In many cases, music, culture and the neural result are inextricable (Huron, 2008).

The way humans use musical instruments extends well beyond providing entertainment. We choose to employ a drum to speak to our own minds and the minds within others (more about this later), even if we are doing so with no knowledge of it, or intention in manipulating brain waves. But essentially, we mustn't forget that these instruments are tools at our disposal, they are also technology whether old or new. However we use the net, it is also a tool. Its ultimate product "cyberspace", shares many important features with music and cultural cohesion as well. Thus our big question becomes: if culture informs the web and visa versa, what neurological impact is it being used for?

In other words, asking a tribal member why a certain drumbeat is used in a ceremony gets you one answer. It is not at all the wrong answer (Narby, 2001). But asking an anthropologist or neurologist who studies the effects of "deep listening", gets a very different answer. The beat ultimately is used to hold the culture together hypnotically.

Let's consider the modern equivalent of a drum though. It is a tool, one that has neurological effects, which may be one reason we use it. Oddly though, we often do not employ computers as tools but as human substitutes. Instead, the tasks for which computers are commonly employed are strangely inappropriate. We pretend they function as specialized brains.

Human brains accomplish most thinking (perceptive and conceptual) by means of switching logically between inductive and deductive reasoning (Dewey 1910; Fodor, 2000; Hawkins, 2005). Computers, with no means of comprehending or creating anything remotely like context, accomplish tasks using only a limited version of

deduction[1]. Some inductive reasoning can be accomplished with a computer by iterating through every single possibility (by making the question deductive). But in real life, this is absurd. Real problems have either infinite unknown possibilities or at least unpredictable ones. Computers simply can't solve things humans can. And we still have no clue as to how we do it.

John Dewey (usually required reading for educational studies) published a very good account of "How We Think" in 1910. It happens to stand as a very good description of how computers don't think. Boolean Logic (Hillis, 1998) and *modus ponens* have been common subjects in Philosophy, Logic, Psychology and Cog Sci for much longer than computers have been on the open market. But no one questions that there is more to human brains and thought than these formalizations. In other words, this is old news from rather common sources. So why have we resisted what should plainly be ingrained into our habits of thought, just to bang our heads against the wall?

On the flip-side, though most of us may want computers to accomplish human functions (as in security facial recognition or recognition of written words), we aren't all actually working on these things first hand. Instead, most of us are using these same machines primarily to store and send strings of text (email, the web, word processing, spread sheets, …). The processor is minimally involved in just delivering a copy from one terminal to another. This is hardly a harmful or bad use. But it certainly doesn't warrant the fancy hardware. Not even close.

The Bigger Picture

An oft posed question: are we guided by technology? is certainly a valid, common sense approach to the issue. However, the answers it produces are necessarily misleading. It is a question like 'When did you stop beating your wife?' There may well be an accurate way to answer, but no satisfying response. Why?

Every question belies a symptom of one particular perspective. From our traditional point of view, the question is rather logical. But in looking at this issue, we need to divorce ourselves from reductionism. Specific views, whether positive or negative vectors, typical of disciplines, cancel each other out. Such margins of error and signal noise are negligible when we get far enough away to view the bigger picture. Specificity is a good thing, but sticking to one small area is not. We need to continually ricochet to a much broader range of subjects. Not adhere to any exclusively. Thus, we will be drawing from neurology, mythology, anthropology, art history, and music theory alike. Applying the ideas of Eric Hoffer on political revolutions to Jeff Hawkins' modern perspective about computers and neurology, musically induced trance states to occurrences of mass hysteria, the psychological theories of Jung to Chomsky's linguistic theories, prove all the more enlightening when taken together.

Why do we choose to delineate between "science", "art", "social work", "exercise" at all and not make all disciplines one? After all much can be culled from one "expert", when applied to other subjects. Obviously, these delineations are breaking down. They are habits of traditional academic organizational systems that we are shedding.

There actually is a deeper rhyme to this reason, albeit a bit arbitrary. Since we humans can only focus on smaller numbers, we justifiably have used somewhat artificial schemes to break up the overwhelming sea of data into categories and subcategories, until the pieces are more easily digestible to the majority of human minds. *Chunking* is a function of our perceptivity and not an a priori reality in nature (Solso, 2003). Incidentally, we tend to project our own limitations onto our inventions. Computers do also have harsh limitations, but this is certainly not one of them[3].

We are finding there is no "normal" when it comes to our mental abilities (Gardner, 1983). Why this is so slowly seeping into our (Western?) culture is certainly a peculiar phenomenon. Despite what we may have been taught, all of us probably actually prefer to organize in a very personal way, though are often pressured into "doing it right". Others of us, who appear to "organized" people as being "disorganized", are often really "pile people". Stacks of papers are arranged not alphabetically, topically or chronologically, but in an ongoing process of convenience. In several ways, a computer is the ultimate pile person. Organization, merely a lens, does not even exist outside our minds (or at least we can never prove it) and can only really be seen by indirect means.

Piaget explained that children assume their thoughts are the only thoughts animating and motivating everything around them (toys, the rain, the moon, etc.) We adults do so too, not out of egocentrism, but often because we do not fully distinguish between our compulsive tendencies toward reductionism and our projection of reductive organization onto chaotic sensory input (Piaget, 1951). Taxonomies are really just more chaos lumped onto chaos – accept (sometimes) to us.

Pina Bausch is a choreographer. In a performance, the dancers appear on stage with piles of sand and begin making castles. (Bausch, 2002) Of course they didn't take classes to learn to play with sand, yet if they did, would this course "Sand Castles Building 101" be found under dance, architecture or child development? Labels, being another element in internal mental organization, are readily classifiable. But icons don't always fit nicely. A behavior pattern as a metaphor for play, childhood, exploration, discovery, all rolled into one, is an icon. The castle is hardly the end goal, but the activity is. The end is a verb, not a noun and not an end at all but an ongoing process. We traditionally categorize in terms of nouns, at least on paper, but not in our subconscious (Calvin, 1996; Sacks, 1996), and it doesn't happen in nature (Gordon, 1999; Johnson, 2001), only in our bloated cortexes.

Figure 1. Nature doesn't classify animals, we do. And we do it impulsively. This phylum diagram was surprisingly easy to compile because the hundreds of such charts throughout the world are all organized almost identically. Even cultures that do not share this biological data, arrange animal classifications this hierarchical way according to easily-observable classifications.

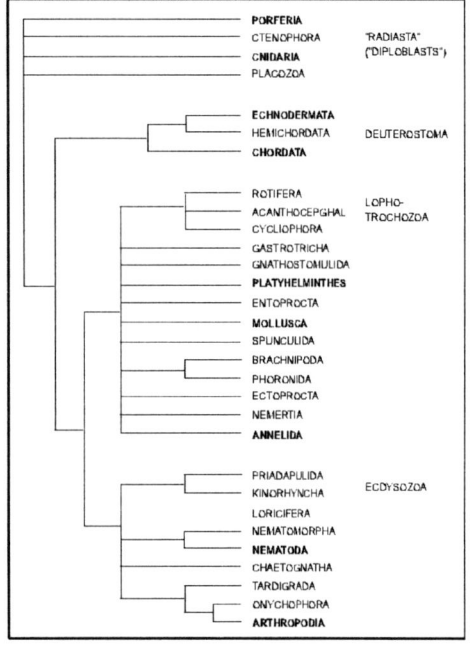

Our use of the web casually confuses the label and the icon. Labeling icons, we often miss more important aspects of the data. The molecules of vapors, as well as their undulating interaction, that defines a cloud, are not organized according to the Dewey Decimal System. Placing them in some sort of linear framework renders the entire thing no longer a cloud at all.

Blind conviction goes much deeper than egotism or even ego-ism. In his history of psychology, Morton Hunt describes the birth of Gestalt psychology. In particular, a few researchers (Duncer is one) began looking at how we solve problems using tools.

In one situation, for instance, the subject was asked to mount three small candles on the door at eye level, ostensibly for 'visual experiments'. On the table were some candles, a few tacks, paper clips, pieces of paper, string, pencils, and some other objects, including the crucial ones: three small empty cardboard boxes. After fumbling around, every subject eventually restructured his view of the things at hand and saw that the boxes could be tacked to the door and used as little platforms to mount the candles on.

But in another version of the problem, the three boxes were filled, one with little candles, the second with tacks, and the third with matches. This time fewer than half of Duncer's subjects solved the problem. They had seen the boxes being used for a specific purpose, and that made it harder to see them as usable in an un-boxlike way. (Hunt, 1993, p. 298)

I. LOOKING BEYOND

The Collective Unconscious

The term "Collective Unconscious" refers to a shared iconography across cultures, time, and distance (Campbell, 1972; Opler, 1994). The term was coined by Carl Jung (Jung, 1919)[2] who writes "In addition to the personal unconscious generally accepted by medical psychology, the existence of a second psychic system of a universal and impersonal nature is postulated. This collective unconscious is considered to consist of preexistent thought forms, called archetypes, which give form to certain psychic material which then enters the conscious. Archetypes are likened to instinctual behavior patterns." (Jung, 1968, p. 451). Jung's points are well taken, but his context is the milieu of psychology in Europe in the first part of the 20th century. He wasn't talking about technology at the beginning of the 21st. Yet, much can be culled by revisiting his ideas.

Jung begins by ruminating about "simple" instincts like thirst, shows that simplicity/complexity are entirely subjective human projections. He soon focuses on the *archetype*, a type of instinct. This may seem problematic, since archetype seems to necessarily imply analysis by the consciousness. But he seems to be discussing archetypes in some abstracted (*pre-conscious*?) form. Joseph Campbell explains this way.

All my life, as a student of mythologies, I have been working with these archetypes, and I can tell you, they do exist and are the same all over the world. In the various traditions they are variously represented, as, for instance, a Buddhist temple, medieval cathedral, Sumerian ziggurat, or Mayan pyramid. The images of divinities will vary in various parts of the world, according to the local flora, fauna, geography, racial features, etc. The myths and rites will be given different rational applications, different social customs to validate and enforce. And yet the archetypal, essential forms and ideas are the same – often stunningly so. And what then are they? What do they represent?

The psychologist who has best dealt with these, best described them and best interpreted them, is Carl G. Jung, who terms them 'archetypes of the collective unconscious,' as pertaining to those structures of the psyche that are not the products of merely individual experience but are common to all mankind. (Campbell, 1972, p. 216)

The Collective Unconscious has since taken on far broader applications. Other books that talk about the Collective Unconscious include *The Universal Collective Unconscious and Metaphysical Utopia, The Complete Idiot's Guide to Being Psychic, Buckland's Complete Book of Witchcraft, Mastering Astral Projection: 90-Day Guide to Out-of-Body Experience, Ouija: The*

Most Dangerous Game. It may seem easy to be disparaging of these titles. However, like Jung's, they are driven by a core idea, and a very astute one. Where they take it is a detail, informed by a chosen culture.

While witches, evangelists and self-help authors have all taken the term Collective Unconscious to far broader definitions than the original conception. Collective Unconscious may not necessarily be limited to mythology per se, but need not be extended to speaking with the dead. Is there a similar a repository, broadly accessible to many, containing shared cultural information? Of course. The web.

Music's Role in Ritual

Now let's turn to the question of how we are effected by various kinds of collective unconsciousness-es. To begin to answer, we will look at how we humans determine and define our cultures, a matter of iconography and ritual. Why exactly do all cultures, often independently of others, make songs?

Perhaps you are familiar with such studies where music is the tool, used in rituals. Natives beat drums not just to entertain. In the West, *The Mozart Effect* (Campbell, 1991) is an interesting illustration of this phenomenon, but even war chants (Sacks, 2007) are fitting examples. Aside from its aesthetic role, music indirectly influences behavior on several levels. Prolonged attention to a beat, tends to excite the thalamus (or so many suspect), and/or bring about a *trance state* in the brain (Levetin, 2006). The neural activity creates subtle, yet complex electric frequencies. In particular, trance states marked by increased Theta waves, from 7-9 MHz. (Robbins, 2000). This occurs to varying degrees both in the performers and the individuals partaking in ritual.

Hypnosis is often seen as un-masculine and shunned in Europe and the US, but in most of the world and throughout most of human existence, it continues to play a leading role in social interaction. However, it still sneaks into our lives in many forms, from opera to rock stars (Hunt, 1993; Rouget, 1985). Like babbling language, perhaps we just can't help it (Crystal, 2005).

A good example of the integration of music and trance is the Balinese Bebuten ceremony. In it participants/dancers plunge daggers into themselves. They do so, acting out a fierce battle between mythical beasts and a witch who has placed a curse on the community (Becker, 2004). They are not so much intending to hurt themselves though. Even under hypnosis, they are fully aware. Maybe even more so. As such, they are transformed into limbs of the community atoning for the tribe, not as representatives, but as appendages of the given social system. By acting out this scene, they are hoping to suggest a new fate for the community and scaring the spirits that are responsible for their luck. The self is entirely enveloped into the "world" (environment) or nether worlds (as defined by the culture) at this point.

The actual logic behind this enactment is not really important, and that is precisely where many foreign observers studying these ceremonies become missionaries rather than scientists (Sells, 1996; Daniélou, 1991; Narby, 2001). Many seemingly obvious factors don't make perfect sense to a removed observer. Their desire to create and adopt shared goals overrides "logic". It actually appeals to a different memory system (LeDoux, 1996; Sacks, 2007). Though emotional memory and episodic memory can influence each other, often they don't. Thus we don't always realize we are reacting to memories. In effect, these performers are accessing a parallel mind. A mind unlocked by music.

One may decide that the participants' interpretations of their motivations are unrelated to the result. For example, that solidarity as a result of these highly emotional rituals may have more influence on defending against invaders than appeals to spirits. This would be a condescendingly smug view. If the enemy invaders are thwarted (or aren't) does it really matter why on a neurological level? Couldn't solidarity with ones peers be inseparable from favor from the gods? Perhaps the brain is merely a radio antenna tunable to the gods? The cynic very importantly chooses to be missing just as much as the fanatic.

In other cases, the subject undergoing the trance is the performer or shaman (Narby, 2001). A more modern example is the young James Brown (Lethem, 2006) whose act intentionally referenced Baptist revival services, which, in turn, borrowed heavily from trance-inducing rituals around the world, particularly from Africa in the 1700's. Brown's ritualistic performance, with his spasmodic yet intricate footwork, ultimately breaking down and covered by a cape, modeled on older rituals, was quite literally a religious experience for many.

Psyching the self up for these rituals is what gives them their effective power. Expectations and acoustics work in tandem. Though trance states are clearly a bio-mechanical function between the brain and the environment, seldom are the rituals effective without a learned expectation of effectiveness[4] (Becker, 2004; Alderage & Fachner, 2006). A ceremony is designed to drive the "witch doctor" to a frenzied state, where the audience is moved sympathetically (Levetin, 2006). As with James Brown, you might begin as an observer, enjoy the show, get more into it, and finally let loose dancing yourself as a very active participant. Perhaps even inspiring the dancer next to you to more enthusiasm. This process happens in many forms. Anecdotes about moving post-9/11 memorials tend to begin with the individual's surprise at their strong emotional reaction and go into how swept up the group became, usually involving singing or music. This is not to say that they weren't actually swept up. But to be swept up, often involves being carried away with others, often employing a hypnotic medium, such as song. Is hypnosis a relinquishing of control? One could just as easily say, this unleashes the "real" self, that's been suppressed.

These rituals serve a dual purpose. By yielding to the community, individuals' problems are, at least temporarily, put aside, for the sake of solidarity. They can also help curb out of control emotional feed-back loops on a neurological level[5], containing unleashed cycles. Fewer cases of pent up rage is certainly good for everyone in the community.

As individual desires are yielded to the will of a team (even if it is never spoken, articulated), we elect to hone the priorities of that culture, like healthier crops, appeasement of a spirit, etc. and develop strategies for working toward them. Common needs create a sense of community, a oneness, where one's super ego is dissolved into a greater life force.

The *deep listener* of pounding tribal drums or the healer's maracas becomes not so much at one with a god per se, but one with a culture (Becker, 2004). Dancing to the pounding bass frequencies of modern music, we may forget we are in a subway car. Individualism is (temporarily) lost, a drop that has been added to the sea. That drop may only wield so much influence alone, however as part of a tide yields unlimited influence. The same is true for the web.

"Teamwork" has a positive ring to it. In small scales, likely groups under 50 (Gladwell, 2000), it probably does more good than harm. Fusing an identity from a small group is hardly infallible, for example the Manson Family, but usually eliciting good (though rarely great, as we'll see later) results in the workplace. On a larger scale however, there may be more examples like the Moonies than Martin Luther King. But who could say? Certainly Gandhi, Lincoln, Roosevelt, etc had positive impacts, though also lead a lot of followers enthusiastically to their deaths.

The point here isn't that conformity or groupthink is always bad, but that only individuals seem capable of assessing if the final decision warrants action. And we can't be both completely self-aware individuals and completely faithful members. Why do we gravitate toward such dual polarism

(Goffman, 1959)? For some reason, we are fed the belief that individuality is situated at one end of a spectrum and often mutually exclusive from cohesion. But thus linearly arranged, the median of opinions as a result has no real value. There is no useful center point. Often the only way to maximize both positions is to move freely among these extremes, respecting and supporting each, but always with an eye open to abuses and harm that can come from either. While unwavering unconditional faith can be fool-hearty, a constantly critical stance would prohibit true immersion and lead to a false sense of understanding.

II. THE MIND

Linguistic Issues

There are an infinite number of alternative ways to convey very specific information. Data saved digitally is one such method, a protocol of binary *bytes*. Humans concoct languages. Each digit is independent of its neighbors. Humans concoct languages. Words are very dependant on their neighbors, not just for grammar agreement but for the general meaning of the sentence.

Importantly, the human method doesn't translate into digital form nicely (Fodor, 2000). But the distinction is far more complex, but the solution is alarmingly simple. We just have to fully grasp it and take advantage of it. As an example, smoke is not essential to friction, but when we see smoke coming from a mechanical object, our first guess is to look for what parts might be rubbing each other wrong. We don't really need machines to create more meaning (to determine how the smoke got there). They aren't good at it, but we do that just fine ourselves. Instead, what computers might be able to do is mimic the key that unlocks our language recognizing function (create virtual smoke, or rather generate fresh air free of smoke).

The software Eliza is a prime example of this. It was created to carry on a conversation, something computers can't usually do convincingly. But it uses a psychological trick, repeating parts of the human's previous replies at random. So the person conversing sees significance in the computer's words, even if the machine doesn't.

The Turing Test says if a panel of experts on a subject can carry on a conversation with an unknown entity, and they are unable to or inaccurately determine the entity's human-ness, the machine passes. Probably more informatively, I would propose the test should go something like this: Select 10 random high school students (I am assuming a Western audience, but any background would surely do). Play them 5 songs, chosen unpredictably on the spot: a pre-18th century Classical piece, a 20's recording of early jazz (including the crackle), a pop song from the 50's, the theme song from a 70's sit-com, and a current (but yet unreleased) rap tune, and ask them in under 15 seconds to determine the chronological order. The question would be if any machine could consistently score better than the high schoolers. Pretty much impossible.

Essentially, the high school students needn't recognize the actual songs, but the genres. Furthermore, one can usually tell at a glance (and we do channel surfing), if an image was filmed or videotaped and in what era, what the level of seriousness is, roughly if it would be interesting to us. Compare a gardening program from 1970 on BBC to a modern reality TV show set in a garden. Color tonality, transition frequency, and how close are the close-ups are all big clues – for humans.

But if those same high school kids were asked to compose a few seconds of early European classical music, that might convince the next group, in under 15 seconds, my money would be on the machine. In fact, the machine could come up with hundreds in a under second. In other words, there are two very distinct skills. Humans induce effortlessly. Computers stink at it. But computers

assemble from deductions effortlessly. Humans get bogged down in possibilities.

Amazingly, nearly every known human language in the world and throughout time can be analyzed into remarkably identical parts of speech, take the form *subject-object-verb*[6]. Some use *subject-verb-object* and only a handful more variants of this order occur very rarely (Hunt, 1993; Pinker, 1995; Calvin, 1996) but many schemes never occur. Why?

Noam Chomsky's theories of an LAD (Language Acquisition Device) and "auto-generative" language demonstrate that an infinite number of variations could be created and still understood without explicitly understanding the rules and exceptions of grammar (in whatever native language) (Chomsky, 1977). Children often make mistakes in trivial ways (saying "foots" instead of "feet", mussing the *surface grammar*) but consistently maintaining, comprehending and employing the culturally defined organization of sentences (*deep structure*). A child may say "My foots hurt!" but never "Hurt feet my!". We understand "Jane said: see Spot Run", "See Spot run toward Dick.", "Jane told Dick about Spot running." without ever literally experiencing all of these specific, seemingly infinite variants. We recognize the deep structure. Marc Hauser describes Chomsky's theories further.

This combinatorial machinery [a module in the human (but not other species) brain for LAD] provides algorithms for specifying the details of our communicative utterances. The fact that we use this machinery for communication is, however, incidental. ... Put simply, even if Australopithecus afarensis – the famous "Lucy" – had a language organ, we would not be in the position to definitively assess its role in communication or, for that matter, any other behavioral expression. This point should be swallowed completely, especially in light of Chomsky's (1986, 1990) own belief that our use of the language organ for communication was quite accidental. (Hauser, 1998, p. 34)

If the brain simply applies this pattern recognition function to whatever stimuli is input, distinguishing relevant organization from randomness, it would stand to reason that the brain could be using the same module for differentiating say music from the sound of leaves in the wind.

Musical Grammar

No matter what culture, recognition of communication is a matter of projecting. The French hear vowels Americans can't, who hear L's in ways Japanese can't (Pinker 1995; Solso, 2003). Music is no different. The strategies are culturally specific, learned, articulated in retrospect. But they are absolutely concrete, not psychological effects. Sounds absolutely do not exist in the air. They exist as sound only in the brain of the listener (Levetin, 2006). Furthermore, composers, musicians and listeners need never even be aware of the rules they apply consistently.

In Western culture, the number of pitch and rhythmic options for a 2 bar melody is huge, but not at all infinite. In fact, we use a remarkably tiny fraction of the possibilities[7]. Which of these subsets of musical options we adopt, is defined by culture. Amazingly, across cultures, our music sounds unique yet identifiable to that culture, in part because our pools are roughly the same uncanny size.

Asian music and Middle Eastern music obey surprisingly similar rules, ones the listeners come to adopt subconsciously (May, 1980). The Westerner tends to hear sounds that do not correspond to normal Western music theory rules. The Eastern ear looks for anomalies according to Eastern Music Theory rules. These rules need not be articulate-able. The individual may not even be aware they know that their brain is using these guidelines as a filter of the chaos of sensory data. Westerners tend not to notice subtleties in more complex rhythms but do in melodies. Whereas for Easterners the reverse is generally true (Aldrige & Fachner, 2007). This is important in that to see from one culturally informed perspective blinds us from others. We are compelled to choose a perspective, but there is absolutely never a "right" one.

Chunking (mentioned earlier) is the term for the brain's tendency to dissect incoming data into digestible bits (7 digit phone numbers, 7 note scales, 7 Wonders, 7 days a week, 7 dwarves. Seven is about average for people around the world, throughout time). This extends far into a myriad of unexpected non-verbal cues. Our brain applies the filter it obtains from experience and culture (in every part of the world, throughout history) to every detail, from understanding how complex Chinese characters can survive the centuries (Coe, 1999), limiting the pool of vocabulary of specific musics (Levetin, 2006) and dance (Shay 1999), recognizing someone in cartoon by their mannerisms (Fliecher, 1932; Linklater, 2002) and even Richard Gregory's *redintegration* experiments (Hunt, 2003; Carter, 1999). We invent "facts" according to what we see (Sully, 1881) revising the surprisingly innate and uniform assessment of beauty we are born with (Etcoff, 2000). Chunking and culture work together (creating icons) to make comprehension easier for us.

III. THE EMPEROR'S NEW PARADIGM

Shifting Between Perspectives

Individuals are free to apply the tools as best they know how to their own very personal situations. Just as the monkey may use a stick to reach a banana, the desire for the banana always precludes the decision how to use the stick. Needs change how we see the tools, not the other way around. In a movie, an ex-special ops mercenary poses as a substitute teacher and lectures the class about "perception".

"What do you think this is? [holds up a yo-yo]

It's a simple toy, man. [drinking from a bottle]

To you it is, but back in 1500 in the Philippines it was a weapon. [lashes out with the yo-yo and shatters the bottle] (Pearl, 1998)

We don't learn by rote that a screwdriver can be used as a hammer. We pick it up and try it. Nonetheless, once a method works at all, it is incredibly difficult for some to discard. Hence, more precisely useful methods are often met with irrationally vehement and disdain, in favor of impractical yet more immediately learn-able alternatives. The resulting perception is then that the object has one very specific purpose and that purpose is defined by a *stable state* (Dawkins, 1978; Gladwell, 2000).

It comes as no surprise that contributors don't want to see the issues complicated out of reach or lost in unfamiliar territory. Web designers want to continue proudly designing pages as if they are fixed layouts for print. More importantly, they have an industry and are paid by it. In recent years, the designer's role has shifted, but has not escaped the goal of making content easier to digest. We employ buffers from being thwarted by unfamiliar technical information in the forms of blogs (Weinberger, 2007). This static-ness is intrinsic to print, not the web at all.

Browsers get away from static-ness with easy to read defaults and allowing user customizations of the general appearance. Still designers insist on over-riding these features in favor of static-ness, especially using CSS (Castro, 2003). Personalized options offered are trivial. Customizing means presenting pages in our favorite colors. "Read more of this article" doesn't purport to alter or update the article in any way.

Though HTML technically does support interaction in the form of hyperlinks, it is the very lowest level of interactivity (Gleick, 2002). Compare the highly interactive question - "What was the most interesting thing you did today?" and the equivalent of hyperlinks - "Do you want fries with that?". You really have three ways to answer: "Yes, go ahead and continue whatever you have planned." (click), "No, wait to continue just yet." (don't click) or supply an answer that gets no meaningful response. However, a computer has no problem "listening for" millions of possible responses. Why would web pages so uniformly ignore what they are capable of, for limitations of linear design? HTML is hardly a bad thing, but it is a cherry on the top. CSS (generally flavorless) and JavaScript may be the colorful sprinkles, but there's still a lot more to a sundae. Yet we regularly throw the rest away.

Design is unavoidable (Tufte, 1990). Arranging any data on any page involves design. But there is design for design's sake (as in logos, marketing or dressing fashionably) to convey an essence by creating an icon and design for function's sake (as in architecture, form design, and even typography) where the graphics are a subservient means to the an unrelated end goal. Will it withstand rain? Does it need to convey specific words? These are questions that (can) trump aesthetics.

Though not entirely separable, design and function can be viewed distinctly. The problem arises, when design limitations are placed on function artificially. If designers of web pages only think in terms of static pages, that leaves very little function. But the page is still essentially static, serves no non-design function, utilizes neither the computer processor nor any significant peculiarities of a vast network. Later we'll talk about how the web can be used, but for the most part, we settle for a 300 billion gallon billboard soup.

We even go so far as to invent a mythological expertise, in order to defend our higher as well as lower techno-caste system (Goffman, 1959; Marvin, 1988). What's wrong with a little ignorance? Nothing, at all! But once we have invested so much in a bad idea, we would look ridiculous (to ourselves) to admit the farce. We need someone to play expert and tell us we are right. How much in time, money and effort would you estimate we have invested in the web? Add up the costs of every computer, not to mention connection hardware, hosting fees, etc. We simply must justify it to ourselves.

The same could be said of cars. Modern ones may work differently than the original models. But drivers benefit from changes not being too drastic, as with solar powered cars. Drivers want to remain safe knowing the steps for finding a mechanic will work every time. One can imagine that the car owners not even be drivers though. They may be using a car exclusively as a couch, while still insisting on their mechanics. After a few generations of this, any mechanic would reasonably claim the only definition of a mechanic is someone who fixes upholstery. The cars will certainly do the job, no harm could come of this. It is a blatant waste - but actually only if you have experience driving. Owners just argue that's the going price of a couch and resent being told they were fooled.

Most users of the web, may be content as infant passengers in the back seat. They have no idea there is interaction happening, but just enjoy just looking out the window (hands in lap). To them riding in a car, is nearly an identical activity as sitting on a couch watching TV. So, it seems fitting the web must be like that too. After all, the web isn't like anything we don't make it like. If we make a "hands-in-lap" web, the web really is like that (Greene, 2004).

Allowing for the widest possible re-purposing, is often a key to survival. VS Ramachandran says this about animal evolution:

A trait may represent a further refinement (through natural selection) of another trait that was selector for a completely different purpose. Feathers evolved from reptilian scales to keep birds warm but have since been co-opted and transformed into wing feathers for flying; this is called preadaption. (Ramachandran, & Blakeslee, 1998, p.209)

If the net bears any similarities to past technologies, it is absolutely only because this is how we have applied the tool. The computer is nothing like a phone. The computer doesn't work similarly to a TV. But we use computers mostly as substitutes for two tools that still work fine[8]. The net is only like the telegraph because we eventually understood the telegraph and applied that understanding to the net.

People called the railroad the 'iron horse' and the automobile the 'horseless carriage'. For decades the telephone was viewed in the context of the telegraph, something that should only be used to communicate important news or emergencies; it wasn't until the 1920s that people started using it casually. Photography was at first used as a new kind of portraiture. And motion pictures were conceptualized as a variation of stage plays, which is why movie theaters had retracting curtains over the screens for much of the twentieth century. (Hawkins, 2004, p. 205)

DIY

The acronym, normally used as an adjective, could well be turned around as a command. Do it your (own damn) self. The desire to bypass years of discipline to become an instant expert is everywhere. Judging by popular titles, we can learn or accomplish anything in 30 days (Shenk, 1999). Often even that much wait proves too excruciating (Glieck, 1999).

DIY grew out of a punk rejection of the influences of "the Man"/big brother. What's ironic is that in all likelihood, punk was probably born in a music industry board room when they decided to team Sid Vicious (he doesn't actually play on studio recordings, had almost no actual musical ability, but looked the part) and Johnny Rotten (also more for his act) with Malcolm MacLaren (who provided stylistic fine tuning rather than anything compositional) (Heylin, 2007). Punk was essentially the discovery that a cool attitude sells, while skills we once assumed were essential are optional.

The Sex Pistols were great! No question there. But the reason the meme spread all over the world depended on distribution. The reason punk was on sale in the big stores, was certainly not because it rejected the industry, it's because marketing departments guessed how to package it. "The Great Rock 'n' Roll Swindle" (Sex Pistols, 1993) was certainly fictionalized for the sake of drama, but like all mythologies there's factualness in the mythological-ness of it. Though most post-punk enthusiasts would be horrified to entertain the thought, there wasn't actually anything idealistic about it. We were all duped. We can argue angrily or lighten up and laugh at ourselves. In the case of the web, we tend to argue.

Looking into the machines themselves, and not the ideals obscuring them. People still commonly think employing a computer can be a "time saver". In cases like musical scoring with a professional MIDI sequencer or now user-friendly software like GarageBand, it can for most. A person fluent in reading and writing on staves will probably loose time. But for others it can make a helpful difference not to need to go through and decipher each note. Classically trained composers may argue that this brings the average quality of composing down, making it is more open to amateurs (Marvin, 1988; Hopkins, 1982). This view is hardly worth arguing for or against, as it misunderstands the essential shift in what is required of the "expert".

Neither side is more right than the other. Both are completely valid from their respective perspectives. By not recognizing schemas, computer technology works - as long as we don't insist on any correct point of view. Computers and a need for truth are simply incompatible. Computers can only provide hypotheticals and alternatives. Often attributed to an age/generation thing (which isn't really true of course), some folks get frustrated with machines, some with the search for fact, but both camps spend their money all the same.

While it is an arguable issue whether a Puritan work ethic results in more invested artists, and thus more profound art, this is not the only issue by far. It is often the only one addressed by fanatic proponents from either side of the argument though. Often extreme zeal and reverence are invoked to describe both the traditional values and the new shift to the new technology (Glieck, 2002; Tapscott & Williams, 2006) as if they are advertising a show on Broadway. The greatest hyperboles ever to appear on Earth!!! Rhetoric aside, one side argues for patience in acquiring skill. But such a work ethic is only interpreted as outdated elitism by others. The other side argues for the freedoms of expression newly available to the common man[9]. Mozart can be boring and therefore of no musical value to kids who want to dance. And Kanye West can be just noise, revealing no interesting musicality to Classicists. This point gets argued to death and leads nowhere.

Nonetheless, the sheer numbers of works, that explode like a burst dam, do have a very real negative effect. No curator (using the term loosely) of artworks or employer listing job openings can realistically expect that, where once they could manage to cull the best candidates out of 200, they can make any such claims when there are 200,000 of them. There just isn't time, nor the attention spans. Doing some quick math, it's plain that they can initially only give each candidate a .2 second glance. Obviously, some ideas take a few seconds to sink in, and those will surely be lost. In fact, which they see at all is either random or limited to the applicants they are already familiar with. By round two, the remaining survivors inevitably all start to look the same any way. And that does bring the quality down (Shermer, 2002).

Let's look at an illustration of this concept. Say in gym class you were the team captain, picking all of your players before anyone else from 100 students, you might pick the best team, and may thereby win more games. But picking from millions the odds of noticing who the key players are no longer significant. Among that huge crowd may be the Yankees, but who could ever notice when there are so many others. Every other captain has roughly the same slim odds. Likewise, ill-chosen HR employees are thereby theoretically less and less adept at what they do, at choosing the next generation of employees. So begins the exponential growth of mediocrity. Averages on the web, do not behave like the bell curves we expect. Nearly every aspect behaves like a steep slope.

Not that the web needs censorship, but why do we choose to "shit in our own bed"? We piously insist there is a new ultra-accessible frontier for gathering information and then set about to transforming it into a hiding place for information. There's a subtle difference. The messages on the net (whether spam or blog diaries) are actually often only meant for a small audience. For grandma, mention of our green socks may bring a little amusement. But it is as if we think we're still using previous media. These messages are not for everyone in the world. Nonetheless, we know full well that's where they go and brush this aside with no further thought of the result.

In targeting our messages, we might as well use a sawed-off shotgun to catch a fly. The question is, why do we pay $1000+ to have access to information that could be sent more efficiently to our target with a 40-cent stamp? Why spend more when we already have a more effective way to reach our specific audience? Skywriting is no way to create intimacy. Moreover, we've known this and seen this developing for decades now and only make the situation worse.

But is any kind of quality on the web even necessary for it's greater function? Maybe not? For example, who care's if someone in class says something other than "here" at role call, so long as their name is checked off, the result is successful. The existence of content is necessary but the content of the content is arbitrary (Calvin, 1996). Has a comparable situation come up before?

IV. DEVOLVING TENDENCIES

Time is a Wheel, not a Line

History always repeats. I am only glancing as far back as the Industrial Revolution, but as you probably already know, every promise the net makes, has been made over and over. What's remarkable is that technology again and again makes these claims as if it were offering a never-before possible opportunity, a new brighter horizon (Barbrook & Cameron, 1995).

In the 1890's, advocates of electricity claimed it would eliminate the drudgery of manual work and create a world of abundance and peace. In the first decade of the 20th century, aircraft inspired similar flights of fancy. Rapid intercontinental travel would, it was claimed, eliminate international differences and misunderstandings. ... Similarly, television was expected to improve education, reduce social isolation, and enhance democracy. Nuclear power was supposed to usher in an age of plenty where electricity was 'too cheap to meter'. (Standage, 1998, p.211)

There is no horizon, neither brighter nor dimmer. And we know it. It's hard to hear the word "utopia" now without sniggering. But we rarely appreciate precisely why? It is because the promise is made obsolete by the development itself. Not by solving any conflicts, but by re-framing them. For instance, let's look at the last century's obsession with signal-to-noise ratio.

Once upon a time, recording audio onto wax cylinders proved too noisy. Progressing to the 60's, things improved and soon technology was such that we could hardly notice tape hiss above the recorded sound without listening for it (Spence & Swayne, 1981). Video, also recorded on magnetic tape, was subject to visual noise as well. This noise also accumulated with age. So at first eliminating it was a genuine problem to solve, even if just for archiving[10]. And soon enough we pretty much

Figure 2. The illusion of the bell curve has been reinforced by centuries of experience. Things do behave in such a way. But suddenly, a lot more things don't fit that model. We assume an average popular site will get an average number of visitors. Not at all. A handful of sites get a phenomenal number of visitors, some get very few, but most get next to none (glancers, who spend 20 seconds or less determining if your site fits with their search). The spread is more like an exponential curve. It may seem like anyone can climb to the top. A handful of anecdotal evidence are all the success stories we need (like Linux and Wikipedia) to ignore the billions of sites that were forgotten before they were ever remembered. The average Joe/Jane is still a serf doing the grunt work to support a remote aristocracy, it's just a very different looking aristocracy. The web certainly is communal, but it is an absolute mirage that everyone is represented. Our web projects never stand the same chances as the ones on top. In the same way. In music, the industry invests a lot of time and money determining what will be popular. Moreover, they have access to channels to test and market their products. Imagine trying to compete by simply playing loudly in the basement. Now imagine there are millions of basements next door (physically iumpoossible but that's the web).

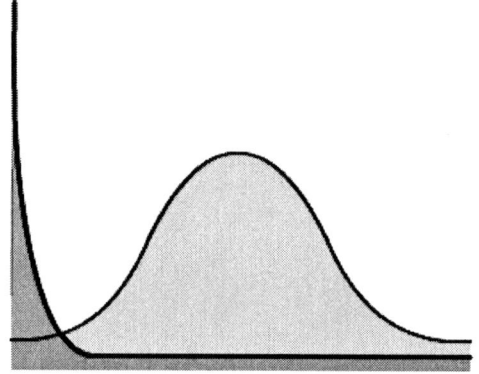

solved it. Nonetheless, we remained in the habit of facing that particular problem. We held on to it, long after the benefits hardly justified the efforts. Moreover, we didn't just want noise eliminated in recordings and edits made in studios fitted with carpeted walls and pro equipment, but wanted to do it at home, too.

We now rarely store mechanical or magnetic reproductions (like clay in a mold, on a macro- or microscopic level), we instead store the instructions to duplicate a subject (like recipes passed down for generations or DNA). In a picture, these instructions take the form of long lists of numbers which define grids of pixel values for red, green, blue and transparency, theoretically accounting for all possible colors an image could be (which is hardly true, such as warmer grays, florescent shades)[11]. A digital image (or sound) is a checkerboard of hues (or sine waves) (Aldrich, 2003; Myler & Weeks, 1993).

The resolution of the reproduction is no longer entirely in the quality of those pixels, but more importantly the sheer number of pixels (or sine waves). Hence "dots per inch"[12]. Despite claims, the noise to video/sound ratio has not been improved at all. Noise is just no longer much of an issue since the copy and the original[13] are now only related by coincidentally having similar colors/sounds in similar places. In fact, once recorded, the "original" (which was really only a list of electrical charges) is discarded forever and playback of any kind absolutely requires constructing an entirely new copy,

At first, the resolution of this clone was limited.[14] In audio one may still notice the sample rate is four times lower for phones than CD recordings, about 11 thousand samples (MHz) and 44.1 (Aldrich, 2005). But this is most apparent in video, where you can often detect blocks of pixels and perpendicular edges in video. This was still a huge obstacle until a few years ago. So in retrospect, one has to ask, why did we support this tool that didn't work all that well for so long (Gleick, 2002)? It's hard to say if digital media (both audio and video) now, after years of improvement, offer better "imaging" than analog recordings. Trying to solve one thing, a new paradigm can often brings other things down. So you'd think there'd have to be a really compelling reason or really strong results. Nope.

Moreover, why choose to make things harder on ourselves with no apparent benefit (at least at first)? We are often comfortable knowing how exacto knives and paste-up boards work, and struggle with layout software. When the print doesn't look like the screen, we assume there must be something done wrong on the computer, without realizing it made exactly the mistake they told it to. That's what computers are good for.

Professional Art Directors, with years of experience and big salaries may not fully understand the impossibility of optics, or down/up-sampling (resolution, blockiness and size anomalies) are often to blame. Traditional methods work. Computer methods can but it requires patient un-learning. However, combining them (even attaching a printer to a computer) brings about the worst of both worlds. Most folks (especially people used to a paper mindset) are just not making the full transition, remain voluntarily stuck with baffling compromises.

Quality does not improve, criteria shift. A common view once was that acting on film for a silent movie was a legitimate expression, but the kind of acting in talkies was not. At first glance, this seems laudable. But remember, at this point, the camera still was considered rather like a replacement of the view of a live theater audience. The camera was not primarily used to focus attention, rather to show the whole scene, where

the projection screen was a direct replacement for the live stage. For several years, the movie was yet another vaudeville act. Only gradually the live performance aspects faded away (Stokes & Maltby, 1999). (Though our computer screens are still influenced by the frame proportions.)

In his traditional manner, ... Elman [a violinist, subject of an early talking short features] chinned his violin, leaned back, and focused on a spot at the top of the balcony. The technicians explained he didn't have to do that anymore, that if he looked directly at the camera it would put his face in front of every member of the audience no matter where they were sitting. Elman couldn't grasp the concept. 'But I always look at someone in the top balcony when I give a concert,' he complained. He agreed to look at the camera, and as soon as the cameras began rolling, he reverted to looking at the balcony. (Eymann, 1997, p.79)

The advent of the talkie also drastically shifted the expertise. Acting was about muscle movement (like dance). Actors used a vocabulary of grandiose gestures to convey the grand emotions in scripts of the era. There even existed formal "code books" for each appropriate emotion and how to portray it (Lutz, 1927). Method actors like Marlon Brando would certainly only appear to them as a half-hearted, perhaps shy, understated attempt at "real" acting (Stanislavsky, 1936).

Adeptness came with years of study. A pleasant voice was simply something any layman had or didn't have, but acquiring it hardly seemed like a skill at all, just luck of the draw. Hence, success in this new paradigm was not based on traditional expertise but a random gift (Eymann, 1997). However, we still have acting schools. The "expertise" did not actually get supplanted, but merely shifted in specifics.

The casual use of the telephone seems so plainly obvious today, but wasn't always. Along the bumpy road to the perspective that we have now (which still stretches before us), telephone calls (like telegrams) were reserved for important or official news. However, an earlier idea folks tried was very different. There was a time you could subscribe to something like a radio station. You would be mailed a schedule, and say at 3:00, they would be playing a Mozart symphony (Kern, 2003, Fischer, 1992). Can you imagine sitting in your kitchen holding one of those old style phones[15] for say 20 minutes? But surely there are things we do with technology now that will seem just as absurd in years to come.

Precedents in Waste

Nowadays, stoves are rather the norm in 1st world households, but this was not always the case. When the gas stove was invented, there was already a strong social "ritual" in place, sharing the cooking area, the communal hearth. At the time when they first appeared on the market, Americans spent (relative) fortunes to have these cooking factories in their own homes (Fischer, 1992). Gradually, social functions shifted and eventually it seemed like an obvious necessity to have a stove in every home. "Fire is revered generally as a deity to this day. The lighting of the household fire is in many cultures a ritual act." (Campbell, 1972, pp 248-249).

Light bulbs had a similar effect. And though now we scarcely notice the costs, originally equipping a house for electric light was a major investment. The investment also dissolved the need for families to gather in the same room. In other words, we paid willingly for privacy.

The cost could not eclipse the desire. Similarly, we spend huge amounts on computers, only to use them insistently for a very minimal purpose. A purpose that often other tools, common household items, do much better and far more cheaply. Not to circumvent an identified inconvenience, but to add several new ones. Moreover, by bringing computers into our homes, we have hardly improved beyond older inventions. The video effects we use now at our desks are the same ones that

were available to pin the 80's. Is this wrong? Of course not, it is up to each individual and their bank account. But it is an odd phenomenon that indicates cost is not nearly the deciding factor in this particular market.

There was once a time on the ARPANET (Hauben, URL), back when graphic editing was too much work for most machines, computers could connect to a server that was the workhorse (Abbate, 1999). We could stand to learn from those days.

If you wanted to create your programs interactively, he reasoned, and if nobody was going to give you your very own 704 [a big computer for back then] to do it with, then the obvious answer was to get together with a bunch of other users to share a machine. ... Give everybody a remote terminal so they could tap into the big computer through telephone lines whenever they liked, wherever they liked. ... And finally, when the users needed some actual processing power, dole it out to them via an artful trick. (Waldrop, 2001, pp.163-164)

In terms of processing muscle, current cell phones are far ahead of where computers were when they became a public commodity (Greene, 2004). So what's a computer got, that a phone ain't got? The QWERTY keyboards (Bluetooth or built in, as in a Blackberry) are a clumsy but usable option. Allowing full size versions would hardly be a technical wonder. Phone screens are small but there could easily be monitors that attach to less powerful computers and supply their own VRAM. Do folks remember duo docks[16]? Software is not a hindrance. With a phone version of a plain text editor (like BBEdit) and FTP (for uploading files)[17], one could replicate most any computer function. But these 3 obstacles are hardly enough to prevent folks from adding their content to the web. It's as if the Great Wall of China were made 3 inches high all over and that was actually enough to keep invaders out.

This has been going on for years though, around the late 90's, we could purchase a webTV system or stand-alone email unit, for a fraction of the cost of a computer. Yet apparently we didn't often enough and the merchandise flopped. We can again send email, take snapshots and even look up web pages, all from our pocket-sized devices, but are dissuaded from giving up on computers. If we were all helpless victims of the industry, stuck with whatever they offered, that would be understandable. But we could hack (and we even have the tools sitting under our finger tips). Creating software is free (even creating software to run on cell phones[18]). We can even benefit from other hackers, as in Bit Torrent. Why doesn't that save us?

One interesting thing about the net is that people can cluster to solve problems, Open Source. Linus Torvald's project Linux is everyone's favorite example (Tapscott & Williams, 2006). But even these instances are subject to shared social influences. Linux would not exist if the need for it wasn't already common. Needs (even if they aren't fully understood or recognized) always preclude effort. Linux could not have happened without a strong selling point and a bit of trend following. Culture, not just some of the individuals in it, needs to agree to the need first, for openness to thrive. In a disappointing but usual case, we upload an Open Source project and nothing happens. The need, even if erroneous, has to be adopted by culture first, before the droves of supporters clamor onto the bandwagon blindly. Though one can sometimes see how a figure like Hitler or even Bill Gates could have profound influence on developing a new culture, it has to be a culture already hungry. It is just waiting for someone to say "I can provide beyond your wildest dreams".

The Nature of the Wave

People upload the concerns at the forefront of their thoughts. If images are uploaded of refrigerators,

it is not because these people are devoted to the appliances. They may be concerned about selling them, repairing them, etc. but the refrigerator itself is a tangential idea. However, the refrigerator is like the stuff of dreams. It can appear to the subconscious out of context. Thus it is an icon (Jung, 1919). It indicates something, but not always the significance or application of that thing. And this describes how, with enough amassed conscious concerns, you can re-contextualize and draw from a subconscious mythology.

Though very unconsciously, we can also choose perfunctory social banter, not unlike role call, over in-depth intimacy. This may seem deplorable if we assume the net is used as a socializing tool (Reeves & Nass, 1996). But if the net is simply a means for everyone to share an iconography, this chatty banter in social software may actually serve a very un-social function. It allows people to say "count me in with everyone else" and weeds out contributions that serve to dissuade cohesion (Jacobs, 1961). Such is the psychological force behind social software. To join the club, the requisite initiation is to encourage replies like "oh kool dood". Insightfulness and real passion may require longer posts, but only really serve to rock the boat, putting esteemed members at risk of drowning.

It would seem there must be an unfathomable tidal wave of ideas to which to give our attention on the web (Popper, 2007). We could be drowned in ideas if we resisted this wave, but we dive in head on. Which actual molecules make us wet? Without the tidal wave pushing from behind, those individual molecules would have no effect. It is the sea of them, many more than we perceive directly, that makes swimming possible. On the web, that means that the ideas that end up at the forefront are there somewhat arbitrarily.

We need not see the millions of pages uploaded to feel their effect. In fact, most of them are seen by under 50 people (and over 200 bots). A few sites are the hubs, seen by millions, the molecules that touch us, and are often only on in a position to touch us because of the combined effects of the other viewers (Barbási, 2002). Sites like Boing Boing may be the molecules making us wet, are only able to, because they reach us before so many ever can. They are the surface of the water, the foam on the shore.

Obviously, not all of the participants in a ritual are the shaman (or James Brown). A concert hardly has the same wild effects on us in an empty arena. The fame, visibility and attention given to the "star" in no way indicates the rest of the crowd is any less effective, just less identifiable. Which brings us back to the idea that anything label-able is really interchangeable. Categorization is essentially *surface grammar*.

Following the Fold

We tend to mistake the notion that we are simply "social creatures" for the underlying idea that we are creatures of opportunity. We often just adopt the first solution that doesn't immediately hurt us. In many cases, latching onto a culture proves far more practical way to go. Do I need my umbrella? Look outside and see if others are carrying theirs (Weinberger, 2007). We (aspire to) contribute symbiotically to the system that helps us. How much of this strategy is genetic is debatable. Some insist humans are social by nature, but that doesn't fully explain why and how we go grocery shopping.

Being social we are also vulnerable to succumbing to mass hysteria. Wacky but harmless notions are common (Heuer, 1951; Simons, 1993; Clancy 2005), but extreme examples include Holland's "tulipomania" of 1634-1636, witch hunts, the Florida real estate frenzy of 1924-1926, the panic caused by *the War of the Worlds* radio broadcast in 1938 can spiral into real trouble. Modern tech phenomena include decades of virus alerts (which ironically spread like viruses), the dot. com bubble burst, the Y2K scare. We seem to be drawn to distractions like OJ Simpson (Rushkoff,

1994), not because of any personal impact, but to revisit our social bonds.

No person of any economic strata, age-group, educational level or time period is exempt from this behavior (Bulgatz, 1992), though often the poor and uneducated are hit the hardest in the mad rush to join the frenzy (Vronsky, 2004). In many cases, as with Jonestown, those people are specifically targeted with tragic results. Feelings of vulnerability often fuel the mass hysteria, but threats and fears can impinge on any classification of people. Stanley Milgram's famous Obedience experiments demonstrate that any of us, given the right external conditions, relinquish our own rationality for our best guess at what is expected of us by "authority" in our chosen culture (Blass, 2004).

Blind faith tends to create impenetrable illusions that ignorance has effortlessly been replaced with expertise (the "money for nothing" bait of successful cons) (Weill, 1948). When it comes to using the web, we are defiantly ignorant. Though many claim to know better, this is fool-hearty. Apparently, our ignorance does not prevent us from unconscious uses of the network though. We just may not realize why we do what we do. Rather, not fully understanding, we may be freer to use the tools to save our sanity. Unconsciousness often protects us from our conscious decisions to over-ride unconscious ones. We need to feel protected in a big social group and simultaneously can be overwhelmed with the demands from so many strangers. The two dynamics are at odds and would make us crazy, if we didn't devise buffers. Walking down a crowded street, talking on a cell phone is one strategy.

Revolution

The history of the net has been something of a "mass movement" as articulated by Eric Hoffer in 1951. In a mass movement, such as the Bolshevik, French, and Nazi revolutions, even Christianity, the status quo is upset by discontent with freedom. Though "more freedom!" may be their chant, freedom also puts responsibility on the shoulders of the individual. What revolutionaries actually want is less responsibility as an individual and broader anonymity. The effort to blend in, generally becomes an effort to make everyone the same, discrediting notions of elitism (whether imagined or real).

This is in stark contrast to the notion that revolution is a reaction to oppressive poverty and despotism. Of course it also echoes tenants of DIY-ism. In many cases, such as "lurkers" (often the vast majority by far) in on-line discussion groups, we voluntarily do not wish to actually be involved. We don't actually want the interactive benefits of the web to replace magazine subscription. We'd often rather just keep tabs on the movement, the discussion that serves to dictate how to behave. Furthermore, people intent on survival see no use in diverting attention to matters of no immediate consequence. In effect conservatism (in all its forms) is the petri dish that inflicts oppressive paranoia[19]. It is people with idle time and disposable incomes that spend their resources to maintain the computer industry. Hopefully, in this essay, you will see, I am not casting blame, I am part of this trend, but asking why.

The ARPANET's transition from a tool for scientists was gradually pried open from institutions who could spend large sums buying into it for large faculties, often with funding, rather than out-of-pocket money, to the hard core techies who weren't academically affiliated has been well documented (Rushkoff, 1994; Shirky, 1995; Waldrop, 2001; Greene, 2004). Then the internet unlocked the door at first to the basement "nerds" and gradually aspired to welcome the functionally technophobic with the advent of web 2.0. Though the cost was at first an obstacle, it has become a given, an expectation. For instance we may still buy Microsoft's Office, though OpenOffice has been free for years. It is also unusual, and thus uncomfortable for the new user to think free software could be better than expensive software. The decisive factor at every step seemed to be, not minimizing cost, but minimizing unfamiliarity (even at the expense of confusion from horribly baffling and inflexible design).

As with any mass movement, opening to a wider base means generalizing for broader appeal. That means that once they have joined, everyone will be reciting the same simplified mantra. In effect, we all must share the same lowest common denominator intelligence, or rather suppress our tendencies to think too far "outside the box". This is not to say greater accessibility is a bad thing, nor a good thing, but that opening admission does result in letting individuals get "lost in the crowd" and "hidden in plain sight", and the "dumbing down" that is often a symptom can easily be mistaken for the disease.

It is not simply the oppressed versus the oppressor that brings out revolutions. The person who harbors no hope of improving their misery, is not a revolutionary prospect, would rather things stay the same. Life is hard enough as it is. Moving the horrible but consistent land mines all around them is a scary proposition. Miserable conditions become easier to live with as long as they remain familiar. Revolutionaries offer disruption only to those who are eager to gamble everything, and feel confident that whatever the outcome, their lives will remain fairly comfortable. Paradigm shifts that upset stability are brought about by folks with leisure on their minds[20]. But the specifics of a doctrine are pretty much arbitrary.

It is futile to judge the viability of a new movement by the truth of its doctrine and the feasibility of it's promises, What has to be judged is its corporate organization for quick and total absorption of the frustrated. Where new creeds vie with each other for the allegiance of the populace, the one that comes with the most perfected collective framework wins. (Hoffer, 1951, p. 41)

We have computers, connected to a network. The capabilities are enormous and yet email and static web pages remain their most common use? Fundamentally, communication over the web can never possibly work ideally on so many levels, but we obediently adjust our criteria to "good enough" (Reeves & Nass, 1996). We discontented yet hopeful seekers of change can all join the faceless crowd with Facebook. Not that Facebook ever intended to promote this anonymity, quite the opposite surely, but its success may stem from a popular need to be nullified. The pressures to subscribe to mass movements "produce both readiness to sacrifice the self and a willingness to dissolve it by losing one's individual distinctness in a compact collective whole." (Hoffer, 1951, p. 59)

Looking at ex.plode.us (a site that compiles the membership of several social software sites) may prove an ideal demonstration of this. Here are six of hundreds of pages retrieved from a single search.

V. TECHNOLOGICAL PERSPECTIVE

Using the Computer

Before going any further, I want to mention some positive examples. There is nothing wrong with buying things that we don't end up needing after all. We also own things, not because they fill any utilitarian purpose, but just because we like having them. Fun clothing, photos and artwork are particularly conspicuous examples. However, here is a case where we see a cool pair of bright red shoes, spend a significant sum on them, decide it is crucial to wear them for as many hours as practical (as if they provide essential nutrients by osmosis?) and wear them every night to sleep in, but remove them each morning. After a year, though the shoes are still quite fine, we replace them with an even more expensive pair. Huh? Let's look at alternatives.

Prominent sites that take advantage of the network of computers (like Amazon or Ebay), rather than solely providing sedatives in the guise of eye-candy, are rare but not impossible to find. More obscure are the personal examples, things we could all accomplish alone in an afternoon. The web can be used simply as a feature of software, a way to reach solutions, an ingredient to be fluent with in wielding this technology. The web is another folder on your desktop, but one so full of files, it has great uses.

Dan O'Sullivan, a professor at NYU's Interactive Telecommunications Program is certainly wielding the tools at hand. One fine afternoon, as a curator of a computer art show of about 150 pieces, he created a very web savvy utility to solve a very real problem, how to arrange these works in the finite space, given finite resources. Pieces often had very specific space, lighting and sound requirements. They all had extensive equipment requirements, needing various things from carpeting and green screens to a water supply, not to mention all the hardware like projectors, CPUs, speakers, cameras or things like wall space, tables, windows, web access.

So in the registration process, he included a software map of the gallery and fixtures. Each artist could place a marker on the map. Artists could return to see the latest version of the marked map. However, you could only edit your own marks (though O'Sullivan could edit all of them as needed). If artist clustered in one popular spot, others could decide to move to a less populated area. The solution to the problem was thus accomplished, finding an the ideal configuration, particularly given the limited efforts expended by everyone involved.

Another example happened when researchers looked into the game "Six Degrees of Kevin Bacon". In the game, a person mentions a movie, someone suggests a pair of actors that appeared in that movie and also were in a second movie together. More movies are mentioned, until the chain of paired actors includes Kevin Bacon. The goal is to see, from any initial movie, how few other movies can be linked to reach a co-starring role with Kevin Bacon.

Using the *IMDB* (Internet Movie Database), Duncan Watts was able to create these tables indefinitely and much more reliably than by poling humans. One thing he found is that Kevin Bacon is hardly the center hub of the Hollywood universe. In links to other actors he rates a little above average. In this way, Watts wielded the tools for the sake of his inquiry (Watts, 2003). He did not look for a pre-packaged research utility or even modify an existing one, he created the software from scratch that accessed the web.

Human Tasks vs. Computational Task

Many of us assume a computer can't consistently create masterpieces (Hawkins, 2004; Carter, 1999; LeDoux, 1996). A few remain convinced that a computer could theoretically create something like pop music and best sellers (Tapscott & Williams, 2006; Jenkins, 2006). How? On the net, one can randomly generate a song-worth of notes, or randomly put sentences together until 200 pages are reached. Then, as with Evolution, generate a variant. Thanks to the enormous number of potential judges available[21] on the net, humans can decide which iteration is the better one, and that becomes the new template for the next generation. Because there are so many, their votes will end up reasonably representative of the group consensus. Theoretically. This concept was made most

famous with the (theoretical) *Turing Machine*, but even *Information Processing*[22] utilizes the computer in this way (Hunt, 1996).

Even if an individual voted "incorrectly", the overwhelming volume of votes would push the outcome, not towards the "better" (as if such a concept were even viable) but toward more popular. This is also why it would never create a "masterpiece". The masterpiece, steps well beyond the expected. A tiny minority may recognize it's "greatness", but most do not at the onset, without other leaders to follow[23]. Einstein and Monet are obvious examples, where public opinion gradually shifted in their favor, but didn't start as comprehension without a few model believers to imitate (Gardner, 1993; Solso, 2003).

The common needs (spoken or not) are what create deep comradery, and not the content of the messages. The net can transmit messages. Many propose that this can engender *social capitol*, a trust between members (Watts, 2003). But trust is not emotion. Membership and kinship are not synonymous. Certainly, the messages are no more faulty for conveying empirical matters (including chit-chat, but don't make sentiments too wordy or require more than a quick glance). Moreover though, the medium simply doesn't work for more complex messages. We don't get an email alert when visitors' to our web pages eyes glaze over. A face-to-face conversation is interactive on hundreds of levels (at least) and not one or two. If we were all Autistic maybe communication mediated by computers (as we do now) would suffice, but so far what the net does and what people do is very distinct (Boyd, 2005).

Figure 3. Looking at ex.plode.us (a site that compiles the membership of several social software sites) may prove an ideal demonstration of this. Here are 6 of hundreds of pages retrieved from a single search.

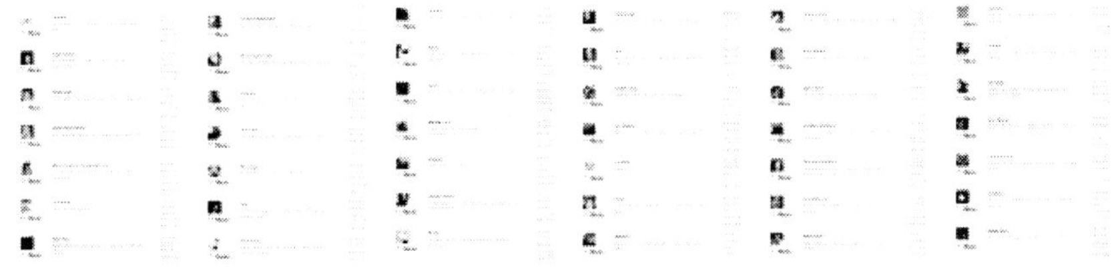

But there are two ways we use computers. One is the hands-on way but the other is the way we imagine computers in the abstract. For instance, we vaguely imagine computers at NASA or even at the bank are for. Most of us don't know or really even care in any detail, yet we have a sense of what we think their role is. We tend to assume our subconscious attitudes are informed by conscious ones. Hopefully not. Movies like *2001:A Space Oddessy*, *War Games* or *The Terminator* series reiterate remarkably similar anthropomorphic messages about our tools. This is the same technophobia as when people became afraid Nikola Tesla was out to destroy the human race (Tesla & Childress, 1996). It's silly in retrospect, but is an inevitable stage in any development.

An example of high-end computing is the *neural network*. At first glance, it would seem this idea would be of no use at all[24]. At second glance, you see it of use, if only it didn't have that preposterous name. A basic model of a neural network may look at a face with 4 binary features, one of two types of eyes, ears, nose, etc. (Adbi & Valentin & Edelman, 1999). Presented with randomly assorted faces of this kind, it guesses if a known face is the same as a new face, perhaps which features are more likely to appear together. The limited number of options is essential to "learning."

As the system "learns", it gives weight to it's predictions, called *Hebbian learning* (the *Widrow-Hoff* method is similar but adds a feedback loop. "Fuzzy logic" is similar but without necessarily networking (McNeal, 1994)). This is obviously a simplification to refer to the concepts, but there are countless pages written about this. However, that's not how brains work. We make unpredictable adjustments to the odds on the fly. "That might be Janet who might be wearing his clown make-up." Like language, there are infinite possibilities, but we use schemas to narrow our choices drastically (likely to chunk-sized options with every recall).

By creating context, we can pool sketchy *top-down* and *bottoms-up* knowledge and derive useful conclusions. A more comprehensive description of the *frame problem* (and the *conduit metaphor*) would require a whole book (Fodor, 2000; Ortony, 1993) so forgive me for trying to sum it up here. Computers can't know about anything beyond what they've been programmed explicitly. Without telling them about 4, they can't add 2 and 2. Thus there is a limit to what they can learn. Computers can only do an amazingly thorough job of re-combining facts they know. They can't consider of things outside of the frame they have been given. Computers just don't "frame" things and can't employ our method. And it ends up, this method is essential to how we conceive of problems (Dewey, 1910; Hawkins, 2004).

Again, this is hardly to say we are wrong in trying (for a while). It is just very peculiar behavior on our part. It is not that we should "correct" our use, as much as we should wonder about how we landed on it, and why we have so tenaciously clung to it, for so incredibly long, when we should know these issues fully well. The points here are not recent or arcane discoveries at all. No industry or advertising is to blame. We choose what to buy and how to use the tools freely. But by clearly identifying our motives, we may be able to shed light on how best to achieve them, particularly by discovering ways to act in our own favor, and no longer in a veiled form of self-abuse.

CONCLUSION

There is a big difference between web 1.0 and 2.0. Its the difference between going to the lumber yard to get wood cut to build shelving and going to the ... store for shelving. In one case its ground up and custom and in the other its glitzy but you're working with pre-defined units. Its a different way of thinking - what gizmos/widgets can i find & add? – Katharine Staelin, Web designer since 1995.

Very true.

However, let's take another step and compare a 100-year-old oak bureau and a unit from Ikea. The Ikea model has benefits that most folks (including the retailer, manufacturer and shipper) need. It's cheap and more applicable to different (but pretty similar) situations like storing books, clothes, flatware, etc. What's worrisome is subtler. Where people once saw funny uses for peculiar furniture, now they just don't want to, are trained not to. They see an old bureau that is obviously a bedroom piece doesn't belong in a kitchen.

That said, most of us are mediocre carpenters. Our pre-fab furniture can only end up so bad. But it can only end up so good, too. At an insightful, yet perfunctory level, it's only the range of quality that is changed. And that's what has happened with the all-purpose look of blogs. But looking at it in reverse, where a page of no interest can be clicked away and immediately forgotten (has minimal effect), the odds are now remarkably slim that we'd land on a page that knocks our socks off.

I am picking on blogs and the web, but these are symptoms. There is a much deeper and more profound dynamic I am referring to here. In this process, we are being trained as to what we need to do to get more expected result, not more useful ones. Part of that training is selective attention. We smooth over differences between what we hope to find and what we are offered. Without noticing our habits shifting, our brains (subconsciously) re-work our inquiries to fit the answers given so things make sense to us (Gallison, 1997). We ask Google for things that cause Google to reward us or not punish us with confusion. This is the wave and we are already drenched.

By subscribing to this wave, we also downplay our individualistic differences. But here we must take caution. An emergent property of this cohesion is that the web now contains mechanisms for enforcing a cultural harmony, even amongst traditionally opposing views. Acceptance is a slippery slope. Ideally, we cast away bigotry

but we can also give child molesters the stamp of approval. This dilemma is not a dilemma at all. It's simply that we are used to judging right/ wrong, true/false, good/bad and the **net** is a way to discard these things. Truth doesn't matter. From a traditional perspective, the net absolutely is lawless anarchy. But the net is not about truth, it is about collecting points of view. The only fact lies in the idea that someone, somewhere wants to make a point of view more accessible. And that is completely trivial, beside the point. Authors and their beliefs are arbitrary because they have deliberately chosen to be so.

Whether a collective unconscious, a gas oven, a cultural ritual, a verbal language or a popular tune, these are all tools, we created at our disposal. We have a choice in how to employ them. As every one of us, in big cities and out in the sticks, is encroached on with greater the pressures of population density and *overload*, the more prone we become to use these tools as a shield, protecting our intimate individuality. Such a scheme is hardly a bad one. Quite the contrary, we may be desperately saving what little sanity is still salvageable.

Democracy is not freedom, but a strategy to allow anonymity. Or if you'd rather believe it is, freedom's key feature is to absorb most of the blows to individuality, by distributing them to the masses in imperceptible little inconveniences. Spam is an obvious illustration, but even the fact that people use blogs that don't precisely fit their needs, though are willing to adjust those needs, is a far more profound case. Or that people eagerly use web pages, a monthly fee and pricy computers instead of filling notebooks and leaving them on the street or sending postcards. What are we anxiously paying for?

Ultimately, we made this collective unconscious. The net is our resource, at our machines' disposal (and not at all visa versa). It is unprecedented-ly useful because it continually evolves, because no individual, consortium or law really has the effect on it of the totals of us, because we feed it, look after it and nurture it. What we created was certainly impressive. Technology is ours to transform, to some degree obviously at the content level, but much more importantly at the perspective level. We are heading further into aimlessness and danger, but aren't on a fixed course at all. Hopefully, a splash in the face, wakes us up. We can now choose to steer.

REFERENCES

Abbate, James (1999). *Inventing the Internet.* Cambridge, MA: MIT Press.

Adbi, Hervé & Valentin, Dominique & Edelman, Betty (1999). *Neural Networks: A Sage University Paper.* Thousand Oaks, CA: Sage Publications.

Alderage, David & Fachner, Jorg (2006). *Music and Altered States.* London, England: Jessica Kingsley Publishers.

Anderson, Laurie (album) (1984). *United States Live.* Warner Brothers.

Arbabi, Freydoon (2000). *Classical Persian Music (Radîf).* San Francisco, CA.

Barabási, Albert-László (2002). *Linked.* Cambridge, MA: Perseus Publishing.

Barbrook, Richard & Cameron, Andy (essay) (1995). *The Californian Ideology.* Alamut.

Bausch, Pina (dance performance) (2002). *Für die Kinder* (For the Children of Yesterday, Today and Tomorrow). Brooklyn NY: BAM.

Blass, Thomas (2004). *The Man Who Shocked the World: The Life and Legacy of Stanley Milgram.* New York, NY: Perseus Books.

Boyd, Dana (essay) (2005), *Autistic Social Software.* SuperNova Conference (reprinted in *The Best of Software Writing*).

Becker, Judith (2004). *Deep Listeners.* Bloomington, IN: Indiana University Press.

Brown, James (liner notes) (1991). *Star Time.* Rykodisk.

Bulgatz, Joseph (1992). *Ponzi Schemes, Invaders from Mars & More Extraordinary Popular Delusions and the Madness of Crowds.* New York, NY: Harmony Books.

Calvin, William (1996). *How Brains Think.* New York, NY: Basic Books.

Calvin, William (1996). *The Cerebral Code.* Cambridge, MA: MIT Press.

Campbell, Don (1997). *The Mozart Effect.* New York, NY: Avon Books.

Campbell, Joseph (1972). *Myths to Live By.* New York, NY: Bantam.

Campbell, Joseph & Moyers, Bill (1988). *The Power of Myth.* New York, NY: Doubleday.

Castro, Elizabeth (2003). *HTML for the World Wide Web.* Berkeley, CA: Peachpit Press.

Carter, Rita (1999). *Mapping the Mind.* Berkeley, CA: University of California Press.

Cavalli-Sforza, Luigi (2000). *Genes, Peoples and Languages.* New York, NY: North Point Press.

Chomsky, Noam (1957). *Syntactic Structures.* Berlin, Germany: Walter Gruyter GMBH.

Chomsky, Noam (1977). *Language and Responsibility, Linguistics and Politics.* New York, NY: The New Press.

Chomsky, Noam (2000). *New Horizons in the Study of Language and Mind.* Cambridge, MA: Cambridge University Press.

Clancy, Susan (2005). *Abducted: How People Come to Believe They Were Kidnapped by Aliens.* Cambridge, MA: Harvard University Press.

Coe, Michael (1999). *Breaking the Maya Code.* New York, NY: Thames & Hudson.

Crystal, David (2005). *How Language Works: How Babies Babble, Words Change Meaning and Languages Live or Die*. Woodstock, NY: Overlook Press.

Dawkins, Richard (1978). *The Selfish Gene*. New York, NY: Oxford University Press.

Daniélou, Alain (1991). *The Myths and Gods of India*. Rochester, VT: Inner Traditions International.

DeLanda, Manuel (2006). *A New Philosophy of Society*. New York, NY: Continuum.

Dewey, John (1910). *How We Think*. Boston, MA: Dover.

Doige, Norman (2007). *The Brain that Changes Itself*. New York, NY: Viking.

Duchenne, GB (1862). *Mécanisme de la Physionomie Humaine*. Paris, FR: Jules Renouard, Libraire.

Ekman, Paul (2003). *Emotions Revealed: Recognizing Faces and Feelings to Improve Communication and Emotional Life*. New York, NY: Owl Books.

Etcoff, Nancy (2000). *Survival of the Prettiest*. New York, NY: Anchor Books.

Eyman, Scott (1997). *The Speed of Sound*. Baltimore, MD: Johns Hopkins University Press.

Fischer, Claude (1992). *America Is Calling*. Los Angeles, CA: University of California Press.

Flietcher, Max (cartoon). (1932). Betty Boop in Minnie the Moocher. *Out of the Inkwell*. New York, NY: Fleicher Brothers Studio.

Fodor, Jerry (2000). *The Mind Doesn't Work that Way: The Scope and Limits of Computational Psychology*. Cambridge, MA: MIT Press.

Gallison, Peter (1997), *Image and Logic*. Chicago, IL: University of Chicago Press.

Gardner, Howard (1983). *Frames of Mind: The Theory of Multiple Intelligences*. New York, NY: Basic Books.

Gardner, Howard (1993). *Creating Minds*. New York, NY: Basic Books.

Gladwell, Malcolm (2000). *The Tipping Point*. New York, NY: Little Brown and Co.

Gleick, James (1999). *Faster: The Acceleration of Just About Everything* New York, NY: Vintage Books..

Gleick, James (2002). *What Just Happened?* New York, NY: Pantheon.

Gordon, Deborah (1999). *Ants at Work*. New York, NY: Frcc Press.

Goffman, Erving (1959). *The Presentation of Self in Everyday Life*. New York, NY: Anchor Books.

Greene, Rachel (2004). *Internet Art*. New York, NY: Thames & Hudson.

Hauser, Marc (1998). *The Evolution of Communication*. Cambridge, MA: MIT Press.

Hawkins, Jeff (2005). *On Intelligence*. New York, NY: Owl Books.

Heylin, Clinton (2007). *Babylon's Burning: From Punk to Grunge*. New York, NY: Canongate.

Heurer, Kenneth (1951). *Men of Other Planets*. New York, NY: Pellegrini & Cudahy.

Hillis, W. Daniel (1998). *The Pattern on the Stone*. New York, NY: Basic Books.

Hinde, RA (Ed.) (1972). *Non-Verbal Communication*. Cambridge, England, Cambridge University Press.

Hoffer, Eric (1951). *The True Believer: Thoughts on the Nature of Mass Movement*. New York, NY: Perenial.

Hoffman, Donald (1998). *Visual Intelligence*. New York, NY: Norton.

Holzner, Steven (2007). *The AJAX Bible*. Indianapolis, IN: Wiley Publishing.

Hopkins, Anthony (1993). *Sounds of the Orchestra*. New York, NY: Oxford University Press.

Hunt, Morton (1993). *The Story of Psychology*. New York: Doubleday.

Jacobs, Jane (1961). *The Death and Life of Great American Cities*. Nerw York, NY: Modern Library.

Jenkins, Henry (2006). *Convergence Culture*. New York, NY: New York University Press.

Johnson, Steven (2001). *Emergence: The Connected Lives of Ants, Brains, Cities and Software*, New York, NY: Touchstone.

Jones, Jim (transcript) (1978) *The final speech of Reverend Jim Jones*. Brighton, England: Temple Press Ltd.

Jung, Carl (essay) (1919). Instinct and the Unconscious. *British Journal of Psychology* (reprinted in *The Portable Jung*, 47-58).

Jung, Carl (essay) (1935). The Concept of the Collective Unconscious. *St Bartholomew's Journal* (reprinted in *The Portable Jung*, 59-69).

Jung, Carl (essay) (1936). The Relations Between the Ego and the Unconscious. *Rascher Verlag*, (reprinted in *The Portable Jung*, 70-138).

Jung, Carl (1966). *The Spirit of Man, Art and Literature*. Princeton, NJ: Princeton University Press.

Kern, Jerome (2003). *The Culture of Time and Space 1890 – 1912*. Cambridge, MA: Harvard University Press.

LeDoux, Joseph (1996). *The Emotional Brain*. New York, NY: Touchstone.

Levin, Golan (essay) (2006). *Computer Vision for Artists and Designers: Pedagogic Tolls and Techniques for Novice Programmers*. Flong.

Levitin, Daniel (2006). *This Is your Brain on Music*. New York, NY: Penguin Books

Linklater, Richard (movie) (2002). *Waking Life*. Fox Searchlight Pictures.

Marvin, Carolyn (1988). *When Old Technologies Were New*. New York, NY: Oxford University Press.

May, Elizabeth (Ed.) (1980). *Music of Many Cultures*. Berkeley, CA: University of California Press.

McNeal, Daniel (1994). *Fuzzy Logic: The Revolutionary Computer Technology that is Changing the World*. New York: Simon & Schuster.

Myler, Harley & Weeks, Arthur (1993). *The Pocket Handbook of Image Processing Algorithms in C*. Upper Saddle River, NJ: Prentice Hall.

Narby, Jeremy (Ed.) (2001). *Shamans Through Time*. New York, NY: Penguin.

Norman, Donald (1988). *The Design of Everyday Things*. New York, NY: Basic Books.

Nørretranders, Tor (1998). *The User Illusion*. New York, NY: Penguin Group.

Opler, Morris (1994). *Myths and Tales of the Jicarilla Apache Indians*. New York, NY: Dover.

Ortony, Andrew (Ed.) (1993). *Metaphoir and Thought*. Cambridge: England: Cambridge University Press.

O'Sullivan, Dan & Igoe, Tom (2004). *Physical Computing*. New York, NY: Thomas Course Technology.

Packer, Randall & Jordon, Ken (Ed.). (2001). *Multimedia from Wagner to Virtual Reality*. New York: Norton.

Pearl, Steven (movie) (1998), *Substitute II: School's Out*. Dynamo Entertainment.

Piaget, Jean (1929). *The Child's Concept of the World*. New York, NY: Rowman & Littlefield Publishers, Inc.

Pinker, Stephen (1993). *The Language Instinct.* New York, NY: Harper Perennial.

Popper, Frank (2007). *From Technological to Virtual Art.* Cambridge, MA: MIT Press.

Prata, Fabio (2002). *Internet Art: Digital Culture.* São Paolo, Brasil: FILE.

Pursell, Carroll (2007). *The Machine in America.* Baltimore, MD: Johns Hopkins University Press.

Ramachandran, VS (2004). *A Brief Tour of Human Consciousness from Imposter Poodles to Purple Numbers.* New York, NY: Pi Press.

Reeves, Bryon & Nass, Reeves (1996). *The Media Equation: How People Treat Computers, Television, and New Media like Real People and Places.* Cambridge, MA: Cambridge University Press.

Robbins, Jim (2000). *A Symphony in the Brain.* New York: Grove Press.

Rushkoff, Douglas (1994). *Media Virus.* New York, NY: Random House.

Sacks, Oliver (1996). *An Anthropologist on Mars: Seven Paradoxical Tales.* New York, NY: Vintage.

.Sacks, Oliver (2007). *Musicophelia.* New York, NY: Alfred A Knopf.

Sells, Michael (Ed.) (1996). *Early Islamic Mysticism.* Mahwah, NJ: Paulist Press.

Sex Pistols, The (album). (1993). *The Great Rock 'n' Roll Swindle.* EMI International.

Sheldrake, Rupert (1999). *Dogs That Know When their Owners Are Coming Home.* New York, NY: Three Rivers Press.

Schwartz, Jeffrey & Begley, Sharon (2002). *The Mind and the Brain.* New York, NY: HarperCollins.

Schwartz, Steven (1994) *Visual Perception: A Clinical Orientation.* Norwalk, CT: Appleton.

Shay, Anthony (1999). *Choreophobia: Solo Improvised Dance in the Iranian World.* Costa Mesa, CA: Academic Publishers.

Shenk, David (1999). *The End of Patience.* Bloomington, IN: Indiana Press.

Shermer, Michael (2002). *Why People Believe Weird Things.* New York: Owl Books.

Shirkey , Clay (1995). *Voices from the Net.* Emryville, CA: Ziff-Davis Press.

Shanahan, Murray & Baars, Bernard (article) (2005). *Applying Global Workspace Theory to ther Frame Problem.* Volume 98, Issue 2, December 2005, Cognition.

Simons, Sarah (Ed.) (1993). *No One May Ever Have the Same Knowledge Again: Letters to Mount Wilson Observatory 1915-1935.* Los Angeles, CA: Society for the Diffusion of Useful Information Press (The Museum of Jurassic Technology).

Slater, Lauren (2004). *Opening Skinner's Box: Great Psychological Experiments of the Twentieth Century.* New York, NY: WW Norton & Co.

Solso, Robert (2003). *The Psychology of Art and the Evolution of the Human Brain.* Cambridge, MA: MIT Press.

Spence, Keith & Swayne, Giles (Ed.) (1981). *How Music Works.* New York, NY: Macmilllan.

Stanislavski, Constantin (1936). *An Actor Prepares.* New York, NY: Theater Arts Books.

Standage, Tom (1998). *The Victorian Internet.* New York, NY: Walker & Co.

Stokes, Melvin & Maltby, Richard (1999). *American Movie Audiences.* London, England: Brittish Film Institute.

Surowiecki, James (2005). *The Wisdom of Crowds.* New York, NY: Anchor Books.

Tapscott, Don & Williams, Anthony (2006). *Wikinomics.* Mew York, NY: Portfolio.

Temperley, David (2007). *Music and Probability.* Cambridge, MA: MIT Press.

Tesla, Nikola & Childress, David (1993). *The Fantastic Inventions of Nikola Tesla.* New York, NY: Adventures Limited Press.

Tenner, Edward (1997). *Why Things Bite Back.* New York, NY: Vintage Books.

Tufte, Edward (1990). *Envisioning Information.* Cheshire, CT: Graphics Press.

Vronsky, Peter (2004). *Serial Killers: The Method and Madness of Monsters.* New York, NY: Berkeley Berkley Books.

Walls, Jeannette (2005). *The Glass Castle: A Memoir.* New York, NY: Scribner.

Watts, Duncan (essay) (2003). *Six Degrees.* New York, NY: WW Norton.

Weinberger, David (2007). *Everything Is Miscellaneous: The Power of the New Digital Disorder.* New York, NY: Henry Holt and Co.

Weizenbaum, Joseph (article) (1966). ELIZA--A Computer Program For the Study of Natural Language Communication Between Man and Machine. MIT (Project MAC).

Weill, JR & Brannon, WT (1948). *Con Man: A master Swindler's Own Story.* New York, NY: Broadway Books.

Zimmer, Carl (2001). *Parasite Rex Inside the Bizarre World of the Nature's Most Dangerous Creatures.* New York, NY: Free Press.

Zimmer, Heinrich (1972). *Myths and Symbols in Indian Art and Civilization.* Princeton , NJ: Princeton University Press.

KEY TERMS

Consciousness: This word (often called "the C-word") gets people all flustered and surely needs definition. Many want to believe it is uncannily elusive. I disagree. Jeff Hawkins seems to as well and articulates it quite nicely. "Consciousness is not a big problem. I think consciousness is simply what it feels like to have a cortex." (Hawkins, 2004, p. 194) Of course "feels like" is even more problematic. I would alter this slightly. Consciousness is the result of whatever the neo-cortex does. We don't precisely know all the details, but whatever they end up being we can just call "consciousness."

Digital: This has to be one of the most abused buzz words ever. It has a very definite meaning that gets lost in a hazy fog. People often think it means something like "newer and better". Actually, it just means "less accurate (for a good reason though)". Theoretically, analog is any value from 0 to the highest, an infinite gradation of grays. In reality, we are limited by human perception and frustrated by mechanical limitations. Digital means that the same reading is now only 0 and 1, black or white.

There is no question this is an imperfect alternative. What is useful though is that when done enough times, we can surpass previous limitations (though we don't always). Creating copies is far more reliable since there are so few possible states per *pixel* (see below). In fact, whereas before "copy" and "original" were useful ideas, now they really don't apply. The thing we play back now is always a copy. The original, which only existed as electrical pulses, is long discarded as soon as it is saved. But there is no longer any reason to distinguish, since they are identical clones.

Network Behavior: Individual entities (nodes) can be grouped to communicate with each other in a myriad of configurations. Two cans and a string is the simplest networking model, where there is only one link between the *nodes*. The phone system is a complex network, allowing any node to relay to any other. Or you can think of social networks. In cases where "she told two friends, then she told two friends and so on, and so on." In networks, tiny events can take on enormous proportions.

Where a germ is passed between nodes, the epidemic charted over time, does not look like a gradual slope, but stays low, then suddenly explodes. The features of the network beyond the sum of its nodes are the *emergent* properties. For instance, telephones are only so handy in isolation but by connecting them in a network. That's a simple emergent idea, but there are more complex ones. For instance, you can scream at one end and raise a person's pulse at the other.

Neural Net(work): A computer system is often designed to mimic the workings of the brain, primarily for the task of statistical learning. However, this term is a bit misleading, since it is not at all likely that the brain learns in even a similar way to the way computers (even networked ones) save and retrieve data from memory. Nonetheless, whether accurately depictive or not, it is a remarkable programming.

In very generalized terms, data comes from several distinct sources (often with their own processors analyzing *input*. Each source is called a *neuron* and has a weight of influence. The data is scaled according to the neurons current weight. The central computer determines the successfulness of this pool and updates the weights accordingly. Thus it seems to learn, which processes influence the outcome more heavily in complex tasks. Of course, this implies fore-knowledge of the goal, which brains don't consider and computers can't do without, However, the pooling and dynamic weighing process is no less effective once the analogy to the brain is discarded.

Pixel: Usually this word is used in the technical sense (and probably one you are well familiar with). But the general concept is often a hazy one. The term being popularly bandied about with careless excitement, refers digital-ity.

It is the smallest indivisible element in a collection of like elements that comprise an entity. "All in all it's just another brick in the wall." You could say the places on a checkerboard are pixels. But they needn't even be square. City blocks may be considered pixels. But they needn't be configured in any orderly fashion. Cells in a cluster of tissue could be pixels. The audio equivalent of resolution is called the *sample rate*.

Usually to call something a pixel, there needs to be an encompassing (explicit or implied) system. A colored dot of ink is not necessarily a pixel, until you consider it part of a larger picture, say in a newspaper.

Surface Grammar and **Deep Structure:** These are terms coined by the linguist Noam Chomsky. Probably this entire essay is predicated on an understanding of these terms. It hardly matters if this theory proves correct for language, it is nonetheless a very useful way of looking at the issues. But if you aren't familiar with the terms, and many aren't, we'll try here.

According to Chomsky, (non-human) animals certainly communicate but insists that these communications can not really be called language because the animals lack essential mental features. He has a point. It's a fine one and could easily be overturned some day. But it is based on the notion that there is a unique mental apparatus. A *Language Acquisition Device*. A bee dance or a bird call serve linguistic purposes, but bees' and birds' brains don't have enough parts.

Each human language obeys a somewhat unique set of conventions. "Noam talks about language." is not the same as "Language talks about Noam." The difference in meaning is derived from their specific *surface grammar*. But the fact that we can put words together in some order and it will convey a message is *deep structure*. Though it is clearly impossible all the specifics of a language would be in our genes (no one is born speaking fluently, we need to learn), it is also unlikely we could learn so many detailed rules, so proficiently, with almost no trial-and-error for many aspects, in the few years we acquire language.

"Colorless green ideas sleep furiously." Chomsky's famous example feels like it should make sense but does not. Possibly because, while we may figure out by "looking up" the references of our learning, that this is meaningless, we are "hardwired" for the rules by which a languages rules are concocted. The brain doesn't actually know syntax from chaos, it simply applies this pattern recognition to whatever stimuli is input. Over the ages we have gradually adapted this mental tool to the task of communication via words.

Turing Machine: Probably, if you bought this book, you are someone who knows about Alan Turing's ideas more than I. It is a very popular concept in computing and for good reason. Turing's original example has been applied in countless variations, in countless disciplines over the years. But what the Turing Machine, from 1936, refers to is a proposed concept (never actually built), where a computer could use feedback with a conditional assessment function to "learn". Almost all computer learning is based on this idea in some way.

ENDNOTES

[1] Object Oriented programming mainly differs from Procedural Programming (now all but extinct) in that there is a short hand method of grouping behaviors into "Classes". If the object in code is a member of the Cat class, any specific instance *inherits* all of the variables and functions of that Class. Thus if Mittens is of the Cat class, and Princess has a variable for breed, and a function purr(), Mittens probably has those things too. However, it would be convoluted to detect if there were purring happening anywhere in the program and then impossible

to determine if it is a Cat doing it. At least without setting up very externally specific and reasoned decisions. There is no identifyCatsByWhateverMeans() function.

[2] Higher level programming languages have *data types*, which may seem comparable to chunking. There are huge, essential differences, but the main ones are that chunks are not a fixed size, and can expand/contract. There is no precise point at which they fail. And there is no analogy to *casting*, or switching between data types.

[3] Jung notes that "resistance to its [archetypical] expression may result in neurosis", which, as we read on, resonates eerily. Sanity may depend on a degree of conformity" (*ibid.*) This notion, that archetypes are not created to serve luxury or interest, but from a genuine psychological need for our mental health, has great ramifications as we continue.

[4] Before the Beatles and Elvis, during Frank Sinatra's early performances audiences, mostly of teenage girls, often screamed and fainted. His manager noted the phenomenon and made it more likely to occur by planting 16 young women who had been instructed to initiate this behavior. They were soon accompanied by authentic screamers and fainters. The antics spread to the thousands of others. As with the laugh track of television sit-coms, people will often follow the herd, even when the leader is obviously a fraud. Shows are funnier to us if the (artificial) crowd seems to enjoy them. We are constantly on the look-out, hungry, for cues to appropriate behavior, even when we "know" better.

[5] Keeping control of neurological feedback loops is what drugs like Prozac are all about. LeDoux shows how potential fear stimuli triggers the amygdala, which sends two signals, one to create visceral effects and one to the pre-frontal cortex to assess the

threat. There, the signal either is abandoned or sent again to the amygdala, completing a loop. There really is often nothing to fear, but fear itself. This is also why feelings of apprehension can take hold, without much antecedent. It is a runaway feedback loop. The amygdala keeps sending the signal to the rest of the body, increase the pulse rate, speed up breathing, dish out more adrenaline. But meditation and other like forms of trance states, has the ability to assuage those loops.

[6] English only relatively recently switched from this structure.

[7] Only about 100 of the thousands of frequencies are commonly used (Levetin, 2006). Though by tuning off concert pitch these dozen notes (times about 8 octaves) are relative and not absolute. Of the dozen tones, they are often further divided into patterns called scales (an example is all the white keys on a piano are C Major, the black keys are "accidentals" of that scale). Cultures tend to stick to the same 3 or so 7 note scales (Temperley, 2007). Though that leaves about 2.5 million possibilities for every interval. Quite a lot but hardly infinite.

Likewise, rhythms tend to be either on the beat, or a division of that by some factor of two (half a beat, half of half, half of half of half, etc). Occasionally factors of 3 or 5 (or even 7, 9 or 11) are used sparingly in Western music. Eastern and Middle Eastern music is more open to factors greater than 2, and tends to limit melodies to 5 of the 7 note scale (May, 1980; Arbabi, 1997). Again with regard to rhythm, the possibilities are enormous, but drastically fewer than infinite.

[8] However, cell phones and TVs are now quite imitative of computers, this is not essential to their actual functions. A peculiar phenomenon. Traditionally, successful need fulfillment, drives manufacturing and thus marketing. This appears to work the other way around. However, as we'll talk about later, a need may actually exist yet.

[9] Ironically, but only examples disruptive to systems that have always worked are ever heard. Napster and Bit Torrent type file sharing is more about the fun of getting things for free. The resulting anarchy is an afterthought, and hardly as detrimental as if it were a bona fide strike against the industries. We literally could upload every song and movie as it comes out, but this may actually be self-defeating.

[10] I would say, an important transition for us now, is to let go of the idea of archiving. While digital storage may be semi-permanent, the means to play it back is in hyper-flux mode. Digitization is paradoxically ephereral-ization. Or rather, the more non-linear and computer-centric the thing, the less likely it will be available for long. But calling archiving an obsolete game is akin to blasphemy.

[11] Actually colors outside of the visible spectrum (like Infra-red and Ultra-violet) would be useful for many applications.

[12] Though "inch" varies slightly, making matters even more complex for traditional designers.

[13] There really is no original. The usefulness of digital-ness is like replacing an assembly line with a hall of mirrors. The fundamental way computers and the net operate, renders issues like copyright, moot points. Looking is "stealing" or rather there just is no useful way to talk about "stealing" anything that can be digitally recorded. This bothers a lot of folks who depend on things like royalties. Prolonged arguments simply conceal the issue. We can try to weave convoluted smoke-screens and offer excuses, but the concept is a simple one – if you understand how digital technology works, you understand that the word "original" has no real

relevance. Without that word, any argument collapses like a house of cards.

[14] Folks today complain that modern audio CD's suffer from a standard set decades ago, limited older capabilities.

[15] Many arms must have grown sore. For a few decades, the phone featured one cup to speak into and another one held to the ear. Later, those two pieces were consolidated. The "French" style one piece receiver, that we still see sometimes in land lines, and ubiquitous until about 2000.

[16] These were early Mac notebooks (1992-1994) with a garage-like box that acted as a monitor base and peripheral hub. I have a few and they still can be networked (LAN or internet). They have browsers, can send email, cost less than those $100 computers for Africa that every talks about. So why do I, along with so many of us, keep buying new machines?

[17] Both programs are surely out there already but which would take a Python or J2ME programmer a day to write from scratch. Often these features are built in to programs such as Dreamweaver and many professional web designers simply aren't always aware they exist (much more usefully) outside of a specific program.

[18] Free IDEs for cell phones are offered by Google (Android) and at Mobile Processing.

[19] Paranoia from schizophrenia (and other psychological problems) may be awful for that individual, but that individual is seldom also able to organize unwilling masses to comply with their defensive strategies. Conservatives (again, loosely defined) tend to project their needs onto others who would be fine if they weren't being so regulated.

Often the minority, they devise very calculated means of getting others to adopt their strategies. However, if conservatism implies a need to keep the status quo from change, the only way they can possible manage to keep things comfortable (or even bearable) for themselves would be to convince the majority that change is a bad thing.

[20] Oddly, this is the exact effect that the adoption of the light bulb had as well. People could read by the fireplace with the family in the living room, but chose to move to secluded bedrooms. People could see by moonlight (and stars back then!) but street lights made it feel less scary to go out for an evening stroll. Note that most places on which livelihood depended, were closed by that hour. Certainly this was not a universal case, but you see how most of the light bulb's cultural impact came more from luxury than dire need.

[21] Note that these are people with idle time and attention. Perfect revolutionary fodder.

[22] A precursor to Artificial Intelligence, where Sociology and Computing initially met.

[23] Recall the plants at Sinatra shows from footnote 4.

[24] The problem can be summed up like this. "[neural nets] do not differ *essentially* from standard statistical models. ... Networks usually have several layers. The first layer is called the *input* layer, the last one called the *output* layer." (Adbi & Valentin & Edelman, 1999, p. 2). Thus you see, the task at hand is not actually accomplished entirely asynchronously but remains essentially linear. And though we don't know much about how brains do what they do, we are sure of many ways they don't work. Brains are definitely not linear.

Section IV
Creativity in Virtual Worlds and Artificial Spaces

Chapter XXI
Theatre in Second Life® Holds the VR Mirror up to Nature

Stephen A. Schrum
University of Pittsburgh at Greensburg, USA

ABSTRACT

As creative people inhabit virtual worlds, they bring their ideas for art and performance with them into these brave new worlds. While at first glance, virtual performance may have the outward trappings of theatre, some believe they don't adhere to the basic traditional definition of theatre: the interaction between an actor and an audience. Detractors suggest that physical presence is required for such an interaction to take place. However, studies have shown that computer mediated communication (CMC) can be as real as face-to-face communication, where emotional response is concerned. Armed with this information, the author can examine how performance in a virtual world such as Second Life may indeed be like "real" theatre, what the possibilities for future virtual performance are, and may require that we redefine theatre for online performance venues.

It's like a Hall of Mirrors, where you keep bumping into reflection after reflection of things that aren't there in the first place!

—Melissa Perreault, a resident of Second Life.

Are we avatars or people performing this particular dance?

—Meghamora Woodward, a resident of Second Life.

PROLOGUE

I dance in the middle of a floor that pulsates with lights from below, and am flanked by a half dozen other moving bodies: shirtless, tattooed men with bulging musculature, and scantily-dressed women perched atop impossibly high Frederick's of Hollywood style shoes. All of us move in synch with choreography copied from *Pulp Fiction*, dancing to an electronic beat emanating from huge speakers at the front of the room. The floor then changes to display a rotating earth so that we seem to be floating in space above it. Colored disco lights flash around us, and suddenly a fog machine by the floor begins pouring out a bluish smoke that engulfs us, and momentarily obscures our view of the surroundings.

We are immersed in the sights and sounds of a disco. But there are no smells or touch sensations here. This is a dance club that exists only as binary information and in our imaginations, one of many in the online virtual world known as Second Life.

INTRODUCTION: WHAT IS SECOND LIFE?

Second Life® is, according to *Second Life: The Official Guide*, "a 3D online virtual digital world imagined, created and owned by its residents" (Rymaszewski, et al., 2007, p. 4). Anyone (with the required computing power) can access this world using the Second Life client, available as a free download from the website, secondlife. com. The user creates an avatar, "an interactive representation of a human figure in a games-based three-dimensional interactive graphical environment" (de Freitas, 69), and then moves through the virtual world. The user travels about, either by walking, flying, or—in the case of more distant locations—teleporting. One may tour shopping areas, art galleries, or other "builds" designed and created by Second Life residents.

At first glance, Second Life (SL) appears to be mostly about shoes, shopping and sex. Almost everywhere one travels, there are shoes and clothing available for sale. If so inclined, one can purchase a parcel of land, and drop a pre-designed house on the parcel. One can then furnish the house completely, with everything from living room sets, to detailed kitchens and bathrooms, to beds that have hidden boxes of animations in order to allow your avatar and another's to engage in simulated sex.

Upon further exploration, however, one finds that the world offers much more. While a casual player or observer may never leave that dance club, the social networking possibilities are widespread. Since there are people connecting from many countries around the world, SL allows for global connectivity. Providing a virtual location for a variety of real world corporations and businesses (such as IBM, Toyota and Coldwell Banker), SL also provides educators with a venue for networking and collaborating with distant colleagues, or for creating a virtual classroom. In-world (inside Second Life) locations such as EduIsland and the New Media Consortium (NMC) offer teaching and learning resources for academics at all levels. For a variety of fields, SL is a rich laboratory, ripe with research topics. Psychologists (Suler, 2007) have analyzed behavior within Second Life; computer scientists (Schweller, 2007) have created virtual interactive objects for student use and experimentation.

For creative artists, there are a variety of small galleries to exhibit works created both in SL and the real world (Real Life, or RL), and numerous venues featuring live music and other, more theatrical, performances. Recently, in increasing ways, theatre practitioners have worked within the SL environment. As performances occur in SL, we learn more about how the virtual environment can or might work as a mirror of RL theatre, and what the possibilities are for future virtual performance. Another question arises: how "real"—meaning how much like RL live theatre—can these performances be?

THEATRICALITY IN SECOND LIFE

In an article entitled "Begin with A Single Step: Adding Technology to a Course" (2000), I compared the use of email and electronic or computer-mediated communication to wearing a mask in an acting class:

While email is no replacement for direct, face-to-face communication, it can lead to more direct communication. For students who are shy and reticent, or who are afraid to ask questions in class (including the fear of looking stupid or, heaven forbid, appearing interested in a class), email allows them to ask their questions in safety. Indeed, the "anonymous" mask of email is much like a mask used in an acting class. With the latter, the student actor, "hiding" behind the mask, feels secure and free to do things he or she would not do without the mask, because of a perceived concealment. In the same way, a student will write very open and informal email notes to the instructor; even though the instructor knows full well who ShyOne@nodename.com is, the student is still masked. If the instructor answers these notes in a friendly and nurturing way, the student often will become comfortable enough with the instructor to venture into more face-to-face dialogues. (Schrum, 2000, 56-57)

The intensely graphic-based and three-dimensional world of Second Life is inherently theatrical, taking the aforementioned example of donning a mask one step further. When you first access the SL environment, you make certain decisions about your avatar: your gender, physical appearance and dress. By then infusing this full-body mask with your personality (or the personality you wish to project to the world), you satisfy the first requirement of acting—taking on the role of an *other*. As you navigate the virtual world, you encounter people from all over the globe and interact with them—again, either as yourself, or as the person you wish to be perceived to be. In either case, the theatricality of role-playing constantly comes into being as you associate with others around you.

To examine the theatrical nature of Second Life, we may begin with the ancient Greeks' conceptions of the theatre, not only because our notion of Western theatre originated with them, but also because, in her seminal study, *Computers as Theatre*, Brenda Laurel (1993) applied the "deep, robust and logically coherent notion of structural elements and dynamics" (36) of Aristotle's *Poetics* to human-computer interaction. This return to Aristotle's description of what theatre can or should be illustrates the parallels between theatre and this virtual world.

Armed with this perspective, Second Life can be perceived as Aristotelian in many of its aspects; the six elements of theatre listed in *The Poetics*—plot, character, verbal expression, thought, visual adornment, and song composition (Aristotle, 26-27)—are in evidence everywhere. The most obvious is that of character; as soon as you first log in, you must create your character, an avatar, which will serve as the agent of the actions you will perform in SL. Connected with this is costume, one of the areas of visual adornment that, according to Aristotle, "can have a strong emotional effect" (Aristotle, 29). One is "assigned" a set of clothes during initial avatar creation, depending on the type of person and gender one chooses to be; there are sixteen choices, ranging from basic shapes to the more exotic, such as "City Chic," "Cybergoth," or "Furry" (a sort of human-animal hybrid). As these choices of costume are always available in the Inventory Library, you can change your basic identity at a later time. Of course, shopping malls are ubiquitous, so with an infusion of the in-world currency called Lindens, or Linden™ dollars, you may purchase almost anything you desire to add to your wardrobe. (Some clothing templates also exist on the World Wide Web for download and editing, so that you can create your own clothes, if you wish.)

The settings of Second Life are another area of visual adornment. It appears that every possible natural and man-made creation has been re-created in SL. Bliss Gardens allows an amusement park-like tour past dizzying heights and beautifully designed flora, waterfalls and walkways. As stated before, the modern commercial mall is well represented, in many shapes, sizes and themes, from pirate ship to ancient Roman forum.

A visit to the Spacestation Alpha International Space Museum allows the visitor to ride a rocket into space, and then teleport to a representation of each of the planets, complete with information about size, year duration, etc.

Aristotle calls song composition "the greatest of sensual attractions" (Aristotle 29). Music can be found throughout SL, in a variety of ways. The owner of any parcel of land can enable the sound from an internet-based stream (such as a SHOUT-cast or Nicecast server) to play through a user's computer speakers. Live music performances, which occur almost constantly somewhere in SL on an hourly basis, provide the consumer with an almost-limitless supply of performers, playing either live music or recorded music with live voiceover announcements.

While SL is a virtual world without hard and fast rules about physics and gravity, thus allowing flying, teleportation, and immediate solar/lunar shifts from midnight to noon, people tend to select signs of reality in order to ground their experience. There is of course no real need for chairs, because avatars do not grow tired standing around and chatting. However, people often feel awkward standing around, and so chairs are provided for them. SL music performance also uses many trappings of visual adornment that are found in RL. A DJ sitting in a room in her home anywhere in the world is represented in SL as an avatar bent over a console replete with turntables and a microphone. A singer-songwriter's avatar sits on a stool behind a microphone stand and holds a guitar, the latter loaded with animations that allow her to sway, rock strum, or shoot flames from the guitar's neck.

To further demonstrate this seeking of a simulation of reality in a virtual world: at one performance, my avatar stood on a stage in front of a microphone, my voice playing through the audio stream. Even though my announcements had been recorded days earlier and emailed to the DJ, and were now being played from an unknown location through the audio stream, the avatars standing in front of the stage turned to watch and listen to my speech. Even though the users could have allowed their avatars to continue to dance while swiveling their cameras to watch my avatar, they chose to stop dancing and stand still, facing the stage. In another instance, a colleague of mine stated that she wanted to arrive early for a performance of a play so that she could sit in the front row. She could have had her avatar sit anywhere (including *behind* the set) and then manipulated her view to a close-up of the actors, but instead chose a front row seat as she would in RL. Clearly an object that signifies its real world counterpart is an important ingredient of the virtual world.

Also in the realm of visual adornment, SL can provide a vast array of eye-popping, awe-inspiring, spectacular (and, of course, inexpensive) effects. At a "live" performance by guitarist Silas Scarborough, the performer stood on a vast stage, overshadowed by towers of speakers that would likely weigh tons in the real world. As Scarborough played his electric guitar to accompanying backing tracks of bass and drums, two dancers swirled and flew about like virtual Ariels (the airy spirit from Shakespeare's *The Tempest*), soaring through a series of dance moves and instantaneous costume changes, metamorphosing from earthbound women to winged creatures in an eye blink.

While character, song composition and visual adornment are Aristotelian elements easily found in SL, verbal expression (language) is a ubiquitous yet problematic area. One of the keys of an environment that is capable of social networking and linking citizens from around the globe, much of the communication and verbal expression within SL still occurs primarily by users typing text into a chat box. The text then shows up on the screen as dialogue, such as, "Phorkyad Acropolis: O brave new world!" One can also communicate with others through an instant message window. Emotes, nonverbal textual signs of a performed action, are also possible. When one types "/me" followed by a phrase in the format of verb and descriptors, others see on the screen a description of the avatar emoting, or doing something. "/me looks around in wonderment" would appear on the screen as "Phorkyad Acropolis looks around in wonderment."

Avatars are also capable of other visual non-verbal communication. Built in to every avatar are certain gestures. With an easy point and click combination, or with a quickly typed command (such as /laugh), an avatar may execute a variety of actions and poses, sometimes with accompanying sounds. For example, as one finishes presenting a poem (by typing it, or copying and pasting it, into the chat window), one may "/bow," after which many of the audience members may choose to "/clap." One can always add more gestures to one's inventory as desired.

Objects in SL can also be scripted to communicate with avatars in their immediate area. Beyond greeter robots, there are interactive objects of a more intimate nature. Perhaps, while in a dance club, two people become friendlier and more intimate as they dance together. If a woman has attached a special appliance to her avatar, as she becomes aroused, either from clicking or from text commands, her vagina, monologuing, would tell observers of her growing excitement.

More recently the element of voice has been introduced. Residents of SL can now communicate to others around them by speaking into a microphone or headset (without using an audio stream or server software, as before), or by speaking directly to one other person through the instant messaging window. One no longer need type what one wants to say; one can merely speak it. While this innovation has widened the possibilities for performance, it has been met with some resistance for a variety of reasons. First, one is not always certain who is talking until one looks around, and sees which avatar has a green waveform (indicating speech is on and functioning) above her head. Second, typed chat allows hearing-impaired persons to communicate on par with others in a text-only environment. Third, typing allows a

slower pace of communication for those who are not native speakers of a language; when people do not feel sufficiently facile with a language other than their own, they prefer to take more time with text, rather than spoken, communication. Also, products in SL such as the Babbler take typed text and translate it, displaying it in the chat window, so that someone unfamiliar with a language (whether French or Japanese) can speak to others, even if the others do not have such a device. There is currently no software solution in SL to handle rapid *spoken* translation.

Perhaps more importantly, however, is that voice may reveal more about a speaker than the speaker wishes. Some, who wear their avatars like masks that allow them to be someone whom they are not, simply feel more comfortable typing, not wanting to reveal their true voices. Some don't like how their voices sound (a common complaint, dating from the advent of telephone answering machines), and so for them, typing is also preferable. For those people whose avatars are a different gender, or perhaps even a different personality in SL than they are in real-life (such as a strong and harsh SL master seeking submissive slaves), they choose not to speak in order to maintain the illusion of the gender or façade they present in the virtual world.

To conclude this coverage of the Aristotelian elements, the elements of plot and thought are much less obvious. Plot, or the structure of story elements, is considered by Aristotle to be of the greatest importance in tragedy (Aristotle, 27). In an environment that resembles a large floating improvisational conversation, with events that occur without any imposed dramatic structure (such as dance club costume contests or live music performances), plot is rarely present. While it may be an element in role-playing sims (simulations, or locales within SL), where there is a greater emphasis on structured behavior, SL does not have a need for beginning/middle/end storytelling, in the way a quest might in other MMORPGs (Massive Multiplayer Online Role-

Playing Games), such as World of Warcraft. In Second Life, events unfold and occur much as in residents' first lives, in a seemingly random way. As for the element of thought: it appears to be entirely absent from SL, unless one examines the world from a more sociological or philosophical point of view regarding the implications of such virtual environments. While I leave larger commentary on that area of study to researchers in those fields, I turn now to a discussion of the relationship of reality to a virtual world, and how that may vary as one spends more time in SL.

THE REALITY OF SECOND LIFE

In his Introduction to the novel *Mother Night*, Kurt Vonnegut (1966) posited that, "We are what we pretend to be, so we must be careful about what we pretend to be" (v). In this seemingly unreal and cartoonish world, one's Second Life can become "real" unexpectedly. Though sometimes compared to other online role-playing games, such as World of Warcraft (WoW), there is one fundamental difference. As explained on the WoW "Introduction to World of Warcraft" webpage:

World of Warcraft is an online role-playing experience set in the award-winning Warcraft universe. Players assume the roles of Warcraft heroes as they explore, adventure, and quest across a vast world. World of Warcraft is a "Massively Multiplayer Online Role Playing Game" which allows thousands of players to interact within the same world. Whether adventuring together or fighting against each other in epic battles, players will form friendships, forge alliances, and compete with enemies for power and glory.

In WoW, players may work together in battle, communicating and cooperating to kill enemies in a fantastic and supernatural milieu. While both the WoW and SL environments feature self-identification with one's avatar, it is likely that,

"Because of the simulation of actual life leading to emotional needs replacing task-oriented needs" (Moran, personal communication, June 25, 2007), combined with SL's interactive nature and the lack of necessary conflict (though there are role playing combat areas in Second Life), residents may find themselves discovering that their virtual relationships are quite real. Unlike WoW, where one may encounter artificial intelligence-controlled non-player characters, every avatar in SL has a human being guiding it, and one can easily sense presence behind the avatar. One SL resident, Iris Ophelia, is quoted in *Second Life: The Official Guide*, as saying:

The most important aspect of how I interact in this virtual world is in emotions. There's a heart beating behind every avatar; including yours. We're not that far from the real world here, and we all have to make a choice to respect that—or ignore it. (208)

Research has shown that emotions can be stirred easily when dealing with other humans via computer-mediated communication. Sherry Turkle illustrated this in *Life on the Screen* in 1995; her example of the relationship of Robert and Kasha and its "easy intimacy" in a MUD (multi-user dimension) environment demonstrated how deeply emotional involvement can occur, even when communicating entirely through a text-only interface (p. 206). Andrew Wood and Matthew Smith describe a similar situation in their textbook *Online Communication* in regard to "self-presentation," "The process of setting forth an image we want others to perceive" (52). Participants of a CompuServe discussion group eventually discovered that Julie, a "mute, paraplegic victim of a car crash who had wrestled with suicidal depression" turned out to be a male professional psychologist doing some research by posing as a woman online (51).

People meeting others through the Internet often believe a person is who they say they are, and accept explanations of identity without question. Perhaps this exchange from the film *The Matrix* (1999) summarizes this tendency:

NEO: I thought it was real.
MORPHEUS: Your mind makes it real.

Users of an online or virtual immersive world tend to accept the reality of that world. Researchers exposing subjects to dangerous heights (a high balcony, or "a rope bridge stretched precariously between two tall buildings) modeled in virtual reality have discovered that those prone to acrophobic responses react to the virtual heights as they would heights in the physical world" (Bolter and Grusin, 163). The conclusion drawn here is that:

This reaction would seem to vindicate the hopes of the cyberenthusiasts: that virtual reality can disappear as an interface and give the viewer the same emotions that she would feel in the real world. If virtual reality can evoke emotions, how can our culture deny that the experience of virtual reality is authentic? (Bolter and Grusin, 165)

In a lecture entitled "'Second Life:' What do We Learn If We Digitize EVERYTHING," Philip Rosedale, founder and CEO of Linden Lab and of Second Life, spoke of virtual identity, not as an anonymous thing, but as a projection of identity in the virtual world "in a more facile and detailed way than you typically do in reality" (2006). This specific presentation can lead others, with their imaginations engaged, to accept one's proffered mask.

If we then couple the freeing of inhibitions through the wearing of this carefully constructed virtual mask (cited before), with a concomitant acceptance of an other's presence in an online world and a willingness to become immersed in the virtual environment, it is easy to see why people find themselves experiencing real emotions as a result of online interactions.

Though some express the opinion in their SL profiles that, "It's just a game," this perspective is sometimes replaced by the notion that, "It gets real very fast." Reading profiles of longer-term residents (over six months, since SL time moves more quickly than RL time), one will find comments suggesting that residents fell prey to "real emotions" in an environment that, for some, can be described as an "interactive unscripted soap opera" (Moran, personal communication, June 25, 2007).

The unscripted, improvisatory nature of SL interaction has often taken residents by surprise, and even those who feel they are well balanced emotionally, with RL situations as workers, spouses and parents, sometimes suddenly discover they feel true romantic attachments to SL avatars (or to the operators behind them). Whether this is a true emotional response, or a projection of one's desires onto the situation, is unimportant for this study; what is more important is the idea that SL residents infuse other avatars with presence, as real humans, and develop true attachments for them. This notion will return later in the context of SL theatre and audience/performer interaction.

WEDDING CRASHERS

One of the ways in which Second Life clearly displays the virtual world as both a mirror of real life and as an environment for theatrical role-playing is in the realm of SL weddings. Residents of SL who fall in love can choose to be partnered, and pay a fee to have their names listed on each other's profiles. However, some residents choose to formalize this linking with wedding ceremonies that emulate the RL event.

In a typical SL wedding, all the usual components of a wedding are present. The bride and groom dress appropriately in wedding gown and tuxedo, or may select garb for a Japanese-themed ceremony. Music plays, and the bride enters, processing down the aisle (perhaps in a jerky fashion, depending on the walking animation her avatar is using, or the lag imposed by a high number of wedding guests present). She clutches a bouquet, which she has attached to her left hand. Her immediate goal is the poseball (an orb that contains scripts that animate an avatar) at the end of the aisle; the groom stands there already posed on his ball, awaiting the arrival of the bride. When she reaches her destination, the couple face each other holding hands, a pose provided by other animation scripts.

At this time, an officiant will type or speak a greeting and provide some comments regarding the couple and their long-term commitment to one another. (I have no statistics on this, but many SL marriages seem to last only a few months.) The two then exchange their vows and slip rings on each other's fingers. A processional follows, and friends join the couple immediately afterward for a reception.

At the reception, events also follow the normal RL pattern. Virtual champagne is available for the guests to sip, and a live performer or DJ provides music for dancing. If the couple has selected the full package, a wedding planner will be standing around demanding that they move on to the next phase, and a photographer will capture images of the newlyweds cutting the wedding cake and dancing their first dance together (again, actions provided by poseballs). It is also likely that the couple will eventually teleport off to their honeymoon chamber for their first virtual encounter as a married couple.

There are of course, some differences between SL and RL. If the bride or groom's software "crashes" during the wedding, they may suddenly disappear and have to log back in, reappearing where they had stood before. However, other than this SL-specific problem, everyone plays his or her assigned roles. The officiant (who may be appearing in the form of a RL cleric, a Wiccan priestess or a hellish, hornéd demon holding a Bible) speaks eloquently of the sanctity of marriage, and the bride and groom pledge their unswerving love. The wedding guests also follow the roles written for them by society; everyone comments on the radiance and beauty of the bride, women will emote the shedding of tears for the happy couple and the event, and everyone will applaud the conclusion. Overall, there seems to be an attempt on everyone's part to follow established social norms and make the event as "real life-like" as possible.

PERFORMANCE AT OTHER SL EVENTS

As stated before, because people interact through avatars, role-playing is a vital part of activities and events that occur in Second Life. One area, known as Mill Pond, featured (as you might suspect) a large pond, European-styled side streets, small shops and cafes, and a reproduction of the Parisian Shakespeare and Company bookstore. On the pond the owners erected a stage for spectators to sit and listen to musicians performing there. In December 2006, the pond was "frozen over"—an ice texture was applied to the water, and the surface was made physical and solid, rather than phantom (which would allow avatars to pass through it). Residents could then don ice skates that were scripted to allow the wearer to glide across the surface, and perform a variety of maneuvers, such as jumping, backwards- and forwards-skating, and leaving trails of sparkles and particle trails when desired.

In SL the user experiences a third-person perspective, with the camera hovering behind the avatar, providing a point of view familiar to those playing certain videogames such as *Tomb Raider*. It would seem that this view does not allow the user to be a direct participant with a "for your eyes only" viewpoint. (You can access a first-person perspective if you choose "mouse look," and focus crosshairs with the mouse.) Instead, you are slightly removed from the action, watching yourself perform the action through your avatar.

In spite of this, residents of Second Life often report emotional reactions to their adventures in SL. I have referred to emotional attachments with friends or virtual lovers. In my own case, I can attest to having a profound feeling of freedom and *joy* while skating around the frozen Mill Pond. I never ice- or roller-skated in my RL childhood or adulthood, because I had never learned how. Here, I was not hampered by age or inexperience; the scripted skates allowed me to be a near-perfect skater. I still had to steer my avatar with relative accuracy, and try not to crash into snow banks or other skaters. However, gliding in and among the others, the R2D2 unit serving drinks, and the ice-fishing igloo and jumping ramps, I—with my imagination in full complicity with the graphics, audio, hardware and software—felt as if I was experiencing skating.

Of course, imagination is a vital ingredient in any theatrical enterprise; for an audience member to be immersed in a theatrical production, we require an actor to create a role, but also require the spectators to engage their imagination and to willingly suspend disbelief, thus creating and accepting the fiction that transpires before them. SL allows us to be the actor and the spectator simultaneously, performing an action (such as skating) and watching ourselves, vicariously or voyeuristically experiencing what we guide our avatars to do.

AT THE BALLET

Within Second Life, there is a ballet company, under the direction of Inarra Saarinen, which produced a piece entitled *Olmannen* during February of 2007. Second Life Ballet says in its program that the company "produces professional works that creatively utilize the Second Life environment; we produce neo-classical, contemporary, and eclectic ballets for and with the residents of Second Life" (Saarinen, 2007, p. 2). *Olmannen*, according to the in-world program:

Is the story of a couple in love, how evil in many forms can divide, and how true love can overcome. It is the story of souls fighting for love.

Centuries ago, Namon agreed to become a Devang to save his one true love, Seraphette, from evil. Namon has waited centuries for her soul to be reborn to reclaim their love. But now that she is reincarnated, Seraphette has changed. Can Namon still love the new Seraphette? Can Seraphette possibly accept the love of Namon, a creature of the Overworld? (Saarinen, 2007)

Dancers accomplish the choreography with "individual gestures and animations, and [they] respond to one another" in real time (Dibou, 2007). This is unlike what one experiences at dance clubs; when you click a dance animation poseball or sign, the script makes your avatar dance with pre-loaded sequences. (At some points, there may be a pause in the dance, as everyone stands around, waiting to resume.) Places such as Ayumi's Geisha House also have preprogrammed dance sequences. A musician "plays" an instrument at the front of the raised, wooden stage, as kimono-clad avatars, amidst a taiko drum and large gong, move in unison through a series of geisha dances.

For the ballet performances, as in RL, the audience sat in the audience and watched the events unfold on a stage in front of them. Audience members were also instructed to "Force Midnight," or change the ambient lighting to darkness, so that the theatrical lighting effects may be seen. The Second Life environment provides some additional advantages to theatrical producers. An opaque curtain can quickly become transparent, or disappear entirely, to reveal the action. Set pieces can also materialize or dematerialize for set changes, and a solid oak wall can be replaced instantly by a huge waterfall. Lavish costumes can include traditional dancewear, as well as fanciful furries (avatars that are animals, such as anthropomorphic foxes or wolves), assuming any shape regardless of the avatar beneath the costume. As for ballet, the dancers can execute movements that take them higher off the ground than any RL, gravity-bound dancer could manage.

For all of these advantages, however, I found the performance of *Olmannen* lacking real interest. Advertised as a full-length ballet, the three acts ran just over twenty minutes, with the three acts timed at five minutes each, and two set changes that seemed longer than necessary in a virtual world. While the performers may have indeed been controlling their movements in response to and along with each other, the overall effect for me was a virtual marionette show. While there were representations of performer's bodies, I personally felt no expression of physical artistry from the avatars, with no distinction between what might be directed movement or recorded animation. Nor did I feel a real connection to the performers, the story or the performance as a whole.

As with SL weddings, audience members played their assigned societal roles to the hilt, shouting "Bravo," applauding, and speaking in hyperboles about the creative vision that had just displayed on the stage before them.

POETRY PERFORMANCE: A STEP TOWARD SL THEATRE

As people extend themselves into their Second Lives, they of course infuse SL with creativity, which is a component part of being human. One aspect of creativity that seems natural in SL is poetry. Before the implementation of voice chat, poetry was the easiest and most accessible literary art form for SL. While storytelling or verbal improv did not work well when typed or copied-and-pasted into the chat box—since one easily can become confused or lost, and has to scroll back, reread, then hurry back and catch up—the short form of poetry allows others present to experience the poem as *readers*. If, when the poet is finished, the audience wishes to review the poem, they

need only click on history, scroll up, and reread. By presenting a poem as a traditionally reading piece, SL allows the poem to be presented and enjoyed in real time; also, feedback is possible: audiences may clap, compliment and critique.

While the text-only presentation of a poem may seem somewhat artificial I recall having intense stage fright when I read my first poem in SL in late summer 2006 at Shakespeare and Company. I infused the surrounding avatars with presence, and thought of them as real people, and real critics. (Fortunately my poems were—and continue to be—positively received.) Thus, the immediacy of the poetry "reading" with people seeing the poem, and reading it synchronously with my typing, added that level of "reality" for me that prompted actual performance anxiety. Since that time, I have attended numerous poetry readings in SL; often I still find that newcomers to a poetry venue will express their nervousness and feeling of vulnerability at the prospect of sharing their feelings and creative works to the audience present.

Because it works so well in SL, many poetry venues have been created. They include:

- The Blue Angel Poets' Dive, run by Persephone Phoenix
- Cruiz Control's Cruizin' Neo-Soul Def Poetry Jam
- Steorling Heron's StarGaze Poetry Reading

Other poetry venues and events have sprung up and disappeared quickly. The ones that endure have also continued to evolve, as different venues offer different experiences. For example, Persephone Phoenix at the Blue Angel reserves the first half of the evening for people to present their poems. This is then followed by the Poetry Challenge; participants must write different types of poems on the spot, and then present them to the rest of the audience. The challenges vary in number of words, the number of lines—and even one type where the poem can be read either forward or backward. The immediacy increases the level of interactivity, with attendees being both reader and spectator.

Before the official implementation of voice chat, there were some experiments using audio as a means of reading poetry aloud. Jilly Kidd of Written Word had a method where a reader would use a phone and dial into a tele-conferencing system that would be connected to the land's audio stream. (Any user who has enabled "play streaming audio" can then turn on the audio with a button on their SL screen.) In this way, the host of an event can read others' poems that they submitted on notecards. The major drawback to this is that the poet is not reading his or her own works; someone else is reading and interpreting them, and this may be different from how the poet intends the work to sound.

Another method, used by Cruiz Control, at his Cruizin' Neo Soul Poetry Jam, was to use audio for "Spoken Word Poetry." Poets would submit mp3s via email, and then at Cruiz's Cotton Club, the host would introduce a poet whose avatar would come to the microphone. Cruiz would play the mp3 through the audio stream, and some poets—with shorter works—would type or copy-and-paste the poem at the same time into the chat window. This allowed the audience to hear the poets themselves, sometimes with musical accompaniment behind the words.

Still others connect to an audio stream directly, using audio server applications. Audio is sent from the performer's microphone through their computer and the server software to a location on the internet. That location then streams the audio, and the land owner in SL sets the "receiver" to that stream. In this way performers can set up background music, as Serene Bechir does for her SNAP performances. She defines SNAP as "spoken narrative and poetry," which can take a variety of forms. In my identity as the poet Phorkyad Acropolis, I use audio tracks created from music loops as background to reading my works.

Sex, Death and Religion: Poetry and Theatre

In January 2008, SL poets Paggles Whitman and Ada Radius put together a poetry performance. Entitled *Sex, Death and Religion,* the two took turns reading each other's poems through voice chat in the Greek Theater on Cookie Island. However, the piece went beyond a simple poetry reading.

Clad in classical costumes, Paggles played a Greek scientist (referred to as Archimedes early on), and Ada played a Greek slave who scrubbed the floor, carried various ancient masks, and danced seductively as she read. A third performer, Upo Choche, wandered about the performance space in a lion avatar, interjecting roars between poems. The lion seemed a mere observer until the very end, when he attacked the slave and the scientist. Animations provided his pounce as well as the victims' bloody sprawls on the stone circle. The scientist then grew angel wings and drifted heavenward; the slave metamorphosed into a gigantic butterfly that soared aloft as well, while singing what sounded like a religious hymn.

While it had no definite conflict until the end, the performance was quite a theatrical experience. The use of costumes, props and movements, along with the performers' spectacular arrival in a hot air balloon from high above, provided the audience with much more visual adornment than a poetry reading usually presents.

One notable event occurred after the final performance on January 30. An audience member instant messaged ToryLynn Writer, who had been running a Line Reader to provide a "closed captioning" text of the poems. The sender thanked ToryLynn for providing that service, since she (the sender) is hearing-impaired and could not hear the poems as spoken by the performers. This proved to be a revelation, as we had been working on voice presentation in SL, and had been neglecting the textual side. Those involved noted this for future presentations, hoping to include simultaneous text scrolling.

"O'ERSTEP NOT THE MODESTY OF NATURE"

The works of William Shakespeare continue to captivate producers of virtual theatre. One of the earliest experiments, *Hamnet*, in 1993, featured the characters of *Hamlet*, an "ASCII-only keyboard symbols [version of] a toy Elsinore castle" (Dixon, 2007, p. 487), and a text "filled with humor, irreverence, Net slang, and carnivalesque wordplay associated with collaborative online textual environments" (Dixon, 2007, p. 486). The production was presented on an IRC (Internet Relay Chat) channel, the precursor of modern instant messaging. In Second Life, the works of Shakespeare continue to entice performers and directors; the Act Up Theatre Company plans to stage *A Midsummer Night's Dream* in SL in August of 2008. In November 2006, I attended a rehearsal of a group of SL theatre actors, under the direction of Takeshi Kiama, preparing for a production of *Romeo and Juliet*.

When I arrived at the rehearsal space, I found the actors sitting on various chairs around a device called a Cardspeaker. This object, loaded with a programming script and a play script, provided the name and lines of dialogue for each character. (Note the similarity between this and the Line Reader used for *Sex, Death and Religion*, cited before.) The intention behind this was to provide the text of the play for the actors so they wouldn't have to type them in, thus freeing them to move about the stage and enact physical poses and expressions.

The physical and gestural actions of the characters would be accomplished with the help of another device, the Anim8tifier. Created by one of the technicians attached to the production, each actor would carry one, and be able to assume a variety of poses as needed. Director Kiama demonstrated this by dropping to his knees, with arms held up in a melodramatic pleading gesture. The device, I was told, would be available on SLExchange (www.slexchange.com), a marketplace

for SL accessories, though to date I have been unable to locate it.

For the production, the physical setting of the performance would provide scenery, and the audience would see objects and hear sound effects and music created specifically for the performance. But would this primarily text-based presentation, with its visual and auditory components, vary that much from *Hamnet*? Also, would this SL Shakespeare production (or *Hamnet* for that matter) be considered theatre, applying the commonly held definitions of the art form? Without hearing the actors speaking, the performance would seem, like *Olmannen*, to be a silent marionette show, with avatars acting to subtitles and accompanying music. Would the performers be truly acting, again, as we commonly define that term? Is there interpretation of roles, or of the text, beyond basic staging?

As a postscript to this, in Spring 2007 I instant messaged Takeshi Kiama, and asked what had happened to the production, but received no reply. I also checked in with one of the actors present at the rehearsal I attended. She was unaware if the production had ever taken place. She gave me a name to contact about an acting HUD that was being developed for the show by Lucas Gealach. A HUD is a heads-up display, an attachment to the screen that allows you to click buttons and manipulate objects, while leaving the rest of the screen visible. In response to an instant message, Gealach wrote:

I created a fully-functional acting HUD with a separate Director object, but I fear it will soon be obsolete: it's designed for the actor to be able to deliver lines through chat from a notecard script, so that the actors don't have to type or cut-and-paste the lines. It has a variety of other features, including a warning light when one has a line coming up, loading a scene for all actors from a central Director box, etc. The one thing I could imagine it being useful for in future (unless someone wanted to do chat acting despite voice) would be as a kind of automatic prompter, with a few adjustments. (L. Gealach, personal communication, May 29, 2007)

A vital step in the evolution of performance in Second Life has been the addition of voice. While some residents have raised legitimate concerns about adding voice (cited before), voice chat allows for more "life-like" performances. Early experiments in poetry, including the phone conferencing system and playing audio files (also cited before), indicate that the truly *spoken* word could be a compelling method of online communication, and performance.

TAKING THE STAGE

Because of its performance potential as a venue for theatre, various groups have been attempting theatre production in Second Life. On youtube.com, there is a video, produced by an SL group called Millions of Us, entitled, "A Night at the Theatre in Second Life." The end titles say it is a presentation of extracts from the play *From the Shadows*, written by Enjah Mysterio, and directed by Osprey Therian. A blog entry on the group's webpage proclaims the presentation's success:

Too tired and inspired to write much more than that after nearly a month of anticipation, the play that was performed was ... highly entertaining. We had a full house and the very idea of sitting here in a virtual theatre in a virtual world with people from around the real world is just mind-blowing. As for the cast and crew, they could very well have all been on different continents which is just stunning. (Millions of Us, 2006)

According to another reviewer, the play itself lasted 10-minutes, with a "traditional three act sequence" (Ixchel). The reviewer adds that, "The mise-en-scene in each act was carefully staged, with a curtain between each to allow for scene and costume changes" (Ixchel).

During the month of May 2007, Dan Zellner, as his avatar Dan Undertone, offered a series of improvisational workshops in SL. His idea was to use:

Games and exercises to further understand the potentials of Second Life for the creation of theatre and similar experiences. Experiences meaning creation of scenarios in which the spectator's role will be different than many other theatre situations in RL. (Zellner, personal communication, May 2, 2007)

Incorporating ideas from Viola Spolin (author of *Improvisation for the Theatre*) and Chicago's Second City, Zellner/Undertone took participants through such exercises as follow-the-leader and passing tie-dyed balls back and forth and over each other's heads, to demonstrate the fundamental idea of the avatar as simultaneous spectator, actor and creator (Zellner, June 2007)).

I attended the first workshop, located in the New Media Consortium (NMC) theatre building, and found these first steps somewhat intriguing, occasionally enhanced and hampered by the environment. While playing the follow-the-leader game, we had some additional abilities not usually found in the real world: we could fly to the rafters, balance on high beams and slowly float to the floor. However, with a somewhat limited visual perspective, following the leader was not always easy; at times, the leader would fly up, out of my visual field, and I would have to stop, search the rehearsal hall for the leader, and then attempt to follow the path I had missed. The workshop did allow those who had not previously experienced these ideas within Second Life to experiment with them, and served as a positive beginning. (Due to time constraints I was unable to attend the remaining sessions during the rest of the month.)

ACT UP, a theatre group organized in SL, decided to stage a Potpourri of One Acts as part of the Second Pride Festival from June 24 through June 30, 2007 (SecondPride.org, 2007). Originally planned as four one-act plays, with rehearsals starting in early May 2007, a blog regarding the theatre production entitled "Derek's SL Report" listed only two one-acts as of June 6, 2007.

1. "One Little Indian Boi", by Christo Larson and Upo Choche and directed by Upo Choche & Kimmi Jewell;
2. "The Thinkerer", by D.F. Dansereau and S.H. Evans (a new take on Aesop's fable "The Tortoise and the Hare") and directed by Marin Mielziner (Hotger, June 6, 2007).

The blog also lists some of the "multifold" challenges for Second Life theatre, which include:

*   The uncertainty whether the voice capability in the SL main grid will be available on time (or the need to alternatively use prerecorded voice);
*   The question which stage to use;
*   The task of finding competent set designers
*   The chore of locating animations which will make the avatars stand, move, and gesture like RL (= Real Life) actors (Hotger, June 6, 2007).

The performance that I attended on June 30, with an audience of approximately 45, began with a play about the creation of Second Life (which doesn't seem to fit either play description). At first, the audience saw only a large garage door hiding the stage. As we would discover, the door/curtain was scripted to tilt up in the manner of a RL garage door. Unfortunately, audience members kept clicking it (whether deliberately or not, is unknown) and it kept opening too soon.

Following an announcement not to IM the actors during the show (the equivalent of using a RL cell phone or text messaging during a production) and not to click the stage, the one act began as the garage door opened, and an angel appeared, hovering above and to the front of the stage. While some audience members realized she represented the narrator, others thought it was a rude person teleporting in, in full view of the audience, to interrupt the performance. (While theatre in SL can be quite fanciful, it may be difficult to determine whether an odd creature is part of the show or of the audience.)

The voice of the narrator (provided by pre-recorded audio played through the stream at the location) explained to us that the characters onstage were the creators of Second Life, working on programming their new invention. We then watched three avatars move around the set, which consisted of a picnic table and trash can far stage right, three desks with chairs at center, a filing cabinet and metal cabinet up center, and a couch with duct tape patches stage left. (In RL terms, the set was quite wide, and had little additional set decoration, other than an "I Want to Believe" poster from the *X-Files*.)

While the avatars were all costumed differently—one in a dress shirt and tie, one in a Linux shirt and cap, and the third in a *Star Trek: The Next Generation* uniform—there was no way to connect the avatars with the voices on the pre-recorded audio. The voices were as distinctive from one another as were the costumes, but nothing provided the audience with hints as to which matched which. There was also no focus on a speaker; the avatars would stand, walk, and sit seemingly unrelated to the audio. One of the advantages of SL performance is that an audience member, regardless of location, can use the camera and zoom in or change perspective as the action unfolds. Here, however, we were given no directorial or acting cues as to whom we should focus on. One of the audience members, Tyrol Rimbaud, later observed that, at this stage, SL theatre is mostly about the audio, and that "the actor's voice is more important than any of the stage directions...the 'acting' here is more of a sideshow..." (T. Rimbaud, personal communication, June 30, 2007). The physical movement of the avatars added little to the performance.

The next piece, a monologue, seemed to be about the experiences of women in SL, but spoken with a man's voice, suggesting the female avatar was a man presenting as female in SL. As people listened to the audio stream, they responded by a series of gestures, such as laugher and applause.

Following this monologue, Upo Choche, a member of the company, appeared in front of the audience and said, "Due to an unusual degree of technical difficulties we will have to postpone the rest of the show." He explained that the problems were "unusually high amounts of lag," resulting in the actors "crashing left and right." As more performers, audience members and objects crowd a sim, the slower things run—hence, lag. Choche invited the audience to return to the next performance that evening.

While the need or desire to create theatre appears as strong in SL as it does in RL, the environment does not appear to have served the medium well in this instance. In the end, this performance more resembled a radio drama with marionettes than what we are accustomed to seeing in RL.

In an attempt to get a first-hand look at the demands of staging theatre in Second Life, I agreed to direct Zayante Hegel's play, *The Perm*. Set in the Cut 'N' Curl beauty salon, Dinah the hairdresser tells her customer Sandy about seeing a woman die on the roadway after being hit by Dinah's husband's new truck. The dialogue neatly captures the essence of two women in a salon sharing both opinions of Hollywood stars' likely "endowments" and a revelation of personal beliefs. The short one-act, with its single setting and only two speaking roles, made it a perfect vehicle for this experiment.

One of the concerns I had about theatre in SL grew out of attending a rehearsal for the other theatre group's "Potpourri of One Acts." Meeting them on the Beta Voice Grid (set up for early voice trials), I saw one actor vanishing and returning (crashing, like the bride and groom of an SL wedding), one actor having difficulty with walking and

navigating, continually crashing into walls, and a third unable to hear the other actors, though she herself could be heard by them. While crashing may be inevitable (you come back and resume) and voice problems are solved (at least partially) by casting people with access to technology that works, the problem of out of control avatars falling off the stage or running into the audience caused me to consider how the particular problem of stage movement might be solved.

In the theatre, we have an actor cross from up left to down right. They begin at one place onstage and end at another; it's a simple matter of blocking, or stage movement. In Second Life, that direct path may be slowed by lag, and an avatar may overshoot the destination "mark" (to use the TV or film term). However, I noticed that, if I was on a stage and needed to go sit down in the audience, I could go directly to the chair by right-clicking the chair and selecting "sit" from the popup menu. There was no stumbling blindly over people as I walked; the software took me in a straight line from where I was to where I intended to be. So the answer, for me, was to install poseballs in the performance floor. They can be hidden to the audience, and visible only to the actor needing to move. Clicking on the next mark, the SL actor would move directly to her next step in blocking, or from one pose to another.

This leads to the topic of Intermediality, "the blurring of generic boundaries, crossover of hybrid performances, intertextuality, intermediality, hypermediality, and a self-conscious reflexivity that displays the devices of performance in performance" (Chapple and Kattenbelt, 2006, p. 11). As Philip Auslander suggests in *Liveness*, each new medium imitates media that came before (11-ff). For example, cinema began by merely filming theatre productions, which resulted in a screenful of small, hard-to-see actors gesticulating and soundlessly mouthing words. When film overcame these initial limitations (a single camera position) and developed its own conventions and language (intercutting among long, medium, and close-up shots), it became a new medium. In the same way, SL theatre has to use what came before, but not just grafted on; it requires some adaptation to make it work. In the end, *The Perm* still resembled a radio drama with marionettes, though the voices of the actors were accompanied by the avatars moving with more focus and intentionality, with Dinah falling to her knees at one point, to demonstrate how she knelt at the side of the dying woman. Audience member Epyllion Basevi remarked that the "emotional expression in voice seemed in ironic tension with the avatar doll-like look" (E. Basevi, personal communication, July 14, 2007).

However, the production still lacked any interaction occurring between the simultaneously speaking and pointing-and-clicking actors and the observing audience.

FUTURE PLANNED THEATRE PRODUCTIONS IN SECOND LIFE

Marin Mielziner spoke of a planned SL production of Eve Ensler's *Vagina Monologues* in February 2008 that was abandoned due to issues over performing rights. While theatrical organizations can produce the *Vagina Monologues* without royalties, HBO—which owns the broadcasting rights—prevented the group within SL from a performance since the company considers SL a "broadcast medium," and not a live venue (M. Mielziner, personal communication, January 30, 2008). This is not only an interesting viewpoint from the perspective of copyright law, it also has implications for defining SL theatre. Can SL theatrical performances be considered theatre in the accepted definition of the performing art?

In an attempt to answer that question: following *The Perm*, in the summer of 2007, I applied for and received a grant from the Foundation for Rich Content. The SL-based FFRC "is a group in Second Life that exists to foster diversity and richness in content and events in Second Life" (FFRC, 2007). I outlined the project as follows:

My project is to create a reproduction of an ancient Greek theatre (such as that of Epidauros). At preferably full-scale, this theatre could serve as a permanent installation to be used by theatre educators, to visit, bring classes, etc. Using plans of Greek theatres (and the notes I accrued in the late 1990s when I began the project using VRML—abandoned as I waited for better technology), Jenene Lemaire, an experienced SL builder, will create the theatre structure. I will then contact SL clothing and accessory designers (such as Angelina Burali and Zayante Hegel) to provide costumes for the characters. Also at this stage, in a new course that I will be teaching in the Spring, entitled Theatre Technology, I will be encouraging the students to contribute objects and music. The premiere use of the space will be a production of Euripides' THE BACCHAE.... I hope to use SL voice-chat for audio, which should be in place by that time.

I chose *The Bacchae* partially because it is a Greek tragedy that will adapt well to the SL environment; notices for the production advertise "Dancing! Godlike powers! Cross-dressing!" Transgendered residents will feel right at home as Pentheus dresses in drag to spy on the Bacchants, and an animation script will provide particles streaming from Dionysus' ivy-covered spear, displaying its magical properties. At the same time, this play returns us to theatre's roots (as did examining SL in light of Aristotle's *Poetics*, as discussed before), but with a modern reinterpretation, as each new era reinterprets those that preceded it.

Builders Jenene Lemaire and Talliver Hartnell of the SL company primeMovers completed the Greek Theater in December 2007. It can be found on Cookie Island at the co-ordinates Cookie 58,28,32. More of a Hellenistic structure than strictly Greek, it is a to-scale reproduction of a classical theatrical performance space. While the production of *The Bacchae* is in rehearsal as of this writing, the theater served as the site for the *Sex, Death, and Religion* performance, and proved an excellent classroom for my Theatre Technology students.

CONCLUSION/FUTURE TRENDS: IS IT THEATRE?

In July 2007, as part of a panel on "Interactive Media in Immersive Performance" at the Association for Theatre in Higher Education (ATHE) conference, I presented a proposed taxonomy of digital performance, in order to categorize media-rich, computer mediated, and virtual performances. I began with the notion that, when we delve into theatrical production in a virtual world, we are talking about performance, and not theatre, as scholars have long defined it. Theatre, traditionally, occurs only when there is an interaction between an actor and a spectator.

In SL, avatars can move about and actors can speak with their voices played through spectators' computer speakers, but can the spectators respond beyond simple clapping gestures and applause sounds? There are SL residents looking into this question. The owner of a comedy club, Atrus Hyun, is "working on a device that lets audiences vary the intensity of their applause. If no one or only one or two people click on it you get polite patta patta sounds. If everyone likes the material you get a roaring ovation" (B. Goode, personal communication, June18 2007).

At the same time, theorists such as Helen Varley Jamieson of New Zealand, continue to experiment with other methods of online performance, including what Jamieson calls "cyberformance." She assisted in developing UpStage, "a web-based platform specifically designed for cyberformance, where the interaction between audience & performer is real & there is most definitely a willing suspension of disbelief" (Jamieson, personal communication, May 15 2008). Jamieson believes the "low-tech, accessible & raw" use of chat applications offers an alternative to virtual worlds such as Second Life ("Cyberformance").

With the greater frequency of computer-mediated performance, such as those in SL, and the changing perception of the audience, perhaps we need to revise our definition of theatre. I say this because, as discussed before, SL is an immersive environment that, when residents fully engage with their imaginations, they become emotionally involved in the virtual world surrounding them. People rapidly attach themselves, as friends and as lovers, to other avatars because they perceive them in particular, and one might say, *real* ways.

In a blog-based condemnation of virtual worlds, writer Malcolm King called for taking out "the intellectual trash." He claimed that, "Many 'cyber-academics' make the astounding claim that the medium of online virtual worlds, such as Second Life, is reality" (King, 2007). While one may contend that a virtual reality is *a* reality, I am unaware of anyone who has asserted that it *is* reality. The addition of the term "virtual" would tend to negate the scientifically measurable plane of reality we share.

Again, I leave further exploration of this question to sociologists, psychologists and philosophers. However, it appears that the virtual performance created by the interaction between performer and the audience could be as real as these online relationships (given imagination and the willing suspension of disbelief of the spectator). People have fallen in love with chat partners, even when they had only text to attract them. Early virtual performances required imagination on the part of the spectators, connected to the hosting MOO by their home or office computers, to fill in the blanks suggested by the text appearing on the screen. Nevertheless, spectators observed or interacted textually with virtual actors in such performances as Rick Sacks' "The *MetaMOOphosis*," based on Kafka's *Metamorphosis* (2000), Twyla Mitchell's *A Place for Souls* (1997), and my own *NetSeduction* (1996). (During one performance in *NetSeduction*, two spectators wandered off for their own private "textual" interaction.)

In 3D online worlds such as Second Life, the user is presented with richly detailed audio-visual components (animation, sound and music). Along with this richer media environment, imagination and a willing suspension of disbelief—vital parts of RL theatre—still play a large part in the creation of the experience. Perhaps as we continue to experiment with theatrical forms in Second Life, we will also continue to expand out traditional definition of theatre to include performances that presume presence even if the performer and spectator are not in a shared physical space. We can then create theatre (more than simply performance) in online and cyberspace venues, and make virtual theatre a true mirror of real-life theatre.

AUTHOR'S NOTE

"Second Life ®" and Linden Lab are trademarks of Linden Research, Inc. This author of this chapter is not affiliated with or sponsored by Linden Research.

REFERENCES

Aristotle. (1970.) *Poetics.* (Gerald F. Else, Trans.) Ann Arbor: University of Michigan Press. (Original work published c. 335 B.C.).

Auslander, P. (1999). *Liveness: Performance in a Mediatized Culture.* New York: Routledge.

Bolter, J.D., & Grusin, R. (2007). *Remediation: Understanding New Media.* Cambridge, Massachusetts: The MIT Press.

Burslem, J. (March 24, 2007). Coldwell Banker Opens Office in Second Life. *Future of Internet Marketing. Retrieved* June 14, 2007 from http://www.futureofrealestatemarketing.com/coldwell-banker-opens-office-in-second-life

Chapple, F., & Kattenbelt, C. (2006). Key Issues in Intermediality in Theatre and Performance. In F.

Chapple & C. Kattenbelt (Eds.), *Intermediality in Theatre and Performance.* New York: Rodopi.

De Freitas, S. (2006, October). Learning in Immersive Worlds: A Review of Game-Based Learning. *Joint Information Systems Committee e-Learning Programme.* Retrieved 18 May 2007 from http://www.jisc.ac.uk/media/documents/programmes/elearninginnovation/gamingreport_v3.pdf

Dibou, A. (2007). *Speaking at a performance of Olmannen in Second Life.* February 2007.

Dixon, S. (2007). *Digital Performance: A History of New Media in Theater, Dance, Performance Art and Installation.* Cambridge, Massachusetts: The MIT Press.

FFRC (n.d.) SL Foundation for Rich Content. Retrieved February 8, 2008 from http://groups.google.com/group/sl-ffrc

Hotger, D. (June 6, 2007). Broadway meets Second Life. *Derek's SL Report.* Retrieved June 27, 2007 from http://derekhotger.blogg.com/2007/06/06/broadway-meets-second-life/

Introduction to World of Warcraft. *World of Warcraft Game Guide.* Retrieved June 15, 2007 from http://www.worldofwarcraft.com/info/beginners/index.html.

Ixchel, Anya (n.d.). All the (Virtual) Worlds a Stage. *Slate Magazine.* Retrieved 21 June 2007 from http://www.slatenight.com/index.php?option=com_content&task=view&id=75&Itemid=40

Jamieson, Helen Varley. "Cyberformance." Retrieved May 17, 2008 from http://www.cyberformance.org/

Jamieson, H. V. UpStage: An open source venue for online performance. Retrieved May 17, 2008 from http://upstage.org.nz/blog/

King, M. (2007, December 5). Virtual Worlds—It's time to take out the intellectual trash. *ON LINE Opinion.* Retrieved December 7, 2007 from http://www.onlineopinion.com.au/view.asp?article=6714

Laurel, B. (1993). *Computers as Theatre*. Reading, Mass.: Addison-Wesley Professional.

Mitchell, T. (March 12, 1997). *A Place for Souls*. Retrieved June 30, 2007 from http://moo.hawaii.edu:7000/4008/

New Media Consortium (2007). Retrieved June 15, 2007 from http://www.nmc.org/

Millions of Us blog (August 25, 2006(Retrieved June 21, 2007 from http://millionsofus.com/blog/archives/28

Millions of Us (August 29, 2006). Performing a play in Second Life. Retrieved June 10, 2007 from http://www.youtube.com/watch?v=6cWF438HgJA

Rosedale, P. (November 30, 2006). Second Life: What Do We Learn If We Digitize EVERYTHING? The Longnow Foundation. Baldwin, Chris (Director). USA: Whole Earth Films. Retrieved January 15, 2008 from http://www.wholeearthfilms.com/rosedale_philip.html#

Rymaszewski, M., Wagner, J.A., Ondrejka, C., Platel, R., Gorden, S.V., Rossignol, J.C. et al (2007). *Second Life: The Official Guide*. Indianapolis, IN: Wiley Publishing.

Sacks, R. (2000). The MetaMOOphosis: A Visit to the Kafka House—A report on the permanent installation of an interactive theatre work based on Franz Kafka's Metamorphosis. In *Theatre in Cyberspace: Issues of Teaching, Acting and Directing* (pp. 159-174). S. Schrum (Ed.). New York: Peter Lang Publishing.

Saarinen, I. (2007). *Olmannen* program/playbill (notecard). Second Life, February 2007.

Schrum, S.A. (2007, July 31). A Taxonomy of Digital Performance. MUSOFYR. Retrieved July 31, 2007 from http://www.musofyr.com/taxonomy/taxonomy.pdf

Schrum, S.A. (2007, July 26). A Taxonomy of Digital Performance. Presented in the panel, "Interactive Media in Immersive Performance and a Taxonomy of Digital Performance." Podcast retrieved February 1, 2008 from http://www.cda.cmich.edu/ATHE%20Podcasts/Interactive%20Media/Interactive%20Media%20(Taxonomy)%20-%20edited.mp3

Schrum, S.A. (2000). Begin with a Single Step: Adding Technology to a Course. In. Schrum (Ed.), *Theatre in Cyberspace: Issues of Teaching, Acting and Directing* (pp. 53-63). S. New York: Peter Lang Publishing.

Schrum, S.A. (1996). *NetSeduction*. Digital Performance Archive. Retrieved January 16, 2008 from http://ahds.ac.uk/ahdscollections/docroot/dpa/authorsdetails.do?project=39&author=61&string=SRational%20Mind%20Theatre%20Company

Schweller, K. (2007). *Personal Webpage*. Retrieved June 15, 2007 from http://web.bvu.edu/faculty/schweller/

SecondPride.org (June 18, 2007). Retrieved June 27, 2007 from http://secondpride.org/main.htm. (No longer available.)

SLExchange Marketplace. Retrieved June 12, 2007 from www.slexchange.com

Suler, J. (2007). Second Life, Second Chance. *The Psychology of Cyberspace*. Retrieved June 13, 2007 from http://www.rider.edu/~suler/psycyber/psycyber.html

Turkle, S. (1997). *Life on the Screen*. NY: Simon and Schuster.

Vonnegut, K. (1966). *Mother Night*. New York: Dell Books.

Wachowski, A., & Wachowski, L. (Directors). 1999. *The Matrix*. USA: Groucho II Film Partnership.

Wood, A.F., & Smith, M.J. (2005). *Online Communication: Linking Technology and Culture.* Mahwah, NJ: Lawrence Erlbaum Associates, Publishers.

Zellner, D. (n.d.). *Theatre and Second Life.* Northwestern University. Retrieved June 27, 2007 from http://door.it.northwestern.edu/sheridan/dms/zellner/zellner-modes.rm

KEY TERMS

Avatar: Originally, the incarnation of a Hindu god. In computing, a representation of a human figure in a computer game, simulation or virtual world.

Computer Mediated Communication: Any communication that occurs through the medium of networked computers, including email, online chatting, instant messaging, etc.

Cyberformance: A performance with remote users meeting online in real time using internet chat applications.

Emote: Presenting text as an action rather than as dialogue.

HUD: Heads-up display; an attachment to the screen that allows you to click buttons and manipulate objects, while leaving the rest of the screen visible.

MMORPG: A Massive Multiplayer Online Role-Playing Game, in which players interact with other players and non-player characters within a virtual world.

Presence: The concept of being in the same space as another. Telepresence allows for presence through various technological tools, such as video conferencing, or meeting in virtual worlds.

Theatre: A form of live performance that relies on the interaction between an actor and a spectator, and that involves a story and dramatic conflict.

Virtual World: A computer-based simulated environment in which users interact as avatars.

Chapter XXII
Machinima in Second Life

Stephany Filimon
Linden Lab, USA

ABSTRACT

This chapter provides a brief history of machinima, films created by computer users within virtual worlds, and focuses on machinima produced within the social virtual world of Second Life, on how to create machinima in Second Life, and on highlighting select examples of Second Life machinima. This chapter also connects user-produced content, like machinima, with the openness and rules of the platforms in which content is created. The chapter concludes with a brief overview of legal thinking surrounding user-created content, including machinima, and points to the rise of the player-producer in these systems.

INTRODUCTION

Worldmaking as we know it always starts from worlds already on hand; the making is a remaking. Anthropology and developmental psychology may study social and individual histories of such world-building, but the search for a universal or necessary beginning is best left to theology. My interest here is rather with the processes involved in building a world out of others.

Nelson Goodman, Ways of Worldmaking (1978)

When we watch a film, we appreciate it emotionally and aesthetically. When we watch the Coen Brothers' *No Country for Old Men*, for example, we are too chilled watching Javier Bardem as a psychopathic serial killer and too absorbed in the horrifying yet beautiful frame of boot scuffs on a police station floor to think deeply about the details of how the scene was shot, the editing tools used, or any legal issues the Coen Brothers may have encountered in adapting a novel to a screenplay. Like a "real" film, machinima films work aesthetic wonders and move us emotionally: try to watch the machinima film *Watch the World* without at least a few tears sneaking out. When watching a machinima film as we would any other, we may not think deeply about the characteristics of the virtual world (the platform) in which it was produced.

When we think about the creation and distribution of machinima specifically, however, as this chapter does, it can be difficult to separate the final work from the virtual world in which it was created. This is because the platform a machinima artist selects strongly influences the genre, point of view, perspective, set, lighting, characters, and objects visually apparent in the final cut. A machinima film created in a first-person shooter (FPS) massively multiplayer online role-playing game (MMORPG) like Halo looks quite different than one created in Hello Kitty Online (Sanrio Town), with pale pink landscapes and places named Flower Kingdom. Ford 2007 used two different game engines, The Sims and Half Life 2, to produce the same film script. Ford notes that each game engine effects genre and thereby intrinsically changes the dynamics of the "same" story being told. In addition, as for most films, the production process and the constraints of time and budget inevitably influence the type of machinima produced and its level of sophistication (Ford 2007).

The virtual world selected for a machinima production also influences more than we can see. Less obvious but just as, if not more, important than aesthetic impact is the degree of a virtual world's openness: the content creation tools it provides (or doesn't) to users, such as camera controls, avatar customizations, in-game video recording, and its terms of service (TOS) and end-user license agreement (EULA), which usually describe what content users should or should not create, how content should or should not be modified, what content (if any) users own, and consequences for content modification.

This chapter aims to contribute to the awareness of machinima and related cultural, technical, and legal issues by describing what machinima is and who creates it, as well as a brief history of how machinima evolved along with virtual worlds and tools that enabled users to create their own content within these virtual environments. This chapter also focuses on the history, content creation tools, machinima, and TOS of Second Life, a social virtual world, and provides a brief tutorial on how to create machinima in Second Life. This chapter concludes with a brief overview of recent and emerging legal questions on user-created content, copyright, and machinima specifically.

1.0 WHAT IS MACHINIMA?

"Machinima" combines "machine," "animation," and "cinema" to describe computer-animated films that are shot within video games or social virtual worlds and primarily distributed online. Machinima is also defined as "a computer movie made using a real-time, 3D game/virtual-world engine instead of a special application dedicated to making computer movies" (Rymaszewski 2007). These definitions make an important distinction between MMORPGs and virtual worlds, like Second Life, that are not games but are more social virtual worlds.

Though MMORPGs and social virtual worlds like Second Life may look like the same 3D spaces occupied by avatars, these terms are not interchangeable or synonymous. Put simply, *social virtual worlds are not games*. The primary purpose of social virtual worlds is the creation of meaning through the manipulation of the world and *communication with others* within the world, while *structured play is the primary purpose* of game play worlds (Damer 2007). Throughout this chapter, the term *social virtual world* will be used as shorthand for "a 3D, online, collaborative, virtual world that is not a game."

At present, machinima means more than just "a specific type of film:" Machinima has evolved into a "player-as-producer open culture phenomenon" as well (Salen 2003), primarily because of the identity of its creators. Players or users of games and social virtual worlds produce machinima within these worlds. Though machinima could be commissioned or in some way used by the company that created a virtual world, machinima is not usually produced by platform creators but by the platform users (or players) themselves. Machinima creators are often referred to as "machinimators," an adaptation of the word "animator." The literature contains a few other terms used to describe content creators, all of which convey the idea that one person is simultaneously the player, or user, and content producer. These are:

- Player-producers (Salen 2003)
- Pro-Ams, short for "professional amateurs" (Sharp 2006, citing Leadbeater and Miller who named a social trend the "Pro-Am Revolution")
- Produser (Sharp 2006)

Accessibility and affordability have contributed, in part, to machinima's rapid growth and increase in popularity during the past four to five years. Compared to professional animation techniques and tools, machinima costs much less to produce: The tools needed to record, encode, edit, and distribute films are often available for free or fairly inexpensively (i.e., less than $150) online, and/or included as part of the virtual world platform, and/or included with the user's computer operating system (OS) like Mac. Finally, YouTube has provided an easy way for users to distribute a variety of films, including machinima, to an enormous audience for no charge.

1.1 A Brief History of Virtual Worlds and Machinima

Virtual worlds and MMORPGs preceded machinima, and it's safe to say they are a necessary condition for machinima. The rise of machinima also correlates roughly with the availability of in-world tools to help users create machinima, namely camera controls, avatar customizations, and in-world video recording.

Text-based role-playing games, like the first Multi-User Dungeons (MUDs), were first seen in the 1970s and 1980s (Damer 2007). These text-based virtual environments evolved into visual ones as the age of affordable graphical computing dawned in the 1980s and 1990s, and expanded to include other genres: first-person shooters, fantasy role-playing games, simulators, shared board and game tables, and social virtual worlds (Damer 2007). User modification of these early virtual environments wasn't far behind, and some MUDs and MOOs (like LambdaMOO) gave some users permissions to create and modify objects (Cherny 1999).

In the spring of 1995, a company called Worlds Incorporated launched Worlds Chat, a 3D space station where users "teleported," could navigate through a rich environment, and could exchange text chat (Damer 2007). Three months later, Worlds Incorporated also launched Alphaworld, the first known social virtual world that allowed users to build objects and create content in-world using prefabricated objects. Alphaworld became and is known today as Active Worlds (Damer 2007).

In the mid-1990s, rich MMORPG environments became extraordinarily popular, much more so than social virtual worlds (Damer 2007). Almost as soon as Quake was released in 1996, gamers began to try and play through its levels as fast as possible and to share recordings of their feats with others (Salen 2003). To make raw footage of gameplay more compelling, players started setting their clips to music, adding their own commentary and voice tracks, and even staging elaborate maneuvers that involved multiple players (Chien 2007). Players discovered the value of elements now common in most games and virtual worlds: strategic editing, soundtracks, and snappy wisecracks (the ability to comment in written text or spoken words) amplify the drama of game events (Chien 2007). Although these early machinima films were clearly a form of retelling play, the films quickly took on a very different role: telling stories. It was only a matter of time before someone made the leap to film (Salen 2003).

Salen 2003 tells us that, according to Quake lore, that leap came in August 1996 when a clan known as The Rangers recorded a demo that exploited the built-in moviemaking capabilities of the game. Rather than restrict their demo recording to play within the game, The Rangers used Quake as a filmmaking tool. This decision transformed the game space into a virtual movie set. The Rangers used their characters as virtual actors and recorded their movements on a death-match map, while typed text messages represented speech. The completed film, *Diary of a Camper*, established the genre of machinima.

In addition, first-person shooter games such as Quake and DOOM were some of the first to offer an open source editor to players, which allowed them to design and program their own maps (environments), skins (character avatars), weapons, and tools for game play. Quake players also wrote Quake code modifications and posted them online for other enthusiasts to download and use (Salen 2003). This pioneering approach, a direct outgrowth of open source software culture, offered players unprecedented power to modify game play by altering the forms and spaces of interaction (Salen 2003).

From the late 1990s and early 2000s on, MMORPGs like Everquest (1999) and World of Warcraft (WoW, launched in 2004) saw their numbers of players grown from the thousands into the millions. In 2003, the U.S. invaded Iraq. This highly controversial act of war combined with the enormous popularity of MMORPGs resulted in the development of what is possibly the most popular and distributed machinima online today, *Red vs. Blue: The Blood Gulch Chronicles* (Chien 2007). Filmed within Halo, a first-person shooter game

produced by Microsoft, the first episode of the *Red vs. Blue* series was released online shortly after the U.S. invasion of Iraq. As Chien 2007 describes, *Red vs. Blue* features two soldiers who don't do much of anything, and certainly don't leap into violent action as we expect them to. The soldiers simply stand around talking, and the content of their conversation makes it clear that they have no idea what they are doing there nor why they are fighting (Chien 2007). The first four seasons of *Red vs. Blue* have been released as feature-length DVDs, and Season Five is available online as of the time of this writing. The cult success of *Red vs. Blue* is due primarily to Internet distribution across online gaming communities (Chien 2007). New episodes are released weekly via the *Red vs. Blue* (Roosterteeth) website.

Though MMORPGs grew by leaps and bounds during the late 1990s and early 2000s, most social virtual world companies proved too soon for their time: By the late 1990s, most of them had lost financial backing, changed hands, or vanished (Damer 2007). This was when the social virtual world of Second Life emerged. Philip Rosedale, founder of Linden Lab and Chairman of the Board, conceived of and began work on the concept that would become Second Life in 1991 (Rymasznewski 2007). Second Life released in beta in 2002 and commercially in 2003. Another social virtual world called There released in beta in 2001 and, like Second Life, released commercially in 2003.

Then and today, both of these social virtual worlds allow users to create their own content using tools provided in the interface, and both worlds have their own economies. The Linden Dollar (L$, also called "Lindens") is the currency of Second Life, which has an average exchange rate of $265-$272 L$:1$ USD at the time of this writing. Second Life users are known as Residents. Second Life Residents use L$ to buy and sell virtual objects from other Residents in-world, and L$ can then be exchanged for U.S. dollars.

In 2004, Linden Lab added the video capture command to the Second Life client (Viewer), and the release of viewer 1.4 provided Residents with the ability to create custom avatar animations and upload them in world (Call 2005). These three additions created a virtual space more amenable to machinima creation, and most early (and available) Second Life machinima online dates from this period, approximately one year after the launch of the commercial version of the product.

As author Wagner James Au points out, however, other virtual environments allowed video capture; unlike them, the world of Second Life was and still is built almost entirely by its users. With "the robust 3D building tools and the internal scripting language available to them, the Second Life community has created nearly all the buildings, vehicles, weapons, clothing, artwork, and mini-games that they enjoy and play with" (Au 2005). Au goes on to write that these factors created the potential for "a new kind of immersive Machinima, with full control over sets, special effects, costumes, and characters that do not ever require external middleware tools... where a large cast and crew can simultaneously work together."

Au also called attention, in 2005, to Linden Lab's policy on intellectual property, which "allows residents to retain ownership over all works they create in-world, including films, to use as they wish, including commercially." The most high-profile proof of this was given in September 2007 when Second Life Resident Molotov Alva, avatar of a man named Douglas Gayeton, sold the North American rights to his machinima film *My Second Life* to HBO. At the time of this writing, Second Life is still the only virtual platform that explicitly allows creators to retain their copyrights (Marcus 2007). The following are relevant but greatly abbreviated excerpts from the Second Life TOS:

3.2 You retain copyright and other intellectual property rights with respect to Content you create in Second Life, to the extent that you have such rights under applicable law. However, you must make certain representations and warranties, and provide certain license rights, forbearances and indemnification, to Linden Lab and to other users of Second Life. Users of the Service can create Content on Linden Lab's servers in various forms. Linden Lab acknowledges and agrees that, subject to the terms and conditions of this Agreement, you will retain any and all applicable copyright and other intellectual property rights with respect to any Content you create using the Service, to the extent you have such rights under applicable law.

4.2 You agree to use Second Life as provided, without unauthorized software or other means of access or use. You will not make unauthorized works from or conduct unauthorized distribution of the Linden Software. "...you may copy the Viewer that Linden Lab provides to you, for backup purposes and may give copies of the Viewer to others free of charge. Further, you may use and modify the source code for the Viewer as permitted by any open source license agreement under which Linden Lab distributes such Viewer source code."

Like Quake, DOOM, and other platform/apps before it, the Second Life Viewer source code was released on January 7, 2007. In a blog post on the Second Life website, Phoenix Linden wrote: "Stepping up the development of the Second Life Grid to everyone interested, I am proud to announce the availability of the Second Life client source code for you to download, inspect, compile, modify, and use within the guidelines of the GNU GPL version 2." With this decision, Second Life exhibited what Salen 2003 calls the three essential qualities of open source games: "the games are open systems that can be modified by a community of players, rather than a single developer; the games are freely shared among players and developers; and the source code is made available."

1.2 Examples of Machinima from Second Life

Selecting just one or two examples of Second Life machinima for this chapter was challenging. Usually, choosing the first of anything makes this task easier, but it's difficult to determine with any accuracy what the first machinima film produced in Second Life was. We do know with some certainty, however, that *Silver Bells and Golden Spurs* was one of the earliest.

Eric Call, an employee of Linden Lab at the time, created a short Western machinima film based on an old poem titled "Silver Bells and Golden Spurs." Stylistically, the film is a typical Western, with old-time saloon player-piano music, a mining town set, and avatars dressed in their Wild West best. In *Silver Bells and Golden Spurs*, we watch a new arrival to town defeat (or does he?) the reigning quick-draw champion and killer of many men, while music and the narrated poem accompany the action. Figures 1-4 show selected stills from *Silver Bells and Golden Spurs*.

A more recent and perhaps more well known machinima is *Watch the World*, created by artist Robbie Dingo, the avatar of Rob Wright. *Watch the World* is an example of advanced machinima in Second Life. Robbie Dingo writes about his inspiration for this film:

Ever looked at your favorite painting and wished you could wander inside, to look at it from different perspectives? Spend a single day in one of mine, from early sunrise on a new day, to dusk when lights come on in cosy homes; through a peaceful night, till morning. Shot on location in Second Life then post-produced, this was an idea I had a while ago. The Sim in this work was on temporary loan so it's all been swept away now, leaving only the film behind. It was always intended however that the video would be the end product, not the build. This work is dedicated to the many weird and very

Figure 1. An opening scene: The camera sweeps down the main street of a small mining town in the West

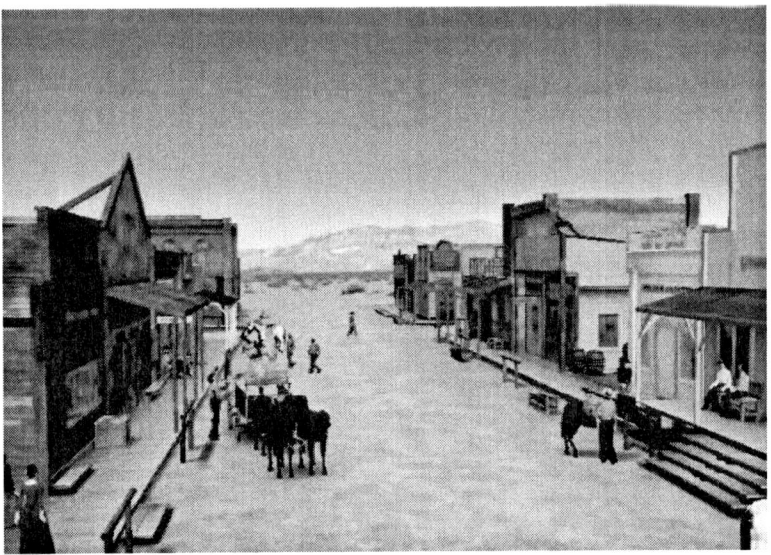

The accompanying narration is:
Twas a mining town called Golden Gulch
While the West was yet untamed.
There two bad men met, made a bet,
And the winnings never claimed.

Figure 2. The villainous Dandy, quick-draw champion and killer, with his back to the saloon room. The music fades as the camera focuses on him

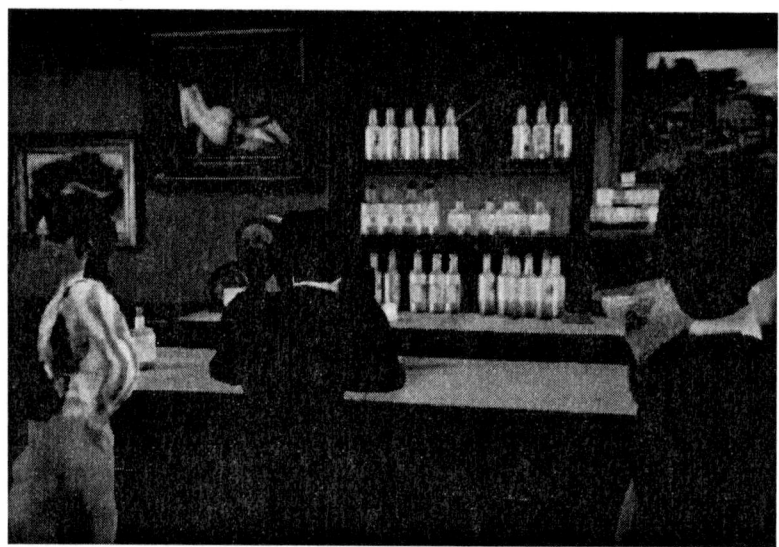

Accompanied by this narration:
Now the Dandy was an onry cuss
If by chance you made him sore,
His only law was the lightning draw
Of the heavy guns he wore.

Figure 3. Onlookers gather for the draw between the Dandy and the stranger

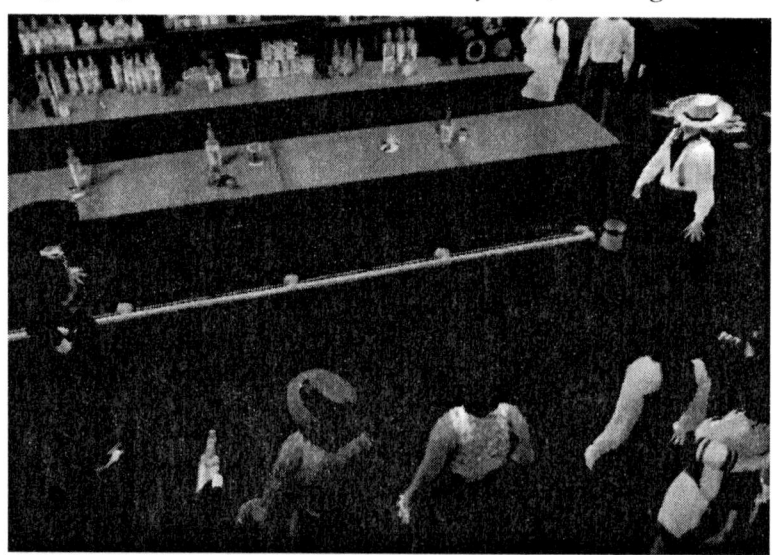

The accompanying narration is:
The stranger watched with narrowed eyes,
The time had passed for talk.
He hadn't drawed but his hands were clawed,
Like the feet of a diving hawk.

Figure 4. The Dandy has fallen, but did he lose? Here we see the stranger's feet

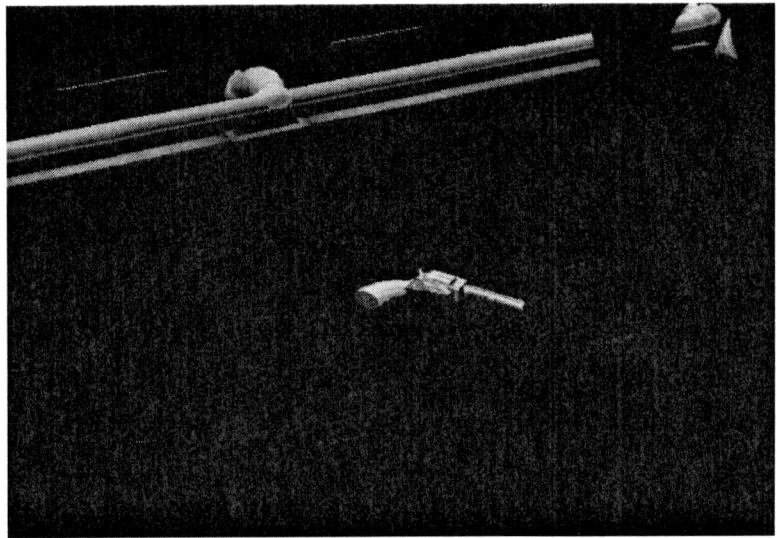

And the narration tells us:
The stranger stood at the end of the bar,
Apparently unhurt,
Except for a spot of red that slowly spread
Beneath the left pocket of his shirt.

Figure 5. Three-dimensional hills appear to bubble up from flatland, filling the outlined areas we see on the screen

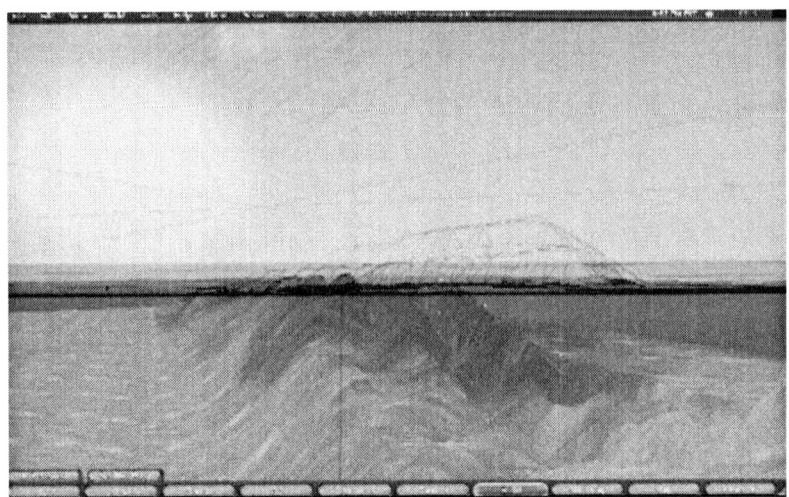

The accompanying song lyrics at this point in the machinima are:
Shadows on the hills,
Sketch the trees and the daffodils

Figure 6. While Dingo builds more 3D elements of the painting, the Second Life viewer has taken on the texture of a pencil sketch. As the formerly grayscale pencil sketch begins to fill with color, as shown here

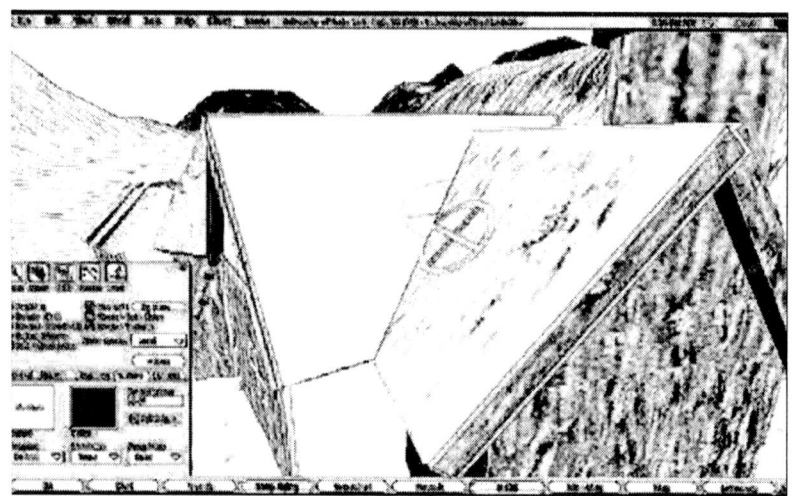

The accompanying song lyrics are:
Colors changing hue, morning field of amber grain,
Weathered faces lined in pain,
Are soothed beneath the artist's loving hand.

wonderful strangers from around the globe I have met, but have never really met.

Watch the World (Figures 5-9) allows us to do just that. Though impossible to see from the paper format of this chapter, Dingo's machinima talent

Figure 7. Here, the music hits a crescendo and the machinima does as well, as the texture of Starry Night begins to fall from the corners over all of the walls, creating brush strokes of the painting we're so familiar with

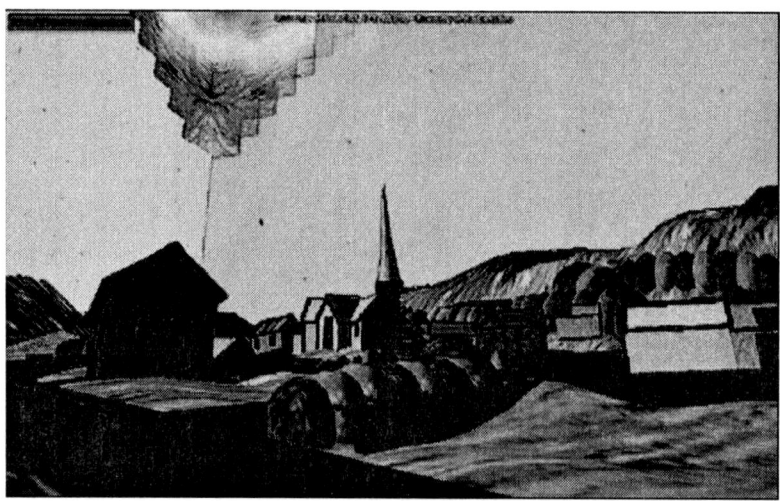

And when no hope was left in sight
On that starry, starry night
You took your life, as lovers often do.

Figure 8. After the avatar, Robbie Dingo, has placed glowing cubes (windows) in the buildings and luminous spheres in the sky, a rotating camera view shows us the nearly complete landscape

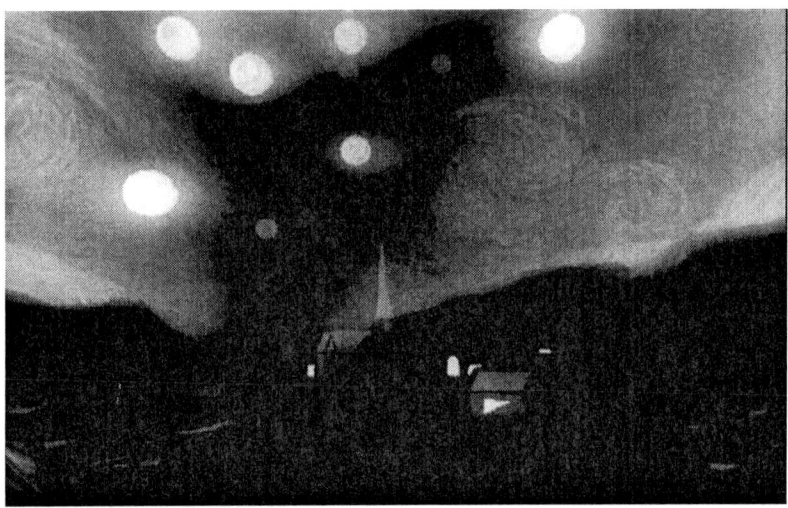

Don McLean sings:
Now I think I know what you tried to say to me,
How you suffered for your sanity,
How you tried to set them free.

is evident in the speed of the film, presented in a sort of rapid time-lapse format, and the pairing between what is shown in film while the song "Vincent" by Don McLean plays. Dingo also uses textures to show "early" 3D shapes in pencil sketch texture, like the early sketches of artists that

Figure 9. With a final flourish, Robbie Dingo frames the painting and his avatar turns, flies in to the village, and is gone

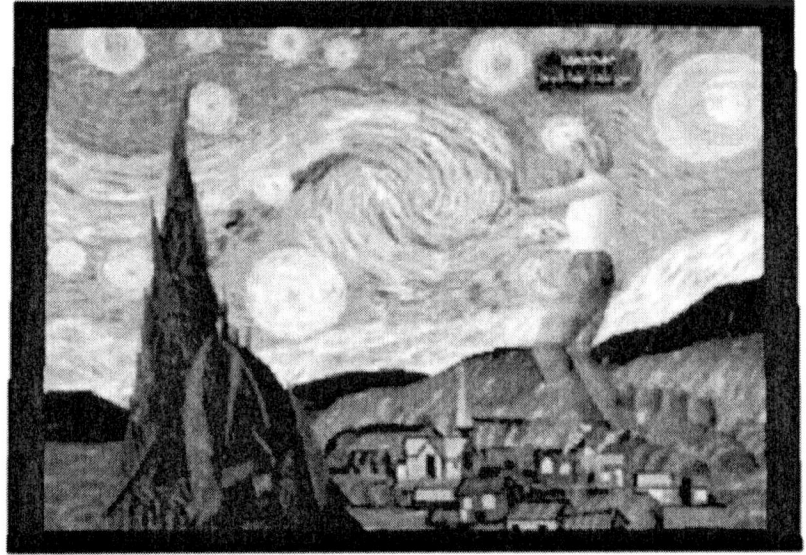

Accompanied by the final lyrics of Don McLean:
They would not listen, they're not listening still.
Perhaps they never will...

contribute to the creation of a painting. Unlike many machinima, this film highlights the building and creation processes in Second Life.

During the past few years, the art of machinima has become more widely recognized and legitimized by the creation of organizations dedicated to machinima and film festivals, both machinima-specific festivals and more traditional film festivals. The Academy of Machinima Arts & Sciences supports and promotes the art, as does the Machinima Artist in Residence program by Millions of Us, a company devoted to creative marketing in virtual worlds. In addition, machinimists participate in projects like the 48 Hour Film Project, a "wild and sleepless weekend" in which teams of film creators in dozens of cities around the world write, shoot, edit, and score a machinima film in just 48 hours. More information about the 48 Hour Film Project completed in Second Life is available at: http://www.48hourfilm.com/secondlife/.

2.0 PLATFORM OPENNESS AND USER-CREATED CONTENT

According to Salen and Zimmerman in their book *Rules of Play: Game Design Fundamentals*, "A game designed as open culture allows players in some way to access the game structure and directly change its meanings." They also point out that "Sometimes, the player-as-producer paradigm takes the modification of a game so far that the invented activity no longer resembles the play of the game at all." The brief history of virtual world platforms presented here identifies some things that different virtual worlds have done to provide a friendly groundwork for machinima creation. Many of these choices reflect a belief on the part of the platform creators in games and social virtual worlds as culturally emergent systems (Salen 2003). This cultural belief is evident in one or more of the following technical attributes of a platform itself. An open platform is one that:

- Contains tools created to support user content creation
- Provides help, tutorial or other documentation to assist users with content creation
- Allows users to import content they've created with another application
- Offers TOS or EULAs that explicitly permit user-created content and distribution (like the Halo 3 license from Microsoft, for example)
- Allows users to retain ownership (intellectual property rights) of their creations
- Enables emergent behavior (rule changes, new in-world avatar roles, non-game focused gatherings, and more)
- Provides client source code to encourage programmers to modify the underlying code itself, changing the product in profound ways and perhaps to look and function very differently than originally intended.

Salen 2003 mentions game modifications that bear little resemblance to original game play, focusing on the degree of technical modification involved in different kinds of user-created content, from in-game object creation to open source platform modification.

Baldrica 2007 proposed a similar spectrum, which he calls the Spectrum of User Contribution. The Spectrum of User Contribution was created in a legal context, to understand different kinds of content creation and modification; understanding where content falls within the spectrum may help to better define the level of protection different types of modifications should receive within the existing legal framework of artistic appropriation and expression. Though created to explore and aid understanding of legal protection, Baldrica's Spectrum of User Contribution is extremely useful for expressing the degree to which content can be or has been modified in a particular system. Baldrica writes that, "On one end, this may include content which results from merely playing the game, or content which is created using user-accessible features within the game. On the other, the content might involve altering the game itself, or using the game to create entirely independent expressive projects."

Baldrica's Spectrum of User Contribution is described in an abbreviated form here:

a. **User-Contributed Content *Comprising* the Game:** Content contributed strictly through users' time spent "playing" or "laboring" within the rules of the game.
   Examples: Unique in-world items like avatar clothing, avatar hair, and tools; virtual real estate (regions in Second Life; buildings).

b. **User-Contributed Content *Exploiting* the Game:** Users exploit the game environment or user tools as a forum to speak or create other expressive content. This is especially relevant to MMOs where the persistent virtual world allows a social interaction.
   Examples: In-world rallies, political protests, and theater productions, which lead to additional questions of both free expression and copyright (because rallies and political protests are protected by free speech laws in the U.S.). Virtual objects (created or imported) by users, like virtual t-shirt designs, and offered for sale to other players for virtual currency that can be exchanged for real funds.

c. **User-Contributed Content *Re-Defining* the Game:** Extending beyond the bounds of the existing game by re-engineering the game engine to create a different type of gameplay. This is where "modding" (modification) generally falls.
   Examples: Simple changes include changing graphics and sounds. More complex changes are altering the very rules by which the game is played. In Second Life, this includes modifying the Viewer itself, which modifies the ways in which users can interact with Second Life. One recent example is a custom Second Life Viewer produced by

the Electric Sheep Company (ESC) for use with a *CSI: New York* television episode. The ESC Viewer contained two things standard Second Life Viewers did't: 1) a Back button (akin to those found in web browsers) and 2) better-looking default avatars than those offered by Linden Lab. By adding a Back button to the Second Life Viewer and different default avatars, ESC offered new users a different experience of Second Life than users otherwise would have had. Being able to return to a previous in-world location via the Back button, for instance, is something Second Life residents cannot usually do, and it changes their experience of movement and navigation in the 3D environment.

d. **User-Created Content Re-purposing the Game:** Using virtual world technology in unanticipated expressive ways in a "meta" creative process. Using the game engine, graphics, or other elements of the virtual world to create a new expression that is no longer a game.
   Examples: Baldrica explicitly places machinima in this range of the spectrum, because it is an alternative use of game technologies to create, and sometimes even commercially release, animated short films.

Baldrica's Spectrum is also useful in examining a range of content modification within a single platform, like Second Life. Content created in-world, using tools like the building tools in Second Life for example, falls on one end of the Spectrum of User Contribution. Custom viewers produced by ESC, made possible by the openness of the Second Life Viewer source code, are even farther along on the Spectrum, while machinima, entirely independent expressive projects that stand on their own outside of the virtual world of Second Life, are at the far end of in-world content creation.

2.2 Emergent Behavior

As noted earlier in this section, one attribute of open systems is that they enable emergent behavior. Salen 2003 gives the example of players modifying existing player professions in the game Ultima Online by creating a new profession, that of prostitution. The authors describe the behavior that emerged among Ultima Online players after the player profession of prostitution was added, citing the appearance of avatars, the selection and use of new spaces in which prostitutes and customers met, and changes to the economy.

Examples of emergent behavior in Second Life include:

- **Land rushes,** in which Linden Lab released a specific quantity of virtual land for purchase after a certain period of time, and "land baron" Residents in Second Life purchased large quantities of it and rented or re-sold it at profit, creating a shortage. Land rushes were part of Second Life's earlier days and have not occurred for a few years now.
- **Architectural standardization:** To the extent that one can claim standardization occurs at all in a constantly changing virtual world that is difficult to explore in full, some common architectural features (and thus behaviors, since residents in Second Life design and build all of the architecture) have emerged over time. Large interior spaces, with extremely wide doorways and very tall ceilings, have most likely evolved due to avatars flying into, out of, and within, enclosed spaces. Large, spacious, open areas with a dozen or more seats are common and support the ways in which wandering avatars often encounter small clusters of people and pause for a while to see what's going on or meet new residents.
- **Gambling:** Though banned in 2007 by the Second Life TOS due to U.S. government restrictions, gambling (and the creation of casinos in which to gather, and 3D games to play) was popular in Second Life for a couple of years.

This last example, gambling, illustrates how—even when emergent behavior is allowed and encouraged to, well, emerge—response behaviors in the form of changes in a platform's TOS and governance might also emerge. The behaviors a platform allows to emerge are not necessarily behaviors that will or can continue to be permitted.

2.3 How to Create Content in Second Life

The Second Life viewer contains many tools, but some tools explicitly enable users to create content and modify the world of Second Life itself, by building content in Second Life. Other tools, while not created only with machinima in mind, assist with and ease creation of machinima in Second Life. Second Life contains the following content creation tools:

- **Building tools** that create 3D objects from **basic shapes called "prims,"** short for "primitives." Cubes and cylinders are two examples of prims.
- **Linden Scripting Language (LSL),** the programming language of Second Life with which Residents can control the behavior of in-world objects. An LSL script, for example, can make a donation box at a virtual food bank accept donations of L$.
- Built-in **video recording** capability
- **Content ownership and permission settings.** Once a Second Life Resident has created an object (like the aforementioned virtual donation box), the creator controls permissions that allow or disallow other users from certain types of activity including subsequent transfers, modifications, and identical copying (Marcus 2007). These

Figures 10 and 11. The Object Editor panel in Second Life, with content permissions illustrated at the bottom: "Next owner can: Modify, Copy, and/or Resell/Give Away"

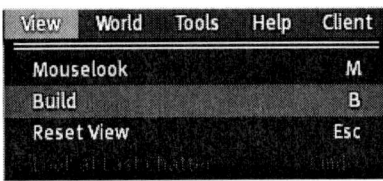

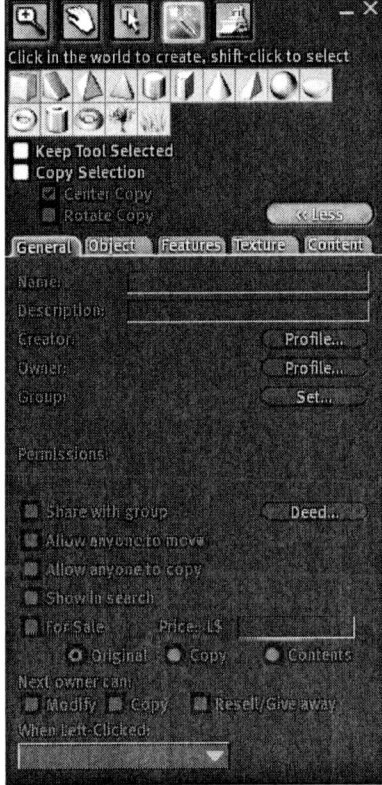

permissions are enabled through click-boxes that appear as part of the editing menu on the Object Editor panel (View > Build), shown in Figures 10 and 11.

- **Camera controls** that enable mouselook (first person) and follow modes (fslightly above and behind your avatar with your avatar in view).

Building Basic 3D Objects in Second Life

The following instructions describe how to build a basic 3D object in Second Life, using some of the tools described before:

Selecting a Basic Shape ("Prim")

1 Right-click the ground your avatar is standing on and choose "Create" in the pie menu that appears. You can also press Ctrl+4. Choosing "Create" will open the Object Editor. It's worth noting that this process will only work if your avatar is standing on land that permits the creation of objects. Not all land in Second Life does, depending on what the landowner wants. Avatars can only create objects on land that permits the creation of objects. Land on which object creation is *not* permitted is designated as No Build. You can identify No-Build land by looking for an icon (a beige cube with a red circle beside it) that will appear at the top of the Second Life viewer window if your avatar is standing on no-build land.

2. With the Object Editor open, you can choose the type of basic shape you wish to create. Basic shapes are called "primitives" or "prims" for short.

3. In the Object Editor, make sure the wand button is selected, as shown in Figure 11. Once it is, you can select the prim you wish to create (cube, sphere, and so on).

4. Next, click the location in-world where you wish to build your object. The prim you selected in the Object Editor should appear in the in-world location you selected (typically with a resounding "whoosh" sound).

Editing and Modifying Prims

How to Move Your Prim

- When the white hand button is selected in the Object Editor (shown in Figure 11), you can use your mouse or trackpad to click and drag your prim, moving it around.

- Dragging moves the prim along the horizontal (X/Y) plane.
- With your prim selected, hold down the Ctrl key to drag the prim vertically (Z) as well.
- Clicking on the red (X), green (Y), and blue (Z) arrows (shown when you hold down the Ctrl key) enables you to drag the prim only along those axes. Again, make sure the hand button is selected in the Object Editor first.

How to Change the Size of Your Prim
- With your prim selected, hold down the Ctrl+Shift keys simultaneously to bring up the sizing box. White, green, red and blue colored cubes will appear.
- Using your mouse or trackpad, click and drag one of the white corner boxes to scale the entire object proportionally.
- Click and drag the red, green or blue sizing boxes to resize a prim's length, width or height, respectively, without changing the other dimensions of the prim.

Creating Machinima in Second Life

Though machinimists may create 3D content to include in their films, such as buildings, furniture, and other objects they need in order to create the film sets they have in mind, creating machinima in Second Life is quite different from creating in-world 3D objects.

Before building 3D sets and recording video, a machinimist needs what any other filmmaker needs: creativity. A storyline, character sketches, and physical context in which the story takes place (i.e., in the Wild West, in the woods, underwater, and so on) will determine what 3D objects need to be created or purchased, what actors (avatars) need to be hired or convinced to volunteer their time, what the avatars should look like (Victorian apparel, 1920s gangster Zoot suits), what the ava-

tars will say (the script), how the actor-avatars will deliver their lines (in written or spoken words, or both), and time of day (morning, noon, sunset).

In addition, a machinimist should check his or her computer to make sure ample hard drive space is available, and make sure s/he has the codecs. Cubes are multimedia compression algorithms that enable the shift from proprietary formats to formats like MPEG, which film editing software can open. Window size also affects the final file size, and high resolution video takes up a lot of hard drive memory. This means machinimists should *shoot* video at a lower resolution than that at which they would normally *play* video, and start with the 640x480 resolution in their computer system's display preferences for settings.

Once these preparation steps are complete, machinimists should understand how to find and use helpful tools available in the Second Life Viewer. These include:

- Camera controls that can change perspective on a scene. Camera controls include mouselook (first-person view) and the default standard, "follow" mode, with the camera behind and slightly above one's avatar. Views can be changed through the "View" item on the toolbar.
- Machinimists should also be aware of settings that can be changed via Edit > Preferences > Input & Camera, which affect how smoothly content can be recorded in Second Life.
- Finally and very importantly, machinimists should know the shortcut for capturing video, which enables Residents to set frame size, codec, and to start/stop the capture process saving footage onto their hard drive. Typing Ctrl-Shift-A or Apple-Shift-A provides a pop-up menu that asks where machinimists would like save Second Life Movie.mov, a default file name that can change.
- After naming this file and deciding where it should be saved, a different dialog appears that lets machinimists choose how they

wish to compress the video. The easiest is Full Frames (Uncompressed), which is not lossy (i.e., will not cause a loss in image quality).

The last time I attempted the described process, it crashed my Mac, recording in Second Life is, at the time of this writing, still known to crash Macs at least some of the time. Even if it didn't, however, using in-world video recording can be difficult because the client is attempting to render Second Life and record video to the hard drive simultaneously.

There are essentially four steps to machinima creation and distribution:

1. Recording
2. Encoding
3. Editing
4. Distributing

Depending on the software application selected, one or more of these capabilities may be included. Discussing them separately, however, better illustrates the distinct steps of creating machinima.

Step One: Recording

Many machinima creators in Second Life opt to record in-world video using an external tool, because it saves the Second Life viewer from having to simultaneously render a 3D world and save a massive video file to a hard drive, two high-demand computer processes. Machinimists who use external video capture tools often use FRAPS, a screen capture and real-time video recording tool that costs just $37 at the time of this writing. Other tools in similar price ranges are available online, but FRAPS is considered best in class by Second Life machinimists.

Step Two: Encoding

Machinimists don't need to understand encoding algorithms. They need to know what file type their film needs to be in so editing software can read it. Many video recording and screen capture tools, including FRAPS, encode their video files in a proprietary format. This means that, after recording a film, a machinimist may be panicked to find s/he cannot open the file in the editing tool s/he would like to use, like iMovie or Adobe Premier. Codecs provide the requisite "container format" to move video from recording to editing software. It is helpful, then, to know which file types the recording software produces, and which file types can be read by editing software. If a machinimist were working on a Mac and thus wanted to use iMovie, for example, s/he should find out what file types iMovie can work with and encode the film accordingly (in this case, to the MPEG-4 format). Popular encoders in the machinima community include Cleaner XL ($125), VirtualDub (free capture with encoding in AVI/MPEG), and TMPGEnc (free shareware).

Step Three: Editing

Editing is both a creative and technical progress, in which the artist omits or corrects mistakes, organizes the narrative and shots that convey the narrative in a certain way, scores the film, and does any other work necessary to get the final film s/he has conceived of. Mac computers include iMovie and Windows systems include Movie Maker, both popular and free (included) editing tools, Animoids (free trial or $24.95), MediaEdit (free trial or $49.95), and similar editing software is available for download online.

Step Four: Distributing

YouTube is currently the go-to destination for posting, sharing, and discussing user-created films, including machinima, and distribution is free. Other machinimists distribute their films in specific online communities of interest, such as websites and forums focused on the platform in which the machinima was made. Still other films are featured on dedicated websites, like *Red vs. Blue*, in which new episodes are posted weekly and old episodes archived.

The goal of the four steps described here is to show how inexpensively machinima can be created. A Basic account in Second Life is free, and a Premium membership $9.95/month. Content creation tools are free and, if a machinimist prefers to purchase L$ to purchase, from other Residents, virtual goods that subsequently s/he doesn't have the skills or time to make, the average exchange rate is 276 L$: $1 USD. In other words, $1 provides a machinimist with approximately 276 L$ to spend on virtual goods in Second Life. As discussed earlier in this chapter, video recording is also free in Second Life, with external video recording software like FRAPS available for $37. Video encoders are available for free, or available for purchase at an average of less than $125. Video editing tools like iMovie and Movie Maker are part of computer operating systems and thus free of charge, while similar applications are often available for $25-$50. An aspiring machinimist can create a machinima for free or for about $40-$200 all told, much less than the cost of traditional animation tools, which can be prohibitively expensive for individual purchasers (versus professional animation houses).

3.0 USER-CREATED CONTENT AND THE LAW

The legal literature that explicitly refers to machinima contains the following major themes:

- Creative expression and copyright
- Possible analogies between physical and virtual spaces (i.e., meaning) analogies to the law governing the "real" physical spaces in which creative or expressive activities take place. If free speech is protected when it is produced or performed in a physical company-owned town, it should arguably be similarly protected in a game-developer-owned virtual town (Baldrica 2007).
- The increasingly questionable power of TOS and EULAs

3.1 Creative Expression and Copyright

As noted earlier in this chapter, Second Life is the only virtual platform that explicitly allows creators to retain their copyrights (Marcus 2007). This policy on copyright retention points to possible extrinsic rewards for an artistic pursuit (machinimia creation) that at first glance may only appear intrinsically rewarding. Kurtz describes the Internet this way, as a place "where people do things, try things, find things, and share things without expecting a financial reward. The explosive creativity shared among millions on the Internet, from musical and video mash-ups to fan fiction and machinima, puts obvious strain on the incentive theory of intellectual property" (Kurtz 2007).

On the other hand, although allowing users to retain rights to their creative works adds the complication of applying copyright to virtual objects, the incentive structure of copyright is actually preserved (Marcus 2007). If virtual objects continue to have physical-world monetary value, then more systems will have an incentive to follow Second Life's example, and shift their own rule systems (Marcus 2007).

Although Second Life Residents can assign permissions to the content they create (by way of Next Owner can: Modify, Copy, and/or Resell/Give Away for example), Marcus also describes the copyright that applies to digital creations in virtual worlds under U.S. law, rather than under the Second Life TOS. He tells us that digital creations in virtual worlds are the subjects of copyright because the software code that underlies a work, or the "script" embedded in the virtual object, is considered a literary work, one of the enumerated copyrightable subject matter categories. A virtual object can also be protected as an audio-visual work because it meets the "fixation" requirement by being fixed in the read-only memory (ROM) hardware of a computer, and can then be perceived with the aid of a machine or device. Machinima, too, seems to fit within this category (Marcus 2007).

For machinima, Baldrica advocates fair use and expression-related legal inquiry: "A more expression-related inquiry is clearly sensible in the case of game-related projects like "machinima," where game engines are put to work to create entirely non-game derivative works, such as short films, and even talk shows produced from within virtual-worlds." Baldrica goes on to write that, with the advent of distribution venues such as You-Tube, such projects are becoming more prevalent and some creators have begun releasing their work commercially, such as *Red vs. Blue*. Furthermore, in addition to their long practice of encouraging the development of mods, some game developers have, as described earlier in this chapter begun sponsoring contests and online festivals for the creation of machinima, with a business rationale that user-created machinima spurs both awareness and ultimate market lifespan of the games used to create them (Baldrica 2007).

Baldrica 2007 also feels that machinima, content modified to the extent that it no longer resembles the original game, "seems to fall most closely in line with the rationale of protected artistic expression embodied in cases such as *Mattel v. MCA*, in which the protections of a copyrighted work are balanced against the social value of transformative fair use of that copyrighted work." In other words, mods are a unique form of collaborative art, and that the current legal regime - in which mods are deemed uncopyrightable derivative works - is inappropriately narrow (Baldrica 2007).

3.2 Analogies Between Real and Virtual Spaces

When we think about real and virtual spaces, it often seems that real-world laws on content usage suit the virtual world, such as laws concerning things like being out in public versus in one's home, and wearing a t-shirt featuring work copyrighted

by another artist. But Marcus 2007 points out important ways in which real and virtual works differ according to the law. In sum, copyright vests in virtual works *fewer* rights than physical works, because virtual creations can be viewed *inherently* as visual or audio-visual works. Therefore, content creators have arguably broader protection over virtual creations and subsequent derivative modifications of those creations than they would have over physical creations (Marcus 2007).

This is because of a limitation like the useful article doctrine, which applies specifically to "pictorial, graphic, and sculptural works." Under the useful article doctrine, if a work is created primarily for functional purposes, then the creation is *not* a proper subject for copyright, except as to the non-utilitarian elements of the creation (Marcus 2007). Although an object may be considered a useful object, those elements of the design that can still benefit from protection are those that can be identified separately from and are capable of existing independently of, the utilitarian aspects of the article. Because virtual objects are inherently visual or audio-visual works, the useful article doctrine does not appear to limit what works can be protected, leading to a greater amount of protected virtual works than physical works (Marcus 2007).

Another possible real-world parallel relates to existing law for artistic creations like photos, always potentially comprised of content created (and copyrighted by) others. Baldrica wonders whether a user's taking of a screenshot is the equivalent of taking a photograph. If it is, he asks, "Would the logic of Burrow-Giles Lithographic v. Sarony apply? Would it matter if the screenshot incorporated architectural or graphical features, such as logos, that were separately trademarked, or if the screenshot rather was simply an image of the virtual wilderness?"

Jankowich 2006 points out some additional differences between real-world law and virtual world law in terms of EULAs, agreements that he feels represent an important crossover point between real world law and virtual world law. Under the terms of their EULAs or TOS, game developers of MMOs in the United States generally claim ownership of everything that exists or occurs within the game servers, though continuing developments may ultimately put the effective scope of these EULAs into question, particularly where they may conflict with constitutional rights of expression (Jankowich 2006).

End-user license agreements, governed by real world law, are the primary instrument of law employed by proprietors (i.e., *platform* creators or owners) in the virtual world, and Jankowich argues that they impose a range of limitations on virtual world participants that would not stand in the real world. Property rights are limited, for example, and a wide range of speech and behavior is also restricted. Jankowich believes that the scope of EULAs should not be extended to claim ownership or dominion over expressive or creative activities undertaken in a virtual environment that would otherwise count as protected activities in a real physical environment, even a privately owned space.

Property rights, mentioned before, relate to the question of who owns the virtual space itself, or any of the virtual items, virtual characters, or other game objects within it. This is an important distinction, because it suggests a conceptual point of differentiation along Baldrica's Spectrum of User Contribution, from user-contributed content comprising the game environment (such as the virtual items that a player's character has earned by playing the game) to content that merely exploits the existence of the game environment - such as, say, an in-game screenshot, which the player consciously composes by moving his character inside the virtual world to capture an image of a particular, original view of the virtual landscape. Baldrica believes that it is entirely plausible to imagine that differing legal analyses can and should be employed when evaluating the validity of EULA claims of ownership over these varying types of user-created content.

Perhaps seeing the writing on the wall, Microsoft recently changed the license associated with Halo 3, a major platform for machinima creation, to explicitly become more lenient and reasonable about it. As Wired Magazine and the Electronic Frontier Foundation Reported, in August 2007 Microsoft published new guidelines governing how its intellectual property could be used for works like machinima.

According to Wired magazine, which provided a helpful summary of the legal changes, the updated Halo 3 rules prohibit:

- The creation of anything "pornographic or obscene ... or otherwise objectionable." The vagueness of this rule upset many creators.
- The use of soundtracks or audio effects from original games. Microsoft said it often licenses music and sound effects and is unable to pass on usage rights.
- The sale of any works that use Microsoft intellectual property. The company said it's OK to sell ads on sites that host machinima, but that's it.
- The reverse-engineering of Microsoft games
- The creation of anything that adds to a game "universe" by expanding upon its story.

Wired points out that "Predictably, the document provoked an uproar. But then two things happened: Legal experts as well as machinima movers and shakers examined the rules and decided they weren't so bad, and Microsoft proved amenable to refining the guidelines and working to clarify misunderstandings." The digital rights advocacy group the Electronic Frontier Foundation signed off on the rules, and a few weeks later Blizzard Entertainment, the developer of World of Warcraft (WoW), came out with its own machinima guidelines (Phan 2007). The consensus among machimists seem to be that both Microsoft's and Blizzard Entertainment's machinima guidelines give players more rights rather than take them away, and that they should be commended for helping the machinima community that previously had received little to no guidance from game companies and platform owners.

ACKNOWLEDGMENT

The author would like to thank Rob Wright (Robbie Dingo) for creating Machinima that makes Lindens cry, and for providing an interview; Torley Linden for providing contacts with Machinima makers; and Ian Linden and Jeska Linden for early Linden Lab history.

REFERENCES

Au, W.J. (February 7, 2005). Second Life – Hooray for Slollywood. *New World Notes.* Retrieved February 8, 2008, from http://www.Machinima.com/article/view&id=432

Au, W.J. (September 4, 2007). HBO Buys U.S. Rights to Second Life Machinima Series, Promotes it as Oscar Nominee Contender. *New World Notes.* Retrieved February 8, 2008, from http://nwn.blogs.com/nwn/2007/09/second-life-mac.html#more

Baldrica, J. (2007). Mod as Heck: Frameworks for Examining Ownership Rights in User-Contributed Content to Videogames, and a More Principled Evaluation of Expressive Appropriation in User-Modified Videogame Projects. *Minnesota Journal of Law, Science & Technology 8*(2), 681-713.

Call, E. (2005). White Paper: *Making Machinima in Second Life.* Retrieved January 13, 2008 from http://s3.amazonaws.com/static-secondlife-com/_files/making_Machinima.pdf

Cherny, L. (1999). *Conversation and Community: Chat in a Virtual World.* Stanford, CA: CSLI Publications.

Chien, I. (2007). Comparisons: Deviation, Red vs. Blue: The Blood Gulch Chronicles. *Film Quarterly, 60*(4), 24-29.

Damer, B. (2007). Meeting in the Ether. *ACM Interactions14*(5), 16-18.

Ford, D. (2007). Virtual Limitations: A Comparison of SIMS 2 and Half Life Games Engines for Machinima Narrative. *ACM International Conference Proceeding Series: Vol. 274. 2nd International Conference on Digital Interactive Media in Entertainment and Arts (DIMEA).*

Goodman, N. (1978). *Ways of Worldmaking.* Indianapolis, IN: Hackett Publishing Company, Inc.

Jankowich, A. (2006). EULAw: The Complex Web of Corporate Rule-Making in Virtual Worlds. *Tulane Journal of Technology & Intellectual Property, 8.*

Kurtz, L.A. (2007). Copyright, Creativity, Catalogs: Copyright and the Human Condition. *U.C. Davis Law Review 40*(3);

Linden, P. (January 8, 2007). *Embracing the Inevitable.* Retrieved December 30, 2007 from http://blog.secondlife.com/2007/01/08/embracing-the-inevitable/

Marcus, T.D. (2007). Fostering Creativity in Virtual Worlds: Easing the Restrictiveness of Copyright for User-Created Content. *New York Law School Law Review, 52,* 67-92.

Phan, M. (September 2007). *Machinima Licenses Spell Out New Rules for Creators.* Wired Magazine, retrieved on February 2, 2008 from http://www.wired.com/culture/art/news/2007/09/machinimalicenses

Rymaszewski, M., et. al. (2007). *Second Life: The Official Guide.* Indianapolis, IN: Wiley.

Salen, K., & Zimmerman, E. (2003). *Rules of Play: Game Design Fundamentals.* Cambridge, MA: MIT Press.

Sharp, D. (2006). Participatory Cultural Production and the DIY Internet: From Theory to Practice and Back Again. *Media International Australia: Practice-Led Research, 118*, 16-24.

KEY TERMS

Avatar: A graphical, usually 3D representation of a computer user, or a computer user's "digital persona" or alter ego.

Codec: A device (hardware) or software program that encodes and/or decodes digital signals or data.

EULA: An acronym for End User License Agreement, which describes digital content users should or should not create, how content should or should not be modified, what content (if any) users own, and consequences for content modification.

Governing Agreements: An umbrella term that encompasses the following user agreements to software use, including virtual worlds: EULAs, terms of service (TOS), rules of conduct, posting policies, and naming policies (Jankowich 2006).

GPL: An acronym for the Gnu General Public License, a free license for software and other kinds of works that enables its users to share and change all versions of a program. For more information on the GPL and its specific terms, visit www.gnu.org.

Machinima: A term that combines "machine," "animation," and "cinema" to describe computer-animated films that are shot within video games or social virtual worlds and primarily distributed online. Machinima is also defined as "a computer movie made using a real-time, 3D game/virtual-world engine instead of a special application dedicated to making computer movies" (Rymaszewski 2007).

MMORPG: An acronym for massively multiplayer online role-playing game (MMORPG) in which a large number of players interact with each other online in a three-dimensional or text-based virtual world.

Mods (also Modders): Programmers who extend the bounds of an existing game by re-engineering the game (**mod**ifying code) engine to create a different type of gameplay (Baldrica 2007).

Open System: A platform that has one or more of the following technical attributes:

- Contains tools created to support user content creation
- Provides help, tutorial or other documentation to assist users with content creation
- Allows users to import content they've created with another application
- Offers TOS or EULAs that explicitly permit user-created content and distribution
- Allows users to retain ownership (intellectual property rights) of their creations
- Enables emergent behavior (rule changes, new in-world avatar roles, non-game focused gatherings, and more)
- Provides client source code to encourage programmers to modify the underlying code itself, changing the product in profound ways and perhaps to look and function very differently than originally intended. (Salen 2003)

Residents: The users or customers of Second Life, a social virtual world.

Social Virtual World: Shorthand for "a 3D, online, collaborative, virtual world that is not a game." The primary purpose of a social virtual world is the creation of meaning through the manipulation of the world and communication with others within the world, while structured play is the primary purpose of game play worlds (Damer 2007).

Chapter XXIII
Player Motivation and Understanding Game Dynamics

Andrew Jinman
Twofour Learning: Immersive Learning Producer, UK

ABSTRACT

Massively multiplayer online role-playing games (MMORPGs) are becoming an increasingly popular recreational activity for social engagement. Transporting players to fantasy realms where they bridge the void between reality and the virtual world, via the creation of their online persona. Since the introduction of "table-top" games in the 70's, social interaction has been shaped by various game dynamics. Following the evolution of the MMORPG genre, the author investigates the implication of these dynamics on in-game social interaction and personal creativity. Identifying the key drivers in a player's motivation allows us to understand how different players are affected by the game's dynamics through research by Bartle (1996) and Yee (2005). Retrospectively these games are being described as immersive due to game dynamics and social content, with a direct intention to increase play. The author also discusses anticipating the occurrence of problematic behavior and addictive nature of the technology.

INTRODUCTION

Virtual worlds are becoming a novel new reality for the establishment of communities and social interaction. In this chapter we will discuss the evolution of MMORPG game dynamics, which constitute these environments. Arguing that game dynamics have evolved to maximize and encourage social interaction and engagement. These fundamental characteristics have become a foundation for identifying player motivation in these virtual-spaces. I intend to use World Of Warcraft® (WOW) (created by Blizzard Entertainment) as a primary example. WOW is one of the most popular MMORPG's to date. I propose to investigate WOW's flexibility in character creation and understand how a large proportion of player types and demographics are catered for, with this innovative system. With each player selecting from different attributes to meet an established requirement (game play or otherwise) as identified by Hartas (2005).

We will identify that the Massively Multiplayer Online (MMO) environment satisfies Oldenburg's (1999) criteria for becoming a "third place", an environment that provides a deep sense of emotional and social engagement. Continuing from research by Steinkuehler (2005) in "The New Third Place: Massively Multiplayer Online Gaming in American Youth Culture", exploring the affect of these game dynamics in MMO environments, defining whether it increases the playability and addictiveness of online games.

Identifying dynamics that have been introduced into a game's design to motivate players to continue playing, create new interactions and social phenomena. However the introduction of these dynamics may cause problematic behavior in respect to addiction, player immersion and social interaction. Guilds are a prime example of game dynamics, designed to increase social interaction between players. These in-game communities provide an effective platform to investigate the occurrence of "bridging" and "bonding" between players as described by Steinkuehler (2005) and Oldenburg (1999).

BACKGROUND

When we think of MMORPG's we usually restrict our thoughts to the last decade where this genre of gaming has really expanded. Predating these preconceptions I explore the origin and history MMORPG's. Understanding the evolutionary timeline and the context of these games allows us to draw on current gaming functionality, examining the gaming dynamics, their introduction over time and their impact on traditional game design.

TRACING THE EVOLUTION OF MMORPG'S

Role-playing is where participants take on and act out the role of a character that usually has differ-ent motives, personalities and backgrounds from the person adopting the role (Waskul, 2006). In the context of Role-playing Games, participants assume the role of a character and collectively create and play out stories. Participants define the actions and behavior of their character within the group. Whether the participant's actions succeed or fail depends on the rules set out and defined by the games dynamics, which in turn shapes the continuing story for the characters dependent on their actions.

This style of game play was truly defined by Table Top games, in particular Dungeons and Dragons (D&D) in 1974 (Wikipedia, 2008) , however this fantasy role playing game wasn't the first of the table top genre. The most obvious precedent to D&D was war games such as Avalon Hill's Tactics II (1958) (Barton, 2007) and sport simulation games like Strat-o-Matic (1961) (Barton, 2007). Nevertheless in 1974 the public was introduced to something completely new. Instead of replaying historical battles or playing out sporting events, they were introduced to a complete fantasy world populated entirely by fictional characters. The player took on the role of a specific race or class and they played out that character's life within this fantasy realm. There are certainly game play features adopted from these old games, an emphasis on calculation for example, but also the question arises about how profoundly the development of D&D was based around J.R.R. Tolkien's Lord of the Rings and traditional fantasy literature as suggested by Barton (2007).

The adventurer's life is structured by guidelines provided by the "core rulebooks". Playing against the Dungeon Master (DM) the players journey through his domain filled with magic and uncertainty. The adventurers continue through the world; fighting monsters, stealing treasure and outwitting enemies (Waskul, 2006). "Dice are important in these role-playing games; they are the principal means of simulating chance and probability maintaining an element of tension and uncertainty, a key characteristic of play (Hunizinga; 1950).

MUD'S AND MUSH'S

Inspired by tabletop games and the creativity of gaming dynamics, the fantasy genre started to evolve along side the technological advances of that time; computers were the next logical step. During the mid 1970's single player adventure games started to appear, "Adventure/Colossal Caves" (Wikipedia, 2008) and "Zork" (Bartle, 1999) were single player text only games, which took you through many puzzles and challenges similar to D&D's. There are no graphics, it's all within the players imagination and has been described as an interactive novel, using simple syntax that is extremely easy to learn. For example, if one wants to fill a bottle with water, just type "FILL BOTTLE WITH WATER".

The imagery of these worlds existed in the player's mind alone, facilitated by the text written on the screen. The next significant steps in the evolutionary timeline following this text only format were MUDs (Multi-User Dungeon, Domain or Dimension) and MUSH's (Multi-User Shared Hack or Habitat). These spaces emerged due to the expansion of the internet, and moved away from standalone servers. Described as a text-based social medium, where multiple users log into the game at the same time and read descriptions of rooms, objects, events, other characters and computer controlled characters know as Non-Player Characters (NPC's) within this virtual world. These games can be classified as the first type of MMORPG's to come about in the mid to late 80's. MUD's are more classically thought of as the original MMORPG and were designed intentionally to include elements of role-playing, described as a combination of hack and slash gaming and elements from social chat rooms. "Co-operation is an important element of survival on adventure MUDs. In many cases, players need each other

to survive" (Reid, 1994). Gemstone II (Genie network) (Mulligan, 2002) was one of the most significant and one of the first MUD's to appear on the market, it allowed users to take on roles, kill monsters, chat with friends and buy clothes and items. It was found to be incredibly addictive. In comparison, MUD's were fundamentally oriented around gaming where as MUSH's were primarily for role-playing and socializing.

Dune's MUSH was one of the most popular titles of the time (With Dune III MUSH still running today (Dune III, (2002)). Strictly oriented around role-play and character interaction, you assume a character from Frank Herbert's award winning 1982 Novel "Dune". Governed by the political views of your Royal House, you embark on an adventure, storylines and plots arise due to the different House politics and intrigue. Allowing players to explore through interactive Role-play (reading text descriptions and giving text commands) the universe described within the book. Players were able to set up there own building, modify room descriptions and create objects. As well as holding events you played to gain power and earn respect from fellow players within your royal house hold. The most compelling differences between MUD's and MUSH's is that within MUSH's players are capable of extending the world by creating new objects or rooms, even defining there behavior using scripting language. Also there is a complete lack of administrative hierarchy, allowing players to have the ability to modify nearly every aspect of the games database.

THE GOLDEN AGE

Ultima Online released in 1997 by electronic arts was branded as a milestone within the progression of MMORPG's (Gustav, 2002). One key innovation was the introduction of a flat rate monthly subscription, in comparison to previous "pay as you play" payment models. This opened the market up, attracting a broader audience instead of the more "hardcore" gamers who would rack up extensive fees on the old system. The game also introduced a more complicated 3D 3rd person graphics engine and included a more engaging plot and story in comparison to its predecessors.

The next commercially successful and most popular game that contributed to the evolution of MMORPG's, was EverQuest online in 1999 by Sony Online Entertainment. EverQuest online was a huge commercial success, and drove the concept of MMORPG into the Western cultures entertainment mainstream. Due to its success, fourteen consecutive expansions followed, with the fifteenth planned for October 2008. Many other games adopted a similar style to EverQuest online and the majority even adopted the popular game dynamics, which had now largely evolved from early MUD's and MUSH's.

However one important factor in its popularity was the media attention EverQuest generated, "*TIME* magazine and other non-gaming press featured stories on EQ, often focusing on the controversies and social question" (Wikipedia, 2007). This was one of the first times computer game culture was reported in mainstream media. Even though the non-gaming press focused largely on the more controversial and social aspects, it still helped propell EverQuest into the mainstream.

EverQuest presented many new ideas that had not been used in MMORPG before. Similarly to previous MMORPG's, users created and customized their character, gaining experience by killing mobs and looting the corpses, collecting equipment for personal and financial gain, whilst following the story set out by the NPC's, in the form of quests.

The major difference between EverQuest and its immediate descendents, such as Neverwinter Nights and Ultima Online, was a completely 3D environment. EverQuest moved away from the more traditional 3D 3rd person view, allowing players to be immersed within the games environment. Users were presented with a world encompass-

ing over 400 different areas from murky forests to alternate planes of reality; gamers had never experienced a MMORPG on this scale before (Wikipedia, 2007). EverQuest also encouraged the social aspect of the game. While many areas could be explored by yourself, you were strongly encouraged to form groups to explore other areas due to their increased difficulty level.

Finally EverQuest used a complicated trade skills system, which allowed players to choose a particular trade skill (such as blacksmithing and tailoring). By collecting the correct raw resources, players were capable of fashioning new items and weapons. This is an extremely time consuming and expensive process, but practice increased your trade skill level. If you were capable of producing expensive and powerful items, players could attempt to sell items online via auction websites. Jakobsson and Taylor (2003) identify that, improving tradeskills is seen as achieving a player's personal goal. But also benefits the greater good of the community, if for example a player chooses to join a guild. This suggests a simple in-game function can constitute towards social interaction between players.

CURRENT MMORPG'S

Bringing us to the modern day, many different genres and titles have been attempted by the massively multiplayer online gaming platform. In the current wave of modern MMORPG's such as Eve online, EverQuest 2, Matrix online and Guild wars, however one game has dominated the genre; Blizzard's 'World Of Warcraft' with over 10 million reported registered users (Blizzard, 2008).

Since the days of games being looked upon as "geeky" for people with no real social lives the perception of participants in this type if game has been overwhelmingly negative, "fantasy gamers have been characterized and caricatured in popular media as socially inept, psychologically unstable, or occultist." (Williams, Hendricks, Winkler, 2006) deeming "fantasy gaming" as a sub culture. However the increasing awareness, integration and appreciation of fantasy in main-stream popular culture is allowing gamers to shed the original misconceptions.

These modern equivalents have completely transcended the mainstream barrier and situated themselves within a wide audience who are aware of these games. Even the impact of popular films have been appearing in the MMORPG genre, such as Matrix Online and Lord of the Rings Online, illustrating the overlap produced between differ-ent types of media and nevertheless increasing awareness and popularity in mainstream culture. Continuing my research I aim to examine World of Warcraft in detail, as this title has dramatically influenced the MMORPG genre. Identifying initially the motivations behind players selecting their class, I investigate the game's dynamics and the influence of creative player motivations, including player immersion.

AN EXPLORATION INTO WOW

World of Warcraft (also known as WOW)

Is an online role-playing experience set in the award-winning Warcraft universe. Players assume the roles of Warcraft heroes as they explore, ad-venture, and quest across a vast world. World of Warcraft is a "Massively Multiplayer Online Role Playing Game" which allows thousands of players to interact within the same world. Whether adven-turing together or fighting against each other in epic battles, players will form friendships, forge alliances, and compete with enemies for power and glory. Blizzard Entertainment (2005)

It was in development for over six years and was highly anticipated in 2004, WOW is now set to release its second expansion pack 'Wrath of the Lich King', where Blizzard expect an increase in the total number of registered users (Blizzard, 2008). Its simplistic gaming style and its appeal to the casual player ensured its survival in the set of MMORPGs. From the onset its simplicity is conveyed to the gamer; by answering a few questions an appropriate server is suggested to the player. Otherwise one can manually select either a PVE (player verses environment) or PVP (player verses player) server. PVP is sometimes seen as more difficult because you are pitched against the environment of the game as well as other gamers. The player is able to create a persona/character on that server, choosing between the games two warring factions the "Horde" and the "Alliance". They are able to decide on their character's gender, race, and class then assign the persona a name. However by selecting your characters race in World of Warcraft; you are affecting where your character may explore and what quests he or she might adventure upon.

Selecting a class in which to belong can dra-matically affect the player's enjoyment of the game and style in which the game is played. There are nine classes to choose from, but they can be bro-ken down into three main groups: healing based classes (druid, priest and paladin shaman) magic classes (mage and warlock) and fighting classes (warrior, hunter and rogue). Each class has its individual strengths and weaknesses, many of which overlap and create hybrid classes, which will be discussed later. Healing classes are able to withstand high levels of damage, when capable of continued healing, but there damage per second (DPS (Damage Per Second); a system in which to measure a characters ability to administer dam-age) is very low. Whereas magic classes are less able to withstand large amounts of damage but themselves deal massive DPS. Finally fighting classes deal a good amount of DPS and are capable of withstanding large amounts of damage, but are unable to heal themselves. This differentiation in skills results in some classes being less capable of playing the game effectively alone, either due to lack of healing ability or low DPS. This greatly alters the games experience, if for example a player wanted to play the game solely on their own they would chose a "hybrid" class e.g. paladin, shaman or druid, a class that offers both reasonable healing powers and DPS. Whilst players that prefer to stick together in a group might choose a class that specializes in either high DPS or more effective healing powers in order to better support the rest of their group. This illustrates how class selection by gamers gives us a great insight into the initial motivations of a player as described by Tazman ODevilsun (2004).

Creating greater flexibility, Blizzard introduced the talents system to WOW, allowing player customization as they progress up until level sixty. Therefore each class fundamentally has the opportunity to become a hybrid class, Rio (2007) changing a players entire play style depending on how the player customizes their persona/character throughout the game. Traditionally the main hybrid classes are the paladin, druid and shaman. Hybrids are classes that have abilities and characteristics that are more normally associated with another class, for example, the paladin is a combination of a warrior and a priest, taking on their fighting and healing capabilities respectively. This dual capability makes hybrid classes more effective at dealing with a variety of situations. (Rio, 2007).

The issue arises when you attempt to examine the balance of power between classes, the hybrid classes become somewhat underpowered, because they lack the effectiveness in comparison to the core classes' abilities. This makes hybrids a problem for group definition, whereas a damage taking class and a healing class both know their roles within the group. This also hinders the amount of group social interaction a player may experience throughout the game. A warrior/healer hybrid, for example, won't be able to take the damage of a core warrior or heal as well as a core healer. A party would prefer to take a warrior and a healer, which each can perform their singular duty most effectively, concluding that hybrid classes are

preferable soloing classes. (Sylene, 2006). Nick Yee a PhD student from Stanford University has extensively researched the demographics of World of Warcraft. In a recent survey, Yee discovered that females are more likely to prefer priests, hunters and druids (Yee, 2005). Whereas men were identified, as more likely to prefer playing more aggressive classes (e.g. rogues, warriors and shamans). The results suggest that women prefer to play more supportive nurturing roles, in essence the caregivers of the game, whereas men take on the primitive aggressive roles.

In this situation neither men nor women have any obligation to take on any role within the world, but we find that both sexes resort to a traditional primitive ideology, when selecting their playing class. Due to Yee's (2005) research, it is also apparent that women tend to associate themselves with the more attractive classes. Males appear to prefer orc's, taurens and the undead; this is most significant within the night elf class, the gender divide is approximately 34% of all women select a night elf character compared to 21% of males (Yee, 2005). Suggesting the night elf race was more attractive/appealing to female players. The survey was simplistic in design and illustrates some motives behind which sexes play which classes. However the simplicity is its downfall; for example it lacks any identification between hybrid classes, in respect to player customization via the talents system. The survey also brings to light two inconsistencies with the conclusion: the shaman and the paladin, both of which are nurturing care giving roles, but are both played more by males than females, suggesting that player motivations are more complicated than this survey suggests.

GAME DYNAMICS & SOCIABILITY

In recent years there have been multiple instances where home media (e.g. TV and games) and the Internet have been criticized for replacing social institutions and community.

...Increasingly concerned with the possible negative social and civic impacts brought on by the diffusion of both traditional media like television and cable and new media such as videogames and the Internet. (Steinkuehler, Williams; 2006)

However MMORPG's have been changing perspectives since launching into the mainstream. These games are being described as sociable environments, where communication is one of the main activities. "While it is true that players spend a significant amount of time in combat... they spend even more time simply communicating with other players." (Ducheneaut, Moore, 2005). Communication and social interaction between players is influenced by the games dynamics. These are implemented into the environment by the game developers. These are intended to increase social interaction (e.g. the functionality to create guilds). In the following section I will explore several main in-game dynamics of World of Warcraft, and its influence on communication and social interaction between players. Game dynamics are constantly improved to increase popularity and game immersion, which in turn affect player motivations. Understanding a games dynamics also allows us to identify player types and motives and problematic behavior, for example addiction.

CMC IN MMORPG'S

CMC or Computer-Mediated Communication is the communication between two or more people via the use of computers connected to a network or the Internet. In the last decade MMORPGs have been viewed as an innovating new form of CMC, designed to encourage social networking. There are two different types of CMC: synchronous and asynchronous (Dix, Finlay, Abowd & Beale, 1993). Synchronous takes place in real time for example in this instance taking place within game. It allows a player to communicate either

to the whole population or to a specific player on a one to one basis. Asynchronous CMC takes place most commonly on websites and forums over time. When we look at World Of Warcraft we observe both forms of CMC in constant use, but to what extent does this have on player motivations? The implication of in-game chat is the primary medium for interaction between players. In a survey that went out to users of several online gaming communities, they found that 39% "reported that the social experience was their primary reason for playing" (Seay, 2004). Within WOW social interaction takes place constantly with several text base communication channels available in game: "private" (one-to-one), "tell" (group chat), "spatial" chat (heard by all players within a certain radius)... "zone" chat, which reaches all the players in a given zone of the game. Zone chat is further subdivided into four channels: general, trade, local defense and "looking for group" (Ducheneaut, Yee, Nickell, Moore; 2006). However social interaction taking place between players cannot be compared to traditional face-to-face interaction; where facial expressions and other non-verbal communication are used heavily. "The absence of meta-communicative features like facial expression, posture and tone of voice encourages users to find other ways of making communication as complete as possible" (Giuseppe, Riva, Carlo Galimberti; 1998). WOW allows avatars to perform expressive gestures such as clap, dance, laugh and cry. But these offer only basic imitations of human facial expressions and gestures. This affectively encourages the development of the games own social norms and structure shaped by the implemented technology "CSSN's [computer-supported social networks] is developing norms and structures of their own. WoW players for instance regularly adopt the use of external software to enable Voice over Internet Protocol during "end game" instances/ dungeons. These emerging communities who's resourcefulness of multiple technologies "are not just pale imitations of Real life. The Net is the Net"

(Wellman, 1996) and communities are evolving independently. Seen as an additional element to the communicative function, it extends the functionality of the game making social interaction easier, faster and more natural in comparison to synchronous text based chat IRC (Internet Relay Chat). These forms of social interactions between players can also become an attractive element to in-game immersion as identified by Nick Yee (2005) and Nicole Lazzaro (2004).

Finally the use of forums within communities plays a vital role within social interaction between guild members. Establishing a sense of belonging within the community is greatly satisfying, however "such collaboration will not automatically occur simply because peer-to-peer interaction is supported and facilitated" (Murphy, 2004). Illustrating that encouragement of social interaction is not solely defined by the facilitated technology to communicate. We must look at how the Massively Multiplayer Online environment encourages this interaction and the motivations of the individual players.

ENVIRONMENT

The Massively Multiplayer Online Environment or social arena where interaction takes place, can't be overlooked. It is important to understanding the type of space and its implied social context. The creative space of the environment gives us a better insight into why social interaction takes place and allows us the means to define it. As we have already noted MMORPG's are in fact extremely sociable places, but at present we do not have the means to define them. Ray Oldenburg is an urban sociologist, who in his 1999 book *The Great Good Place*, writes about the importance of informal spaces within society. Defining three very different social spaces: first, second and third places. First places, are somewhere you feel comfortable and relaxed, for example your home. Second places are where you spend the majority

of your time when your not at home, for example work. These places provide social interaction and a sense of community. The final type of space is the Third place, where you feel relaxed and comfortable in addition to providing social interaction and a sense of community (Oldenburg; 1999). For example, general stores, bars and coffee shops. Ray Oldenburg consequently defined each social space with a set of characteristics, eight of which were identified to characterize a third place (Steinkuehler; 2005). Subsequently MMORPG's have been described as a modern day third place, but how do these eight characteristics translate to the Massively Multiplayer Online environment? With reference to Constance Steinkuehler research; I aim to explore the concept of MMO environments as "Third places" in reference to Oldenburg's, (1999) orginal definition.

THE THIRD PLACE CHARACTERISTICS IN MMORPG'S (STEINKUEHLER, 2005)

Neutral Ground: Players are able to come and go as they please from the game. There is little or no obligation to play, except occasionally the social obligation whilst grouped with a party, to complete the quest.

- **Leveler:** During play social status or rank within society are rarely invoked, to players this is insignificant and unimportant. Avatar-mediated social interaction allows players to shed any reservations they have over their real life appearance or personality.
- **Conversation is Main Activity:** Whether this is the main activity of MMORPG's is debatable, However several media scholars identify conversation and social interaction as the "main activity" of Massively Multiplayer Online environments (Steinkuehler & Williams, 2004; Yee, 2005).

- **Accessibility and Accommodation:** MMORPG allow players to log in and play at any time, coming and going when they please. Inside the MMORPG environment there will always be someone else logged in and playing.

- **The Regulars:** MMORPG's accommodate two types of regulars; 79% of players join a guild (Yee, 2006) Therefore "Guild members" who play together regularly and "Squatters" players who occupy a particular region regularly, become the regulars of the MMO environment.

- **A Low Profile:** MMORPG's visual form does not fit Oldenburg's (1999) criterion "low profile", but the social function does. The population of gamers follows a parabolic curve, maximum population on release and then the regulars persist while others leave for more recent titles.

- **The Mood is Playful:** The atmosphere and general mood of MMO environments are playful and witty. Players mock and make fun of each other and perform amusing actions. If a serious Real Life (RL) issue is raised the conversation is usually changed into a more amusing situation. A Home Away From Home: MMORPG's become a regular daily activity for players, a sense of rootedness and homely warmth surround MMO environments, the absence of a guild member or friend would cause concern.

MMORPG'S THIRD PLACE OBSERVATIONS

Fundamentally MMO environments satisfy Oldenburg's (1999) eight criteria, but the question remains "are virtual communities really communities, or is physical proximity necessary?" (Steinkuehler: 2006)

"Bridging" occurs when two people from different backgrounds come together via the means of a social network. And "Bonding" "occurs when strongly tied individuals, such as family and close friends, provide emotional or substantive support for one another" (Steinkuehler, 2006). Within MMORPG's players come from many diverse backgrounds therefore the occurrences of social bridging between social capitals occurs often. However it is usually less frequent for bonding to take place. "While deep affective relationships among players are possible, they are less likely to generate the same range of bonding benefits as real-world relationships because of players geographic dispersion" (Steinkuehler, 2006). In reflection MMO environments and MMORPG's satisfy Oldenburg's criteria for being classified a "Third place" But 'gamers become more involved in long-term social networks such as guilds and their activities become more intense. The function of MMOs as "third places" begins to wane' (Steinkuehler: 2006). Illustrating that gamers who experience social bridging and bonding, highlight the implication of social motivations. In contrast gamers who express a desire to advance and who are dedicated to their characters progression through the game, highlight the implication of "hardcore" player traits and motivations.

GUILDS AND GROUPING

"While role-playing games and shooter games give birth to clans, tribes and guilds, sports games are played in local and global teams and leagues. Here, we move towards the "next level"" (Sotamaa, 2002). Within MMORPG's and the World Of Warcraft environment in particular, the functionality of the game allows players to form guilds. A collaboration of players that band together to play regularly and achieve shared objectives. However in WOW at player level 60 guilds play an even more important factor.

Much of the content of World Of Warcraft for level 60 players relies on the ability to accomplish feats only possible with a large group operation. Because of the nature of "instances/dungeons" in World of Warcraft, many options are only available to characters, who group with others. (Brendan, 2005)

In the following section I aim to outline the difference in motivations between "end game" players (level 55+) who participate in guild activity, low-level players who are in guilds and those that are not in guilds at all. Subsequently discussing the social structure and sense of belonging created within a guild. Taylor describes "...The vast variety of guild types in existence" (2006), if different player types and motivations exist, there must exist the equivalent range of guilds. Ensuring this level of flexibility in the game design allowing the dynamic to create guilds throughout the game satisfies the variety of player motivations.

Players gaming style and objectives vary considerably between players of different of age, sex, ethnicity and moral beliefs; this is reflective of the broad demographic attracted to MMORPG's. Therefore a players gaming style must be reflective of the other guild members play styles, similarly to guild objectives and player objectives. "A guild serves to unite players of common interest and motivations, such as raid cooperatives, economic cooperatives or associates, players who share common playing habits" (Brendan, 2005) For example a level 60 player wishing to experience "end game" instances and dungeons, wouldn't join a solely Player versus Player guild. "A hardcore endgame raider would have little place in a guild whose social norm was to place a priority on player vs. player content. The presence of such a player would harm the overall cohesion of the guild" (Chris Taylor, 2006) Once a guild is created; including; naming of the guild, recruiting a minimum of ten players and designing a tabard (An item of clothing worn by guild members. The graphic design and style signifies which guild they

belong to). The guild leader creates a hierarchy of command "Guild dynamics highlights classic themes in political theory while providing insight into the formation of political communities" (Castranova, 2001). Players are elected to take charge of particular elements of the guild, for example class leaders for recruiting new members of a particular class. This illustrates the sharing of responsibility within the community, e.g. Usually a paladin will know more about his class than a completely unrelated class, for example a mage so it makes sense to make a paladin a paladin leader. Guilds quickly generate their own "code of conduct" how guild members are expected to act and behave. Members feel a sense of pride as the guild's name is displayed below the player's name. Consequently a player can easily be brought to the attention of the guild leader. If there behavior within game is unacceptable or the player illustrates exceptional skill, building a reputation for the guild. "Clan members depend on one another's strengths and exploits for there own individual success in the game by cultivating a shared clan reputation" (Steinkuehler, 2006).

A notable difference in playtime activity is also present between players that are in guilds and those who are not. "Playing time is more stable after level 40 for guilded players and fluctuates more for non-guilded ones" (Ducheneaut, Yee, Nickell, Moore; 2006) this may illustrate the social motivations of players within guilds. It could also be explained by the activities organized by the guild leaders, which regrettably creates negative social affects to the stability of the guild. "It has been proposed that guilds put "social pressure" on their members to play longer... guilds often organize raids and other events requiring planning, which could create a sense of obligation for the members" (Ducheneaut, Yee, Nickell, Moore; 2006) and in-turn may constitute towards problematic play. Players were found to be more likely to group with members of their guild. 'Our second type of social network connects players who are observed to be in the same zones of the game, excluding the major cities. Such a network highlights players who are spending time together, grouping with guild mates to run quests and visit dungeons.' (Ducheneaut, Yee, Nickell, Moore, 2006).

Admission to high-level guilds is also a rigorous process, unlike low-level guilds where the admission process usually consists of a conversation with the guild leader. High level admission is much more selective, dependent on the guild's requirements, the applicants typically have to apply via a website specifying equipment, play times, past guilds and whether there avatar is attuned to "end game" instances/dungeons (obtained the keys to access certain dungeons). "In order to be considered, you first have to play a character class the guild is currently in need of, and be able to enter endgame dungeons which require special keys" (Chris Taylor, 2006). This acts as an exclusive club, strengthening the sense of pride and belonging players feel.

Analyzing the affect of the guild functionality of MMORPG's we see a distinct trend in the observed social aspects of player motivations. However not all players participate in a guild or any such organized guild events, so some player social activities are unaffected by the implementation of guilds. Finally the shift in player motivations of high-level players and the functionality of guilds transforms attracting players that are motivated by exploring & advancement. Research presented by Ducheneaut, Yee, Nickell, Moore (2006) also suggests high-level players (55+) felt they were obligated to progress to end game instances/dungeons and subsequently join a guild to progress with the game, an observation shared by Brendan (2005) "Participating in guild activities – such as raids, trade, or player vs. player (PvP) competition – is the thing to do". However research suggests it isn't up until level 55+ that the majority of grouping takes place. This could be explained by the same previous explanation, players are too engrossed with exploration and personal advancement.

ECONOMICS & TRADE

Economics and in-game trade isn't a new game dynamic, it can be traced back to MUD and MUSH style games. However in MMORPG's the economics and trade systems have been designed to maximize social interaction between players. "...The professional system and the economy in Star Wars Galaxies are both structured so that players have to interact" (Ducheneaut & Moore, 2004). Within WOW there are three main types of trading that occurs, these all incorporate an element of social interaction between players and include: Auction house trading, Direct trading and Quests & Crafts. Each major city within WOW contains its own auction house, which allows players to buy and sell items. Items are purchased by placing a bid and letting an auction finish or by using the "buy it now" function similar to many online auction sites. The social interaction that occurs during this process is limited, however players regularly contact the "sellers" directly and attempt to buy the items privately at a cheaper rate. Direct trading occurs when two players decide to trade directly, contacting each other usually via the "Trade IRC channel" and then through private messages. They discuss whether to barter (trade items for other items of the same value) or to trade using the games in world currency. They arrange a meeting spot and conduct the transaction; this illustrates the social aspect of trading. Finally Quests and Crafts, throughout the game players will be presented with quests which, in order to be completed, require the use of a particular item, crafted by another players character. Players that are "Leather workers" can only produce the relevant leather equipment, at the beginning of the game you are able to specialize and select two professions that allow you to produce goods. Therefore for the quest to be completed the player must either find himself a leather worker capable of making the equipment, implementing the occurrence of social interaction, or the player must be capable of making them themselves (impossible for all quests), requiring in game social interaction.

MOTIVATIONS

Contrary to popular belief, playing computer games is not a solitary activity but more and more a social experience. Starting with MUDs (Multi-User Dungeons), players and designers quickly took advantage of the capabilities offered by the Internet to build complex online social worlds where people could meet and play. (Ducheneaut, 2004)

MMORPG's attract such a wide audience due to their increased mainstream popularity; it is important to understand what motivates so many people to play these games. Defining differentiation between player types is fundamental to further research. Providing an understanding for these motivations would allow us to explore concepts such as gamer preferences (e.g. gender swapping, class selection, player behaviors like moral issues and identifying problematic usage). Freud (1990) suggests that all action or behavior is a result of internal, biological instincts, which are classified into two categories: life (sexual) and death (aggression). Research has found this to be prevalent within the context of MMO games. "Role-playing's ability to affirm the player's fundamental drives, encouraging a sense of self-worth and power while indulging male erotic desire" (Nephew, 2006). In the following section I aim to examine these ideologies, by investigating the following models; Bartle's Player Types (1996) and Nick Yee's Player Motivations (2005). By design MMORPG's are extremely sociable places, progressively throughout the game the difficulty increases. Causing players to form groups or teams, to conquer particular enemy or complete a particular quest. As we have seen, many MMORPG's such as World of Warcraft have the facility to allow permanent groups of

players to form, known as guilds; first introduced in Never Winter Nights (AOL, 1991). Such 'in-game' dynamics and mechanisms are intended to increase sociability between players.

RICHARD BARTLE

Bartle's Player Types were initially proposed within his paper "Hearts, clubs, diamonds and spades: Players who suits MUD's" (1996) Bartle's model is specific to MUD's; this renowned study can be applied to modern day games, due to their similarity, as we have previously described MUD's were the original MMORPG's. This model describes four different player types within MUD games; These "Player Types" do not describe the motivations of players. What they do describe is the type of players that play, and we believe there is a relationship between the two.

These four player types have been defined from the interrelationship of two dimensions of playing style, action versus interaction, and world-oriented versus player-oriented.

1. **Achievers:** Players who play primarily to achieve goals and beat the game (world-orientated), collecting masses of gold or resources for example. Achievers prefer to beat the game world and its mechanisms.
2. **Explorers:** Players who play primarily to explore and interact with the games environment (world-oriented), seeking to "know" everything about the fantasy world (e.g. back story, history and the NPCs who populate it).
3. **Socializers:** Players who play primarily to interact with other players (player oriented), by using the game's communicative facilities., maintaining either their real world persona, or their in-game fantasy persona.
4. **Killers:** Players who play primarily to cause distress to other players (player orientated). By using aggressive and intimidating be-

havior, with the intent to kill other players or disrupt their gamine experience.

Bartle's research provides some very credible evidence, the model is very simple and precise. His research into player motivations of MUD players has acted as a foundation for many other researchers and academics.

NICK YEE

Nick Yee's research declares that Bartle's Player model gives a powerful insight into player motivations but suffers from several limitations. "Bartle suggested that different Player Types influenced each other in certain ways. But unless we have a way of assessing and identifying players of different Types, theories built on top of Bartle's model are inherently unfalsifiable." (Yee, 2005). Subsequently Nick Yee devised his own set of ten subcomponents that coexist and together explains the motivations of the players. In an extensive study of close to 7000 responses via an on-line open-ended questionnaire, he was able to define the subcomponents that make up a player's motivations. He stresses that these together identify the motivations of a player and the subcomponents are not player types directly. The ten subcomponents are:

The Achievement components:
(http://www.nickyee.com/daedalus/archives/001298.php?page=5)

1. **Advancement:** Players who gain satisfaction from achieving goals within the game environment, such as collecting large amounts of resources and gold, similar to Bartle's observations (1996). However Yee expands to identify player's striving to collect experience, leveling there characters and skills extremely quickly.

2. **Mechanics:** Gamers who gain satisfactions from understanding and calculating the in-game numerical mechanics of the game. (For example, working out the precise combination of attributes and equipment to maximize damage and defense. Striving to understand the entirety of the inner mechanics of the game, usually to optimize there characters skills).

3. **Competition:** Players who receive satisfaction from beating other human players. This may include lawful and fair challenges such as dueling or PVP battlegrounds. Yee reveals that this may however include scamming and griefing behavior (intentionally disrupting another players gaming experience).

The Social components:
(http://www.nickyee.com/daedalus/archives/001298.php?page=6)

4. **Socializing:** Gamers who enjoy and gain satisfaction from socializing with other players, a similar observation to Bartle's research.

5. **Relationships:** Gamers that intentionally play the game to form and sustain meaningful relationships with other players. Sharing real life issues and problems with there online friends. When tough situations occur in real life they will usually turn to their on-line friends for help and support.

6. **Teamwork:** Gamers who gain satisfaction from playing as part of a team or in a group. Players that don't demonstrate this sub-component are described as displaying "soloing characteristics".

The Immersion components:
(http://www.nickyee.com/daedalus/archives/001298.php?page=7)

7. **Discovery:** Players who gain satisfaction from exploring the world discovering new;

locations, quests, NPC's and treasures.

8. **Role-playing:** Gamers that gain enjoyment from taking on their fantasy persona in the game world, participating in the on-going story and reading all back story associated with the game world.

9. **Customization:** Gamers that enjoy playing to customize there character usually concerned with personal appearance and categorized by the adoption of a unique style. These gamers enjoy games with a large breath of equipment variation, initially noticeable in Gemstone II (Mulligan, 2002).

10. **Escapism:** Gamers who gain satisfaction from using the MMO environment for relaxation (Ray Oldenburg; MMO's as a '3rd place', discussed previously). Used as a way of escaping from real life issues and problems. This subcomponent is also typically associated with conventional additive behavior (Wesley, 2007).

At a glance you understand the sheer complexity in comparison to the Bartle's model, but there are definite similarities between Nick Yee's model and the Bartle's player motivation model. For example; the Discovery element of the Immersion component is very similar to the Explores of Bartle's model. Nevertheless Nick Yee identifies several key differences between his own research and Bartle's model. "However, the questions he asked were based on his own observations of possible player motivations, and this may have limited the discovery of other possible new motivations" (Foo, 2004). Nick Yee declared this, but somewhat overlooked its importance "Open-ended responses from earlier surveys" (Yee, 2005). Subsequently basing this research on previously composed survey's where Yee had already made observations on player motivations, might have limited his discovery of new motivations, as stated by (Foo, 2004).

IMMERSION

As previously stated, Bartle didn't identify Immersion as a player type, or as motivation towards play. However Nick Yee identified immersion as an important factor within player motivations. Immersion is intertwined with the game dynamics, and they are dependant upon each other. In the following section I aim to identify the close relationship between game dynamics and player immersion and determine whether improvements of the game dynamics influence the immersive effect of the game. Previously I highlighted the effect of game dynamics on social interaction; nevertheless I aim to draw on the immersive aspects of game dynamics. Social implications as identified earlier, also provide an attractive form of immersion; this is especially true of the fantasy genre.

NARRATIVE, ROLE-PLAY AND THE AVATAR!

Narrative is heavily rooted within the constructs of Role-playing. In MMORPG's the narrative and role-play is an important factor of immersion, situating the player within the construct of a story "As a means of creating emotional engagement..." (Hayes, 2005). Narrative is implemented into a MMORPG via two game dynamics: NPC's (Non playing characters) and through the game worlds "back-story." Both of which originally appeared in the Dungeons and Dragons Table-Top games. Continuing their importance throughout the evolution of the fantasy genre, these two elements have been prevalent throughout. NPC's, opposite to PC's (playing characters); refers to a character within the MMORPG that are controlled by someone that isn't playing the game (this could be human or otherwise). Typically AI (artificial intelligence) intertwined with the game to supply merchandise, increase game play, progress the plot of a game (e.g. the delivery of the games back-story or quests to the PC's). NPC's populate the entire virtual environment with intertwining narratives and plots; they build up an entire history of the world. Showing emotion and conveying their feelings about the virtual world to the playing characters. These Dungeon Masters of the MMORPG genre act as locked boxes of information, waiting until a playing character utters the correct saying within the vicinity of the NPC, to release the NPC's information or reward (Stern, 2002). As in Tabletops, MUD and MUSH games still required the back-story and narrative to occur in the player's imagination "Narrative is a certain type of mental image, or cognitive construct which can be isolated from the stimuli that trigger its construction" (Klastrup & Tosca, 2004). However contributing factors (cinematic titles, characters, history, plot etc) and the 3D environment reinforce the context of the back-story, positively increasing immersive play. By understanding what role-play fantasy gaming involves, "Seeing it as a way of playing a game, rather than a game in itself, role-playing can be perceived as game playing motivated with narrative desires" (Heliö, S, 2004).

In joining the MMORPG world, participants are playing out a completely fabricated adventure that is not restrained by "realities" boundaries and rules. "In fantasy role-playing games, participants must actively establish symbolic boundaries between player, persona, and person and assume the right role in each situation" (Waskul, 2006). The Avatar is the in-game representation of the real person, allowing the gamer to situate themselves within the back-story and historical events of the virtual world. "When we step through the screen into virtual communities, we reconstruct our identities on the other side of the looking glass" (Turkle, 1995). With this opportunity to completely redefine your personality, players are encouraged to immerse themselves within the context of the environment and social structure.

"You can be whoever you want to be. You can completely redefine yourself if you want. You can be the opposite sex. You can be more talkative. (...)" You don't have to worry about the slot other people put you in as much. Its easier to change the way people perceive you because all they've have got is what you show them...." (Turkle, 1995)

However World Of Warcraft's environment lacks the functionality to create user generated content, this lack of functionality may result in the environment becoming limited and restraining for players.

Neal Stephenson's 1992 novel Snow Crash introduced the term 'Metaverse', a term given to a completely immersive 3D environment where people interact, socialize, work and are entertained. Unlike WOW and many other MMORPG's these are completely player driven (similar to MUSH's) and are representative of the real world. These technologies have the potential to expand into every market imaginable for every day activities; work, entertainment and education, moving away from the traditional player driven environments such as commercial MMORPG's.'...By giving its users the vibrant complexity and dynamics of real-world cities rather than simple, repetitive game-play.' (Ondrejka, 2004). Without the ability for player driven content, immersive experiences will become saturated. Allowing a player to be situated and influential within the worlds architecture and history gives players a sense of ownership over the environment and its content. 'Creation is needed if there is to be any hope of creating an online world that dwarfs the complexity of the real world.' (Ondrejka, 2004)

ADDICTION

Increasingly addiction is becoming even more common within MMORPG, with china recently introducing an anti online game addiction system for children under 18 (Rio, 2007). And with

Korea reporting 6,271 cases of gaming addiction within 6 months (Yong, 2006). 40% of players considered themselves addicted, 30% agreed that they continue to participate in the environment even when they are frustrated with it or are not enjoying the experience (Yee, 2005). Illustrating that addiction is an increasing problem within the MMORPG genre. As we have already seen, game dynamics encourage many motivations, including social and immersive elements. These can become very addictive, but the main focus of game designers is to improve such gaming dynamics, so surely this will increase player addiction, therefore games companies could be accused of making MMORPG's more addictive for financial gains.

FUTURE TRENDS

The research discussed in this chapter highlights the technological change over the last forty years. It is prevalent to assume that technology will continue to evolve. My continued research will investigate whether gaming dynamics are used to encourage play. I will continue to observe the evolution of these dynamics in MMO technologies. For example, I will look deeper into the improvements in future in-game CMC techniques, evolution of convenient communication, and essentially more natural interaction with these worlds. Additional research has identified barriers to entry associated with this technology. Again this will be reduced as the technology evolves. In addition future projections identify that the uptake of virtual world technologies is set to increase between broadband users. "By the end of 2011, 80 percent of active Internet users, will have an avatar in a virtual world" (Gartner, 2007). This suggests that virtual world technologies will be increasingly used in a wider range of serious and creative purposes. However with the growing applications of this technology, come growing fears and concerns. Escapism and addiction are a grave

concern. These and similar fears lead to the Byron review (2008) being commissioned. The review was to investigate the risks associated with video games and surfing the Internet, but preserving the right of young people to engage with this technology in a safe and informed manor. With many more MMORPG's on the cards for release, it is important to continue to research the positive and negative implications of game functionality. It is important we influence game design and address the growing fears and concerns, which occur naturally with new technologies.

Richard Bartle has since published further research into design of MMO environments in "Designing Virtual Worlds" (2004). This book delves in to the sheer complicities of virtual world design, covering a range of subjects. This includes the expansion of his initial research into player motivations. In addition Bartle investigates three further player type models; Social Dimensions (F. Randall Farmer, 1992), Circles (Hedron, 1998) and he goes further to investigate Nick Yee's previous research. The books serve to educate designers and encourage them to develop concepts and gaming functionality that the players actually want and not simply sticking to already successful formulas.

Bartle has also expressed his concerns regarding World of Warcraft. This huge successful title has carved out a massive market share in the online gaming world, and since 2008 has seen massive growth to over 60% of MMOG market share (mmogchart.com, 2008). This has lead to a lack of creativity when considering fundamental game dynamics. Due to WOW's success, many games have merely tried to introduce similar styles to WOW. Sticking to the successful formula, instead of pushing the creative boundaries and introducing new dynamics, Bartle expressed similar concerns. This subsequently sparked a massive debate within the MMO community:

While I thought the original post had a certain aggressive tone to it (which is fine, makes for more interesting reading if nothing else), I do think your points were still valid.

The word 'revolutionary' is greatly overstated when speaking about MMOs and peoples wants for them. WoW improved on the EQ formula greatly, and while perhaps not the definition of revolutionary, it certainly did ENOUGH better to reach a far greater market share, a market share that pre-WoW no one could have predicted could even be reached...(Comment by Syncaine at tobolds.blogspot.com)

Bartle seems to like clarifying his statements. I'm sure this case is a bait for a debate where he gets to score one of those special points on those who seem to hook on to it.

I also think he is trying to actually fight the battle against slow, predictable and boring evolution by hinting that anyone who thinks of making any type of "clone design" will be used as Bartle bait. (Comment by Wolfie at tobolds.blogspot.com)

This illustrates the intersection of technology and creativity and identifies the question; should technology be the main driver in MMO design or should creativity? Shouldn't we really push technology to its limits to fit our creative natures or simply stick with a successful formula? My research will continue to investigate these changes and highlight the introduction of new gaming dynamics and primarily answer a few of the questions discussed in this chapter.

CONCLUSION

Game dynamics are increasingly used to motivate players to continue playing. Throughout the history of MMORPG's we have seen the introduction of several game dynamics, which are still fundamental in current game design.

By investigating game dynamics we were able to identify and observe the affect of game dynamics on player motivations described by Yee and Bartle. For example CMC encourages the various "social elements" behind player motivations. Constantly being used by players whilst: socializing, forming relationships and working together as a team. However during my investigation of gaming dynamics I found that guilds and the MMO environment waver in their ability to encourage social interaction when a player reaches level 55+ "end game". Player motivations seem to change, shifting from a majority of social aspects to more achievement based aspects of motivations. Players wish to progress to end game activities and therefore feel obligated to join a guild. Immersion is also deeply rooted within player motivations, and again game dynamics have been introduced to heighten and maximize the impact on players. However current MMORPG's are limited in their immersive capability, due largely to technological limitations. For example, 3D environments are still represented on 2D monitors and limited NPC AI, which detract from the immersive affect of the narrative. In conclusion MMORPG's still have a massive amount of information to offer us on social interaction within these virtual environments. With current MMORPG game design, the player must invest a great deal of time and effort to increase social interaction and player immersion. "Designing deep game experiences... offers a different avenue to enhance the Player Experience as a whole and by refining them through play testing provides more opportunities for emotion in games" (Lazzaro, 2004). However the occurrence of addiction to MMORPG's is ever more prevalent in today's society, so will increasing emotional, social and immersive game dynamics only increase the levels of addiction? As we have now observed this technological change over the last forty years, seeing the new functionality emerge. It is prevalent to assume that technology will continue to evolve.

REFERENCES

Bartle, R. (1996). *Hearts, Clubs, Diamonds, Spades: Players Who Suit MUDS*. Retrieved January 28, 2007 http://www.mud.co.uk/richard/hcds.htm

Bartle, R. (1999). *A Zork: A Computerized Fantasy Simulation Game*. Retrieved January 28 2007 http://www.mud.co.uk/richard/zork.htm

Barton, M. (2007). *The history of computer Role-Playing Games part 1: The early years (1980-1983)*. Retrieved February 28, 2007 http://www.gamasutra.com/features/20070223a/barton_01.shtml

Blizzard (2006). *Burning Crusade, release*. Retrieved January 28, 2007 http://www.blizzard.com/press/070307.shtml

Blizzard (2005). Retrieved January 28, 2007 www.worldofwarcraft.com

Blizzard (2008). *World Of Warcraft hits 10 million*. Retrieved February 22, 2008 from http://www.mmogchart.com/2008/01/30/world-of-warcraft-hits-10-million/ and http://www.blizzard.com/us/press/080122.html

Dennis, D.W. (2006). The Role-Playing Game And The Game of Role-playing, the Ludic self and everyday life. In J.P. Williams, S.Q. Hendricks, & W.K. Winkler (Eds.), *Gaming as Culture: Essays on Reality, Identity, and Experience in Fantasy Games*. Jefferson, NC: McFarland Publishing.

Dix, Finlay, Abowd & Beale (1993). *Human–Computer Interaction, Prentice Hall, Implementation support*. Retrieved January 28, (2007) from http://www.cc.gatech.edu/computing/classes/cs4753_94_fall/slides/Lecture-13(Implementation).ps.Z

Ducheneaut, N., & Moore, R.J., (2004). *The social side of gaming: a study of interaction patterns in a massively multiplayer online game*. Retrieved January 28, 2007 www.parc.xerox.com/research/publications/files/5223.pdf

Ducheneaut, N., & Moore, R.J. (2005). More than just 'XP': learning social skills in massively multiplayer online games. *A Palo Alto Research Center (PARC) journal*. Retrieved January 28, 2007 from http://www2.parc.com/csl/members/nicolas/documents/ITSE.pdf

Ducheneaut, N.m Moore, R.J., & Nickell, E. (2007). *Virtual "third places": A case study of sociability in massively multiplayer games*. Computer Supported Cooperative Work.

Ducheneaut, N., Yee, N., Nickell, E., & Moore, R.J. (2006). A lone Together? Exploring the Social Dynamics of Massively Multiplayer Games. *In conference proceedings on human factors in computing systems*. Retrieved January 28, 2007 from http://www.nickyee.com/pubs/Ducheneaut,%20Yee,%20Nickell,%20Moore%20 %20Alone%20Together%20(2006).pdf

Dune III (2002). Retrieved January 28, 2007 http://www.dune3.net/

Filiciak, M. (2003). Hyperidentities: Postmodern identity patterns in massively multiplayer online role-playing games. In M.J.P. Wolf & B. Perron (Eds.), *The video game theory reader*. New York: Routledge.

Foo, C.Y. (2004). *Grief Player Motivations*. Curtin University, Nokia Research center. Retrieved January 28, 2007 from http://www.tu-chemnitz.de/phil/medkom/mn/spive/index.php?option=com_docman&task=doc_download&gid=25

Freud, S. (1990). *Beyond the pleasure principle*. New York: W. W. Norton & Company.

Gartner media (2007). *Gartner Symposium/ITxpo 2007 Emerging Trends*. Retrieved January 28, 2007 from http://www.businesswire.com/portal/site/google/index.jsp?ndmViewId=news_view&newsId=20070424006287&newsLang=en

Gustav, T. (2002). *Guilds: Communities in Ultima Online*. Retrieved January 28, 2007 http://cid.nada.kth.se/pdf/CID-167.pdf

Hartas, L. (2005). *The art of game characters.* Lewes, East Sussex: Ilex Press.

Heliö, S. (2004). Role-Playing: A Narrative Experience and a Mindset. In M. Montola & J. Stenros (Eds.), *Beyond Role and Play,* (pp. 65-74). Solmukohta. Vantaa, Ropecon. Retrieved January 28, 2007 from http://www.ropecon.fi/brap/ch6.pdf

Jakobsson, M., & Taylor, T.L. (2003). The Sopranos Meets EverQuest Social Networking in Massively Multiplayer Online Games. *fineArt forum, 17*(8). Retrieved January 28, 2007 http://hypertext.rmit.edu.au/dac/papers/Jakobsson.pdf

Klastrup, L., & Tosca, S. (2004). Transmedial worlds - rethinking cyberworld design. In *Proceedings International Conference on Cyberworlds 2004.* IEEE Compuater Society, Los Alamitos, California, 2004. Retrieved January 28, 2007 from http://www.itu.dk/people/klastrup/klastruptosca_transworlds.pdf

Lazzaro, N. (2004). *Why We Play Games: Four Keys to More Emotion Without Story.* Retrieved January 28, 2007 from http://xeodesign.com/xeodesign_whyweplaygames.pdf

Mmogchart.com (2008). Retrieved April 5, 2008 from http://www.mmogchart.com/Chart7.html

Mulligan, J. (2002). *Biting The Hand #17: Talkin'' bout my... Generation.* Retrieved January 28 2007 http://www.skotos.net/articles/BTH_17.shtml

Murphy, E. (2004). Recognizing and promoting collaboration in an online asynchronous discussion. *British Journal of Educational Technology, 35*(4), 421-431. Retrieved January 28, 2007 from http://grail.oise.utoronto.ca/blog/karaisko/files/2007/12/murphy_recognizing_and_promoting_collaboration.pdf

Nephew, M. (2006). Playing with identity: unconscious desire and Role-Playing Games. In J.P. Williams, S.Q. Hendricks, & W.K. Winkler (Eds.), *Gaming as Culture: Essays on Reality, Identity, and Experience in Fantasy Games.* Jefferson, NC: McFarland Publishing

Ondrejka, C. (2004). *Escaping the Gilded cage: User Created Content and Building the Metaverse.* Linden Research, Inc., San Francisco. B.S. United States Naval Academy. Retrieved January 28, 2007 from http://www.nyls.edu/pdfs/v49n1p81-101.pdf

Reid, E. (1994). *Cultural Formations in Text.*

Rio, S. (2007) *China to implement anti-online game addiction system.* Retrieved January 28, 2007 http://mmorpg.qj.net/China-to-implement-anti-online-game-addictionsystem/pg/49/aid/89019

Rio, S. (2007). *Now everyone's a hybrid: Tseric on hybridization.* Retrieved January 28, 2007 http://wow.qj.net/Now-everyone-s-a-hybrid-Tseric-onhybridization/pg/49/aid/80618

Seay, A.F., Jerome, W.J., Lee, K.S., & Kraut, R.E. (2004). Project Massive: A Study of Online Gaming Communities. In *Proceedings of ACM CHI 2004* (pp. 1421–1424). New York: ACM Press.

Seay, A.F., Jerome, W.J., Lee, K.S., & Kraut, R.E. (2004). Project massive: a study of online gaming communities. *Conference on Human Factors in Computing Systems (CHI'04),* Vienna, Austria (2004).

Sotamaa, O. (2005). Creative User-centred Design Practices: Lessons from Game Cultures. *A (Eds.), Everyday Innovators, Researching the Role of Users in Shaping ICTs.* Springer, Dordrect, (pp. 104-16). Retrieved January 28, 2007 from http://members.aol.com/leshaddon/Sotamaa.pdf

Steinkuehler, C.A., & Williams, D. (2006). *Where Everybody Knows Your (Screen) Name: Online Games as "Third Places".* Retrieved January 28, 2007 http://jcmc.indiana.edu/vol11/issue4/steinkuehler.html

Steinkuehler, C.A. (2003). *Massively multiplayer online videogames as a constellation of literacy practices.* Paper presented at the International Conference on Literacy, Ghent, Belgium.

Steinkuehler, C.A. (2004c). *Online cognitive ethno¬graphy: Methods for studying massively multiplayer online videogaming culture.* Paper presented at the 17th Annual Conference on Interdisciplinary Qualitative Studies, Athens GA.

Steinkuehler, C.A. (2005). The new third place: Massively multiplayer online gaming in American youth culture. *Tidskrift Journal of Research in Teacher Education.* Retrieved January 28, 2007. http://website.education.wisc.edu/steinkuehler/papers/Steinkuehler_TIDSKRIFT2005.pdf

Steinkuehler, C.A. (2005c). *Styles of play: Gamer-identified trajectories of participation in MMOGs.* Paper presented at the Annual Conference of the Digital Games Research Association (DIGRA), Vancouver. Retrieved January 28, 2007 from http://www.academiccolab.org/resources/documents/Steinkuehler_NEWLIT.doc.

Sylene (2006). *Hybrid vs. Core.* Retrieved January 28, 2007 http://wow.stratics.com/content/features/editorials/hy/

Taylor, C. (2006). Bonds of trust: An in-depth look at social bonding within MMO guilds. *Undergraduate term paper for the course "Games for the Web".* Trinity University. Retrieved January 28, 2007 from http://www.trinity.edu/adelwich/worlds/articles/trinity.chris.taylor.pdf

Tazman_ODevilsun (2004). *Why are good soloing abilities important in a MMORPG?* Retrieved January 28, 2007 http://eq2vault.ign.com/View.php?view=columns.Detail&category_select_id=8&id=228

Turkle, S. (1995). *Life on the screen: Identity in the age of the Internet.* New York: Touchstone.

Wellman, B., et al. (1996). Computer Networks as Social Networks: Collaborative Work, Telework, and Virtual Community. *Annual Review of Sociology, 22,* 213-38.

Wesley, J. (2007). *Overcoming addiction and escapism.* Retrieved November 20, 2007 http://cid.nada.kth.se/pdf/CID-167.pdf

Wikipedia (2007). *Revews MMORPG's- MMORPG.cvr.pl, History.* Retrieved January 28, 2007 http://mmorpg.cvr.pl/history.htm

Wikipedia (2008). *Dungeons & Dragons.* Retrieved January 28, 2007 http://en.wikipedia.org/wiki/Dungeons_%26_Dragons

Wikipedia (2008). *Neverwinter Nights (AOL game).* Retrieved January 28, 2007 http://en.wikipedia.org/wiki/Neverwinter_Nights_(AOL_game) and Medar (2001). Neverwinter nights the original. Retrieved January 28, 2007 http://www.bladekeep.com/nwn/

Wikipedia (2008).*Colossal cave adventure.* Retrieved January 28, 2007 http://en.wikipedia.org/wiki/Colossal_Cave

Williams, J.P., Hendricks, S.Q., & Winkler, W.K. (Eds.) (2006). *Gaming as Culture: Essays on Reality, Identity, and Experience in Fantasy Games.* Jefferson, NC: McFarland Publishing

Yee, N. (2004). *World of Warcraft Scenery.* Retrieved January 28, 2007 http://terranova.blogs.com/terra_nova/2004/12/world_of_warcra.html

Yee, N. (2005). *WoW Character Class Demographics.* Retrieved January 28, 2007 http://www.nickyee.com/daedalus/archives/001367.php

Yee, N. (2005). *A model of player motivations.* Retrieved January 28, 2007 http://www.nickyee.com/daedalus/archives/001298.php?page=1

Yee, N. (2005). *Addiction.* Retrieved January 28, 2007 http://www.nickyee.com/daedalus/archives/000818.php

Yong, K. (2006). *Addiction to MMORPGs: Symptoms and Treatment.* Retrieved January

28, 2007 http://www.netaddiction.com/articles/addiction_to_mmorpgs.pdf

KEY TERMS

Character: The persona or role taken on by a player when he or she enters the game environment.

Fantasy Genre: Is an art style that uses magic and supernatural forms to propel the narrative, theme and or setting. Originally descended from myths and legends these were early examples.

Game: An activity structured with certain objectives, which are shared by participants, usually undertaken for enjoyment. With reference to this thesis "game" is used to describe the MMORPG's, and activities preformed by its participants.

Game Dynamics: Proportions of game functionality or characteristic created to intentionally encourage game play or game enjoyment; for instance, Social activities and Fighting.

MMO Environment: Massively Multiplayer Online environment, or virtual world where the game takes place.

MMOG: Massively multiplayer online game; variety of game, which is capable of supporting thousands of simultaneous players usually within a virtual world or an MMO environment.

MMORPG: Massively multiplayer online role-playing game.

Online Identity: is a social identity that network users establish in online communities, for this thesis this applies to the online identity formed within an MMORPG, and the persona the player takes on. See Introduction to online identity.

Player: The real person sat at the computer participating within the game, with his or her character.

Player Organization: An association of players with common interests and goals within the MMO environment; sometimes called a guild or clan.

Reality: The non-game environment where the player exists, as opposed to the MMO environment where the character, avatar or persona exists.

RPG: Role-playing game, a game where the player takes on the role of another character, acting and behaving like them.

Social Learning: Social learning refers to learning within a social environment e.g. MMORPG exclusively within Player Organizations.

Compilation of References

Abbott, E. (1999, c1884). *Flatland: A romance of many dimensions*. London, UK: Penguin.

Abrams, J., & Hall, P. (2006). *Else/where: mapping new cartographies of networks and territories*. Minneapolis, MN: University of Minnesota Design Institute.

Abrams, J., & Hall, P. (Eds.) (2006). *Else/Where: Mapping*. Minneapolis: University of Minnesota Press.

Adams, C. C. (1995). *Technological allusivity: appreciating and teaching the role of aesthetics in engineering design*. In Proceedings of the Frontiers in Education Conference, 1995. (pp. 3a5.1-3a5.8). Washington, DC, USA: IEEE Computer Society.

Aderson, Eh. (2000). Lightness perception and lightness illusions. In MGazzaniga (Ed.), *The New Cognitive Neurosciences*. Cambridge: MIT Press.

Agarawala, A., & Balakrishnan, R. (2006). Keepin' it real: Pushing the desktop metaphor with physics, piles and the pen. *Proceedings of CHI 2006 - the ACM Conference on Human Factors in Computing Systems* (pp. 1283-1292).

Agre, P. (1997). *Computation and Human Experience*. Cambridge University Press, ISBN 0521386039.

Alexander, C. (1987). *The New Theory of Urban Design*. New York: Oxford University Press.

Alexander, C., Ishikawa, S., & Silverstein, M. (1977). *A Pattern Language*. New York: Oxford University Press.

Algazi, V.R., Duda, R.O., Thompson, D.M. & Avedano, C. (2001). *The CIPIC HRTF Database*. Paper presented at the IEEE Workshop on Applications of Signal Processing to Audio and Acoustics, New York, USA.

Algorithms in School Mathematics. 1998 Yearbook. Reston, VA: National Council of Teachers of Mathematics.

Alinovi, C. (2003). *NET.ART The art of our times- 4/2003 Stedelijk Museum Bulletin* [Electronic Version]. Retrieved 27 Jun 2006 from http://www.stedelijk.nl/oc2/page.asp?PageID=390.

Al-Mualla, M. (2003). Motion field interpolation for frame rate conversion. *ISCAS, 2*, 652-655.

Anderson, N. C. (2005). *The Creative Brain, the Science of Genius*. New York, NY: Penguin Group, 2006.

Andre, T., Cagnazzo, M., Antonini, M., & Barland, M. (2004). Motion-compensated lifting-based wavelet transform. *IEEE International Acoustics, Speech, and Signal, 3*, 121-123.

Anthropology and Mathematics. In B. Greer & S. Mukhopadhyay (Eds.),

Appiah, A. (1992). *In my Father's House*. NY: Oxford.

Argent, L., Depper, B., Fajardo, R., Gjertson, S., Leutenegger, S. T., Lopez, M. A., et al. (2006). Building a game development program. *Computer, 39*(6), 52-60.

Arikan, B. (2008). *MYPOCKET: Predicted Objects*. Retrieved April, 2008 from http://transition.turbulence.org/Works/mypocket/predicted_objects/

Aristotle. (1970.) *Poetics*. (Gerald F. Else, Trans.) Ann Arbor: University of Michigan Press. (Original work published c. 335 B.C.).

Arnheim, R. (1974). *Art and Visual Perception*. Berkeley, CA: University of California Press.

Arnheim, R. (1996). From chaos to wholeness. *The journal of aesthetics and art criticism, 54*(2), 117-120.

Arnheim, R. (2006). *Film as Art*. Berkeley, CA: University of California Press.

Aronilth, Jr., W. (1991). *Foundation of Navajo culture*. Navajoland, USA: Wilson Aronilth, Jr.

Ascher, M. (1991). *Ethnomathematics: A multicultural view of mathematical ideas*. Boca Raton, FL: Chapman & Hall/CRC.

Au, W.J. (February 7, 2005). Second Life – Hooray for Slollywood. *New World Notes*. Retrieved February 8, 2008, from http://www.Machinima.com/article/view&id=432

Au, W.J. (September 4, 2007). HBO Buys U.S. Rights to Second Life Machinima Series, Promotes it as Oscar Nominee Contender. *New World Notes*. Retrieved February 8, 2008, from http://nwn.blogs.com/nwn/2007/09/second-life-mac.html#more

Auslander, P. (1999). *Liveness: Performance in a Mediatized Culture*. New York: Routledge.

Axerold, R. (1984). *The Evolution of Cooperation*. New York: Basic.

Baldrica, J. (2007). Mod as Heck: Frameworks for Examining Ownership Rights in User-Contributed Content to Videogames, and a More Principled Evaluation of Expressive Appropriation in User-Modified Videogame Projects. *Minnesota Journal of Law, Science & Technology 8*(2), 681-713.

Barker, E., Webb, N., & Woods, K. (1999). *The Changing Status of the Artist*. Yale University Press. ISBN 0300077424.

Barnhardt, R., & Kawagley, O. (2005). Indigenous knowledge systems and Alaska native ways of knowing. *Anthropology and Education Quarterly, 36*(1), 8-23.

Barr, A., & Feigenbaum, E., (Eds.) (1981). *The Handbook of Artificial Intelligence, 1*. Morgan Kaufmann.

Barrow, J. (1992). *Pi in the sky: Counting, thinking and being*. Clarendon Press.

Barta, J., & Eglash, R. (2008). Seeing With Many Eyes: Connections Between

Barthes, R. (1977). *Image, music, text/ Roland Barthes; essays selected and translated by Stephen Heath*. New York: Hill and Wang.

Bartle, R. (1996). *Hearts, Clubs, Diamonds, Spades: Players Who Suit MUDS*. Retrieved January 28, 2007 http://www.mud.co.uk/richard/hcds.htm

Bartle, R. (1999). *A Zork: A Computerized Fantasy Simulation Game*. Retrieved January 28 2007 http://www.mud.co.uk/richard/zork.htm

Barton, M. (2007). *The history of computer Role-Playing Games part 1: The early years (1980-1983)*. Retrieved February 28, 2007 http://www.gamasutra.com/features/20070223a/barton_01.shtml

Batteau, D.W. (1967). The role of the pinna in human localization. *Proc. Roy Soc, London, B168*, 158-180.

Batteau, D.W. (1968). Listening with the naked ear. In S.J. Freedman (Ed.), *The neuropsychology of spatially oriented behavior* (pp. 109-133). Homewood, IL: Dorse Press.

Baudrillard, J. (1998). *The Consumer Society: Myths and Structures*. (p. 116). London, England: Sage Publications Limited.

Baumgartel, T. (2001). Net Art. On the history of Artistic Work with Telecommunications Media. In W. Peter, D. Timothy, & Zentrum fur Kunst und Medientechnologie Karlsruhe (Eds.), *Net-condition : art and global media* (pp. 398). Cambridge, Mass: London: MIT Press.

Baxter, W., Scheib, V., Lin, M.C., & Manocha, D. (2001). Dab: interactive haptic painting with 3d virtual brushes. In *'SIGGRAPH '01: Proceedings of the 28th annual conference on Computer graphics and interactive techniques'* (pp. 461–468). ACM Press, New York, NY, USA.

Baxter, W., Wendt, J., & Lin, M. (2004). Impasto: a realistic, interactive model for paint. In *'NPAR '04: Proceedings of the 3rd international symposium on Non-photorealistic animation and rendering'* (pp. 45–148). ACM Press, New York, NY, USA.

Bazin, A. (2004). *What Is Cinema?* Berkeley, CA: University of California Press.

Begay D., & Maryboy, N. (1998). *Nanitáá S Nanitáá Sạạqh Naagháí Nanitáá Bikäeh Hózhóón, Living the Order: Dynamic Cosmic Process of Diné Cosmology.* Unpublished doctoral dissertation, California Institute for Integral Studies, San Francisco, CA.

Ben Youssef, B., Bizzocchi, J., & Bowes, J. (2005). The future of video: User experience in a large-scale, high-definition display environment. *ACM SIGCHI International Conference on Advances in Computer Entertainment Technology,* Valencia, Spain. (pp. 204-208).

Benjamin. W. (1935) *The work of art in the age of mechanical reproduction.* Originally published in Zeitschrift für Sozialforschung.

Bennis, W., & Biederman, P. W. (1997). *Organizing genius: The secrets of creative collaboration.* Cambridge, MA: Perseus Books.

Berdyaev, N.A. (1931). *Self investigation. The attempt of philosophical autobiography.* Paris, 1949 (Chapter VII)(in Russian).

Berghaus, M. (2007). Simulated Chance and Staggered Gear Ratios. *Leonardo Music Journal,* 17, 1-90.

Bertelsen, O. W., & Pold, S. (2004). *Criticism as an approach to interface aesthetics.* In Proceedings of the third Nordic conference on Human-computer interaction (pp. 23-32). New York, NY, USA: ACM Press.

Bestor, C. (2003). Installation art: image and reality. *SIGGRAPH Comput. Graph.,* 37(1), 16–18.

Binkley, T., Entis, G., Maxwell, D., & Smith, A. R. (1994). Computer technology and the artistic process: how the computer industry changes the form and function of art. In *'SIGGRAPH '94: Proceedings of the 21st annual conference on Computer graphics and interactive techniques'* (pp. 494–495). ACM Press, New York, NY, USA: Chairman-Jane Flint DeKoven.

Bishop, A. (1991). *Mathematical Enculturation: A Cultural Perspective on Mathematics.* Melbourne, Australia: Kluwer Academic Publishers.

Biswas, A., & Singh, J. (2006). *Software Engineering Challenges in New Media Applications.* In the Proceedings of Software Engineering Applications (~SEA 2006~) (pp. 7). Dallas, TX, USA: ACTA Press.

Biswas, A., Donaldson, T., Singh, J., Diamond, S., Gauthier, D., & Longford, M. (2006). *Assessment of mobile experience engine, the development toolkit for context aware mobile applications.* In Proceedings of the 2006 ACM SIGCHI international conference on Advances in computer entertainment technology (pp. 8). New York, NY, USA: ACM Press.

Bizzocchi, J. (2008). The Aesthetics of the Ambient Video Experience. *Fibreculture Journal, Issue, 11,* <http://journal.fibreculture.org/issue11/issue11_bizzocchi.html> - viewed July 1, 2008

Blackmore, J., & Miikkulainen, R. (1993). Incremental grid growing: encoding high-dimensional structure into a two-dimensional feature map. In *Neural Networks, IEEE International Conference on, 1,* 450-455).

Blais, J., & Ippolibo, J. (2006). *At the Edge of Art.* New York, NY: Thames & Hudson.

Blassnigg, M. (2005). Documentary Film at the Junction between Art and Digital Media Technologies. *Convergence-The International journal of New Media Technologies, 11*(3), 104-110.

Blauert, J. (1996). *Spatial Hearing, the Psychophysic of Human Sound Localization.* Cambridge, Mass, USA: The MIT Press Cambridge.

Blizzard (2005). Retrieved January 28, 2007 www.worldofwarcraft.com

Blizzard (2006). *Burning Crusade, release.* Retrieved January 28, 2007 http://www.blizzard.com/press/070307.shtml

Blizzard (2008). *World Of Warcraft hits 10 million*. Retrieved February 22, 2008 from http://www.mmogchart.com/2008/01/30/world-of-warcraft-hits-10-million/ and http://www.blizzard.com/us/press/080122.html

Bobbio, L. (2004). *A più voci. Amministrazioni pubbliche, imprese, associazioni e cittadini nei processi decisionali inclusivi*. Napoli: Presidenza del Consiglio dei ministri, Edizioni Scientifiche Italiane.

Boden, M. A. (2004). *The Creative Mind: Myths and Mechanisms*. Routledge; London, 2nd edition.

Bogart, B. (2008). *Memory association machine: An account of the realization and interpretation of an autonomous responsive site-specific artwork*. Master's thesis, Simon Fraser University.

Bollinger, T. (1997). The interplay of art and science in software. *Computer, 30*(10), 128, 125-127.

Bolter, J., & Grusin, R. (1999). *Remediation: understanding new media*, London; Cambridge, Mass: MIT Press.

Bolter, J.D., & Grusin, R. (2007). *Remediation: Understanding New Media*. Cambridge, Massachusetts: The MIT Press.

Boltyansky, V.G. (1984). About one parquet. *Mathematics in School, 1*, 65–66.

Bond, G. W. (2005). Software as art. *Communications of the ACM, 48*(8), 118-124.

Borges, J. L. (1998). *Collected Fictions*. New York, NY: Penguin Putnam, Inc.

Bourdieu, P. (2007). *Distinction: A Social Critique of the Judgement of Taste (*Nice, R trans).Cambridge, Ma: Harvard University Press.

Bowlt, J.E., & Long, R-C.W. (1984). *The Life of Vasilii Kandinsky in Russian art: a study of "On the spiritual in art" by Wassily Kandinsky*. Newtonville, MA: Oriental Research Partners.

Bowman, J. (1990). *Performance Art* (Publication.: http://www.bright.net/~dapoets/performa.htm

Boyd, J. E., Hushlak, G., & Jacob, C. J. (2004). *SwarmArt: interactive art from swarm intelligence*. In Proceedings of the 12th annual ACM international conference on Multimedia (pp. 628-635). New York, NY, USA: ACM Press.

Boyle, P., & Boice, B (2002). Best Practices For Enculturation: Collegiality, Mentoring, and Structure. *New Directions for Higher Education, 1998*, 87-94.

Brakhage, S. (1978). Metaphors in vision. In P. A. Sitney (Ed.), *Avant-garde film*. New York, NY: New York University Press.

Branscomb, L.M., & Auerswald, P.E. (2002, November). *NIST GCR 02–841: Between Invention and Innovation*. Retrieved February 27, 2008, from NIST Advanced Technology Program Web site: http://www.atp.nist.gov/eao/gcr02-841/contents.htm

Brettell, Sue (2007). Communicating Your Personal Brand 2007 Global TeleSummit Creating a Decade of Personal Branding (2007) *http://www.personalbrandingsummit.com*

Brintrup, A., Ramsden, J., & Tiwari, A. (2007). An interactive genetic algorithm-based framework for handling qualitative criteria in design optimization. *Computers in Industry* , (pp. 279-291).

Brogger, A. (2000). *Net art, web art, online art, net art?* (Publication. Retrieved Jun 2006: http://www.afsnitp.dk/onoff/texts.html

Brooks, R. (2004, November 2004). The other exponentials: Moore's law isn't alone. many technologies now improve so quickly it boggles the mind. [Electronic version]. *MIT Review,* Retrieved July 30, 2007 from http://www.technologyreview.com/Infotech/13863/

Brooks, R. A. (1992). Intelligence without representation. *Foundations of Artificial Intelligence, 47*, 139-159.

Brotchie, A., & Gooding, M. (Ed.). (1991). *A Book of Surrealist Games*. Boston, MA: Shambhala Publications, Inc.

Brougher, K. (2005). Visual-Music Culture. In K. Brougher, J. Strick, A. Wiseman, & J. Zilczer (Eds.), *Visual Music: Synaesthesia in Art and Music Since 1900*. New York: NY, Thames & Hudson.

Bryan-Wilson, J. (2003). Sol LeWitt. In Molesworth, H. (Ed.), *Work Ethic* (pp. 158-159). University Park, PA: The Pennsylvania State University Press.

Burgin, V. (1994). *Thinking photography*. Houndmills: Macmillan.

Burnett. K. (1993). *Toward a Theory of Hypertextual Design*. http://www.iath.virginia.edu/pmc/text-only/issue.193/burnett.193 accessed 26/02/2008

Burslem, J. (March 24, 2007). Coldwell Banker Opens Office in Second Life. *Future of Internet Marketing*. *Retrieved* June 14, 2007 from http://www.futureofrealestatemarketing.com/coldwell-banker-opens-office-in-second-life

Buxton, B. (1997). Artists and the art of the luthier. *SIGGRAPH Comput. Graph., 31*(1), 10–11.

Cajete, G. (1999). *Igniting the sparkle: An indigenous science education model*.

Call, E. (2005). White Paper: *Making Machinima in Second Life*. Retrieved January 13, 2008 from http://s3.amazonaws.com/static-secondlife-com/_files/making_Machinima.pdf

Candy, L. (1999). *COSTART Project Artists Survey Report: Preliminary Results*.: Loughborough University.

Candy, L., & Edmonds, E. (2002). *Modeling co-creativity in art and technology*. In Proceedings of the 4th conference on Creativity & cognition (pp. 134-141). New York, NY, USA: ACM Press.

Candy, L., & Edmonds, E. (Eds.). (2002). *Explorations in art and technology*. New York, NY: Springer.

Capellazzo, A. (2000). Making time: Considering time as a material in contemporary video and film. *Palm Beach Institute of Contemporary Art*, Lake Worth, FL.

Capra, F. (1947). *It's a Wonderful Life*. Writers: Van Doren Stern, P (story). Goodrich, F. (writer). Liberty Films II. United States.

Caroline, A. J. (2005). *Sensorium: Embodied experience, technology and contemporary art*. London, England: MIT Press.

Carpenter, G. A., & Grossberg, S. .(1994). *Adaptive Resonance Theory*. Boston University, Center for Adaptive Systems and Dept. of Cognitive and Neural Systems.

Carroll, L. (1893) Sylvie and Bruno Concluded. England. Macmillan & Co.

Carter, K. (1993). The Place of Story in the Study of Teaching and Teacher Education. *Educational Researcher, 22*(1), 5-12.

CASPAR. (2005). *European Project*. http://www.casparpreserves.eu

Chaitin, G.J. (1975). Randomness and mathematical proof. *Scientific American* 232(5), 47-52.

Chapple, F., & Kattenbelt, C. (2006). Key Issues in Intermediality in Theatre and Performance. In F. Chapple & C. Kattenbelt (Eds.), *Intermediality in Theatre and Performance*. New York: Rodopi.

Cheang, S. (2006). *E-mail interview with Shu Lea Chang by Yueh Hsiu Giffen Cheng*.

Chen, G. (1998). Art on the net, is not equal to net art. *The journalist*.

Chen, Y. (2002). The boundary of virtual wisdom and stratagem. *Artist Magazine*.

Chen, Y. (2004). Art relaxed. *Artist Magazine*.

Cheng, S. (2000). Traveling between virtual and reality world- Shu Lea Cheng (Publication. Retrieved Apr 2006: http://goya.bluecircus.net/archives/004358.html

Cherny, L. (1999). *Conversation and Community: Chat in a Virtual World*. Stanford, CA: CSLI Publications.

Chien, I. (2007). Comparisons: Deviation, Red vs. Blue: The Blood Gulch Chronicles. *Film Quarterly, 60*(4), 24-29.

Cho, S.-B. (2002). Towards Creative Evolutionary Systems with Interactive Genetic Algorithm. *Applied Intelligence, 16*, 129–138.

Choi, B., Lee, S., & Ko, S. (2000). New frame rate up-conversion using bi-directional motion estimation. *IEEE Transactions on Consumer Electronics, 46*(3), 603-609.

Chomsky, N. (2006). *Language and mind.* New York, NY: Cambridge University Press.

Chupeau, B., & François, C. (2000). Region-based motion estimation for content-based video coding and indexing. *SPIE Visual Communications and Image Processing, 4067,* 884-893.

Chupeau, B., & Salmon, P. (1993). Motion compensating interpolation for improved slow motion. In E. Dubois, & L. Chiariglione (Eds.), *Signal processing of HDTV* (pp. 717-724). Elsevier Science Publishers B.V.

Cohen, H. (1979). What is an image? In *Proceedings of IJCAI.*

Cohen, H. (1995). The further exploits of aaron, painter. *Stanford Humanities Review, 4,* 141-158.

CollabNet. (2000). *Subversion.* http://subversion.tigris. org.

Conover, D., Czuchra, D., & Caloyanis, N. (2004-2008). *Sunrise earth* [Television series]. United States: Compass Light Productions.

Consumers winners in HD wars. (2008, Feb. 19, 2008). *Vancouver Sun.* (pp. D3).

Cope, D. (1996). *Experiments in musical intelligence.* AR Editions.

Cornwell, R. (1996). Artists and interactivity: Fun or funambulist? In *Serious Games.* London: Barbican Art Gallery/Tyne and Wear Museums.

Cramer, F., & Gabriel, U. (2001). Software Art and Writing. *American Book Review, 22*(6).

Cytowic, R.E. (1989). *Synesthesia: A union of the senses.* New York: Springer.

Cytowic, R.E. (1996). *Synesthesia: Phenomenology and neuropsychology.* New York: Psyche.

Cytowic, R.E. (1997). Synaesthesia: Phenomenology and neuropsychology. In S. Baron-Cohen, & J. E. Harrison (Eds.), *Synaesthesia: Classic and contemporary readings* (pp. 17–39). Massachusetts: Blackwell.

D'Albergo, E., & Moini, G. (2007). Il potenziale trasformativo delle pratiche partecipative : tre casi a confronto. In E. D'Albergo, & G. Moini (Eds.), *Partecipazione, movimenti e politiche pubbliche a Roma.* Rome: Aracne.

D'Ambrosio, U. (1985). Ethnomathematics and its place in the history and pedagogy of Mathematics. *For the Learning of Mathematics, 5,* 44-48.

Damer, B. (2007). Meeting in the Ether. *ACM Interactions14*(5), 16-18.

Danks, M., Geiger, G., Zmölnig, J.M., Clepper, C., & Tittle, J.II. (1995). *Graphics environment for multimedia,* http://gem.iem.at/.

Danziger, M. (2008). *Information Visualization for the People* @ MIT Program in Comparative Media Studies Advised by Nick Montfort. Archived at: http://cms.mit. edu/research/theses/MichaelDanziger2008.pdf

Davis, F.D., Bagozzi, R.P., & Warshaw, P.R. (1989). User acceptance of computer technology: A comparison of two theoretical models. *Management Science, 35*(8), 982–1003.

De Carlo, G. (1973). *L'architettura della partecipazione.* Milano: Il Saggiatore.

De Freitas, S. (2006, October). Learning in Immersive Worlds: A Review of Game-Based Learning. *Joint Information Systems Committee e-Learning Programme.* Retrieved 18 May 2007 from http://www.jisc.ac.uk/ media/documents/programmes/elearninginnovation/ gamingreport_v3.pdf

De Landa, M. (2003). *War in the Age of Intelligent Machines.* New York: Zone Books.

De Pietro, L., & al., e. (2003). *Linee guida per la Promozione della Cittadinanza digitale: E-democracy (in Italian: Guidelines for promotion of digital citizenship: e-democracy).* FORMEZ.

Dempsey, A. (2002). *Styles, schools and movements: an encyclopaedic guide to modern art.* London: Thames & Hudson.

Dennis, D.W. (2006). The Role-Playing Game And The Game of Role-playing, the Ludic self and everyday life. In J.P. Williams, S.Q. Hendricks, & W.K. Winkler (Eds.), *Gaming as Culture: Essays on Reality, Identity, and Experience in Fantasy Games.* Jefferson, NC: McFarland Publishing.

Depocas A., Ippolito J., & Jones C. (Ed.). (2004). *Permanence Through Change: The Variable Media Approach.* New York: Guggenheim Museum Publications.

Deren, M. (1978). Cinematography: The creative use of reality. In P. A. Sitney (Ed.), *Avant-garde film* (pp. 72-73). New York, NY: New York University Press.

Dibou, A. (2007). *Speaking at a performance of Olmannen in Second Life.* February 2007.

Dickens, C. (1910). *A Christmas Carol.* Edison Manufacturing Company. (short story). United States.

Dix, Finlay, Abowd & Beale (1993). *Human–Computer Interaction, Prentice Hall, Implementation support.* Retrieved January 28, (2007) from http://www.cc.gatech.edu/computing/classes/cs4753_94_fall/slides/Lecture-13(Implementation).ps.Z

Dixon, S. (2007). *Digital Performance: A History of New Media in Theater, Dance, Performance Art and Installation.* Cambridge, Massachusetts: The MIT Press.

Donna, J. C. (1991). Interdisciplinary collaboration case study in computer graphics education: "Venus & Milo". *SIGGRAPH Computer Graphics, 25*(3), 185-190.

Douglas, K. (Ed.). (1990). *From Marxism to Postmodernism and Beyond.* Baudrillard, Jean. *The Consumer Society,* (p.109, p. 33). Stanford, CA: Stanford University Press.

Ducheneaut, N., & Moore, R.J. (2005). More than just 'XP': learning social skills in massively multiplayer online games. *A Palo Alto Research Center (PARC) journal.* Retrieved January 28, 2007 from http://www2.parc.com/csl/members/nicolas/documents/ITSE.pdf

Ducheneaut, N., & Moore, R.J., (2004). *The social side of gaming: a study of interaction patterns in a massively multiplayer online game.* Retrieved January 28, 2007

www.parc.xerox.com/research/publications/files/5223.pdf

Ducheneaut, N., Yee, N., Nickell, E., & Moore, R.J. (2006). Alone Together? Exploring the Social Dynamics of Massively Multiplayer Games. *In conference proceedings on human factors in computing systems.* Retrieved January 28, 2007 from http://www.nickyee.com/pubs/Ducheneaut,%20Yee,%20Nickell,%20Moore%20%20Alone%20Together%20(2006).pdf

Ducheneaut, N.m Moore, R.J., & Nickell, E. (2007). *Virtual "third places": A case study of sociability in massively multiplayer games.* Computer Supported Cooperative Work.

Dulac, G. (1978). Visual and anti-visual films. In P. A. Sitney (Ed.), *Avant-garde film* (pp. 33-35). New York, NY: New York University Press.

Dunbar, R.I.M. (1992). Neocortex size as a constraint on group size in primates. *Journal of Human Evolution, 22,* 469-493.

Dune III (2002). Retrieved January 28, 2007 http://www.dune3.net/

Dusinberre, D. (1975). *Avant-garde british landscape films (introduction to programme notes).* London: Tate Gallery.

Ebert, D. S., & Bailey, D. (2000). A collaborative and interdisciplinary computer animation course. *ACM SIGGRAPH Computer Graphics, 34*(3), 22-26.

Edmonds, E., Turner, G., & Candy, L. (2004). *Approaches to interactive art systems.* In Proceedings of the 2nd international conference on Computer graphics and interactive techniques in Australasia and South East Asia (pp. 113-117). New York, NY, USA: ACM Press.

Edmonds, E., Turner, G., & Candy, L. (2004). Approaches to interactive art systems. In '*GRAPHITE '04: Proceedings of the 2nd international conference on Computer graphics and interactive techniques in Australasia and South East Asia*' (pp. 113–117). ACM Press, New York, NY, USA.

Edwards, B. (1986). *Drawing on the Artist Within: A Guide to Innovation, Invention, Imagination and Creativity.* New York, Ny: Simon & Schuster.

Eglash, R., Bennett, A., O'Donnell, C., Jennings, S., & Cintorino, M. (2006). Culturally situated design tools: Ethnocomputing from field site to classroom. *American Anthropologist, 108*(2), 347-362.

Eglash, R., Croissant, J., Di Chiro, G., & Fouché, R. (Ed.) (2004). *Appropriating technology: Vernacular science and social power.* Minneapolis, MN: University of Minnesota Press.

Eiben, A., & Smith, J. (2003). *Introduction to Evolutionary Computing.* Springer Verlag.

Eisenstein, S. (1969). *Film Form: Essays in Film Theory.* Washington, PA: Harvest Books.

Eisenstein, S. (1969). *The Film Sense.* New York, NY: Harcourt Books.

Eisenstein, S. (2006). *The Eisenstein Collection.* Oxford, England. Seagull Books.

Eno, B. (1978). *Music for airports (album liner notes)* (PVC 7908 (AMB001) ed.)

Ernest, P. (1985). *Social Constructivism as a Philosophy of mathematics.* Albany: State University of New York Press.

Esther, L. (2000). *Walter Benjamin electronic resource: overpowering conformism.*

Ethnomathematics. Oxford, UK: Routledge Publishing.

Fauconnier, G., & Turner, M. (2002). *The way we think: Conceptual blending and the mind's hidden complexities.* New York: Basic Books.

Feddersen, W.E., Sandel, T.T., Teas D.C., & Jeffress, L.A. (1957). Localization of high-frequency tones. *Journal of the Acoustical Society of America, 29*, 988-991.

Fels, S., Kinoshita, Y., Tzu-pei Grace, C., Takama, Y., Yohanan, S., Gadd, A., et al. (2005). Swimming across the Pacific: a VR swimming interface. *Computer Graphics and Applications, IEEE, 25*(1), 24-31.

FFRC (n.d.) SL Foundation for Rich Content. Retrieved February 8, 2008 from http://groups.google.com/group/sl-ffrc

Figgis, M. (2000). *Timecode.* Red Mullet Productions. United States.

Filiciak, M. (2003). Hyperidentities: Postmodern identity patterns in massively multiplayer online role-playing games. In M.J.P. Wolf & B. Perron (Eds.), *The video game theory reader.* New York: Routledge.

Fishwick, P. (2003). Nurturing next-generation computer scientists. *Computer, 36*(12), 132-134.

Fishwick, P. (2005). Enhancing experiential and subjective qualities of discrete structure representations with aesthetic computing. *Journal of Visual Languages & Computing, 16*(5), 406-427.

Fishwick, P. A. (2007). Aesthetic Computing: A Brief Tutorial. In F. Ferri (Ed.), *Visual Languages for Interactive Computing: Definitions and Formalizations*: Idea Group Inc.

Fishwick, P., Davis, T., & Douglas, J. (2005). Model representation with aesthetic computing: Method and empirical study. *ACM Trans. Model. Comput. Simul., 15*(3), 254-279.

Fitzmaurice, G.W., Balakrishnan, R., Kurtenbach, G., & Buxton, B. (1999). An exploration into supporting artwork orientation in the user interface. In *'CHI '99: Proceedings of the SIGCHI conference on Human factors in computing systems'* (pp. 167–174). New York, NY, USA: ACM Press

Flagg, M., & Rehg, J.M. (2005). *Oil painting assistance using projected light: Bridging the gap between digital and physical art.* Gvu technical report git-gvu-05-35. Georgia Institute of Technology.

Foo, C.Y. (2004). *Grief Player Motivations.* Curtin University, Nokia Research center. Retrieved January 28, 2007 from http://www.tu-chemnitz.de/phil/medkom/mn/spive/index.php?option=com_docman&task=doc_download&gid=25

Foote, J. (2003). Net Art: A New Voice in Art. Challenging Perceptions of the Virtual and Physical (Publication., from History & Philosophy of Mass Media Final Paper: http://babel.massart.edu/~jfoote/netartpaper.html

Forbes, S., & Ashton, P. (1998). The identity status of African Americans in middle adolescence: a reexamination of Watson and Protinsky - 1991 - response to M.F. Watson and H. Protinsky. *Adolescence, 26,* 963.

Ford, D. (2007). Virtual Limitations: A Comparison of SIMS 2 and Half Life Games Engines for Machinima Narrative. *ACM International Conference Proceeding Series: Vol. 274. 2nd International Conference on Digital Interactive Media in Entertainment and Arts (DIMEA).*

Fowles, J. (1992). *Why viewers watch: A reappraisal of television's effects* (Revised ed.). Newbury Park, CA: Sage Publications.

Freud, S. (1990). *Beyond the pleasure principle.* New York: W. W. Norton & Company.

Fritzke, B. (1991). Unsupervised clustering with growing cell structures. In *Neural Networks. IJCNN-91-Seattle International Joint Conference on, 2,* 531-536.

Fry, B. (2000). *Organic Information Design.* Masters Thesis @ MIT Program in Media Arts and Sciences Advised by John Maeda.

Fry, B. (2004). *Computational Information Design.* Dissertation @ MIT Program in Media Arts and Sciences Advised by John Maeda.

Gabora, L.M. (2002). Cognitive mechanisms underlying the creative process. In T. Hewett & T. Kavanagh, (Eds.), *Proceedings of the Fourth International Conference on Creativity and Cognition,* (pp. 126-133).

Gadamer, H. (1993). *Truth and method* (2nd rev. ed.). London: Sheed and Ward.

Gaertner, M. (2000). In H. Friedel, S. Gaensheimer & U. Wilmes (Eds.), *Moments in time: On narration and slowness* (). Stuttgart: Cantz Editions.

Gamma, E., Helm, R., Johnson, R., & Vlissides, J. (1995). *Design Patterns: Elements of Reusable Object-Oriented Software.* Boston, MA, USA: Addison-Wesley.

Gartner media (2007). *Gartner Symposium/ITxpo 2007 Emerging Trends.* Retrieved January 28, 2007 from http://www.businesswire.com/portal/site/google/index.jsp?ndmViewId=news_view&newsId=20070424006287&newsLang=en

Garvey, G. P. (1997). Retrofitting fine art and design education in the age of computer technology. *ACM SIGGRAPH Computer Graphics, 31*(3), 29-32.

Gerdes, P. (1988). On culture, geometrical thinking and mathematics education. *Educational Studies in Mathematics, 19,* 137-162.

Gerzon, M. (1974). Periphony: With-height sound reproduction. *Journal of the Audio Engineering Society, 21*(1/2), 2-10.

Giangrande, A., & Mortola, E. (2000). *Progettare con la comunità.* Università Roma Tre (DiPSA) and Comune di Roma (USPEL).

Gibson J.J. (1966). *The Senses Considered as Perceptual Systems.* Boston: Houghton Mifflin Company.

Global Ergonomic Technologies (1998). *Comparison of Postures from Pen and Mouse Use.* Guerneville, CA, U.S.A. URL: http://www.wacom.com

Goldberg, D.E. (1989). *Genetic algorithms in search, optimization and machine learning.* Addison-Wesley.

Goldberg, D.E. (1999). *Genetic and Evolutionary Algorithms in the Real World.* IlliGAL Report No 99013, University of Illinois at Urbana, Department of General Engineering.

Goldberg, R. (1988). *Performance art : from futurism to the present* (Rev. and enl. ed.). London: Thames and Hudson.

Goldsworthy, A. (2007). *Enclosure.* New York, NY: Abrams.

Goodman, N. (1978). *Ways of Worldmaking.* Indianapolis, IN: Hackett Publishing Company, Inc.

Gorder, P. F. (2007). Multicore processors for science and engineering. *IEEE Computing in Science & Engineering, 9*(2), 3-7.

Grau, O. (2003). *Virtual Art – From Illusion to Immersion.* Cambridge, Massachusetts: The MIT Press.

Grau, O. (2006). *MEDIA ART HISTORIES.* Cambridge: The MIT Press.

Grau, O. (Ed.). (2007). *Media Art Histories.* Cambridge, Ma.: M.I.T. Press.

Greene R. (2005). *Internet Art.* London: Thames & Hudson Ltd.

Gregory, A., Ehmann, S. & Lin, M. (2000). inTouch: interactive multiresolution modeling and 3d painting with a haptic interface. In *Virtual Reality* (pp. 45–52). IEEE, New Brunswick, NJ.

Gregory, R.L. (1998) Eye and Brain, The Psychology of Seeing. Oxford, England.

Grey, J. (2002). *"Human-computer interaction in life drawing, a fine artist's perspective".* In Sixth International Conference on Information Visualisation (IV'02) (pp. 761-770). Los Alamitos, CA, USA: IEEE Computer Society.

Grice, H.P. (1962). Some Remarks About the Senses. In R. J. Butler (Ed.), *Analytical Philosophy, First Series.* Oxford: Basil Blackwell

Griffin-Pierce, T. (1992). The Hooghan and the Stars. In R. A. Williamson, & C. R. Farrer (Eds.), *Earth and sky: Visions of the cosmos in Native American folklore* (pp. 110-130). Albuquerque: University of New Mexico Press.

Gross, J. B. (2005). *Programming for artists: a visual language for expressive lighting design.* In IEEE Symposium on Visual Languages and Human-Centric Computing, 2005 (pp. 331-332). Los Alamitos, CA, USA: IEEE Computer Society.

Guo, G., Gong, D., Hao, G., & Zhang, Y. (2006). Interactive Genetic Algorithms with Fitness Adjustment. *Journal of China University of Mining and Technology , 16*(4), 480-484.

Gustav, T. (2002). *Guilds: Communities in Ultima Online.* Retrieved January 28, 2007 http://cid.nada.kth.se/pdf/CID-167.pdf

Haahr, M. (1998-2008). *Introduction to Randomness and Random Numbers.* Retrieved January 15, 2008, from http://random.org/randomness/

Haberland, W. (1986). Aesthetics in native American art. In E. Wade (Ed.), *The arts of the North American Indian: Native traditions in evolution.* New York, NY: Hudson Hills Press Inc.

Hajo, D. (2000). *Wassily Kandinsky 1866–1944: A Revolution in Painting.* New York: Taschen.

Halonen, K. (2007). Open Source and New Media Artists. *Human Technology - An interdisciplinary journal on humans in ICT environments, 3*(1), 98-114.

Hammershøi, D. (1995). *Binaural Technique: a Method of True 3D Sound Reproduction.* PhD thesis, Aalborg Universitetsforlag, Denmark.

Hao, G., Gong, D., & Huang, Y. (2006). Interactive Genetic Algorithms Based on Estimation of User' s Most Satisfactory Individuals. *Intelligent Systems Design and Applications, 2006. ISDA'06. Sixth International Conference on, 3.*

Harris, C. (Ed.). (1999). *Art and innovation: the Xerox PARC Artist-in-Residence program.* Cambridge, Massachusetts: MIT Press.

Hartas, L. (2005). *The art of game characters.* Lewes, East Sussex: Ilex Press.

Hayes, B. (2001). Randomness as a Resource. *American Scientist, 89*(4), 300-304. Research Triangle Park, NC: Sigma Xi.

Hayles, N.K. (1990). *Chaos Bound: Orderly Disorder in Contemporary Literature and Science.* Ithaca, NY: Cornell University Press.

Hayles, N.K. (1994). Chance Operations: Cagean Paradox and Contemporary Science. In M. Perloff, & C. Junkerman, (Eds.), *John Cage: Composed in America* (pp. 226-241). Chicago, IL: University of Chicago Press.

Heath, T. L. (Trans.). (1956). *The Thirteen Books of Euclid's Elements, Books 1 and 2.* Euclid. Mineola, NY: Dover Publications.

Helander, M. (2005). *A guide to human factors and ergonomics.* Taylor and Francis.

Held, G., & Marshall, T. R. (1991). *Data compression: Techniques and applications: Hardware and software considerations* John Wiley & Sons, Inc.

Heliö, S. (2004). Role-Playing: A Narrative Experience and a Mindset. In M. Montola & J. Stenros (Eds.), *Beyond Role and Play,* (pp. 65-74). Solmukohta. Vantaa, Ropecon. Retrieved January 28, 2007 from http://www.ropecon.fi/brap/ch6.pdf

Hemmi, K. (2006). *Approaching proof in a Community of Mathematical Practice.* Doctoral Dissertation, mathematical Department, Stockholm University.

Henzen, A., Ailenei, N., Fiore, F.D., Reeth, F.V., & Patterson, J. (2005). Sketching with a low-latency electronic ink drawing tablet. In *GRAPHITE '05: Proceedings of the 3rd international conference on Computer graphics and interactive techniques in Australasia and South East Asia'* (pp. 51–60). New York, NY, USA: ACM Press.

Hermida, A. (2005). *Sony shows off new PlayStation 3.* Last updated May 17, 2005 on BBC News website, http://news.bbc.co.uk/2/hi/technology/4554025.stm

Hoffmann, R., & Krauss, K. (2004). *A critical evaluation of literature on visual aesthetics for the web.* In Proceedings of the 2004 annual research conference of the South African institute of computer scientists and information technologists on IT research in developing countries (pp. 205-209). South Africa: South African Institute for Computer Scientists and Information Technologists.

Holloway, J. (2005). *Promise, paradox and opportunity.* URL: http://moca.virtual.museum/editorial/holloway.htm

Hoover, C. W., & Jones, J. B. (1991). *Improving Engineering Design, Designing for Competitive Advantage.* The National Research Council, Washington D.C.: National Academy Press.

Hotger, D. (June 6, 2007). Broadway meets Second Life. *Derek's SL Report.* Retrieved June 27, 2007 from http://derekhotger.blogg.com/2007/06/06/broadway-meets-second-life/

Hu, S. (2005). Experiencing the Climax of communication. In *Climax- The Highlight of Ars Electroniza* (pp. 204). Taipei: National Taiwan Museum of Fine Arts.

Hunter J., & Choudhury Sh. (2006). PANIC – an integrated approach to the preservation of complex digital objects using semantic web services. *International Journal on Digital Libraries: Special Issue on Complex Digital Objects, 6*(2), 174-183.

Huskey, P., & Korn, G. (Eds.). (1962). *Computer Handbook.* New York, NY: McGraw-Hill Book Co.

Hwang, S., & Park, Y. (2006). *Time delay estimation from HRTFs and HRIRs.* Paper presented at the 8th International Conference of Motion and Vibration Control (MOVIC 2006), Kaist, Daejeon, Korea.

IEEE Visualization. (2006). Call for participation. In *IEEE Symposium on Visual Analytics Science and Technology 2006* . Retrieved 11/14/2007 from http://conferences.computer.org/vast/vast2006/

Inge, T. M. (2001). Theories and Methodologies Collaboration and Concepts of Authorship. *PMLA, 116*(3), 623-630.

Introduction to World of Warcraft. *World of Warcraft Game Guide.* Retrieved June 15, 2007 from http://www.worldofwarcraft.com/info/beginners/index.html.

Ixchel, Anya (n.d.). All the (Virtual) Worlds a Stage. *Slate Magazine.* Retrieved 21 June 2007 from http://www.slatenight.com/index.php?option=com_content&task=view&id=75&Itemid=40

Jaccheri, M. L., & Sindre, G. (2007). *Software Engineering Students meet Interdisciplinary Project work and Art.* In Proceedings of the 11th International Conference on Information Visualisation (pp. 925--934). Washington, DC, USA: IEEE Computer Society.

Jakobsson, M., & Taylor, T.L. (2003). The Sopranos Meets EverQuest Social Networking in Massively Multiplayer

Online Games. *fineArt forum, 17*(8). Retrieved January 28, 2007 http://hypertext.rmit.edu.au/dac/papers/Jakobsson.pdf

Jamieson, H. V. UpStage: An open source venue for online performance. Retrieved May 17, 2008 from http://upstage.org.nz/blog/

Jamieson, Helen Varley. "Cyberformance." Retrieved May 17, 2008 from http://www.cyberformance.org/

Jankowich, A. (2006). EULAw: The Complex Web of Corporate Rule-Making in Virtual Worlds. *Tulane Journal of Technology & Intellectual Property, 8.*

Jenkins, H. (2000). Keynote address. Paper presented at the *Computers and Videogames Come of Age,* Cambridge, MA. Retrieved February 27, 2008 from http://web.mit.edu/cms/games/opening.html

Jennings, P., Giaccardi, E., & Wesolkowska, M. (2006). *About face interface: creative engagement in the new media arts and HCI.* In CHI '06 extended abstracts on Human factors in computing systems (pp. 1663-1666). New York, NY, USA: ACM Press.

Johnson, S. (1997). *Interface Culture: How the Digital Medium - From Windows to the Web - Changes the way We Write, Speak.* New York: Harper Collins.

John-Steiner, V. (2000). *Creative collaboration.* New York, NY: Oxford University Press.

Johnston, E., & Hicks, D. (2004). Speaking in teams: Motivating a pattern language for collaboration. *Interdisciplinary Description of Complex Systems , 2*(2), 136 – 143.

Jones, S. (2005). *A cultural systems approach to collaboration in art & technology.* In Proceedings of the 5th conference on Creativity & cognition (pp. 76--85). New York, NY, USA: ACM Press.

Jütte, R. (2005). *A History of the Senses: From Antiquity to Cyberspace.* Cambridge: Polity Press.

Kahn, D. (2001). The Sound of Music. In D. Kahn (Ed.), *Noise Water Meat: A History of Sound in the Arts.* New York, London: The MIT Press.

Kalish, M.K. (2007). *Navajo immersion mathematics: Culturally grounded 5th grade mathematics curricular and pedagogical materials study.* Unpublished doctoral dissertation, New Mexico State University.

Kamii, C. & Dominick, A. (1998). The harmful effects of algorithms in grades 1-4. In

Kandinsky, W. (1963). *Concerning the Spiritual in Art: And Painting in Particular.* New York: G. Wittenborn.

Kandinsky, W. (1979) *Point and Line To Plane.* Mineola, NY: Dover Publications.

Kang, S. B., Uyttendaele, M., Winder, S., & Szeliski, R. (2003). High dynamic range video. *ACM Transactions on Graphics, 22*(3), 319-325.

Karmel, P. (Ed.). (1999). *Jackson Pollock: Interviews, Articles, and Reviews.* New York, NY: The Museum of Modern Art.

Katz, F.G.B. (1996). New approach for obtaining individualized head-related transfer functions. *Journal of the Acoustical Society of America, 100,* 2609.

Kaye, N. (2000). *Site-specific Art: Performance, Place and Documentation.* Routledge,. ISBN 0415185599.

Kazin, A. (Ed.). (1977). Blake, William. *The Portable William Blake.* (p.150). New York, NY: Penguin Books.

Kelly, K. (1995). *Out of Control.* New York, NY: Perseus Books.

Kelvin, L. (William Thomson). (1901). Nineteenth century clouds over the dynamical theory of heat and light. *The London, Edinburgh and Dublin Philosophical Magazine and Journal of Science, Series 6, 2, 1–40.*

Kesseler, A., & Bergs, A. (Ed.). (2003). *New Media Language.* New York, NY:

Ketner, J., Herbert, L., & Volk, G. (2002). *Roxy Paine: Second Nature.* Houston, TX: Contemporary Arts Museum, & Waltham, MA: The Rose Art Museum, Brandeis University.

Khan, D. (Ed.). (1999). *Noise Water Meat.* Camrdige, Ma: M.I.T. Press.

Khan, M.R., & Al-Ansari, M. (2005). Sustainable Innovation as a Corporate Strategy. *Triz-Journal, 2-1*, Retrieved February 27, 2008, from http://www.triz-journal.com/archives/2005/01/02.pdf

King, M. (2007, December 5). Virtual Worlds—It's time to take out the intellectual trash. *ON LINE Opinion*. Retrieved December 7, 2007 from http://www.online-opinion.com.au/view.asp?article=6714

Kistler, D., & Wightman, F. (1992). A model of head-related transfer functions based on principal components analysis and minimum-phase reconstruction. *Journal of the Acoustical Society of America, 91*, 1637-1647.

Kitchenham, B. (2004). *Procedures for Performing Systematic Reviews*: Keele University Technical Report TR/SE-0401 and NICTA Technical Report 0400011T.1.

Klastrup, L., & Tosca, S. (2004). Transmedial worlds - rethinking cyberworld design. In *Proceedings International Conference on Cyberworlds 2004*. IEEEE Compuater Society, Los Alamitos, California, 2004. Retrieved January 28, 2007 from http://www.itu.dk/people/klastrup/klastruptosca_transworlds.pdf

Klinker, S. (2007). Spinning Form: How to Tell Stories with Product Design. Taken from Core 77 Design Student Guide: http://www.core77.com/hack2school/klinker.asp

Kluszczynski, R. W. (2005). Arts, Media, Cultures: Histories of Hybridisation. *Convergence-The International Journal of New Media Technologies, 11*(4), 124-132.

Knuth, D. (1981). *The Art of Computer Programming, Volume 2: Seminumerical Algorithms (2nd Edition)*. Reading, MA: Addison-Wesley Publishing Company.

Koch, S. (1978). Andy warhol's silence. In P. A. Sitney (Ed.), *Avant-garde film* (p. 165). New York, NY: New York University Press.

Kohonen, T. (2001). *Self-Organizing Maps.* Springer. ISBN 3540679219.

Kraaijveld, M.A. (1992). A non-linear projection method based on kohonen's topology preserving maps. Pattern Recognition, II. *Conference B: Pattern Recognition Methodology and Systems, Proceedings, 11th IAPR International Conference on*, (pp. 41-45), Aug-3 Sep. doi: rm10.1109/ICPR.1992.201718.

Kracauer, S. (1997). *Theory of Film.* Princeton, NJ: Princeton University Press.

Kubey, R., & Csikszentmihalyi, M. (1990). *Television and the quality of life: How viewing shapes everyday experience*. Hillsdale, NJ: Lawrence Erlbaum Associates.

Kurtz, L.A. (2007). Copyright, Creativity, Catalogs: Copyright and the Human Condition. *U.C. Davis Law Review 40*(3);

Kwon, M. (2004). *One Place After Another: Site-Specific Art and Locational Identity.* MIT Press. ISBN 026261202X.

L. J. Morrow & M. J. Kenney (Eds.), *The Teaching and Learning of*

Landow, G. P. (1997). *Hypertext 2.0* (Rev., Amplified, ed.). Baltimore: Johns Hopkins University Press.

Lane, B. (1982) Stoscereopic Displays,"Processing and Display of Three-Dimensional Data". SPIE Proc. 367, 1982, pp. 20-32.

Lanier, J. (2003). One half a manifesto. In J. Brockman (Ed.), *The new humanists: Science at the edge* (pp. 233-262). New York, NY: Barnes and Noble.

Lash, S. (1991). *Post-structuralist and post-modernist sociology.* Aldershot, England ; Brookfield, Vt., USA: E.Elgar Pub.

Lau, A., & Vande Moore, A. (*2007). Towards a Model of Information Aesthetics in Information Visualization.*

Laurel, B. (1993). *Computers as Theatre.* Reading, Mass.: Addison-Wesley Professional.

Lazzaro, N. (2004). *Why We Play Games: Four Keys to More Emotion Without Story*. Retrieved January 28, 2007 from http://xeodesign.com/xeodesign_whyweplaygames.pdf

Le Corbusier & Ozenfant (1920). Purism. *L'Esprit Nouveau, 4*, 369-386.

Lee, S., Yang, S., Jung, Y., & Park, R. (2002). Adaptive motion-compensated interpolation for frame rate up-conversion. *IEEE Transactions on Consumer Electronics, 48*(3), 444-450.

Legrady, G., & Honkela, T. (2002). *Pockets full of memories: an interactive museum installation. Visual Communication, 1*(2), 163-169. http://vcj.sagepub.com/cgi/content/abstract/1/2/163.

Lem, S. (1974). *The Cyberiad: Fables for the Cybernetic Age* (M. Kandel, Trans). New York, NY: Harcourt Brace Jovanovich.

Lenarcic, J (2004). Behavioral Issues in Software Development: The Evolution of a New Course Dealing with the Psychology of Computer Programming. *Journal of Issues in Informing Science and Information Technology*, (pp. 247-252).

Leth, J. (1967). *Perfekte menneske, Det* (The Perfect Human). Demark. Laterna Film.

Levaco, R., (Ed.), (1975). *Kuleshov on Film: Writings by Lev Kuleshov*. Kuleshov, L. Berkley, CA: University of California Press.

Levy, D. (2001). *Scrolling Forward: Making Sense of Documents in the Digital Age*. New York: Arcade Publications.

Lin, x. (2006). 2006 Web 100. *Next Publishing Corp.*

Linaza, T. (2003). *Artnouveau project: Recommendations and Generic Framework*. Brussels: European Commission, Project ID: artnouveau IST-2001-37863, Deliverable ID: D.5.

Linden, P. (January 8, 2007). *Embracing the Inevitable*. Retrieved December 30, 2007 from http://blog.secondlife.com/2007/01/08/embracing-the-inevitable/

Lipka, J. (1998). Expanding curricular and pedagogical possibilities: Yup'ik-Based mathematics, science, and literacy. In J. Lipka, G. V. Mohatt and the Ciulistet Group (Eds.), *Transforming the culture of schools: Yup'ik Eskimo examples* (pp. 139-181). Mahwah, NJ: Erlbaum.

Lipset, S.M. (1963). The Value Patterns of Democracy: A Case Study in Comparative Analysis. *American Sociological Review , 28*(4), 515-531.

Liu, A. (2004). *The Laws of Cool: Knowledge Work and the Culture of Information*. Chicago: University of Chicago Press.

Llorà, X., Sastry, K., & Alìas, F. (July 8–12, 2006). Analyzing Active Interactive Genetic Algorithms using Visual Analytics. *ACM GECCO '06*. Seattle, Washington, USA: ACM.

Löwgren, J., & Stolterman, E. (2007). *Thoughtful interaction design – a design perspective on information technology*. Cambridge, Massachusetts: The MIT Press.

Machin, C. H. C. (2002). *Digital artworks: bridging the technology gap*. In Proceedings of the 20th Eurographics UK Conference, 2002. (pp. 16-23). Washington, DC, USA: IEEE Computer Society.

MacKenzie, I.S., Sellen, A., & Buxton, W.A.S. (1991). A comparison of input devices in element pointing and dragging tasks. In *CHI '91: Proceedings of the SIGCHI conference on Human factors in computing systems* (pp. 161–166). New York, NY, USA: ACM Press.

Maeda, J. (2000). *Maeda @ Media*. New York: Rizzoli.

Maeda, J. (2007). *Aesthetics and Computation Group*. http://acg.media.mit.edu (accessed May 2007).

Maeda, J. (Ed.) (n.d.). *ACG Concepts Volume 01: Elements of Reactive Form*. Retrieved March, 2008, from http://acg.media.mit.edu/concepts/volume01.html

Mahemoff, M. (2006). *Ajax Design Patterns*. O'Reilly Media.

Mamykina, L., Candy, L., & Edmonds, E. (2002). Collaborative creativity. *Communications of the ACM, 45*(10), 96-99.

Manovich, L. (2000). *The language of new media*. Cambridge, Mass.; London: MIT Press.

Manovich, L. (2005). The Shape of Information. http://manovich.com/DOCS/IA_Domus_3.doc

Manovich, L. (n.d.). After Effects, or Invisible Revolution. Presented at Danube Telelectures #4 *Remixing Cinema.* November 8th, 2007. http://www.donau-uni. ac.at/en/department/bildwissenschaft/veranstaltungen/ telelectures/archiv/index.php

Manovich, L. (n.d.). Data Visualization as new Abstraction and Anti-Sublime (2002) http://manovich.com/ DOCS/data_art_2.doc.

Marchese, F. T. (2006). *The Making of Trigger and the Agile Engineering of Artist-Scientist Collaboration.* In Proceedings of the conference on Information Visualization (pp. 839-844). Washington, DC, USA: IEEE Computer Society.

Marcos, A. (2007). Digital Art: When artistic and cultural muse and computer technology merge. *IEEE Computer Graphics and Applications, 5*(27), 98-103.

Marcus, G. (1990). *Lipstick Traces: A Secret History of the Twentieth Century.* Cambridge, Ma: Harvard University Press.

Marcus, T.D. (2007). Fostering Creativity in Virtual Worlds: Easing the Restrictiveness of Copyright for User-Created Content. *New York Law School Law Review, 52,* 67-92.

Marks, U. L. (2000). *The Skin of the Film: Intercultural Cinema, Embodiment, and the Senses.* Durham, NC: Duke University Press.

McConnell, S. (1998). The art, science, and engineering of software development. *IEEE Software, 15*(1), 120, 118-119.

McCullough, M. (1998). *Abstracting craft : the practiced digital hand/ Malcolm Mc-Cullough.* London, Cambridge, Mass: MIT.

McLuhan, M. (1994). *Understanding Media.* Cambridge, Ma: MIT Press.

Mcluhan, M. (1994). *Understanding Media: The Extensions of Man.* Boston, Mass: MIT Press.

McLuhan, M. (2007). Visual and Acoustic Space. In C. Cox & D. Warner (Eds.), *Audio Culture: Readings in Modern Music* (pp. 67-72). New York, London: Continuum.

Medler, D.A. (1998). A brief history of connectionism. *Neural Computing Surveys, 1,* 61-101.

Menegat, R. (2002). Participatory democracy and sustainable development: integrated urban environmental management in Porto Alegre, Brazil. *Environment & Urbanization , 14*(2).

Merleau-Ponty, M. (1969). The Visible and the Invisible, trans. Evanston: Northwestern University Press.

Metz, C. (1990). *Film Language: A Semiotics of the Cinema.* Chicago, Il: University of Chicago Press.

Meyer, J., Staples, L., Minneman, S., Naimark, M., & Glassner, A. (1998). *Artists and technologists working together (panel).* In Proceedings of the 11th annual ACM symposium on User interface software and technology (pp. 67-69). New York, NY, USA: ACM Press.

Millions of Us (August 29, 2006). Performing a play in Second Life. Retrieved June 10, 2007 from http://www. youtube.com/watch?v=6cWF438HgJA

Millions of Us blog (August 25, 2006(Retrieved June 21, 2007 from http://millionsofus.com/blog/archives/28

Minor, V.H. (1994). *Art History's History.* Saddle River, NJ: Prentice Hall.

Mitcham, C. (1998). *The Importance of Philosophy to Engineering.* University Park, PA, Penn State University: STS Program.

Mitchell, T. (March 12, 1997). *A Place for Souls.* Retrieved June 30, 2007 from http://moo.hawaii.edu:7000/4008/

Mitry, J. (2000). *The Aesthetics and Psychology of the Cinema.* London, England: Athlone Press.

Mizrach, S. Talking pomo: An analysis of the postmodern movement (Publication. Retrieved 6 Feb 2007, from Florida International University: http://www.fiu. edu/~mizrachs/academentia.html

Mmogchart.com (2008). Retrieved April 5, 2008 from http://www.mmogchart.com/Chart7.html

Møller, H., Sørensen, M.F., Jensen, C.B., & Hammershøi, D. (1996). Binaural Technique: Do We Need Individual Recordings? *Journal of the Audio Engineering Society*, *44*(6), 451/469.

Moore, B.C.J. (2003). *An Introduction to the Psychology of Hearing*. London, UK: Academic Press.

Morgan, J., & Muir, G. (2004). *Time zones: Recent film and video*. London, UK: Tate Publishing.

Morgan, M. (2003). The Space Between Our Ears. Weidenfield & Nicholson.

Moritz, W. (2007). Retrieved 1/2007, 2007, from http://www.iotacenter.org/visualmusic/articles/moritz

Morris, S. (2001). Museums & New Media Art- A research report commissioned by The Rockefeller Foundation. [Electronic Version] from http://www.cs.vu.nl/~eliens/onderwijs/multimedia/mma1/college/@archive/refs/Museums_and_New_Media_Art.pdf

Mortola, E., Fortuzzi, A., & Mirabelli, P. (1995). Communications Project of Designing with Multimedia Interactive Tools. *ECAADE Conference Proceedings, Multimedia and Architectural Disciplines*. Palermo, Italy.

Mullet, K. (2003, November). *The Essence of Effective Rich Internet Applications*. Macromedia Experience Design Team.

Mulligan, J. (2002). *Biting The Hand #17: Talkin'' bout my... Generation*. Retrieved January 28 2007 http://www.skotos.net/articles/BTH_17.shtml

Murphy, E. (2004). Recognizing and promoting collaboration in an online asynchronous discussion. *British Journal of Educational Technology, 35*(4), 421-431. Retrieved January 28, 2007 from http://grail.oise.utoronto.ca/blog/karaisko/files/2007/12/murphy_recognizing_and_promoting_collaboration.pdf

Muybridge, E. (1955). *The Human Figure in Motion*. Mineola, Dover Books.

Nalder, G. (2003). *Art in the Informational Mode*. In Proceedings of the Seventh International Conference on Information Visualization (pp. 110). Washington, D.C, USA: IEEE Computer Society.

Nephew, M. (2006). Playing with identity: unconscious desire and Role-Playing Games. In J.P. Williams, S.Q. Hendricks, & W.K. Winkler (Eds.), *Gaming as Culture: Essays on Reality, Identity, and Experience in Fantasy Games*. Jefferson, NC: McFarland Publishing

New Media Consortium (2007). Retrieved June 15, 2007 from http://www.nmc.org/

Nielsen, J. (1993, November). *Iterative Design of User Interfaces. IEEE Computer, 26*(11), 32-41.

Noguchi, H. (1990). Mozart - Musical Game in C K. 516f. *Mitteilungen der International Stiftung Mozarteum, 38*(1-4), pp. 89-101.

Nolan, C. (2000). *Memento*. Nolan, J. (short story Memento Mori) Nolan, C. (screenplay). France: Newmarket Capital Group.

Norman, D. A. (2003). Emotional *Design: Why We Love (Or Hate) Everyday Things*. New York, NY: Basic Books.

Norman, D.A. (2002). *The Design of Everyday Things*. New York: Basic books.

Núñez, R.E., & Sweetser E. (2006). Aymara, where the future is behind you: Convergent evidence from language and gesture in the crosslinguistic comparison of spatial construals of time. *Cognitive Science, 30*(2006), 401-450.

Nussbaum, B. (2006). *The Best Product Design of 2006*. Taken from BusinessWeek Magazine *http://www.businessweek.com/innovate/content/jun2006/* ©2008 The McGraw-Hill Companies Inc.

O'Riley. T. (2006). Thinking Through Art, Reflections on Art as Research, (p. 94). Routledge.

OAIS - Open Archive Information System (2002). http://ssdoo.gsfc.nasa.gov/nost/isoas/us/overview.html.

Oates, B.J. (2006). New frontiers for information systems research: computer art as an information system. *European Journal of Information Systems, 15*(6), 617-626.

OECD, Organization for Economic Co-Operation and Development. (2006). *Citizens as Partners: Information, Consultation and Public Participation in Policy-making.* OECD.

OECD, Organization for Economic Co-Operation and Development. (2003). *Promise and Problems of E-Democracy.* OECD.

Ondrejka, C. (2004). *Escaping the Gilded cage: User Created Content and Building the Metaverse.* Linden Research, Inc., San Francisco. B.S. United States Naval Academy. Retrieved January 28, 2007 from http://www. nyls.edu/pdfs/v49n1p81-101.pdf

Oppenheimer, R. (2007). The conversation continues: When artists and engineers first collaborated. In P. Jennings (Ed.), *Speculative data and the creative imaginary: Shared visions between art and technology* (). Washington DC: National Academy of Sciences.

Orek, J. (Director) Matthies, E (Producer) (1995). *Making the Matrix* [Motion Picture]. New York, NY: Home Box Office

Ortny, A. (1993). *Metaphor and Thought.* Essay, Reddy, M. J. New York, NY: Cambridge University Press.

Oxford english dictionary. Retrieved 2/2007, 2007, from http://dictionary.oed.com.proxy.lib.sfu.ca/cgi/entry/50182439?>

Panfosky, E. (1997). *Three Essays on style; Style and the medium in motion pictures*, (p. 93). Boston, Mass: MIT Press.

Panofsky, E. (1972). *Renaissance an Renascences in Wester Art.* New York: Harper & Row.

Parberry, I., Kazemzadeh, M. B., & Roden, T. (2006). *The art and science of game programming.* In Proceedings of the 37th SIGCSE technical symposium on Computer science education (pp. 510-514). New York, NY, USA: ACM Press.

Partecipando. (2006). Retrieved 2 25, 2008, from European Project "Partecipando": http://urbact.eu/projects/partecipando/home.html

Paul, Ch. (2005). *Digital Art.* London: Thames & Hudson Ltd.

Peacock, K. (1988). Instruments to Perform Color-Music: Two Centuries of Technological Experimentation. *LEONARDO, 21*(4), 397-406.

Pearce, C., Diamond, S., & Beam, M. (2003). BRIDGES I: Interdisiplinary collaboration as practice. *Leonardo, 36*(2), 123-128.

Pehkonen, E. (2004). State-of-Art in Problem Solving: Focus on Open Problems. In *ProMath Jena 2003. problem Solving in math Education* (eds. H.rehlich & B.Zimmerman), (pp. 93–111). Yidesheim:Verlag Franzbecker.)

Pehkonen, E., & Rakov, S. (2005). Comparative Survey on Pupils Beliefs of Mathematics Teaching in Finland and Ukraine. *Teaching Mathematics and Computer Science, 3*(1), 13-33.

Peterson, I. (1998). *The Jungles of Randomness: A Mathematical Safari.* New York, NY: John Wiley & Sons.

Phan, M. (September 2007). *Machinima Licenses Spell Out New Rules for Creators.* Wired Magazine, retrieved on February 2, 2008 from http://www.wired.com/culture/art/news/2007/09/machinimalicenses

Picard, R.W., & Klein, J. (2002). Computers that recognize and respond to user emotion: theoretical and practical implications. *Interacting with Computers, 14*, 141–169. URL: http://www.sciencedirect.com/science/article/B6V0D-459BFXM-3/2/1aea6019fe1bb3835dd6e2480658a68e

Picinali, L. (2006). *Techniques for the extraction of the impulse response of a linear and time-invariant system.* Paper presented at the DMRN Doctoral Research Conference, University of London, UK.

Pier, M.D., & Goldberg, I.R. (2005). Designing interfaces for art applications. In *CW '05: Proceedings of the 2005 International Conference on Cyberworlds* (pp. 172–178). Washington, DC, USA: IEEE Computer Society.

Pimentel, T., & Branco, V. (2005). *Dynamic and Interactive typography in digital art.* Computer & Graphics Journal, *6*(29), 882-889.

Pincus, S., & Singer, B.H. (1996). Randomness and Degrees of Irregularity. In *Proceedings of the National Academy of Sciences of the United States of America, 93*(5), 2083-2088. Washington, DC: National Academy of Sciences.

Pine, B. J., & Gilmore, J. H. (1999). *The Experience Economy; Work is Theater & Every Business a Stage.* Boston: Harvard Business School Press.

Pink, D. H. (2004). *A Whole New Mind: Moving from the Information Age to the Conceptual Age.* New York, NY Riverhead Hardcover.

PLANETS. (2004). European project. http://www.planets-project.eu

Pohflepp, S. (2006). Buttons. *Between Blinks & Buttons.* Retrieved February 22, 2008, from http://www.blinksandbuttons.net/buttons_en.html

Pole, R. (1968, January). 3-D Imagery and Holograms of Objects Illuminated in White Light. *Applied Physics Letters, 12*(1), 10 –12.

Polli, A. (2004). *DATAREADER: a tool for art and science collaborations.* In Proceedings of the 12th annual ACM international conference on Multimedia (pp. 520-523). New York, NY, USA: ACM Press.

Pool, R. (1999). *Beyond Engineering: How Societies Shape Technology.* New York, NY: Oxford University Press.

Possiant, L. (2007). *Media Art Histories,* (pp. 229-250). MIT Press.

Poundstone, W. (2005). *Fortune's Formula.* New York, NY: Hill and Wang.

Powell, J.W. (1880). *Introduction to the Study of Indian Languages with Words Phrases and Sentences To Be Collected.* Washington: Government Printing Office.

Pritchett, J. (1988). *The use of chance techniques in the music of John Cage, 1950-1956.* Unpublished doctoral dissertation, New York University, New York.

Puckette, M. (1996). *Pure data.* http://puredata.info.

Pulkki, V. (1997). Virtual sound source positioning using vector based amplitude panning. *Journal of the Audio Engineering Society, 45*(6), 456-466.

Putnam, R. (1993). The Prosperous Community: Social Capital and Public Life. *The American Prospect , 13*(1), 35-42.

Quintas, R., & Dionísio, T. (2005). Displacement: Instalação Musica-Visual Imersiva que Analisa e Retracta a Expressividade Corporal. In A. Marcos, L. Valbom, & M. Meira. (Eds) *Proceedings of Artech 2005 – International Conference on Digital and Interactive Art.* Vila Nova de Cerveira, Portugal: Computer Graphics Center Press.

Quiroz, J., Louis, S., Shankar, A., & Dascalu, S. (2007). Interactive Genetic Algorithms for User Interface Design. *IEEE Congress on Evolutionary Computation,* (pp. 1366-1373).

R Development Core Team. (2007). R: A Language and Environment for Statistical Computing. *R Foundation for Statistical Computing,* Vienna, Austria. http://www.R-project.org. ISBN 3-900051-070.

Rabiner, L.R., & Gold, B. (1975). *Theory and Aplication of Digital Signal Processing.* Englewood Cliffs, NJ, USA: Prentice Hall, INC.

Rakov, S., & Gorokh, V. (1998). Information Technologies in Geometry (an example of generalization of one well known problem about squares). *Bulletin User Group of Derive, 31,* 25–30.

Rakov, S.A. (2005). *Math education: competency approach with ICT support.* Kharkov, Ukraine: Fakt.

Rayleigh, L. (1907). On our perception of sound direction. *Philosophical Magazine, 13,* 214-232

Rehyner, J. (1992) *Teaching American Indian students.* Norman, OK: University of Oklahoma Press.

Reid, E. (1994). *Cultural Formations in Text.*

Reynolds, C.W. (1987). *Flocks, Herds, and Schools: A Distributed Behavioral Model, in Computer Graphics, 21*(4) (SIGGRAPH '87 Conference Proceedings), 25-34.

Rio, S. (2007) *China to implement anti-online game addiction system.* Retrieved January 28, 2007 http://mmorpg.qj.net/China-to-implement-anti-online-game-addictionsystem/pg/49/aid/89019

Rio, S. (2007). *Now everyone's a hybrid: Tseric on hybridization.* Retrieved January 28, 2007 http://wow.qj.net/Now-everyone-s-a-hybrid-Tseric-onhybridization/pg/49/aid/80618

Rius, M., Parramón, J.M., & Puig, J.J. (1985). *The five senses: Sight.* Hauppauge, NY: Barron's Educational Series.

Roberts, E.B. (1988). Managing Invention and Innovation. *Research –Technology Management, 31,* 11-29.

Rokeby, D. (1990). *The Giver of Names.* http://homepage.mac.com/davidrokeby/gon.html.

Rokeby, D.. (2001). *n-cha(n)t.* http://homepage.mac.com/davidrokeby/nchant.html.

Roland, R. (1992). *Globalization: Social Theory and Global Culture.* London: Sage Publications.

Rosedale, P. (November 30, 2006). Second Life: What Do We Learn If We Digitize EVERYTHING? The Longnow Foundation. Baldwin, Chris (Director). USA: Whole Earth Films. Retrieved January 15, 2008 from http://www.wholeearthfilms.com/rosedale_philip.html#

Ross, D. (1999). Net.art in the Age of Digital Reproduction (21 Distinctive Qualities of Net.Art) [Electronic Version]. Retrieved 20 May 2006 from http://switch.sjsu.edu/web/v5n1/ross/index.html

Rymaszewski, M., et. al. (2007). *Second Life: The Official Guide.* Indianapolis, IN: Wiley.

Rymaszewski, M., Wagner, J.A., Ondrejka, C., Platel, R., Gorden, S.V., Rossignol, J.C. et al (2007). *Second Life: The Official Guide.* Indianapolis, IN: Wiley Publishing.

Saarinen, I. (2007). *Olmannen* program/playbill (notecard). Second Life, February 2007.

Sacks, R. (2000). The MetaMOOphosis: A Visit to the Kafka House—A report on the permanent installation of an interactive theatre work based on Franz Kafka's Metamorphosis. In *Theatre in Cyberspace: Issues of Teaching, Acting and Directing* (pp. 159-174). S. Schrum (Ed.). New York: Peter Lang Publishing.

Salen, K., & Zimmerman, E. (2003). *Rules of Play: Game Design Fundamentals.* Cambridge, MA: MIT Press.

Salingaros, N. (2005). *Principles of Urban Structure.* Amsterdam, Holland: Techne Press.

Sapir, E. (1975). *Navajo texts.* New York: AMS Press Inc. (Original work published 1942).

Sardon, M. (2006). *Books of sand.* In Proceedings of the 14th annual ACM international conference on Multimedia (pp. 1041-1042). New York, NY, USA: ACM Press.

Scaff, J. H. Art and Authenticity in the Age of Digital Reproduction [Electronic Version]. *Digital arts institute.* Retrieved 03 Mar 2007 from http://www.digitalartsinstitute.org/scaff/index.html

Schleiner, A. M. (1999). E-mail interview discussing artists and the computer game industry by Jim McClellan (Publication.: http://www.opensorcery.net/jiminterview.html

Schrage, M. (1995). *No more teams! mastering the dynamics of creative collaboration.* New York, NY: Currency Doubleday.

Schrammel, J., & Tscheligi, M. (2006). Experiences evoked by today's technology - results from a qualitative empirical study. In *20th International Symposium on Human Factors in Telecommunication.*

Schrum, S.A. (1996). *NetSeduction.* Digital Performance Archive. Retrieved January 16, 2008 from http://ahds.ac.uk/ahdscollections/docroot/dpa/authorsdetails.do?project=39&author=61&string=SRational%20Mind%20Theatre%20Company

Schrum, S.A. (2000). Begin with a Single Step: Adding Technology to a Course. In. Schrum (Ed.), *Theatre in Cyberspace: Issues of Teaching, Acting and Directing* (pp. 53-63). S. New York: Peter Lang Publishing.

Schrum, S.A. (2007, July 26). A Taxonomy of Digital Performance. Presented in the panel, "Interactive Media in Immersive Performance and a Taxonomy of Digital

Performance." Podcast retrieved February 1, 2008 from http://www.cda.cmich.edu/ATHE%20Podcasts/Interactive%20Media/Interactive%20Media%20(Taxonomy)%20-%20edited.mp3

Schrum, S.A. (2007, July 31). A Taxonomy of Digital Performance. MUSOFYR. Retrieved July 31, 2007 from http://www.musofyr.com/taxonomy/taxonomy.pdf

Schweller, K. (2007). *Personal Webpage*. Retrieved June 15, 2007 from http://web.bvu.edu/faculty/schweller/

Seay, A.F., Jerome, W.J., Lee, K.S., & Kraut, R.E. (2004). Project Massive: A Study of Online Gaming Communities. In *Proceedings of ACM CHI 2004* (pp. 1421–1424). New York: ACM Press.

Seay, A.F., Jerome, W.J., Lee, K.S., & Kraut, R.E. (2004). Project massive: a study of online gaming communities. *Conference on Human Factors in Computing Systems (CHI'04)*, Vienna, Austria (2004).

SecondPride.org (June 18, 2007). Retrieved June 27, 2007 from http://secondpride.org/main.htm. (No longer available.)

Sedelow, S. Y. (1970). The Computer in the Humanities and Fine Arts. *ACM Computing Surveys (CSUR), 2*(2), 89-110.

Sensable Technologies (1993). *The phantom*. URL: http://www.sensable.com

Sharp, D. (2006). Participatory Cultural Production and the DIY Internet: From Theory to Practice and Back Again. *Media International Australia: Practice-Led Research, 118*, 16-24.

Shaw, E.A.G., & Teranishi, R. (1968). Sound pressure generated in an external-ear replica and real human ears by a nearby sound source. *Journal of the Acoustical Society of America, 44*, 240-249.

Shi, Y.Q., & Sun, H. (2000). *Image and video compression for multimedia engineering: Fundamentals, algorithms, and standards*CRC Press.

Shneiderman , B. (1980). *Software Psychology: Human Factors in Computer and Information Systems*. Boston, Ma: Winthrop Computer Systems Series.

Shoniregun, C. A., Logvynovskiy, O., Duan, Z., & Bose, S. (2004). *Streaming and security of art works on the Web*. In IEEE Sixth International Symposium on Multimedia Software Engineering (ISMSE'04) (pp. 344-351). Washington, DC, USA: IEEE Computer Society.

Sivo, G. (1995). Intervention at meeting. *I Laboratori di quartiere nella città di Roma*. Rome, Italy: Orme.

Skopets, Z.A. (1990). *Geometric miniatures*. Moscow, Russia: Prosveschenie.

Skyland, NC: Kivaki Press.

SLExchange Marketplace. Retrieved June 12, 2007 from www.slexchange.com

Smith, O. (2008). *Raphael Di Luzio and The Concept of Time-Based Painting*, (p.16). Wynwood, Florida: Wynwood The Magazine of Art.

Snow, C. P. (1959). The two cultures. *Rede Annual Lecture*, Senate House, Cambridge, UK.

Solari, S. (1997). *Digital video and audio compression*. New York: McGraw-Hill.

Sontag. S. (1977). *On Photography*. New York: Farrar, Straus and Giroux. P154 ISBN.

Sotamaa, O. (2005). Creative User-centred Design Practices: Lessons from Game Cultures. *A (Eds.), Everyday Innovators, Researching the Role of Users in Shaping ICTs*. Springer, Dordrect, (pp. 104-16). Retrieved January 28, 2007 from http://members.aol.com/leshaddon/Sotamaa.pdf

Spaink, K. (1997). Prix Ars Electronica 1997- net Jury statement, July 1997 [Electronic Version] from http://www.spaink.net/english/ArsPrix97.html

Steinkamp, J. (2001). My Only Sunshine: Installation Art Experiments with Light, Space, Sound and Motion. *Leonardo, 34*(2), 109-112.

Steinkuehler, C.A. (2003). *Massively multiplayer online videogames as a constellation of literacy practices*. Paper presented at the International Conference on Literacy, Ghent, Belgium.

Steinkuehler, C.A. (2004). *Online cognitive ethno¬graphy: Methods for studying massively multiplayer online videogaming culture.* Paper presented at the 17th Annual Conference on Interdisciplinary Qualitative Studies, Athens GA.

Steinkuehler, C.A. (2005). *Styles of play: Gamer-identified trajectories of participation in MMOGs.* Paper presented at the Annual Conference of the Digital Games Research Association (DIGRA), Vancouver. Retrieved January 28, 2007 from http://www.academiccolab.org/resources/documents/Steinkuehler_NEWLIT.doc.

Steinkuehler, C.A. (2005). The new third place: Massively multiplayer online gaming in American youth culture. *Tidskrift Journal of Research in Teacher Education.* Retrieved January 28, 2007. http://website.education.wisc.edu/steinkuehler/papers/Steinkuehler_TIDSKRIFT2005.pdf

Steinkuehler, C.A., & Williams, D. (2006). *Where Everybody Knows Your (Screen) Name: Online Games as "Third Places".* Retrieved January 28, 2007 http://jcmc.indiana.edu/vol11/issue4/steinkuehler.html

Stewart, V. (2006). *Math and Science Education in a Global Age: What the U.S. can Learn from China* Published By Asia Society. http://internationaled.org/mathsciencereport.pdf

Stocker, G. (2005). The Art of Tomorrow. In *Climax- The Highlight of Ars Electroniza* (pp. 204). Taipei: National Taiwan Museum of Fine Arts.

Strömberg, H., Väätänen, A., & Räty, V.-P. (2002). *A group game played in interactive virtual space: design and evaluation.* In Proceedings of the conference on Designing interactive systems: processes, practices, methods, and techniques (pp. 56-63). New York, NY, USA: ACM Press.

Sturtevant, W. (1986). The meaning of native American art. In E. Wade (Ed.), *The arts of the North American Indian: Native traditions in evolution.* New York, NY: Hudson Hills Press Inc.

Suler, J. (2007). Second Life, Second Chance. *The Psychology of Cyberspace.* Retrieved June 13, 2007 from http://www.rider.edu/~suler/psycyber/psycyber.html

Sullivan, G. (2006). *Artifacts as evidence within changing contexts.* Working Papers in Art and Design 4 http://www.herts.ac.uk/artdes/research/papers/wpades/vol4/gsfull.html

Sung-dae, H., Jin-wan, P., & Won-Hyung, L. (2006). *Designing Audio Visual Software for Digital Interactive Art.* In 16th International Conference on Artificial Reality and Telexistence--Workshops, 2006. ICAT '06. (pp. 651-655). Washington, DC, USA: IEEE Computer Society.

Sylene (2006). *Hybrid vs. Core.* Retrieved January 28, 2007 http://wow.stratics.com/content/features/editorials/hy/

Tait, W. (1998). The space between: fine art and technology. *SIGGRAPH Comput. Graph. 32*(1), 17–19.

Takagi, H. (2001). Interactive Evolutionary Computation: Fusion of the Capabilities of EC Optimization and Human Evaluation. *Proceedings of the IEEE, 89,* 1275–1296.

Tarantino, Q. (1994). *Pulp Fiction.* A Band Apart. United States.

Tator C. (1998). *Challenging Racisms in the Arts.* Toronto, CA: University of Toronto Press.

Taylor, C. (2006). Bonds of trust: An in-depth look at social bonding within MMO guilds. *Undergraduate term paper for the course "Games for the Web".* Trinity University. Retrieved January 28, 2007 from http://www.trinity.edu/adelwich/worlds/articles/trinity.chris.taylor.pdf

Taylor, R. (Ed.). (2006). *Vsevolod Pudovkin: Selected Essays.* Oxford, England: Pudovkin, V.Seagull Books.

Tazman_ODevilsun (2004). *Why are good soloing abilities important in a MMORPG?* Retrieved January 28, 2007 http://eq2vault.ign.com/View.php?view=columns.Detail&category_select_id=8&id=228

Tekalp, A. M. (1995). *Digital video processing.* Upper Saddle River, NJ: Prentice Hall PTR.

Thorburn, D., Barrett, E., & Jenkins, H. (2004). Series Editors Media In Transition Book series. In L. Gittelman, and G. B. Pingree (Eds.), *New media* (pp. 1740-1915). Cambridge, MA: MIT Press.

Tippett, L.H.C. (1927). Random sampling numbers. *Tracts for Computers, 15*. London: Cambridge University Press.

Tokuno, H., Hamada, H., Kirkeby, O., & Nelson, P. (1996). Binaural sound reproduction in a stereo dipole system. *Journal of the Acoustical Society of America, 100*, 2700.

Torres-Nez, J. (2004). *Beesh Łigaii in Balance*. Santa Fe, NM: Museum of Indian Arts and Culture.

Trifonova, A., Ahmed, S. U., & Jaccheri, L. (2007). *SArt: Towards Innovation at the intersection of Software engineering and art*. Paper presented at 16th International Conference on Information Systems Development, Galway, Ireland.

Tufte, E. (1993). *The Cognitive Style of PowerPoint*. New London, CT. Yale University Press.

Tufte, E. (2001). *The Visual Display of Quantitative Information*. Cheshire: Graphics Press.

Turing, A. (2004). *The Essential Turing: Seminal Writings in Computing, Logic, Philosophy, Artificial Intelligence, and Artificial Life, Plus the Secrets of Enigma, chapter Computing Machinery and Intelligence (1950)*. Oxford University Press, USA.

Turkle, S. (1995). *Life on the screen: Identity in the age of the Internet*. New York: Touchstone.

Turkle, S. (1997). *Life on the Screen*. NY: Simon and Schuster.

Turner, M. (2005, July 12-15). *Mathematics and Narrative*. Paper presented at Mathematics and Narrative, Mykonos, Greece. Retrieved April 5, 2006, http://www.thalesandfriends.org/en/papers/pdf/turner_paper.pdf

Tykwer, T. Director. (1998). *Lola Rennt*, (Run Lola Run). Tykwer, T. (writer). 1999. X-Filme Creative Pool. Germany.

Ultsch, A. (1993). Self-organizing neural networks for visualization and classification. *Information and Classification*, (pp. 307-313).

Ultsch, A., & Siemon, H. (1989). *Exploratory data analysis: Using kohonen's topology preserving maps*. Technical Report 329, University of Dortmund, Germany.

Urbact Project, Partecipando. (2006). *European Handbook for Partecipation*.

Valbom L., & Marcos A. (2005). WAVE: Sound and music in an immersive environment. *Computer & Graphics Journal, 6*(29), 871-881.

Valbom, L., & Marcos A. (2007). Presenting a prototype of an immersive musical instrument. *IEEE Computer Graphics and Applications, 4*(27), 14-19.

Vanechkina, I. (1980). Complex approach to the research on A.N.Scriabin's light-music conception. In *Problem of complex research of art creativity*. - Kazan: Izd. KGU, (p. 107-115).

Varese, E. (2007). The Liberation of Sound. In C. Cox & D. Warner (Eds.), *Audio Culture: Readings in Modern Music* (pp. 17-21). New York, London: Continuum.

Venkatesh, V. (1999). Creation of favorable user perceptions: Exploring the role of intrinsic motivation. *MIS Quarterly, 23*(2), 239–260.

Vertov, D. (1985), *Kino-Eye: The Writings of Dziga Vertov*. Berkley, CA: University of California Press.

Vigosky, L.S. (2004). *Psychology of human development*. Moscow, Russia: Smisl (in Russian).

von Hornbostel, E.M., & Wertheimer, M. (1920). Über die Vahrnehmung der Schallrichtung [On the perception of the direction of sound]. *Sitzungsber. Akad. Wiss. Berlin*, (pp. 388-396).

von Trier, L. Director. (2003). *De Fem benspænd*, ("The Five Obstuctions"). Denmark. Almaz Film Productions S.A.

Vonnegut, K. (1966). *Mother Night*. New York: Dell Books.

Vonnegut, L (1976). Slaughter-House-Five. Laurel/Dell Books paperback.

Vygotsky, L. (1989). *Psychology of art*. MIT Press.

Wachowski, A., & Wachowski, L. (Directors). 1999. *The Matrix*. USA: Groucho II Film Partnership.

Wacom Technology Inc. (n.d.). *Wacom cintiq 21ux*. URL: http://www.wacom-europe.com/int/products/cintiq/whatto.asp?lang=en

Wade, E. (1986). The arts of the North American Indian: Native traditions in evolution. In E. Wade (Ed.), *The arts of the North American Indian: Native traditions in evolution.* New York, NY: Hudson Hills Press Inc.

Wakin, D. (2006, May 5). *An Organ Recital for the Very, Very Patient* [Electronic version]. *The New York Times*.

Walden, K. L. (2002). Reviews: Peter Weibel and Timothy Druekrey (eds), Net_Condition: Art and Global Media. *Convergence-The International journal of New Media Technologies, 8*(1), 114-116.

Walter, B. (1955). *Art in the Age of Mechanical Reproduction, Illuminations*. New York: Schocken Books.

Wang, Y., Ostermann, J., & Zhang, Y. (2002). *Video processing and communications*. New Jersey: Prentice Hall, Inc.

Warr, A., & O'Neill, E. (2007). *Tools to Support Collaborative Creativity*. Paper presented at Tools to Support Collaborative Creativity workshop held as part of Creativity and Cognition conference 2007, Washington D.C., USA.

Webster, M. (2007). *Miriam Webster Dictionary* Online: http://www.merriam-webster.com/dictionary/experience

Wei-Lung, W. (2002). Beware the engineering metaphor. *Communications of the ACM, 45*(5), 27-29.

Welker, G. (1996, February 8). Song of the horses. In *Indigenous people's literature*. Retrieved 2006, April 29 from http://www.indigenouspeople.net/songhors.htm

Wellman, B., et al. (1996). Computer Networks as Social Networks: Collaborative Work, Telework, and Virtual Community. *Annual Review of Sociology, 22*, 213-38.

Wesley, J. (2007). *Overcoming addiction and escapism.* Retrieved November 20, 2007 http://cid.nada.kth.se/pdf/CID-167.pdf

Whale, G. (2002). Why use computers to make drawings? In *C&C '02: Proceedings of the fourth conference on Creativity & cognition* (pp. 65–71). ACM Press.

Whitelaw, M. (2008). Art Against Information: Case Studies in Data Practice. In *Fibreculture*. No. 11, 2008. Retrieved March, 2008, from http://journal.fibreculture.org/issue11/issue11_whitelaw.html

Wijers, G. (2005). Preservation and/or Documentation: The Conservation of Media Art (Publication. Retrieved 04 Sep 2006, from The Netherlands Media Art Institute: http://www.nimk.nl/en/

Wikipedia (2007). *Revews MMORPG's- MMORPG.cvr.pl, History*. Retrieved January 28, 2007 http://mmorpg.cvr.pl/history.htm

Wikipedia (2008). *Dungeons & Dragons*. Retrieved January 28, 2007 http://en.wikipedia.org/wiki/Dungeons_%26_Dragons

Wikipedia (2008). *Neverwinter Nights (AOL game)*. Retrieved January 28, 2007 http://en.wikipedia.org/wiki/Neverwinter_Nights_(AOL_game) and Medar (2001). Neverwinter nights the original. Retrieved January 28, 2007 http://www.bladekeep.com/nwn/

Wikipedia (2008).*Colossal cave adventure*. Retrieved January 28, 2007 http://en.wikipedia.org/wiki/Colossal_Cave

Wikipedia contributors. (2007). Terminology. In *Scientific visualization*. Retrieved November 14, 2007 from http://en.wikipedia.org/wiki/Scientific_visualization

Wikipedia: The free encyclopedia. (2006, February 15th). FL: Wikimedia Foundation, Inc. Retrieved March 13, 2008, from http://en.wikipedia.org/wiki/Chain_Home

Williams, J.P., Hendricks, S.Q., & Winkler, W.K. (Eds.) (2006). *Gaming as Culture: Essays on Reality, Identity, and Experience in Fantasy Games*. Jefferson, NC: McFarland Publishing

Wilson, S. (1995). Artificial intelligence research as art. *Stanford Electronic Humanities Review, 4*(2).

Wilson, S. (2002). *Information Arts: Intersections of Art, Science, and Technology.* Cambridge, Massachusetts: The MIT Press.

Wilson, S. (2002). *Information arts: Intersections of art, science, and technology.* Cambridge, MA: MIT Press.

Wilson, S. (2002). *Information arts: intersections of art, science, and technology.* Cambridge, Mass.; London: MIT.

Wilson, S. (2002). *Information Arts: Intersections of Art, Science and Technology.* MIT Press.

Witherspoon, G. (1977). *Language and art in the Navajo universe.* Ann Arbor: The University of Michigan Press. (Original work published 1977).

Wittgenstein, L. (1994) Tractatus Logico-Philisophicus. London: Routledge.

Wood, A.F., & Smith, M.J. (2005). *Online Communication: Linking Technology and Culture.* Mahwah, NJ: Lawrence Erlbaum Associates, Publishers.

Woods, W.A. (1970) Transition network grammars for natural language analysis. *Commun. ACM, 13*, 591-606, 1970.

Yee, N. (2004). *World of Warcraft Scenery.* Retrieved January 28, 2007 http://terranova.blogs.com/terra_nova/2004/12/world_of_warcra.html

Yee, N. (2005). *A model of player motivations.* Retrieved January 28, 2007 http://www.nickyee.com/daedalus/archives/001298.php?page=1

Yee, N. (2005). *Addiction.* Retrieved January 28, 2007 http://www.nickyee.com/daedalus/archives/000818.php

Yee, N. (2005). *WoW Character Class Demographics.* Retrieved January 28, 2007 http://www.nickyee.com/daedalus/archives/001367.php

Yeh, J. (2002). Net. Art - Exhibition. *Artist Magazine.*

Yong, K. (2006). *Addiction to MMORPGs: Symptoms and Treatment.* Retrieved January 28, 2007 http://www.netaddiction.com/articles/addiction_to_mmorpgs.pdf

Yoshida, S., Kurumisawa, J., Noma, H., Tetsutani, N., & Hosaka, K. (2004). Suminagashi: creation of new style media art with haptic digital colors. In *MULTIMEDIA '04: Proceedings of the 12th annual ACM international conference on Multimedia* (pp. 636–643). New York, NY, USA: ACM Press.

Yost, W.A. (2000). *Fundamentals of Hearing: An Introduction.* San Diego, California, USA: Academic Press.

Youngblood, G. (1970). *Expanded cinema.* Cambridge and London: E. P. Dutton.

Zanni, C. (2004). Interview with Golan Levin. *CIAC Magazine,* 16 June 2004.

Zebra Imaging (2009) Application. http://www.zebraimaging.com/html/industries.html

Zellner, D. (n.d.). *Theatre and Second Life.* Northwestern University. Retrieved June 27, 2007 from http://door.it.northwestern.edu/sheridan/dms/zellner/zellner-modes.rm

Zheng, W., Kanatsugu, Y., Itoh, S., & Tanaka, Y. (2000). Analysis of space-dependent characteristics of motion-compensated frame differences. *International Conference on Image Processing, 3,* 158-161.

Zimmerman, G. W., & Eber, D. E. (2001). *When worlds collide!: an interdisciplinary course in virtual-reality art.* In Proceedings of the thirty-second SIGCSE technical symposium on Computer Science Education (pp. 75-79). New York, NY, USA: ACM Press.

Zmölnig, J.M. (2001). *ann_som: Component of the Artificial Neural Network library for Pure Data.* http://puredata.info/Members/dmorelli/ann/?searchterm=neural.

About the Contributors

James Braman is a lecturer for the Computer and Information Sciences Department at Towson University (Towson, MD). He holds a master's degree in computer science from Towson University and is currently a doctoral candidate in applied information technology. He has been teaching courses dealing with computers and art for the past several semesters. Combining art and technology has become a focal interest over the years. His current research focus includes art and technology, intelligent agents, simulated emotions and education in virtual and immersive environments.

Giovanni Vincenti is in charge of research and development at Gruppo Vincenti, a family-owned company with interests across several fields. His main areas of research include fuzzy mediation, information fusion, emotionally-aware agent frameworks and robotics. He held several positions at Towson University, including a lecturership with the Department of Computer and Information Sciences. He also taught courses for the Center of Applied Information Technology, also at Towson University. He is the author of many publications, and the father of the concept of fuzzy mediation, as applied to the field of information fusion.

Goran Trajkovski is the director of product strategy and design for it/engineering at Laureate Higher Education Group, Baltimore, MD. He was the chair of the Department of Information Technologies of South University and associate professor of IT at its Savannah, GA campus. He was previously the founding director of the Cognitive Agency and Robotics Laboratory (CARoL) at Towson University, Towson, MD, USA. The virtual version of CARoL now exists in Second Life. He also taught at Towson University, West Virginia University, Parkersburg, WV, USA, and the University "Ss Cyril and Methodius," Skopje, Macedonia. His research focuses on cognitive and developmental robotics, and interaction and emergent phenomena in agent societies. He is an affiliate of the Institute for Interactivist Studies at Lehigh University, and a member of the organizing committee of the biannual Interactivist Summer Institutes. He has authored over 200 publications, including ten books and edited volumes. He has chaired two symposia for the Association for Advancement of Artificial Intelligence. Trajkovski is the founding editor-in-chief of the *International Journal of Agent Technologies and Systems*, published by IGI Global. His work has been funded by the NSF, the National Academies of the Sciences, and OWASP (Open Web Application Security Project). Trajkovski hold a BSc in applied informatics, MSc in mathematical and computer sciences, and PhD in computer sciences from the University "Ss Cyril and Methodius," Skopje, Macedonia.

* * *

Annette Aboulafia holds a master of science and a PhD in Psychology from Copenhagen University. Presently, she is a Research Scholar at the University of Limerick, Ireland. Previously, she was an associate professor at Roskilde University, Department of Philosophy and Psychology, Denmark. She has worked on a number of EU projects in the area of work psychology, human-computer interaction (HCI), computer-mediated learning, and laterly in the area of simulation-mediated medical training. She has researched and published in these areas for about 20 years, and has a number of years experience in teaching in a variety of areas within psychology, and lately within interaction design.

Salah Uddin Ahmed is a PhD student in software engineering group at NTNU. He has completed his bachelor's in computer science from University of Dhaka, Dhaka, Bangladesh. In 2005, he obtained MSc in Engineering and Management of Information Systems from the Royal Institute of Technology (KTH), Stockholm, Sweden. His PhD research is part of the SArt (software and art) project which focuses on research issues at the intersection of software and art. He is especially interested in software intensive artworks and art projects with software engineering interventions for the development, maintenance and upgrading of the artworks.

Francesco Altarocca is a software developer and architect. He currently works in the IT field with ISTAT (Italian National Institute of Statistics). He develops many systems that support statistical researchers ranging from electronic and web questionnaires to data warehouse and portal's surveys. He also worked as Web master, back-end and a centralized authentication system development. In ISTAT he wrote some publications regarding design patterns and open source development, the use of new technologies in statistical data catching. His MS degree is in computer science at "La Sapienza" University of Rome and attends a graduate studies course in computer science at "Tor Vergata" University of Rome. His research interest deals with design patterns, software engineering, open source, sentiment analysis, swarm intelligence and Web 2.0 technologies.

Jim Barta, Associate department head of regional campus and distance education at Utah State University has been involved in multicultural mathematical educational research, scholarship, and curricular development for over 15 years, with a particular emphasis in Native American mathematics education. He currently is involved in two related projects; he works to enhance mathematics education with indigenous teachers in the rural highlands of Guatemala and he collaborates with upper elementary and middle school teachers on The Northern Ute Reservation to improve their STEM teaching. He remains interested in developing culturally responsive professional development for mathematics educational leaders.

Eugenia Benelli is an architect and a concept designer. She's currently working on the art direction of various international programs based on unconventional development to valorize the individual and its organizational growth. She is also the creative director of PIMBY (**P**lease **I**n **M**y **B**ackyard) - project, a national prize about people inclusion in the energy projects involved in urban strategies - and Raabeart - multidisciplinary lab based in Rome and Vienna. She has worked on interaction design developing prototypes for e-democracy and collaborative art at Enterprise Digital Architects S.p.A. (ICT company in Rome). Her research has been focused on the public participation processes and on the inclusion of people into the valorization and the "creative" management of the environment and the cultural heritage. Graduated cum laude at Roma Tre University she has attended design masterclasses at Architectural

Association (London), Berlage Institute (Rotterdam), Escuela Tecnica Superior de Arquitectura de Granada, University of Bath and Universitat Internacional de Catalunya, Barcelona. She has worked in the research field at the Architectural Design Department of Roma Tre University and has taken part in different architectural projects, landscape projects and international competitions

Jim Bizzocchi is a researcher, instructor, and video artist. He is an assistant professor in the School of Interactive Arts and Technology at Simon Fraser University. He did his graduate studies in the Comparative Media Studies Program at the Massachusetts Institute of Technology. His research interests include the future of the moving image, interactive narrative and game design. He has taught in all these areas, and is the recipient of a university Excellence in Teaching Award. His academic papers have been published widely, and his video art has been exhibited in Canada, the United States, England, Australia, and China.

Ben Bogart is an artist working in installation, audio-visual improvisation and software development. His installations create content live in response to their sensed environment. He works in an Open Source context and makes all the software he develops, that is of general use, available under the GPL. Physical modeling, chaos, feedback systems, evolutionary algorithms and artificial intelligence have been used to inform and engage in his creative process. In collaboration with the Pure-Data Documentation Project (PDDP), Ben is working on a curriculum for electronic media arts based solely around Open Source tools. Ben holds a Master's of Science in Interactive Arts and Technology from Simon Fraser University. His current work deals with computational implementations of embodied creativity, memory and dreaming.

Pedro Branco holds a PhD in Information Systems from the University of Minho. In 2000, he joined Fraunhofer's U.S. operations as researcher/3d software engineer in the development of virtual reality interaction techniques. Since 2003 he has worked at IMEDIA in Providence, RI, studying user interface usability based on physiological monitoring. In January 2007 he joined the Department of Information Systems at University of Minho, Portugal. His research interests are on monitoring users' facial expressions and the associated computer vision topics, intelligent user interfaces and anthropomorphic interfaces.

João Alvaro Carvalho is full professor at and chair of the Department of Information Systems, School of Engineering, University of Minho, Portugal. He holds a PhD in Information Systems from the UMIST (University of Manchester Institute of Science and technology), UK (1991). His research and teaching interests include: the foundations of information systems, information systems development, meta-modelling, requirements engineering and knowledge management. He is the national representative of Portugal in IFIP TC 8 since 1996 and the President of the Portuguese Association for Information Systems (APSI – Associação Portuguesa de Sistemas de Informação).

Yueh Hsiu Giffen Cheng is a Taiwanese new media artist, designer, researcher and writer based in Sydney and Taipei. She completed a master's of visual arts at the Australian National University and a doctoral degree from the University of Technology, Sydney. She has edited a series of books and papers on new media art and contributed to various exhibitions. Her online-portfolio is available on http:// giffenspace.blogspot.com. "Throughout the life of art creation, countless ups, downs and unpredictable variables await; the constant pursuit of breakthroughs for exceeding thyself will therefore never end. Creating art, writing and educating young people are the three elements that enrich my life. The creation of art allows me to communicate with my own soul and to inspire ideas in me about every trifle in my life. Writing to me is a way of simmering down and sorting out my thoughts. As far as I'm concerned, educating young people is like a farmer irrigating the seedlings. It requires all-time patience and commitment. Although the fruitage might not be perfect, it is definitely worth committing oneself to educating our younger generations."

Stefano De Luca is the CEO of Evodevo s.r.l, an innovation company specialized in semantic web technologies, electronics, sensor networks and homeland security. He realized products as a logic-based middleware, a semantic web enabled CMS, agent-based hazard simulators. He has a degree in philosophy with specialization in mathematical logic and philosophy of science, and teaches Artificial Intelligence and Multi-Agent Systems at Rome University "Tor Vergata." Before founding Evodevo, he worked as R&D manager in multinational companies. He has published books on agent-based modeling and simulation, process optimization in health care, and papers on intelligent adaptive systems. His research interests ranges in nature inspired computing, evolutionary computation, Semantic Web, homeland security, intelligent agents.

Raphael DiLuzio was born on the coast of California. As a child he wanted to make images in any way possible from drawing to experiments with film. When he was nine he began a formal study of drawing. Raphael DiLuzio received a BFA in drawing and painting from California State University Long Beach. He attended graduate school at Cornell for one year then received a MFA in painting at the University of Pennsylvania. In 1990 he moved to New Orleans and began a career as a painter while developing public art programs for at risk children, writing grants and doing public arts projects. In 1997 he worked with emerging digital mediums; exploring the possibilities of the medium. Raphael was creating time-based digital art by 1999 while still painting and drawing. His explorations in time-based media seek to reconnect a traditional praxis in painting with technology, resulting in time-based paintings, installation and digital performances.

Dario Dussoni is a graduate in information technology from University "Tor Vergata" in Rome, where he is studying for a postgraduate. For a short time he has been a systems engineer while his major interests have evolved into object oriented programming, network applications, cryptography, design patterns, genetics algorithm, ontologies and so on. Moreover Dario is a collaborator with Evodevo. He likes very much to practice Jeet Kune Do, a martial art.

Ron Eglash holds a BS in cybernetics, an MS in Systems Engineering, and PhD in history of consciousness, all from the University of California. A Fulbright postdoctoral fellowship enabled his field research on African ethnomathematics, which was published by Rutgers University Press in 1999 as *African Fractals: modern computing and indigenous design*. He is now an associate professor of science and technology studies at Rensselaer Polytechnic Institute. His courses include a studio class on the design of educational technologies as well as graduate seminars on the relations between science and society.

Stephany Filimon. At the time of this writing, Stephany Filimon works at Linden Lab, creators of Second Life, where she manages development of the Second Life voice program. Stephany is also a PhD candidate in technical communication at the Illinois Institute of Technology, at least until she graduates in May 2009. The views contained in this article should not be considered representative of or endorsed by Linden Lab, which did not in any way sponsor, support, or endorse the writing in this article. The views presented here are exclusively those of the author.

Viktor Gorokh is an associate professor of the mathematics department in Grigoriy Skovoroda Kharkiv national pedagogical university; graduated from the mechanics and mathematics faculty of Kharkiv national university (1978); worked as a teacher of mathematics in secondary school; post-graduate student of the geometry department in Kharkiv national university (1985-1988); successfully defended candidate dissertation in Leningrad national university (1989); coauthor of the package DG – free software for modeling in geometry which used in schools and universities in Ukraine; coauthor of textbooks in ICT support of math education.

Lindsay D. Grace is a professor of interactive media, an artist, and a computer programmer. He has presented, written, and taught for a variety of colleges including the University of Illinois, American Intercontinental University and the Illinois Institute of Art. His professional experience includes business systems development, web design, database management, inventory management and video game design. His artistic practice includes electronic art, writing, and photography. Lindsay holds the bachelor's of arts in english and the master's of science in computer information science, both from Northwestern University. He is also a master's of fine arts candidate in electronic visualization at the University of Illinois, Chicago.

Ethan Ham is a sculptor and installation artist who often uses kinetics, electronics, and computers in his artwork. His recent shows include the PS122 Gallery, a showing of Rhizome.org commissions at The New Museum of Contemporary Art, and a group show the Photo Resource Center in Boston. Recent commissions include ones from Turbulence.org, Rhizome.org, and The Present Group. Ethan's background includes stints in the computer game industry as game designer, producer, programmer, and executive.

Andrew Jinman is originally a child of the golden age of gaming. Growing up on a steady diet of Ultima online, Warhammer and Neverwinter nights, he is no stranger to the MMORPG's scene. In 2002 Andrew continued his interests in this genre and enrolled at Plymouth University to study digital art and technology, to expand and open his mind to conceptualism. Later his interest for immersive environments grew as he began to further shape and broaden his areas of study. Andrew now works for Twofour Learning and has gone on to produce several Second Life projects for local schools and educational institutions. Andrew also stands as an advisor in virtual environments, understanding that different virtual world platforms provide different benefits to clients. He continues to write on the theme of the 'virtual classroom' for a number of journals and aims to continue to justify the importance of virtual worlds.

Letizia Jaccheri is professor at the Department of Computer and Information Science at the Norwegian University of Science and Technology (NTNU) since 2002. In the last five years she has been involved in the supervision of at least ten PhD students. She has more than fifteen years experience with research projects, both at National level (Italian and Norwegian) and international. Letizia has been working with software research issues since 1988. She wrote her PhD on software process modeling in

1994. She is interested in software intensive processes with special focus on artistic software and open source software.

Mia Kalish is currently the director of distance education for Diné College in Tsaile, Arizona on the Navajo Nation. She focuses on the cognitive effectiveness of multi-modal learning models that satisfy the intense demands of contemporary learning, especially in the virtual domains typical of distance education, in the fields of mathematics and computer science. Dr. Kalish also uses, creating the tools where necessary, the language and cultural understandings of the people for whom her learning materials are designed. When and where available, she uses centuries-old writings to incorporate this knowledge in contemporary materials to create for learners a visualization of their cultures' scientific, technological and mathematical histories.

Heejoo Kim is a visual+digital artist, graphic/web designer, video producer/editor, and educator. She received a MFA degree in film, video, and new media from The School of the Art Institute of Chicago (2000), BFA degree in Art and Technology from The School of the Art Institute of Chicago (1997), and BFA degree in painting and drawing from Hongik University in Korea (1993). With a diverse academic background in painting, animation, video, film and design, she has been teaching and researching digital media. Currently, she is an adjunct professor in Interactive Arts and Media at Columbia College Chicago and an assistant director at the art school in Glenview.

Adérito Fernandes Marcos holds a PhD in computer graphics from the Technical University of Darmstadt, Germany. He is a Professor at the Department of Information Systems, University of Minho. He is (co)author of more than 40 articles in the area of Computer Graphics and Digital Art. During 8 years he was the executive director of the Computer Graphics Center, a research institute of the University of Minho. He was the (co)founder of the conference series SIACG (Ibero-American Symposium on Computer Graphics) and Artech (Int. Conference on Digital Arts), now in their fourth edition. He participated in more than 20 R&D projects (12 as PI) and has been IT consultant for several institutions including the European Commission and INIGraphicsNet Foundation. He is the founder and actual director of the Master Course in Technology and Digital Arts at the University of Minho.

Gabriele Meiselwitz is an assistant professor in the Department of Computer and Information Sciences at Towson University. She has 15 years of industry experience as a computer engineer in both Europe and the U.S. and 10 years of teaching experience. She earned a MS in computer science and an EdD in instructional technology from Towson University. Dr. Meiselwitz is involved in development and implementation of courses addressing creativity and creative development in computer science. She has published papers on topics including computer science and creativity, information security for non-majors, privacy, web credibility, online learning, and usability/accessibility in online learning.

Kirill Osenkov is a software design engineer in test, Microsoft Corporation, Redmond, WA, USA. Studied applied mathematics at Kharkiv National University, Kharkiv, Ukraine. Graduated from Brandenburgische Technische Universitaet Cottbus, Germany in 2007 with a master's degree in computer science. Together with Sergiy Rakov and Viktor Gorokh, co-authored and developed the DG Dynamic Geometry Package (http://dg.osenkov.com). Interests include dynamic geometry, interactive development and modeling environments, developer tools and programming languages.

Lorenzo Picinali works currently as a Lecturer in Audio/Music Technology at the De Montfort University, Leicester, UK, and as a part-time researcher for IRCAM, Paris, France. He is also enrolled as a PhD student in the Music, Technology and Innovation Research Centre, at the De Montfort University. His research topics focus mainly on 3D representation of acoustic environments using various techniques for sound reproduction through loudspeakers and through headphones, and on spatial sound perception.He is also involved in a research project with GNReSound Italia and LIM (Università degli Studi di Milano) about objective and subjective audio quality evaluation of hearing aids.

Joseph William Pruitt is one of the lead design engineers at *Slingshot Product Development Group* in Atlanta, Georgia. He started out wearing white shirts everyday but switched over to wearing mostly black during his master's studies in industrial design at the *Savannah College* of *Art* and *Design*. These days he typically wears grey shirts and is comfortable doing it. He also has more friends at work than most people.

Nicola Quinn holds a master's of science in computer science from the University of Limerick, Ireland. Her main research interests are within the realm of technology and art. Presently she is working as an User Experience Designer, in London, UK. Previously she worked in the Interaction Design Centre at the University of Limerick, Ireland. She has worked on a number of interaction design, usability and human-computer interaction projects for both industry and academia. She has a keen interest in fine art, and paints regularly as a hobby.

Sergiy Rakov is a professor of the computer science department in Grigoriy Skovoroda Kharkiv national pedagogical university; graduated from the mechanics and mathematics faculty of Kharkiv national university (1971); a lecturer of mathematics in universities of Ukraine; Candidate of physics and mathematics from Moscow institute of electronic engineering (1977), doctor in pedagogy (2005) from Kyiv National pedagogical university; Ukrainian national coordinator of some International Projects in Math education; coauthor of the package DG – free software for modeling in geometry which used in schools and universities in Ukraine; coauthor of textbooks in ICT support of math education.

Martin Richardson joined De Montfort University in 2003 after successfully running his own company The Holographic Image Studio, founded in 1988 the same year he gained the worlds first PhD in display holography. In 1999 he was awarded a Millennium Fellowship by the UK Millennium commission for his work within inner city schools, prompting him to consider a position in full time education. He is currently chair of modern holography at DMU. He has recorded the film directors Martin Scorsese and Alan Parker as holograms, as well as the Fine Artist Sir Peter Blake, and worked with David Bowie on several projects including promotional material for the album 'hours' all of which was documented in his first book authored book 'SPACEBOMB: Holograms and Lenticulars 1984 – 2004' was published 2005 and his second book 'The Prime Illusion: Modern Holography In The New Age Of Digital Media' was published in 2007.

Paul Scattergood works in the fields of fine and applied arts, visual communication and advanced technologies in both academic and commercial settings. His creative practice encompasses many communications platforms, focusing upon creative exploration of understanding and perception. His principal interests lie in the exploration of the material qualities of light, its deployment within holographic optics, and stereoscopic projections. Scattergood initially trained as a Fine Art Painter before undertaking research within the Institute of Creative Technologies in Leicester. He has produced a number of investigations into optics, utilising Holography, Lenticular imaging and digital video, which expand upon his painterly concerns. His continuing research career involves the production of experimental outputs, installation artworks and philosophical writing.

Stephen A. Schrum is currently assistant professor of Theatre Arts at the University of Pittsburgh at Greensburg. With a PhD in dramatic art from the University of California, Berkeley, Schrum begun teaching with technology in 1993, and since then has been writing and presenting on technology, including editing the book, *Theatre in Cyberspace: Issues of Teaching, Acting and Directing* (2000). His other interests include digital filmmaking, virtual performance, and playwrighting. Schrum directed a one-act play, The Perm, in Second Life (SL) in the summer of 2007, and has scheduled a production of Euripides' The Bacchae in SL for the summer of 2008, in a historically accurate virtual Greek Theater. As for playwrighting, Schrum continues to perform his full-length monologue, Immaculate Misconceptions, taking it into the virtual world of Second Life in mid-summer 2008. More information can be found on his website, MUSOFYR— pronounced "muse of fire"— at www.musofyr.com.

Guttorm Sindre (born 1964) is a professor in Information Systems at the Norwegian University of Science and Technology (NTNU). He received his PhD in 1990 at the University of Trondheim, Norway. His main research interests are conceptual modeling, requirements engineering, and early stage techniques for eliciting safety and security threats to information systems.

Greg J. Smith is a Toronto-based designer and researcher with interests in media theory, representation and digital culture. Greg received a master's in architecture from the John H. Daniels Faculty of Architecture, Landscape and Design at the University of Toronto in 2007. Recent work includes "Critical Sections" a web-based database drawing project for Vectors, the Journal of Culture and Technology in a Dynamic Vernacular (fall 2008) and ongoing research into the representation of urban space. Greg has taught courses pertaining to digital culture at McMaster University's Department of Communications Studies and Multimedia and he blogs regularly at http://serialconsign.com

Anna Trifonova graduated from New Bulgarian University (Sofia, Bulgaria) in 1999 with specialty "Information Systems and Technologies - Applications in Business and Office." In 2006 she finished her PhD at the International Graduate School of Information and Communication Technologies at the University of Trento (Italy). Her research topic was "Mobile Learning: Wireless and Mobile Technologies in Education." Since January 2007, Anna has a post-doc position within the Software Engineering group at the Norwegian University of Science and Technology (NTNU). Her research interests in software and art also extend to their intersection with mobile computing and education.

Judson Wright is a computer artist, but more accurately a "Behavioral Artist". He recently completed a faculty fellowship at the Interactive Telecommunications Program at NYU/Tisch, studied as a visual artist and composer (Brown University, graduated in 1991 and at RISD) and has programmed artwork for interactive and multimedia performance, art shows and web galleries for over ten years, around the world. http://funkymomma.org

Belgacem Ben Youssef received his PhD in electrical engineering from the Cullen College of Engineering at the University of Houston (Houston, Texas, USA). He is currently an assistant professor in both the TechOne Program and the School of Interactive Arts & Technology at Simon Fraser University (Vancouver, British Columbia, Canada). His research interests include parallel and distributed computing, computational tissue engineering, visualization, and video signal processing. Dr. Ben Youssef has also two years of industrial experience in software development and technical project management in the telecommunications and business sectors. He is a member of both the IEEE Computer Society and the ACM.

Index

virtual world 395, 415, 416
voting parks 244

W

Widrow-Hoff method 364
World Of Warcraft (WOW) 417